THE NATURE AND PRACTICE OF

BIOLOGICAL CONTROL OF PLANT PATHOGENS

S. D. Garrett (1906–)
Scholar of Soilborne Pathogens and Dean
of Root Pathologists

THE NATURE AND PRACTICE

OF BIOLOGICAL CONTROL

OF PLANT PATHOGENS

R. James Cook

U. S. Department of Agriculture

and

Washington State University

Kenneth F. Baker

U. S. Department of Agriculture

and

Oregon State University

Published by
APS PRESS
The American Phytopathological Society
St. Paul, Minnesota

Cover design by Julie Ann Cook

SB
732.6
C66
1983

Library of Congress Catalog Card Number: 83-71224
International Standard Book Number: 0-89054-053-5

©1983 by The American Phytopathological Society
  Second Printing, 1989

Typeset by the Washington State University Computing
Services Center
Published by The American Phytopathological Society
Printed in the United States of America

The American Phytopathological Society
3340 Pilot Knob Road
St. Paul, Minnesota 55121, USA

CONTENTS

Preface

Our original intention in writing this book was to revise and update our *Biological Control of Plant Pathogens*. However, it was clear already from the first efforts that we were writing a completely new book and that the first volume would therefore still be useful to workers in this field. Furthermore, the quantity of new information on biocontrol was so large that to incorporate it into a second edition was almost impossible. Our decision to build on and extend the principles of the first book, rather than to revise it, was strengthened when the American Phytopathological Society (APS) suggested in early 1982 that they reprint our first book which had been out of print for two and a half years. This solved the problem of continued availability of the first book and led to our agreement that APS would also publish the second book as a companion volume.

This book differs from the first in several important respects. The broad subject of biological control of plant pathogens — whether of aerial or subterranean plant parts, whether viroids, viruses, prokaryotes, fungi, or nematodes — is treated in an integrated, unified framework of concepts and principles. Relevant information is included from soil physics on the water and gaseous environment of soil, and from soil microbiology on microbial biomass and biomass turnover in the soil. Mechanisms of biological control are emphasized and related to current concepts of plant physiology, soils, and microbiology. Although some of the aspects covered are outside our expertise (e.g., cross protection between viruses and the physiology of host-parasite interactions), we hope that our perspective as biocontrol specialists may provide useful insights to the experts of these subjects and motivate them to apply these phenomena in biological control. One of the principal themes of this book, only briefly discussed in our earlier work, is that slight changes in an environmental factor often produce striking effects in interactions among microorganisms, or between them and the crop plant, and provide an effective means of achieving biological control of plant pathogens. Pathogen-suppressive soils, which have been the focus of much new research since our first book, are discussed in the context of the soil ecosystem and in relation to the source fields.

The beginning of a roster of the diverse agents shown to be antagonistic to plant pathogens is presented to stimulate more study of the biology and ecology of the microbial agents of biocontrol as well as of the pathogen to be controlled. Biological control by introduced antagonists, now a promising frontier, was represented 20 years ago by only three examples, and 10 years ago by only six examples, only two of which were used commercially.

The subject is here developed around the broad definition of biological control introduced in the first book, because this broad definition has been widely accepted and has proved to be useful and workable. Principles are again set forth in boldface type, 30 in this book, 51 in the first book. Of the 1081 references in this volume, 60% are since 1974, and 87% were not cited in the first volume. This provides a bibliography of 1,557 references on biological control. Fifteen key examples of the successful application of biocontrol are highlighted as boxed summaries set apart from the text. These summaries give a "thumbnail sketch" of the pathogen and the disease caused as well as the method of biocontrol, and they reveal the variety of approaches now in use against these important diseases.

This book, *The Nature and Practice of Biological Control of Plant Pathogens*, explains how biocontrol works in soil, in crop residue, on the surface of the living plant, and inside the plant. Investigations of this subject should never be futile or sterile laboratory exercises, but should start in the field, "where the clues and the action are," and after a phase of detailed laboratory study, should return to the field for testing and application. This book shows how biological control can be achieved in practice and lists more than a dozen potential opportunities for application in the field that have not been exploited. Biological control is as applicable to the agriculture of developing countries as to the high-production, high-energy practices of the western world, and it provides answers to many problems in the ultimate development of a sustainable world agriculture.

The landmark studies in the development of biological control of plant pathogens are presented, and photographs of 36 contributors from nine countries to these landmark studies are included. Biological control as a research field is still young, perhaps just entering its exponential growth phase. However, in some ways it is a maturing, organized discipline with research contributions dating back 60 years. Certainly, it is today one of the most exciting and rapidly developing areas of plant pathology, drawing on basic information from many related disciplines and source fields. Although no single book can now hope to cover biological control completely, we hope that this volume will provide focus and guidance to this expanding subject, stimulate more and better research in the laboratory and the field, and lead to wider application of biocontrol in agriculture. Although plant pathologists are no longer skeptical about the feasibility of biological control, the task of applying it to field conditions is only beginning.

We are happy to have participated in the growth of biological control as a subject during the "vintage years" of its development, and we look forward with confidence to the fruitful maturation of the subject and to its incorporation into world agriculture.

March 1983

R. James Cook
Kenneth F. Baker

Acknowledgments

Our concepts, philosophies, and interpretations of biological control are a product of discussions over more than 20 years with our colleagues. Particularly stimulating in this regard have been the 29 annual Pacific Coast meetings of the Conference on the Control of Soil Fungi, the 16 annual meetings of the Western Regional Project W-38, the 12 annual meetings of its successors WRCC-12 and W-147, and the 1963, 1968, 1973, and 1978 International Symposia on Soilborne Plant Pathogens. We acknowledge with thanks the many courtesies and chances for discussions extended by plant pathologists of the Plant Breeding Institute and the University of Cambridge Botany School, Cambridge, England, to RJC during the five months that he worked there on this book. We are deeply grateful to the many persons who have thus contributed to our understanding of the subject.

In addition, we acknowledge with pleasure the specific assistance and helpful suggestions of R. R. Baker, J. A. Browning, G. W. Bruehl, W. R. Bushnell, R. G. Gilbert, R. G. Grogan, D. A. Inglis, P. Lentz, R. G. Linderman, M. J. O'Brien, and H. T. Tribe, who read part or all of the manuscript. The authors are, however, wholly responsible for the interpretations and views presented. We also acknowledge with pleasure the generous assistance of G. C. Ainsworth, D. G. Hagedorn, G. A. Hepting, R. J. Leach, K. Maramorosch, W. Sackston, G. A. Zentmyer, the American Phytopathological Society, the Hunt Institute for Botanical Documentation, Lederle Laboratories, and the University Archives of the University of Leeds for generous help in locating the photographs of investigators of biological control.

The photographs for this volume were prepared by J. W. Sitton, and the charts and graphs were drawn by Joy Schroeder. The manuscript and revisions were entered on computer datasets using the text editing program, WYLBUR, by Teresa Arndt and Virginia S. Tremblay. The final typeset pages of copy, formatted by a TeX program, were produced by

Alan Hagen-Wittbecker of the Washington State University Computing Services Center and provided to the publisher as a camera-ready document; Janene Winter provided typeset samples, and Dean Guenther created a rough index with the WYLBUR program. The professional efforts of these individuals are gratefully acknowledged. We also are grateful to Katharine C. Baker, Beverly A. Cook, and Kammy Ragan for the hours of typing and checking during the preparation of this volume and to Edward Bassett, William Howie, and Jerry Sitton for their help with the indexing.

THE NATURE AND PRACTICE OF

BIOLOGICAL CONTROL OF PLANT PATHOGENS

WHY BIOLOGICAL CONTROL?

*All things are connected. . . . Whatever befalls the Earth
befalls the sons of the Earth. . . . This we know—
the Earth does not belong to man, man belongs
to the Earth. . . . Man did not weave the web of life;
he is merely a strand in it. Whatever
he does to the web, he does to himself.*
—CHIEF SEATTLE, DWAMISH TRIBE, 1854

*We have not inherited the earth from our fathers,
we are borrowing it from our children.*
—LESTER A. BROWN, 1981

This, our second book on biological control of plant pathogens, appropriately begins with "Why Biological Control?," the question asked nearly 10 years ago in the concluding chapter of our first book (Baker and Cook, 1974). An attempt is made here to build on the foundation laid in that book, emphasizing progress of the last decade, particularly in the changing viewpoints and approaches. Our earlier work presented evidence to uncertain or skeptical plant pathologists for the existence of biological control; its reality now accepted, interest is shifting from "why" to "how."

The subtitle of the proceedings of the first international symposium on biological control (Baker and Snyder, 1965) phrased this uncertainty simply as *Prelude to Biological Control.* The editors sometimes jokingly were asked when the rest of the symphony could be expected. Perhaps our earlier book was the andante first movement of that symphony. This volume, we hope, may be the rondo second movement, complete with halftones and footnotes. If so, the final coda is for the future to compose.

In the first volume, we held that "a degree of biological control of diseases in crop plants is being achieved, even under present methods, and that by specific and judicious manipulation of environmental factors, perhaps without destroying the pathogen or even preventing it from infecting, satisfactory control of many additional diseases may be achieved." A thesis of the present volume is that biological control offers answers to many serious problems of modern agriculture and is an essential component in the development of a sustainable agriculture capable of continuing without interruption or diminution.

History bears witness that civilizations have rarely developed an enduring agriculture. Ancient Egypt apparently did so, because the land was replenished with silt in each annual flood; now that the Aswân High Dam is in place and the land is not replenished as before, the above statement is again true. The Sumerians in what is now Iraq, the Indians of the Indus Valley, the Mayans of Guatemala, the Aztecs of Mexico, the Incas of the Andes, the Anasazi of southwestern United States, the Singhalese of Ceylon, the Khmers of Cambodia, and others developed strong civilizations based on agriculture. When agriculture became nonproductive because of soil erosion, silting of irrigation systems, soil deterioration from slash-and-burn agriculture, soil salinity, or other causes, these civilizations declined and some vanished. Ancient China had devastating soil erosion problems, but developed a labor-intensive agriculture based on extensive use of organic manurial treatments that has made the People's Republic of China largely self-sufficient in food production, despite having the largest population of any country. The example of China offers hope that extensive adoption of biological control measures may help retard the environmental deterioration brought about by modern agriculture.

BIOLOGICAL CONTROL ANSWERS MANY AGRICULTURAL PROBLEMS

Events of the past decade have reinforced our view that biological control holds the answers to many problems of today's agriculture.

Increasing Crop Production Within Existing Resources

The success of modern agriculture in providing sufficient food for the increasing world population has come largely from the increased use of fossil-fuel energy and from increasing the area of cultivated land. Both of these factors have limits that have been reached or exceeded. The Organization of Petroleum-Exporting Countries (OPEC) has made us acutely aware of limitations on the energy supply, and the moment of truth on the amount of available land and on production ceilings is now recognizably at hand. The challenge facing agriculture today is to discover how to maintain, much less expand, production without making greater demands on limited nonrenewable resources.

The expansion of agriculture into arid and semiarid land by use of irrigation, besides being limited by available suitable land and by a finite supply of water, much of it from ancient, slowly replenishing aquifers, is dependent on expensive energy for pumping. In addition, salts that accumulate in irrigated land can be managed only by increased use of water; the more saline the soil or water, the more water required to leach the salts from the root zone. This involves pyramiding expense for the energy necessary to supply the water.

Increasing cultivable land through deforestation also has limits. The perennial drought problem of once-productive areas of northern Africa is a tragic reminder of the dangers of expanding farmland by deforestation.

High-production western-type agriculture is high-energy agriculture. The cereal crops of the Green Revolution achieve their increased yield potential by increased use of fertilizers and water, made possible because the new cultivars with their short stiff stems are less prone to lodge than standard-height cultivars when heavily fertilized. The day of abundant natural gas or oil for manufacturing cheap nitrogen fertilizer is past, and energy costs can only increase. This imposes an economic limit to achieving the full yield potential of the dwarf cereals.

Meanwhile, much productive land is disappearing beneath cities and pavements, and soil is being lost through erosion. An estimated two billion tons of topsoil is lost from U.S. cropland annually. Although resources are declining, the human population is increasing, necessitating increased crop production.

Ways must be found to increase or at least maintain the level of food production per unit area of land, while protecting and preserving declining and nonrenewable resources. Agriculture must become more productive but must also be sustainable. Toward this end, even more effort must be directed at improving the health and hence the productivity of green plants. Control of plant diseases is among the most significant ways still remaining to increase crop production, and biological control achieves this goal in a way that is economical and, most important, sustainable.

Browning (in Kommedahl and Williams, 1983) describes the present and future as the "Age of Plants." Humans have experienced many ages, including the recent Atomic Age and Space Age. According to Browning, recognition of the Age of Plants arrived in the 1970s with the oil embargoes and resultant energy crises. The total dependence of civilizations on the production of green plants is now abundantly clear. We can no longer take plants, and especially plant health, for granted. Plant health is of more than academic interest.

Plant diseases also play an important direct role in the diminution of natural resources for agriculture. Farming with reduced tillage or no tillage conserves energy and decreases soil erosion but is unacceptable in many areas for many crops because yield is less. Winter wheat in the Pacific Northwest commonly yields 25% less when direct drilled (sown no-till) than with conventional intensive tillage systems. Significantly, yields are the same or better with no-till if the soil is freed of pathogens by soil fumigation

(Cook, *in* Bezdicek and Power, 1983). Therefore, the problems of poor wheat yield with no-till in the Pacific Northwest are biological, not physical. Specifically, the lower yields are caused by poor root health. The shift to less tillage provides ecological niches previously unavailable to certain soilborne pathogens and to soil- and residue-inhabiting pests. Through understanding the biological stresses responsible for lower yields and the reasons why these stresses become more acute with reduced tillage, we can make adjustments to again close these niches for the betterment of plant health.

Crops could use fertilizer much more efficiently if the problem of poor root health caused by soilborne pathogens could be eliminated. *Pythium* spp., nematodes, and many other pathogens responsible for root pruning on most major food crops of the United States greatly reduce nutrient absorption by roots (Cook, *in* Bezdicek and Power, 1983). Growers often try to compensate for poor root health of their crops by applying more fertilizer. The well-known increased-growth response (IGR) of crop plants to soil fumigation (Wilhelm and Nelson, *in* Toussoun et al, 1970) results largely from decrease or elimination of these "root nibblers" and other plant pathogens (Cook, *in* Bezdicek and Power, 1983). The IGR resembles a fertilizer response because the healthier roots are more absorptive and also more effective in manufacturing growth factors for transfer to the plant tops. Soil fumigation and the IGR reveal the yield potential for a crop, but fumigation is not an economical or sustainable practice for most food crops. Genetic modification of crop cultivars, combined with management of microbial populations to obtain biological protection of the roots, offers an acceptable and sustainable alternative.

Although crops probably have a maximum yield that cannot be significantly exceeded by intensified management, the yield potential for a given area and agroecosystem is rarely attained. Perhaps the nearest approach to that level is attained in glasshouses with controlled environment and freedom from root nibblers. The record yields for corn, soybeans, potatoes, and most other food crops for a given area, soil, and climate are commonly double the average for that crop and area. The world record wheat yield of 209 bu/A (182 hl/ha) established 20 years ago under irrigation in the Columbia Basin in Washington still is unbroken. Indeed, the average yield in the same county is only about 100 bu/A (87 hl/ha), despite application of 75–100 cm of irrigation water and 150–200 kg of nitrogen per hectare annually. The yield of a given crop in a given area and season can be considered the product of an equilibrium between the genetic potential of the crop, the climate, and the availability of water and nutrients on the one hand, and the external biological stresses on that crop on the other (Chapter 11). Yields cannot even be maintained without adequate attention to plant health.

Biological control offers a powerful means to increase yield by suppression or destruction of pathogen inoculum, protect plants against infection, or increase the ability of plants to resist pathogens. The recent demonstration (Burr et al, 1978; Kloepper et al, 1980) of yield increases for potatoes following inoculation of seedpieces with *beneficial microorganisms* ("plant growth-promoting rhizobacteria") exemplifies the kind of crop response possible when nonpathogenic microorganisms are used to exclude or displace pathogens on roots (Figure 1.1). Biological control with beneficial microorganisms is in its infancy and has a promising future. It provides a way to increase production of crops without increased energy or land demands and without environmental pollution.

The alleviation or prevention of abiotic stresses on crop plants is another important means of increasing production. However, abiotic stresses rarely act alone when they affect plant productivity. The direct effect of abiotic stresses probably is significantly less important than their role in causing plants to become more susceptible to weak parasites. Winter barley and wheat are markedly more hardy and hence more able to survive severe winters in eastern Washington if they are protected by a fungicide from *Typhula incarnata*, a weak parasite but a potentially destructive pathogen of these crops (G. W. Bruehl, unpublished). In this same area, desiccation of wheat crowns caused by severe freezing for a few days early in the spring of 1982, when the plants had no snow cover, subsequently favored widespread damage from fusarium foot rot. The most winterhardy wheats had the least fusarium foot rot (R. J. Cook, unpublished). Also, in 1982, in plots of winter wheat near Pullman, Washington, plants were stressed by winter injury, growth was slow to resume in the spring, and the average yield was only 63.7 bu/A (55.6 hl/ha). However, in adjacent plots where soil had been fumigated with methyl bromide just before sowing the previous fall, plants were not damaged by the winter, grew vigorously early in the spring, and gave an average yield of 100.8 bu/A (87.8 hl/ha) (R. J. Cook, unpublished). Stress from winter injury predisposes wheat to some pathogens, whereas other pathogens apparently predispose wheat to greater winter injury. Soil fumigation as a research tool can reveal the greater yield that is possible in spite of abiotic stresses, if pathogens favored by the stresses can be controlled (Chapter 11). Adequate crop rotation allows time for biological control of the pathogens in soil, and the crop growth response can resemble that from soil fumigation (Cook, *in* Bezdicek and Power, 1983).

Even the milder frost damage that occurs on sensitive plants at temperatures only slightly below freezing is now known to involve microorganisms. Lindow et al (1975a, 1975b) showed that the bacterium *Pseudomonas syringae* on leaves of frost-sensitive plants provides nuclei for ice crystallization, which prevents water from supercooling below −2 to −5°C and increases frost damage. Plants having this bacterium or some strains of *Erwinia herbicola* on the foliage sustain frost injury, but plants lack-

Figure 1.1. Root and shoot growth from potato seedpieces that were inoculated by dipping into a cell suspension of a *Pseudomonas* strain (left) or water only (right) prior to being planted in a natural soil. (Reprinted with permission from Schroth and Hancock, 1982.)

ing these bacteria do not. Lindow (1979, 1981, 1983) showed that several species of bacteria applied to foliage would decrease, by a type of biological control, the number of ice nucleation-active bacteria and, thus, the frost injury.

Of particular concern are the more subtle forms of abiotic stress that are not apparent by inspection of standard crop characteristics but nevertheless are recognizable by the presence of certain pathogens, which then become more aggressive and cause greater yield loss. Fusarium foot rot of wheat (caused by *F. roseum* 'Culmorum') in the Pacific Northwest became increasingly important in the low- to intermediate-rainfall area beginning in 1962. The region's first semidwarf cultivar, Gaines, was first grown there commercially in 1961. Subsequent experiments showed that intensified management systems, including greater use of nitrogen fertilizer, caused plant water stress that favored the *Fusarium*. Use of cultural practices to avoid stress (Chapter 10), and of semidwarf cultivars less prone to stress (Chapter 6), has provided biological control of this disease. The trend toward more intensive crop management to produce maximum yields is probably accompanied by subtle physiological plant stresses of many kinds, to which pathogens respond. Biological control aimed directly at the pathogen or mediated through adjustments in the host offers unlimited opportunities to reduce losses associated with stress.

Avoiding Development of Pathogen Resistance to Chemicals

The development of pathogen strains resistant to fungicides and bactericides, not yet common in 1974, has now become a major problem throughout the world (Dekker and Georgopoulos, 1982). Chemicals must now be used as mixtures, intermittently, or alternately in sequence to avoid selection for resistant strains. The newer materials affect specific metabolic processes and therefore are more selective in their action. They usually give more selective and effective control of specific pathogens (e.g., carbendazim fungicides against many Ascomycetes, and metalaxyl against Oomycetes) but increase the problem of chemical-resistant strains. The early fungicides (e.g., lime sulfur, Bordeaux mixture, mercuric chloride, formaldehyde) were nonspecific and broad-spectrum in their effect; although they produced undesirable effects on non-target organisms, they presented no problem of development of resistance by the pathogen. Similarly, thermal treatment of soil and of plant propagules has brought about no increase in microorganism resistance to heat. **The more specific the effect of a biological, physical, or chemical factor on a microorganism, the more limited the metabolic processes affected, and the greater the probability of decreasing the effect through genetic shifts in the population.**[1]

The problem of pesticide resistance is not yet as serious in plant pathology as in entomology, where "In fact, only a very few species regularly treated with pesticides have not developed significant levels of resistance" (Smith, *in* Rabb and Guthrie, 1970). Nevertheless, populations of plant pathogens with resistance to chemicals used for their control are known, including: *Venturia inaequalis* (cause of apple scab), *Erysiphe cichoracearum* (cause of powdery mildew of cucurbits), and *Botrytis cinerea* (on several plants) resistant to benomyl; *Phytophthora infestans* (cause of late blight of potato), *P. parasitica* var. *nicotianae* (cause of black shank of tobacco), and *Peronospora tabacina* (cause of tobacco blue mold) resistant to metalaxyl; *E. graminis* (cause of wheat powdery mildew) resistant to triadimefon; *Tilletia foetida* (cause of wheat smut) resistant to hexachlorobenzene; *Helminthosporium avenae* (cause of foot rot of oats) resistant to organomercurials. Other instances of pesticide resistance unquestionably will appear in the future.

Resistance to antibiotics produced by microorganisms also has developed, as in the well-known penicillin-resistant pathogens of humans and animals. Strains of *Erwinia amylovora* (cause of fire blight of pome fruits) with resistance to streptomycin now occur commonly in orchards where this antibiotic has been used intensively, and some strains of *Agrobacterium radiobacter* pv. *tumefaciens* are resistant to agrocin 84 produced by *A. radiobacter*. Because antibiotics, being specific selective pesticides, affect rather limited

---

[1] Principles in this volume are in boldface type. A principle is a concisely stated fundamental proposition or generalization that aids in the integration of related facts and has prediction value.

metabolic processes, it was to be expected that populations with resistance to them would appear. Biological control involving only antibiosis would, therefore, probably be no more stable than control with the new selective fungicides.

However, when antibiotics are formed in situ by microorganisms, more than one antibiotic may be produced or some other biological control process may also be involved. **A population of a given organism necessarily has more than one way of achieving an essential physiological objective.** The tendency among investigators, after one process or substance has been verified as operative in a given situation, is to divert further research efforts elsewhere, rather than to seek additional and alternative explanations. The alternatives may become evident only when the first way becomes inoperative. Thus, for example, *Trichoderma viride* was thought to control *Armillaria mellea* by antibiosis following a carbon disulfide soil treatment (Bliss, 1951). Webster and Lomas (1964) showed that *T. viride* did not produce antibiotics and that the organism thought by Bliss to produce antibiotic probably was *Gliocladium virens*. Chet and Baker (1980, 1981) also found no evidence of antibiosis by *Trichoderma* and showed that *T. harzianum* is an effective mycoparasite that produces β-(1-3) glucanase and chitinase. Ohr and Munnecke (1974) found that methyl bromide soil treatment inhibits production by *A. mellea* of an antibiotic that protects it from *T. viride*. Thus, what seemed a fairly simple and direct mechanism is obviously a complex pathogen-antagonist interaction. A biocontrol thought to be due to a given antibiotic may be due to several processes and will therefore be more stable than a specific chemical control.

Loss of host resistance to a pathogen is another risk in agriculture. Pea cultivars formerly resistant to *Fusarium oxysporum* f. sp. *pisi* have thus proved susceptible to newly developed strains of the pathogen, and previously resistant tomato cultivars are proving susceptible to new strains of *F. oxysporum* f. sp. *lycopersici*. Loss of effective host resistance has been especially common and a serious problem in the cereal grains, e.g., to leaf diseases such as rusts and mildew of wheat and barley, and blast of rice.

Maintaining Relatively Pollution- and Risk-Free Control

Modern agriculture has tended to apply ever larger amounts of pesticides to attain the yield potential set by the climate, the agronomic inputs, and the genetics of the cultivar. Replacing pesticides entirely by biological control is not an attainable or even a reasonable goal, except perhaps under special circumstances (e.g., in glasshouse operations). Meeting the future challenge for a productive but sustainable agriculture will require the use of all available strategies that are effective, economical, safe, and compatible. Growers will always need to use more than one technique to successfully control a plant disease, just as organisms do to achieve an essential objective within the biological balance of nature. The rapid adoption over the past 10 years of integrated pest management as an approach and way of thinking shows clearly that the philosophy of multiple strategies is an integral part of our institutions of research, extension, and teaching. However, some change in thinking may be required, so that biological control will be tried as a first rather than as a last strategy. It is becoming increasingly clear that even multiple chemicals will not permit the grower to economically achieve the yield potential of his crop and that, despite the temporary benefits, the backlash from nontarget effects cannot be accepted indefinitely.

In the 1960s, the public became aware of environmental pollution and of the increasing amounts of toxic chemicals in nature's food chains. Public opinion, once so favorable to the cheap and rapid control of pathogens and insects by chemicals, then moved toward the opposite extreme of favoring serious restrictions on their use. At the same time, plant pathologists realized that upsetting the biological balance can lead to severe disease outbreaks and, more than ever, they have included biological control procedures in their integrated control programs. Most proprietary fungicides have low toxicity to humans and their use has not received as much adverse publicity as have insecticides. Mercurials and formaldehyde, exceptions to this, have been banned. Similarly, soil treatment with live steam (100°C) is being replaced by treatment with aerated steam (60°C) that does not kill many beneficial soil saprophytes and that causes less discomfort and poses fewer hazards to employees.

Most fungicides have only a temporary effect and therefore require repeated applications. Biological control usually lasts longer. Some fungicides may actually increase plant disease by inhibiting or killing antagonists of pathogens. The eutypa canker of apricot trees increased in South Australia after copper sprays were used to control *Clasterosporium carpophilum* (Carter, 1971); applications of phytoactin and cycloheximide to control white pine blister rust adversely affected the mycoparasite *Tuberculina maxima* (Kimmey, 1969); Bordeaux mixture applied to snapdragons, presumably to control rust, inhibited the mycoparasite *Fusarium roseum* (Dimock and Baker, 1951). It is becoming clear that increasing the potency or number of treatments does not prevent the severe fluctuations in pathogen attack resulting from overkill treatments but may actually increase the fluctuations. Biological control often works by gentler, more subtle methods to which there is little resistance. Moreover, disease control is more than merely killing pathogens. In fact, "nudging" the microbiota toward a more stable, better buffered, less disease-prone configuration by mild or specific treatment with chemicals combined with adjustments in cultural practices and by changing the cultivar has often proved better than using eradicative "dynamite" treatments. This is exemplified by the more stable control of *Gaeumannomyces graminis* var. *tritici* achieved with long-term monoculture than with soil fumigation (Chapter 7).

In 1979, an international workshop considered the potential hazards arising from procedures for biological control of insects (Lundholm and Stackerud, 1980). H. D. Burges there reported that *Bacillus popilliae* and *B. thuringiensis* had been in extensive commercial insecticidal use for 30 and 20 years, respectively, without incident or risk of vertebrate infection, toxicity, or allergy and without persistence or multiplication in the environment. D. Pimentel reported that "Biological control using microorganisms (protozoans, nematodes, bacteria, fungi, viruses) has not resulted in any known environmental problem. Of all the biological control methods used, the microorganisms offer the best potential for insect pest control." L. E. Caltagirone and C. B. Huffaker found that "The use of predators and parasites for pest control, when it is the result of a well thought out, carefully executed program, is . . . risk-free." The assembled representatives of the nine participating countries concluded that "from a consideration of the published data and evaluation results, and from the experience so far gained in their use, biological control agents clearly present a lower risk potential than that associated with many chemical pesticides."

The safety factor in biological control of plant pathogens would appear to be at least as great as that of insects, since a closer physiological relationship exists between mammals and insects than between mammals and microorganisms. Indeed, antagonists of plant pathogens will probably not be able to infect mammals. It is also difficult to visualize any health risk or environmental damage that could result from enhancement of resident antagonists through cultivation practices. However, inoculation (introduction) of antagonists conceivably could present some minimal hazards.

One potential hazard of antagonists is allergy of workmen in laboratories mass producing inoculum or applying it in the field. Most microorganisms used in biocontrol probably would be soil or plant inhabitants already present in the normal environment, although in smaller populations. Microorganisms that produce spores in slime are less likely to be extensively wind disseminated. Most of the introduced antagonists used for biocontrol of plant pathogens are slime-spored fungi or slime-celled bacteria in the following genera: *Agrobacterium, Ampelomyces, Ascocoryne, Bacillus, Bdellovibrio, Chaetomium, Coniothyrium, Endothia* (hypovirulent strains of *E. parasitica*), *Erwinia, Fusarium, Gliocladium, Laetisaria, Myrothecium, Nematophthora, Phialophora, Pseudomonas, Pythium, Sphaerellopsis, Trichoderma,* and *Verticillium.* Genera of dry-spored microorganisms used in biocontrol are *Arthrobotrys, Cladosporium, Dactylella, Hansfordia, Penicillium, Peniophora, Scytalidium, Sporidesmium, Streptomyces, Trichothecium,* and *Tuberculina,* and of these, only *Hansfordia, Penicillium, Peniophora, Scytalidium,* and *Sporidesmium* are commonly used. Austwick (*in* Lundholm and

Stackerud, 1980), in a discussion of possible allergic reactions to fungi used for biocontrol of insects, concluded that "considering the long period over which insect pathogenic fungi have been grown in quantity for field experimentation, it is surprising that there are so few reports of adverse effects to laboratory workers or field operatives." Many of the potential hazards can be obviated by careful handling (e.g., not using dry spores, or wearing masks to prevent inhalation).

Poisoning from toxins is another potential hazard of antagonists. Aflatoxins produced by *Aspergillus flavus* and *A. parasiticus* preclude these common insect pathogens from commercial use, and similar restrictions may prove necessary for some antagonists of plant pathogens. However, we know of no reports of significant mammalian toxins produced by antagonists of plant pathogens.

Although there have been no important examples of infections, allergies, or toxicoses of humans resulting from introduced antagonists of plant pathogens, the possibility of this occurring must be kept in mind as a greater variety of antagonistic microorganisms are used. The testing requirements for registering biological pesticides in the United States were discussed by Rogoff (*in* Lundholm and Stackerud, 1980), who reported, "Of all the life forms currently registered as biological pesticides not one has shown positive results in adverse effects testing at any dose level with the minor exception of some skin allergy responses or minor eye irritations . . . " Similar testing requirements in the United Kingdom and in France were discussed by Papworth and by Hascolt and Hurpin (*in* Lundholm and Stackerud, 1980), respectively.

Most cultivation practices that have been used to augment biological control (e.g., addition of organic matter, crop rotation, delaying the plowing-in of stubble, planting of trap, inhibitory, or cover crops, timing of irrigation applications, ringbarking of forest trees before felling, deep plowing) have no deleterious effect and may have a beneficial effect on the environment. An exception is the practice of thorough tillage to fragment and bury pathogen-infested crop residue and thereby control residue-inhabiting pathogens. This approach to disease control is expensive because of the fuel required, and soils are then rendered more prone to erosion. Conceivably, many of the more traditional or conventional methods of tillage have been successful because they help accelerate the death of residue-inhabiting pathogens. Although any approach to biocontrol may become obsolete for economic reasons or because of undesirable effects on the environment (e.g., greater soil erosion), many other biological control strategies are still to be developed and tested.

One type of environmental pollution that has been too little considered is the spread by man of plant pathogens that persist in an area once they are introduced. The sale of nursery stock infected with, or growing in soil infested with *Phytophthora cinnamomi* (Zentmyer et al, 1952; Baker, 1959) is a case in point. Apparently, this pathogen was carried on heather plants from nurseries in southern California to the San Francisco Bay area, and on nursery stock from eastern to western Australia (Zentmyer, 1980). Since eradication of such a soilborne pathogen is not economically feasible, the sale of contaminated or infected nursery stock is a particularly dangerous type of environmental pollution. The suppression of soilborne pathogens in nursery stock by fungistatic drenches such as pentachloronitrobenzene, Lesan (Dexon), or metalaxyl is simply to postpone disease expression until the host is planted in the field, where the loss is much greater and the soil is then permanently contaminated. Allen et al (1980) found that metalaxyl applied (100 g/10 liter of soil) to containerized avocado nursery stock controlled *P. cinnamomi* root rot for four months, but that one chlamydospore per 300 grams of soil remained viable. Fortunately, the lytic antagonists were not eliminated by the treatment. Measures for production of avocado nursery stock certified free of *P. cinnamomi* have been outlined in California (California Department of Agriculture, 1958), New South Wales (Deleney, 1977), and Queensland (Pegg, 1978).

When the pathogen is present, the use of chemical drenches to decrease loss in the nursery, rather than soil treatment to eliminate the pathogen, is irrational and should be illegal. Similarly, removal of the galls of root-knot nematode or crown gall (caused by *Agrobacterium radiobacter* pv. *tumefaciens*) from roots of nursery stock to get by the inspector or purchaser is fraud, not disease control. *The objective in nurseries should be to produce healthy stock free of the pathogen, not merely free of disease symptoms.*

This line of reasoning leads also to consideration of the role of suppression of disease by biological control. For example, it is possible, using antagonists, to suppress phytophthora root rot of avocado, or by altering soil pH to suppress phytophthora root and heart rot of pineapple (Figure 1.2), in soil in which *P. cinnamomi* persists in fairly high population. This is an excellent strategy in infested orchard soil, but it is dangerous in producing nursery stock. The recommendation in Queensland, Australia, therefore, is to plant certified nursery stock grown in treated soil free of *P. cinnamomi* (Broadbent and Baker, *in* Bruehl, 1975; Pegg, 1978). This exemplifies an important point about the use of biological control: it is not an overpowering, self-sufficient, single-shot control method, nor does it overcome and make permissible sloppy accompanying procedures.

Biological control should be regarded as one part of the whole disease-control program. Moreover, its relative importance can be expected to vary with different diseases, dominant in some instances where other measures have failed (e.g., chestnut blight, caused by *Endothia parasitica*; Chapter 5), but of minor importance where other measures provide inexpensive effective control (e.g., heat treatment of soil to control damping-off

Figure 1.2. Pineapple in rows where the soil was amended (left) or not amended (right) with elemental sulfur (preplant). The sulfur is converted to sulfuric acid by *Thiobacillus* spp., which lowers the soil pH, favors *Trichoderma* spp., and suppresses *Phytophthora cinnamomi*. (Reprinted with permission from Pegg, 1977b.)

of seedlings caused by *Rhizoctonia solani* and *Pythium ultimum*; Chapter 7).

Cross protection is the inoculation of a plant with an avirulent or weakly virulent related strain of a pathogen, or with an unrelated pathogen from the same or another host, to protect the plant from subsequent infection (Chapters 3, 4, 6, and 8). Cross protection is being increasingly investigated, but has some risks: 1) The biocontrol agent may mutate to a more virulent pathogen when used in the field; 2) the agent may prove pathogenic to some adjacent crop or to one used in the rotation; 3) the agent may mutate to a strain that is ineffective in cross protection. Inoculation with a plant virus has the additional risks that it may be synergistic with some other mild virus in the plant and thereby become part of a serious syndrome, or it may mutate to a strain that is carried more efficiently by its insect vector. These risks have been considered in tests of the method, and no example is yet known where such changes have occurred. Nevertheless, the experimental use of an avirulent strain to control apple mosaic in New Zealand in 1961 and 1962 was abandoned because of the risks involved, and other such examples could be cited. Perhaps plant pathologists have been unduly concerned about the risks from inoculation of plants with mild strains of a pathogen, to the detriment of progress in this approach to biological control. All human activities have an element of risk, and biocontrol is no exception. However, the record makes

Phytophthora Root Rot

Biocontrol of the pathogen responsible for this disease illustrates the use of resident antagonists and management of the host to eliminate or suppress inoculum. It is accomplished for different crops through: 1) decreasing zoospore formation by inhibiting the bacteria stimulatory to zoospore formation and by lysis of pathogen hyphae, effects favored by high soil organic matter content and ammonium nitrogen; 2) decreasing zoospore formation and favoring *Trichoderma* spp., accomplished by lowering soil pH; 3) changing the forest understory from susceptible plants to resistant, inhibitory plants.

Pathogen: *Phytophthora cinnamomi* (Mastigomycotina, Oomycetes, Peronosporales).

Hosts: 950 varieties and species of plants, mostly woody types.

Diseases: Decay is produced in tips of small absorbing roots, which commonly results in decline of aboveground parts and often causes death. The pathogen also may cause trunk cankers and decay of larger roots. It attacks the growing point of pineapple, causing heart rot.

Life Cycle: Nonseptate mycelia survive saprophytically in soil and in plant residue. Hyphae are able to invade roots to some extent through wounds but are less important than zoospores, the principal infection units. Zoosporangia are formed in field soil but not in sterile soil. A nonvolatile, thermostable, water-soluble organic acid produced by *Pseudomonas* spp. stimulates *Phytophthora* sporangium formation; this material compensates for low light intensity and an excess of available nutrients in soil, thus promoting zoosporangium production underground. Lowering the soil temperature results in release of zoospores, which may be attracted chemotactically to the region of elongation of roots, where they encyst. The cysts germinate, and the germ tubes may form appressoria before direct and rapid penetration of healthy and wounded tissue. Thick-walled, asexual, resting chlamydospores form copiously in diseased tissues and in surrounding soil and serve as survival structures and

clear that, compared with the alternatives, biocontrol is safe to use and worth the risk.

Some control procedures cannot be used under conditions involving risk of exposure of humans. Carbon disulfide for control of armillaria root rot of fruit trees cannot be used in cities or near residences because of the health hazard. Under these conditions, aerated steam treatment (Munnecke et al, 1976) or polyethylene tarping and solarization (Ashworth et al, 1982) to elevate soil temperature may be used to increase susceptibility of the pathogen to soilborne antagonists, but this has been little tested in the field. Soil fumigation with chloropicrin or methyl bromide is banned in residential areas because of health hazards.

soilborne inoculum that may germinate and infect roots. Formation of zoosporangia and of chlamydospores is stimulated by different and antithetical environments. Oospores are formed from intra- or interspecific pairing or in response to aging or exposure to some root extracts or to *Trichoderma* spp. Few oospores germinate (2–10%), and their role is not clear. In Western Australia, the pathogen spreads between *Eucalyptus* trees through root mats of susceptible *Banksia grandis*.

Environment: The pathogen is favored by excessive soil moisture and moderate soil temperature (optima 20–32.5°C, minima 5–15°C, maxima 30–36°C). Pathogen and disease are inhibited by soil pH below 3.9.

Biological Control: For **avocado** in Queensland and New South Wales, Australia, application of abundant organic amendments, poultry manure, and maintenance of nearly neutral soil pH by addition of dolomitic limestone help maintain organic matter, calcium, and ammonium nitrogen at levels comparable to those of the undisturbed rain forest and suppressive to the pathogen. For **pineapple** in Queensland, Australia, root rot and heart rot are controlled by adding sulfur (Figure 1.2) to the soil to keep the pH below 3.9; this decreases zoosporangium formation and favors the mycoparasite, *T. viride*. For **eucalyptus** forests in Western Australia, inoculum density is reduced by replacement of the susceptible *Banksia grandis* understory with resistant and inhibitory *Acacia pulchella* through prescription high-intensity burning, which kills *Banksia* plants and seed and breaks dormancy of *Acacia* seeds but causes some injury to *Eucalyptus*. Perhaps lower-intensity fire and airplane sowing of heat-treated *Acacia* seed would prove effective and less damaging. Mycorrhizae also may be involved in control.

References: Baker (1978), Broadbent and Baker (1974a, 1974b), Malajczuk (*in* Erwin et al, 1983), Malajczuk and McComb (1979), Malajczuk et al (1977), Pegg (1977a, 1977b), Malajczuk (*in* Schippers and Gams, 1979), Schoulties et al (1980), Shea and Malajczuk (1977), Zentmyer (1980).

Adopting Practices Compatible with Sustainable Agriculture

The declining resources of agriculture outlined above make imperative the development and application of new types of management. The problems of soil erosion and increasing energy cost have made reduced tillage a viable alternative to present practices, but reduced tillage has led to new and different problems of disease control. Agriculture must turn from the short-term view and the "quick kill," to the long-term view and sustainable procedures. Practices that promise long-term effectiveness, even if at a somewhat lower level than the short-term procedures, are more likely to be sustainable. *The need for a sustainable agriculture will be met in part by wider use of biological control of plant pathogens,* but except for development of disease-resistant cultivars, research on biological control lags seriously behind that on chemical control of pathogens. Meanwhile, some highly effective biological controls achieved through crop rotation and other cultural practices are becoming obsolete as growers shift to recropping, double-cropping, and other intensive specialized cropping systems.

One reason for the slow development of biological control is that the application of cultural practices has depended on the grower to make the decisions and execute the procedure. Because there was nothing to sell the grower (Baker and Cook, 1974), there was little of the active solicitation or promotion that growers have come to expect with new devices or chemicals. The move toward more sophisticated systems of integrated disease management, where both public and private consultants work closely with growers, is changing this situation. In addition, with the selection and development of microorganisms specific for a given function, and with the legal protection from patents on plants and microorganisms, the stage is now set for companies to promote the sale and use of microorganism cultures, perhaps even with plant cultivars bred for compatibility with them. The use of beneficial microorganisms as seed inoculants and the biological control of crown gall by *Agrobacterium radiobacter* K84 are examples, but these are only the beginning of what can be done with the techniques of the new biology when private and public interests join forces. In other words, biological control is now ready to move from a self-selling phase to one of active promotion by consultants and industry.

The suppressive soils publicized in our first book have now been widely accepted, and many new examples have been discovered (Chapters 7 and 11). We hope and expect that the next decade will see a similar wide exploitation of genetic engineering and the new biotechnology in the service of biological control.

Biological control is a way of thinking about a plant disease problem; it seeks a solution in terms of restoring the biological balance. Biological control works where chemical controls have been unacceptable or inadequate, as with fusarium foot rot of wheat, take-all of wheat, crown gall of fruit trees, and phytophthora root rot of avocado and pineapple. Application of biological control is not restricted to high-value crops, to well-financed farms, or to wealthy nations. Because many effective cultural practices do not require expensive equipment or large investments of capital, they can be applied by growers in developing nations as well as by those in wealthier nations. Perhaps the best large-scale demonstration of effective biological control by cultural practices is the widespread multiple-cropping organic system used in the People's Republic of China (Kelman and Cook, 1977; Williams, 1979; Chapters 10 and 11). The agriculture of that country, which feeds nearly one fourth of the earth's population, clearly demonstrates that farming can be both intensive and sustainable and, if stabilized for years or perhaps centuries, can provide a biological balance and disease suppression similar in effect to the disease suppression that can occur with prolonged monoculture of some crops (Chapter 10).

Biological control must become part of modern large-scale agriculture if it is to assume its rightful role in helping agriculture become more sustainable. Some biocontrol procedures such as an organic manurial treatment, the use of legumes in the rotation, or perhaps even biological seed treatments, are more readily applied on the diversified farm. Large-scale agriculture has moved away from diversity toward specialization in one or two crops. On the other hand, large-scale agriculture may be better able than the small farm to afford the expense of a biological control program that may require several years to become profitable. The eventual cost and energy demands for biological control are less than for chemical control, but at first, when losses may be greatest, such long-term but ultimately more sustainable control may be beyond the means of the small grower. One practice of both large and small growers that is incompatible with biocontrol is leasing or renting a different field each year to grow a certain crop. Such a practice promotes a short-term view because the grower is not concerned with the potential for disease on the next crop grown in that field. Biocontrol in one form or another can be practiced by the large as well as the small grower, but requires a commitment to the long-term view. In this regard, the same philosophy underlies biocontrol and sustainable agriculture.

Furthermore, the disease-control strategy must be part of the whole crop culture. For example, although pseudocercosporella foot rot of winter wheat (caused by *P. herpotrichoides*) in eastern Washington and adjacent Idaho and Oregon is controlled by seeding late (October 1–15 rather than September 15 or earlier), the plant size attained before onset of winter rains is then too small to check soil erosion. Fungicidal control of this disease is, therefore, necessary to permit the early seeding needed for erosion control and must continue until resistant wheat cultivars or other controls can be developed that are consistent with a sustainable agriculture. **A program for disease control, by whatever means, must fit into cultivation practices, or the practices must be modified before the control program can be adopted.**

The long-term gains to society as well as to commercial interests from even a single accomplishment in plant disease control can be enormous, but they may not always be predictable. This is illustrated by the development of the modern pot-chrysanthemum industry. In the 1930s, only chrysanthemum cultivars resistant to verticillium wilt were grown in the fall months for cut flowers. A. W. Dimock developed the cultured-cutting technique in 1943 to obtain pathogen-free propagules, and this was commercialized by an Ohio company. Coupled with soil fumigation, the technique provided effective control of verticillium wilt. Because wilt resistance was no longer needed, the number and types of available cultivars increased. Commercial day-length manipulation to control flowering, beginning in the late 1940s, greatly extended the chrysanthemum season. Development in the late 1950s of the apical-culture technique to obtain healthy propagules strengthened the program. Precise, dependable scheduling of year-round production of the pot-chrysanthemum became a reality, but the public still regarded the plant as a fall flower, and acceptance became general only about 1970. The development of many new cultivars and the publication of tabular schedules reduced culture know-how essentially to the following of these tables. Because production of chrysanthemum cuttings was concentrated in so few places, the national epidemic of viroid stunt disease in 1947–1949 was brought under control by 1950. Pot chrysanthemums are now enjoyed every month of the year, and have become the leading florist crop, profoundly affecting the economics of the entire glasshouse industry (Baker and Linderman, 1979).

BIOLOGICAL BALANCE

While this book was being written, the spaceships Voyager I and II were moving through interstellar space, exploring distant planets and their moons, sending back clear pictures for all to see. When our first book on biological control was written, man had just successfully landed on the moon. Both of these enormously complex space achievements were based on the predictability of natural events that made it possible to fix the position of the cycling planets in both space and time in relation to the moving spaceships. The orbital movement of a planet is affected by the gravitational pull of its neighbors, and it affects their motions in turn, even as organisms affect each other on earth. The universe is thus in equilibrium, each body in relation to its neighbors.

Living things on earth by necessity have adjusted to the diurnal cycle of day/night, the seasonal cycle of spring/summer/fall/winter, and the tides of the oceans, among other reliable events extraterrestrially determined. The built-in physiology of earthly biota is thus in biological balance with itself and with its physical environment. This basic predictability of biotic relationships has given rise to biological balance and has enabled man to achieve dominance of the earth.

One should not need in 1983 to justify the existence of biological balance on the earth. However, the current political changes revoking established legal measures for protecting the biological environment, and the rise of "scientific creationism" in the United States clearly indicate that one cannot assume that the obvious is apparent to everyone. That there have long been skeptics is shown by the 1966 comment of a government biologist that "The balance of nature is not an achievable ideal, if it is an ideal at all." As

Snoopy commented in a recent Peanuts cartoon strip, those "who believe in the balance of nature [are] those who don't get eaten." It therefore seems necessary to briefly discuss biological balance—the foundation of biological control.

The biological world is a vast network of living organisms interacting in their natural environment. The presence of an organism in a given place and time is determined by: 1) its having evolved or been introduced there; 2) the presence of a favorable physical environment; 3) the presence of associated organisms (symbionts) favorable to its development, or of organisms required for its survival (e.g., hosts for parasites); and 4) the inhibition or absence of organisms (disease organisms; antagonists, predators) so detrimental as to cause its extinction. Thus, interaction is the essence of the continued existence of a population, and this continued existence is itself evidence of biological balance having been achieved.

In this dynamically balanced condition, individual species follow cyclical changes normal for the existing environment, or even unusual cyclic variants, without significantly affecting the whole network, because compensating changes induced in other components maintain the balance. This is the result of long evolutionary adjustment, and it conforms to thermodynamic laws. Those organisms unable to tolerate the many slow changes of the environment in the past half billion years became extinct. Those that survived each climatic change struggled among themselves for space, nutrients or substrates, light, and other factors. Organisms that produced metabolites (antibiotics) that inhibited competitors for a given *ecological niche* had an advantage, as did those that caused some or all of their competitors to enter resting stages by modifying some aspect of the environment. Some actively occupied a site during a warm or moist period and became dormant during the cool or dry part of the year. An ecological niche may thus exist in either space or time. In such a manner, organisms developed an interacting system of spatial and temporal succession.

Organisms affect their physical environment, as well as being affected by it. Microorganisms may modify their environment by producing enzymes and metabolites. Anaerobic bacteria may cause acidification or elevated temperature in microsites. Microorganisms may attain such numbers in soil as to temporarily tie up the available nitrogen. As J. E. Lovelock commented in 1981, "The atmosphere is not merely a biological product, but more probably a biological construction: not living, but like a cat's fur, a bird's feather, or the paper of a wasp's nest, an extension of a living system designed to maintain a chosen environment." He suggested that the remarkable constancy of salinity levels in the ocean indicates a sort of biotic control; a similar conclusion was reached by Olson (1982). The lush Amazonian vegetation, by using carbon dioxide in photosynthesis, and releasing oxygen during growth, plays a significant role in maintenance of the world's atmosphere.

There is thus a beautifully reciprocal interaction between living things and the physical environment that results in biological balance. An organism cannot use more than its share of some ingredient because its population will then decline because of the resulting deficiency, and an organism exposed to abundant materials probably will increase its population. Such biofeedback can only result in a dynamic balance. An essential factor in maintaining balance is the existence of a highly heterogeneous, variable population of organisms, so that when one fails to make the grade, for whatever reason, it is replaced by another. As Spinoza pointed out in the 17th century, "Nature abhors a vacuum."

Taylor and Williams (1975) pointed out that to maintain a stable mixed population of several microorganisms in a continuous-flow system, there must be at least as many growth-limiting substrates as there are different species; "environments where a single or a few growth substrates are outstandingly deficient . . . produce populations of low species diversity . . . negative interaction, through a toxin, will decrease stability whereas positive interaction, e.g., where one . . . of the organisms produces a growth compound needed by the other, would create a . . . stable population."

In undisturbed conditions, plant populations are well buffered and have a tough resistance to change. Only the well-adapted and competitive have found a niche, and to maintain possession they have had to integrate their population, temporal sequence, and activities with those of their associates and with the climatic cycle. To be enmeshed in such an association is to gain external strength and protection, as a stone gains stability and strength when built into a wall and imparts these features to associated stones. The population of a single type of organism will fluctuate wildly with environmental changes, but in association with other suitable types, a stable situation develops. As rephrased by Marston Bates from C. S. Elton, **"the greater the complexity of the biological community, the greater is its stability."**

Such a situation ensures maximal use of each ecological niche, and develops an interlocking *ecosystem* in which an alien or newly evolved organism will have difficulty in becoming established. Such an ecosystem is *biologically buffered* and exhibits dynamic equilibrium among its members and between them and their abiotic environment. A fluctuating population density of each organism is maintained within certain definite limits. Establishment of alien plants in some habitats may mean that they are suited to the particular niche as well as or better than are the residents, that they have been introduced in such numbers that they temporarily or permanently swamp the residents, or that they modify the physical environment in some way especially favorable to themselves. It usually means, however, that the natural balance has been upset in some way more favorable to the alien than to the resident.

Plant disease is one of the factors that determine whether a plant becomes established. Plants in the wild state are adapted to their pathogens and insects; those not adapted are so weakened that they are subordinated or displaced by those better adjusted. In addition, plants grow in mixed stands with a natural sequence or rotation, which provides some protection against pathogens. If a plant becomes unusually plentiful because of favorable climatic conditions, its parasites also tend to increase and reduce the fitness and, subsequently, the number of susceptible plants, and thus, in turn, their own numbers through a diminishing food supply. Disease is therefore an important force in maintaining the biological balance of plants (Harlan, 1976). Pressure from parasites tends to select those plants that, through mutations and resultant variability, have a measure of resistance or means of escape. For this reason, plant pathologists and breeders obtain genetic sources of resistance to pathogens from areas of the native habitat of the host. However, the many methods by which a plant may escape disease in a natural ecosystem may be forfeited in an agroecosystem, unless some adjustment in the cultural practice is made.

Selection among microorganisms is also a balancing factor. Some may compete with a pathogen for available energy sources, preventing its increase. Others may flourish on materials that leak from the pathogen, increasing its loss of nutrients either through steeply decreasing the nutrient gradient outward from it, or by production of toxins or antibiotics that increase leakage. This loss of nutrients would tend to weaken the pathogen, diminish its population, and thus enforce the balance. Other microorganisms may produce substances or create an environment that causes the pathogen to go into a resting or dormant state, forming chlamydospores, oospores, or sclerotia, thereby reducing the attack on the host.

If, as occasionally must have happened, all these checks and balances failed to reduce the activity of some pathogen, the host may have been extinguished and the ecological niche may become available for some better-adapted plant. This may have left the pathogen with a lessened substrate or an enforced saprophytic existence. Perhaps for many organisms, the price of destroying the balance is untimely death.

In the final analysis, the sources of microorganisms parasitic on plants and animals are the soil (including water and dead organic matter) and the host (Baker, *in* Holton et al, 1959). The control of plant diseases, therefore, should place major emphasis on 1) procedures that eliminate pathogens in soil or effectively suppress or avoid their pathogenic capacity, 2) use of plant propagules free of the pathogen, and 3) use of resistant cultivars when available. The first two general procedures, plus rigorous sanitation, have essentially eliminated many of the formerly major diseases of glasshouse crops (Baker and Linderman, 1979). Under field conditions, pathogen-free stock should be planted no matter what the pathogen inoculum level of the soil or the level of genetic resistance of the host.

The population of an organism in a given ecosystem is under continuous adaptive selection, through interactions with other organisms, for ever greater fitness. Probably, however, organisms rarely if ever become perfectly adjusted to their environment, which also is steadily and gradually changing. The organisms, therefore, never quite catch up in adaptation. Under natural conditions, the whole complex thus tends toward biological balance through ecological negative feedback systems powered by both abiotic and biotic factors. The result is an equilibrium that is resilient but balanced, complex but interlocking, stable but fluctuating, strongly resistant to change in composition, and generally uncongenial to alien organisms.

The soil is more stable than the aerial environment in practically all respects, but it can slowly be changed by many conditions. Because microorganisms have become adapted to this relatively stable habitat, slight changes in one or more environmental conditions may have profound effects on soil microorganisms (Baker and Cook, 1974). Such a slight change may exert an insignificant direct effect on growth rate and reproduction of any given microorganism, but the incremental effects on the interacting population may be profound. A niche can be opened more widely to a microorganism or completely closed to it. Adjustments in the soil or plant environment that effectively close the ecological niche of that pathogen are the bases for most biological control achieved with cultural practices (Chapter 10).

A plant disease results from interaction between and among the abiotic environment, host, pathogen, and antagonists. In each situation, the host, pathogen, and antagonists interact against a background biota involved minimally or not at all in the interaction. Host resistance may be so high that no disease results, regardless of the physical environment, unless resistance actually depends on certain environmental conditions.

As each factor of the abiotic environment moves outside the optimum range for the host plant, the plant is increasingly stressed and has weakened or senescent tissue. **Damage by facultative necrotrophic parasites usually is favored by conditions that weaken or stress the host.** This does not imply that infection by these parasites is limited to senescent or weakened tissue. *Facultative and often weak parasites may establish and maintain a nonpathogenic colonization of vigorously growing plants until the environment stresses the plant, upsetting the balance between host and parasite. Plant damage, which we recognize as disease, then appears, often rapidly as the parasite aggressively colonizes the weakened, but still living, tissues.*

The interactions of microorganisms may be qualitative, quantitative, or both. Thus the microbiota involved in a given situation, and their ability to produce antibiotics or toxins, often qualitatively determine the success of a pathogen. However, experimental introduction of the pathogen in abnormally large quantities into an ecosystem may quantitatively "swamp" the antagonist population for a time, even though the antagonistic population may eventually regain dominance.

A soil with the requisite kinds and numbers of microorganisms antagonistic to a pathogen and adapted to or favored by the ecological niche of the pathogen will be suppressive to the pathogen as long as the necessary environment is maintained and the balance is not upset. With take-all of wheat, both host and pathogen appear to be necessary for the development of suppressive soil associated with take-all decline, since the infected roots seem to provide the niche for development of bacteria antagonistic to the pathogen, *Gaeumannomyces graminis* var. *tritici* (Figure 1.3). Populations of the pathogen quickly become very small in soil during cultivation of a crop not susceptible to the pathogen, but the antagonists responsible for take-all decline apparently also then sink to ineffective populations. Upon return to monoculture wheat, the cycle must repeat (Chapter 10). Similarly, soils of the virgin eucalyptus rain forests of Queensland are highly suppressive to *Phytophthora cinnamomi* and, if the rain-forest conditions are maintained, continue to protect avocado groves planted there after the forest is cleared.

Apparently, a soil can be suppressive to some pathogen-host associations before the host is grown there or, in other associations in which the pathogen has not previously occurred. Thus, the Queensland eucalyptus rain-forest soils are suppressive to *P. citrophthora* (Broadbent and Baker, 1974a, 1974b), and certain soils in the Salinas Valley of California are suppressive to *Fusarium oxysporum* f. sp. *dianthi* even though carnations probably never have been grown there (R. Baker, 1980). The factor that determines whether a given soil is suppressive is essentialy whether a microflora capable of suppressiveness is qualitatively present, even if quantitatively insufficient. Possibly other *Phytophthora* or *Pythium* spp. in the Queensland soil, or other formae speciales of *F. oxysporum* in the Salinas Valley soil, may have favored development of the antagonists in these soils in the past. The significance of this question is obvious: Would the soil in southern California avocado groves that is now conducive to *P. cinnamomi* become suppressive if the Ashburner cultivation system was applied there; i.e., are the necessary suppressive microbiota present in southern California soils but in deficient amounts?

Effects of Agriculture on Biological Balance

Evolutionary changes and those imposed by environmental shifts are generally slow, although in restricted areas some may be sudden. The slowness provides

Figure 1.3. Demonstration of the effect of suppressive soil on severity of take-all caused by *Gaeumannomyces graminis* var. *tritici*. The five pots on the right contained soil from five different fields, each diluted 1:100 with the same fumigated (methyl bromide) soil amended with 0.3% of oat grains containing the take-all pathogen. Thus, 99% of the soil in each pot was the same. The 1% soil in the third and fourth pots from the right was from fields where take-all had declined owing to wheat monoculture. The 1% soils in the other three pots were from fields not previously or regularly planted to wheat. The three pots starting on the left were, respectively, fumigated soil not infested with *G. graminis* var. *tritici*, nonfumigated soil not infested with the pathogen, and fumigated soil infested with the pathogen but not amended with a nontreated soil. (Reprinted with permission from Cook and Rovira, 1976.)

opportunity for biological adjustment through selection among genetically variable lines and thus for maintenance of a stable balance. Environmental changes induced by farming practices, on the other hand, tend to be major and rapid, permitting little time for readjustment of the biological balance (Thresh, 1982). Moreover, crop cultivars selected for desired specialized features may have little or no survival ability in nature. This has so reduced variability and adaptability of some plants that many are unable to survive, or are likely to be severely stressed, without protection.

To feed and clothe our expanding multitudes, we have intensified the use of fertilizers, pesticides, and other crop management practices, and have selected ever higher-yielding cultivars to increase production. The complex plant community has been replaced with a simple one, the heterogeneous population with single genotypes, and characteristics suited for adaptation and survival with those for yield and quality. We have become completely dependent on our crops, and they in turn more dependent on us. The biological balance that served to protect crops against disease epidemics has diminished.

That agriculture has disturbed biological balance and augmented plant disease in numerous ways is illustrated by the following examples. First, growing vast acreages of a few wheat cultivars (a simplified biological community) favors epidemics, as in rust and other diseases of wheat (Baker and Cook, 1974). Similarly, most of the epidemics of rice blast (caused by *Piricularia oryzae*) have occurred in Asia when large areas have been sown to the same cultivar. The 1970 southern corn leaf blight epidemic (caused by *Helminthosporium maydis*) occurred because essentially all corn hybrids carried the same factor for male sterility and susceptibility to leaf blight. Second, introduction of foreign pathogens into a population of susceptible established plants has led to several epidemics, including chestnut blight (caused by *Endothia parasitica*) in the United States and Europe, white pine blister rust (caused by *Cronartium ribicola*) in the United States, and downy mildew (caused by *Plasmopara viticola*) on grape in Europe. Third, bringing desert land into cultivation with irrigation has favored some diseases (e.g., curly top [virus] of sugar beet, and verticillium wilt and nematodes of potatoes, cotton, and other crops). Fourth, irrigation during the normal warm, dry season may favor disease. Such a practice enables endemic *Armillaria mellea* to kill native oak trees in California and is currently favoring leaf rust (*Puccinia recondita*) on wheat under pivot irrigation in the Pacific Northwest. Fifth, cultivation buries pathogen propagules such as sclerotia of *Sclerotium* (*Athelia*) *rolfsii*, enabling them to survive; they may then resurface with the next cultivation, germinate, and infect. Sixth, prolonged and extensive cultivation causes a loss of organic matter from soil and the soil may then become more conducive to certain soilborne pathogens, as the rain-forest soils in Queensland, Australia become conducive to *Phytophthora cinnamomi* when clean cultivated.

The disease epidemic that sometimes occurs when a new pathogen is introduced often follows the so-called grand cycle of disease — severe for a period, followed by less severe outbreaks, an occasional mild outbreak, and finally perhaps a decline to unimportance. This cycle may represent the establishment of a new balance, in which the pathogen is rarely eradicated but exists in a state of suppression as exemplified by the chestnut blight in Italy. Discovered there in 1938, the disease nearly destroyed the important chestnut industry in a few years. In 1951, near the site where the disease first appeared in northern Italy, cankers were reported to be healing spontaneously. By 1964, the spontaneous healing had spread widely through Italy and southern France, and in many plantations no new canker was evident. Mittlempergher (*in* MacDonald et al, 1979) reported that "the disease is no longer a problem in the cultivation of chestnut in Italy." Further studies provided clues to an important and unique biological control (Chapter 5).

In undisturbed ecosystems, plant-disease epidemics are unknown or rare, pathogen suppression is usual, and health rather than disease is the norm for the population as a whole. Undisturbed associations tend toward stability. Biological control, in this context, is the retention or restoration of such a disease-suppressive natural balance. Usually it is not the original balance but a new one, stable and compatible with the changed conditions. It is doubtful whether agriculture could have developed if such biocontrol were not common even under the disturbed conditions of cultivation. When a pathogen is present or has been repeatedly introduced into an area without the appearance of significant disease, biological balance clearly exists.

The occurrence of a plant disease epidemic indicates that some aspect of the biological network is not in equilibrium: 1) the *pathogen* is genetically homogeneous and highly virulent, in high inoculum density, or not in equilibrium with antagonists; 2) the *abiotic environment* is relatively more favorable to the pathogen than to the host and/or antagonists; 3) the *host plant* is genetically homogeneous, highly susceptible, and continuously or extensively grown; 4) the *antagonists* are absent or in low populations, lack nutrients or the proper environment to function as antagonists, are inhibited by other microorganisms, or their antibiotics are sorbed by soil or inactivated by other microorganisms. Conversely, absence of a disease epidemic indicates that the pathogen is absent, or its population genetically diverse, the host is highly resistant, or at least minimally diverse, the abiotic environment is unfavorable to the pathogen part or all of the time, or antagonists are inhibitory to the pathogen.

In recent years, man has realized that it is better to work with natural biological forces than to ignore or override them. Rachel Carson's 1962 *Silent Spring*, and A. L. Rudd's 1964 *Pesticides and the Living Landscape* expressed this shift in our approach to nature. This shift was further shown by the U. S. Department of Agriculture *Report and Recommendations on Organic Farming* (Papendick et al, 1980), the 1981 American Society of Agronomy symposium on Organic Farming (Bezdicek and Power, 1983), and the introduced but unsuccessful 1982 Congressional legislation on the Organic Farming Act (House version) and Innovative Farming Act (Senate version). The use of broad-spectrum pesticides that cause overkill has given way to subtler, more sophisticated methods and to greater awareness of biological and ecological principles. Modest use of pesticides to decrease populations of insects and pathogens, along with biological control and modified cultivation practices, gives best results.

Economical control of plant disease is rarely achieved by a single procedure. It must be supported by use of planting material that is free of the pathogen, by tillage methods that are unfavorable to the pathogen, favorable to the host and/or antagonists, by sanitation or use of fungicides, and especially by host plant resistance. Such *integrated control* is based on the fact that different methods work best at different times and places or under different conditions, each compensating to some extent for the deficiencies of the others. *Biological control should not be thought of as a single unaided procedure, like chemical spraying.*

In accord with this changing attitude, there has been a shift away from single unaided overkill treatments. Entomologists have turned to more specific chemicals and to integrated control measures. Furthermore, for biocontrol of insects, there has been "a shift in emphasis from the introduction of exotic parasites and predators to the recognition of the importance of naturally occurring biological control agents . . . this approach is gradually becoming one of the major topics in applied entomology" (Brader, 1980). Entomologists and pathologists are thus coming to use the same approach to biocontrol. Plant pathologists are turning to proprietary, specific, and relatively benign materials, rather than to the broad-spectrum materials formerly used.

As pointed out in our first volume, ecosystems are so complex that we can hope to describe and understand only a few of the simplest ones. Appreciation of this complexity caused plant pathologists for many years to be skeptical of the possibility of effective biological control of plant pathogens. However, such a view failed to recognize that similar complexities involved in agriculture itself have been surmounted, and it is now recognized that biocontrol of plant pathogens need not await the millenium of complete understanding. As Samuel Johnson observed long ago, "Nothing will ever be attempted if all possible objections must be first overcome."

HISTORICAL CONSTRAINTS ON THE USE
OF BIOLOGICAL CONTROL WITH ANTAGONISTS

When our first volume was written 10 years ago, we outlined 19 reasons why the development of biological control of plant pathogens had been delayed (Baker and Cook, 1974, pp. 23–25, 50–55). Schroth and Hancock (1981) recently presented an additional 10 reasons "why biological control in plant pathology . . . is undeveloped"; apparently they felt that further explanation is still necessary. Although good and sufficient reasons delayed the start of biological control of plant pathogens, the last two decades have seen a remarkable increase of interest and research, with at least 19 international symposia and conferences held on the subject, and several commercial field applications of biological control by antagonistic microorganisms.

The general subject of biological control by antagonists has been reviewed by various authors in Baker and Snyder (1965), Toussoun et al (1970), Bruehl (1975), Krupa and Dommergues (1979), Schippers and Gams (1979), Papavizas

(1981), Pimental (1981), and Thresh (1981). Other reviews have been written by Baker (1968), Baker (1973), Baker and Cook (1974), Cook (*in* Horsfall and Cowling, 1977), Baker (*in* Ellwood et al, 1980), Mankau (1980), Papavizas and Lumsden (1980), Corke and Rishbeth (*in* Burges, 1981), Baker (*in* Mace et al, 1981), Schroth and Hancock (1981, 1982), Cook (1982), and Baker (*in* Kommedahl and Williams, 1983).

Although the application of biocontrol is easiest when an antagonist is introduced into treated and nearly sterile soil for high-value crops, a situation fully compatible with commercial glasshouse operations, most applications of biocontrol with microorganisms have been through management of resident antagonists of pathogens on relatively low-value crops in nontreated field soil (e.g., cereal and fruit crops, and forests). Although early work by Ferguson (1958) and Olsen (1964) in California showed the efficacy of managing antagonists under commercial glasshouse conditions, it is disappointing that the method has not been applied or the leads followed up (Powell, 1982). Application to commercial glasshouse operations offers a neglected but very promising area of research in biological control. As predicted (Baker and Cook, 1974), biocontrol has been "most used in situations where alternative control measures are unavailable, difficult to apply, or economically impractical, without regard to the agroecosystem in which they appear."

THE POTENTIAL OF GENE MANIPULATION

The development of agriculture traditionally has been restricted by the necessity of finding a crop plant or organism with all of the desired features, or of finding one with a single desired feature, which must then be genetically introduced by hybridization into the preferred germ plasm. Such hybridization is beset with many limitations. We are now on the threshold of an era of gene splicing and manipulation (genetic engineering) at the cellular and molecular level. This modern technique began about 1970, with the demonstration that plants could be regenerated from protoplasts, and makes possible the prescription development of biological entities to fit specific niches or functions.

Details of methods that are being developed can be obtained from Setlow and Hollaender (1979), Ingram and Helgeson (1980), Wu (1980), Hollaender et al (1982), Old and Primrose (1981), Panopoulos (1981), Williamson (1981), California Agricultural Experiment Station (1982), and from the journals, *Practical Biotechnology* (Harpenden, England, 1981 –), *Recombinant DNA Research* (Bethesda, Maryland, 1976 –), and *Genetic Engineering News* (New York, N. Y., 1981).

Some special applications of gene manipulation to biological control of plant pathogens are briefly presented.

1. *Microorganisms* (mainly bacteria and fungi). The DNA molecules of a potential donor microorganism and the DNA molecular vector are enzymatically cut by a specific restriction endonuclease. The cut ends are "sticky" or cohesive and may be rejoined in recombinant molecules to transfer a gene to the desired site in another organism. Thus, it may become possible to transfer a gene for production of an antibiotic effective against a pathogen, from an organism unable to survive in the given habitat, to another organism that survives well in that habitat but which produces no effective antibiotic. Microorganisms may thus be tailored for specific purposes but must still possess other fitness attributes to survive and compete.

2. *Crop Plants.* Cell walls of various tissues are removed enzymatically, leaving naked protoplasts that may be regenerated to produce new plants that exhibit surprising variability of characters, sometimes having greater disease resistance than the source plant. The protoplasts may be fused, or foreign DNA can be introduced into them by a DNA molecular vector (e.g., a plant DNA virus, a bacterial or yeast plasmid, or plant organelle DNA), making possible crosses between incompatible plants and the prescription development of plants with desired characters. It may even be possible in the future to introduce genes from an antagonistic microorganism into the plant genome and thereby produce a resistant cultivar. The new resistant type, however, must still be subjected to the same rigorous testing procedures as other sources of variability. Furthermore, the procedure presents the disturbing potential for decreasing genetic heterogeneity.

Gene manipulation has the potential to greatly expand the use of biological control by making available a greater number of suitable antagonists and by increasing the level of host resistance. The application of these developments should make interesting reading in the next book on biological control.

2

DEVELOPMENTAL HISTORY OF BIOLOGICAL CONTROL OF PLANT PATHOGENS

*More than 100 years after the introduction of soil
disinfestation and more than 50 years after Sanford's classical
publication on biological control . . . we are
frustrated by the large gap between promising results
in the greenhouse and failures in the field. . . .
We no longer aim to achieve absolute control, but rather
an economic reduction in disease level.*
—J. KATAN, 1981

Many empirical practices for increasing crop production have been discovered during the evolution of agriculture in the last 10,000 years, and some probably involved biological control of plant pathogens. The fallowing practiced in the archetypal dry farming in the hilly country of Iraq and Iran increased the amount of water available per crop and may also have provided soil conditions for accelerated attrition of pathogens. When this Sumerian farming moved down to the flood plains of the Tigris and Euphrates rivers about 8,000 years ago and irrigation and drainage systems were developed, the annual deposit by the rivers of fertile silt provided plant nutrients and soil relatively free of pathogens. The practice of intermixed planting of crops, common in this area today and undoubtedly used in ancient times, restricted the activity of plant pathogens, as did intermixed double-cropping (Grigg, 1974).

In the fifth millenium B.C., Egyptian farming that developed along the Nile River was based on irrigation and drainage, the annual deposit of river silt, and intermixed double-cropping (Grigg, 1974). White rot of onion (caused by *Sclerotium cepivorum*) was not important in Egypt until the Aswân High Dam was built about 1965. Because this dam stopped the annual deposit of virgin clay and the destruction of sclerotia of the pathogen resulting from the anaerobic conditions during the three-month flooding, the disease is now important there (Abd-El-Moity, 1979; Georgy and Coley-Smith, 1982). During the summer fallow in Egypt, the soil reaches temperatures of 40–50°C at 10-cm depth (Prescott, 1920). Such temperatures when continued for 29–34 days are lethal to propagules of soilborne pathogens or may weaken the resistance of propagules to antagonists (Katan, 1980; Pullman et al, 1981a, 1981b). Summer temperatures in both Sumer (now Iraq) and Egypt were sufficiently high to provide effective soil treatment.

Chinese farmers practiced crop rotation and application of manure about 5,000 years ago, and they began irrigation about 3,000 years ago (Forbes, 1965). Slash-and-burn agriculture also was practiced in forested areas. Disease control procedures were outlined by 300 B.C. (Chiu and Chang, 1982). In the first century B.C., farmers were advised, "If a field gave a poor crop in the second year, fallow it for one year" (Fan Shêng-Chih, 1974).

The Incas of South America had a highly developed irrigated agriculture on mountain terraces long before the 16th-century Spanish conquest. They used bird guano, llama and human manure, and fish heads extensively for fertilizer. Corn and amaranth were used in mixed plantings (Garcilaso, 1966).

The Romans used varied cultivation in their wide-ranging empire (50 B.C.–476 A.D.). Land was fallowed and cereal crops were rotated with peas, beans, vetches, or lupines. Manuring, irrigation and drainage, and liming also were used on the land, and green manuring and composting for garden use were practiced. On some forested areas, a type of slash-and-burn agriculture was used, supplying wood ashes (White, 1970). Bassus' *Geoponica* (rewritten about 949–956 A.D.) contains the recommendation, "If you would have fungi to grow from the ground, you must select a spot of light soil on a hill where reeds grow; there you must collect together twigs and other inflammable materials, and set all on fire just before the rain is expected. . . . " (Buller, 1915). This is perhaps the first recorded intentional heat treatment of soil and the first successful manipulation of soil fungi. Numerous modern sources report the occurrence of abundant basidiocarps and ascocarps in burned-over areas and old bonfire sites.

The two-field system (alternating winter grain and fallow) was common during the Dark Ages in Europe. The three-field system (alternating winter grain, spring grain or legumes, and fallow) was used after 763 A.D. This system, made possible by the development of the heavy plow, helped to increase yields. In some cases, a two- to three-year rotation with grass replaced the fallow. Turnips for cattle feed entered the rotation after the 16th century.

Numerous obnoxious concoctions were used in the 17th and 18th centuries for treatment of tree wounds. Austen (1657) advocated treatment of fresh pruning wounds with a mixture of cow dung and urine to prevent a canker disease of apple. Forsyth (1791) recommended the application of a mixture of fresh cow dung, lime, wood ashes, and sand to fresh tree wounds. Le Berryais (1785) suggested that pruning wounds of trees should be coated with fresh mud to prevent infection by fungi. Grosclaude (1970) showed that mud so applied reduced infection by *Chondrostereum purpureum* from 100% down to 30%, demonstrating the efficacy of the soil microbiota as antagonists. Weidlich (*in* MacDonald et al, 1979) also showed that soil applied to cankers of *Endothia parasitica* on American chestnut caused cankers to heal, perhaps due to *Trichoderma* spp.

Information on the developmental cycles of fungi and bacteria accumulated from the 17th to the 19th centuries. This emerging science was given a firm base with the development of laboratory culture methods after 1850 and the poured-plate technique in 1881. W. Roberts (1874), the first to note antibiotic action in cultures, introduced the term antagonism into microbiology. He wrote, "the growth of fungi appeared to me to be antagonistic to that of Bacteria and vice versa. I have repeatedly observed that liquids in which *Penicillium glaucum* was growing luxuriantly could with difficulty be artificially infected with Bacteria; it seemed, in fact, as if the fungus . . . held in check the growth of Bacteria. . . . On the other hand, the *Penicillium glaucum* seldom grows vigorously, if it grows at all, in liquids that are full of Bacteria. It has further seemed to me that there is an antagonism between the growth of certain races of Bacteria and certain other races of Bacteria. . . . It would be hazardous to conclude that because a particular organism was not found growing in a fertile infusion, that the germs . . . were really absent from the contaminating media." M. C. Potter (1908) showed that cultures of *Erwinia carotovora* growing in turnip tissue, and *Penicillium italicum* in orange rind, were killed by liquid media in which the respective organism had been grown. This may be the first report of inhibition of a plant pathogen by its own staling products.

F. Viullemin in 1889 applied the word antibiosis to situations where one organism destroyed another to sustain its own life. G. Papacostas and J. Gaté in 1928 used the word for injurious effects of one organism on another in vitro; such an effect in vivo was called antagonism. S. A. Waksman (1947) stated that antibiotic should mean "a chemical substance, produced by microorganisms, which has the capacity to inhibit the growth [of] and even to destroy bacteria and other microorganisms." This definition has been generally accepted (Lamanna et al, 1973). The word antibiotic is very broad and includes widely disparate mechanisms. For example, penicillin interferes with bacterial cell wall formation, streptomycin interferes with bacterial protein synthesis, bacteriocins are compounds produced by bacteria that affect species closely related to the producer, and siderophores are compounds that inhibit growth by sequestering iron needed by the organism affected. Bacteriocins and siderophores might be included in Waksman's broad definition, but some workers regard antibiotic more narrowly. Until sufficient information is avaialable to permit specialists in this field to redefine the term, it seems best for the present to simply use the specific terms. These terms undoubtedly will be clarified by future research, but such clarifications are not requisite to the use of the phenomena as tools of biological control.

"Analysis of the microorganisms involved, and the biochemistry of their relationships, . . . becomes a means of perfecting the result obtained, not a necessary percursor to attempting biological control" (Baker and Cook, 1974).

The inhibition of one microorganism by another, and similar phenomena in laboratory culture was observed subsequently by many others, but use of this phenomenon came only after the discovery of the antibiotic penicillin by A. Fleming in 1928, and its purification and use in medicine by W. Florey, E. Chain, and N. G. Heatley in 1939 (Hare, 1970). This development probably provided the single greatest stimulus to studies of antagonists to plant pathogens. More than 100 other antibiotics were described between 1929 and 1949, 75 of them in 1945 to 1949. Most of them were obtained from saprophytic soil microorganisms.

The first successful transfer of a natural enemy from one country to another apparently was the introduction of the mynah bird from India to Mauritius in 1762 to control the red locust (Doutt, *in* DeBach, 1964). The first control of an insect pest by an introduced insect was that of the cottony-cushion scale of citrus by the vedalia beetle, introduced from Australia to California in 1888 (DeBach, 1964). In 1916, L. O. Howard referred to insect control by natural enemies as "the biological method," and in 1919, H. S. Smith called it *biological control* (De Bach, 1964).

INTRODUCED ANTAGONISTS

The first attempts at direct applications of biological control of plant pathogens were made in 1920-1940. C. Hartley (1921) inoculated forest nursery soils with 13 antagonistic fungi in an attempt to control damping-off of pine seedlings. Pot tests with autoclaved soil inoculated with *Pythium debaryanum* gave 27.5% damping-off. When *Phoma betae, Phoma* sp., *Chaetomium* sp., *Rhizopus nigricans, Trichoderma roseum, T. koningii, Aspergillus* spp., *Rosellinia* sp., and *Penicillium* sp. were added, there was 16.9% damping-off. The antagonists without *Pythium* gave 11.1% damping-off. In another test, *P. debaryanum* added to nontreated nursery soil produced 35.8% damping-off, but in the same soil autoclaved, the pathogen produced 100% damping-off. Hartley concluded that "competition of different fungi is a factor to be considered and . . . the inoculation of treated soil with saprophytes may sometimes prove of value in increasing the efficacy of heat disinfection." This was the first attempt to use introduced antagonists to control a plant pathogen in treated soil.

W. A. Millard and C. B. Taylor (1927) were more successful in attempts to control common scab of potato (caused by *Streptomyces scabies*) by the addition of green grass cuttings plus the antagonist *S. praecox*, an obligate saprophyte, to sterilized soil in pot tests. No antibiosis was exhibited by *S. praecox* in agar culture, but numbers of *S. scabies* were strongly decreased by *S. praecox* in soil, with or without grass cuttings. This was thought to be "probably caused by a starving out of the weaker organisms in competition for the available food supply of the soil." Grass cuttings alone "exerted no inhibitory action on scab." The effect was not due to soil acidification by the grass. The photos of the potato tubers showed excellent control of scab when *S. praecox* was added.

A. W. Henry (1931) found that eight cultures of actinomycetes, bacteria, and fungi from soil, when inoculated in various combinations in sterilized soil, resulted in slightly less disease caused by *Helminthosporium sativum* on wheat.

RESIDENT ANTAGONISTS

K. Bancroft (1912) showed that mycelia of *Rigidoporus* (*Fomes*) *lignosus* were unable to infect young seedings of rubber trees if the root food base of the pathogen was removed and that they would then die in four to five days. This was perhaps the first demonstration of the significance of the food base in infection of host roots, a feature used in many subsequent biological control studies (Leach, 1937; Rishbeth, 1963; Fox, 1977).

The usual control practice for potato common scab in the 1920s and 1930s was soil acidification. The plowing under of green rye plants was thought at that time to decrease scab by increasing soil acidity, but G. B. Sanford (1926) showed that green rye (20.6 metric tons/ha) neither decreased scab nor acidified the soil in 58 days in field tests. Of 27 bacteria inhibitory to *Streptomyces scabies* in agar media, only one, isolated from soil, inhibited the pathogen without acidifying the medium. The number of bacteria and fungi in nonsterilized soil in a laboratory test increased as the amount of green rye leaves was increased from one part in 20 of soil to one in three. He commented that "the same toxicity demonstrated on artificial media may also occur naturally in the soil . . . it may be that certain microorganisms of the greatly increased soil flora are antibiotic to *A. scabies.*" He suggested that "When scab is controlled by green rye crops in some soils . . . the antibiotic qualities of certain predominant soil microorganisms influence the development of *A. scabies.*" Garrett (*in* Baker and Snyder, 1965) credited this paper with having triggered an awakening of plant pathologists to the antagonistic potential of microorganisms by suggesting that saprophytic microorganisms affect activities of plant pathogens, and that the microbiological balance of soil can be changed by addition of organic matter.

Key Contributors to Biological Control of Plant Pathogens

C. Hartley 1887–1968

W. A. Millard 1881–1964

G. B. Sanford 1890–1977

H. H. McKinney
1889–1976

A. W. Henry 1896–

O. A. Reinking
1890–1962

J. C. Walker 1893–

R. Weindling 1899–

R. Leach 1903–1980

Sanford and Broadfoot (1931) provided experimental evidence supporting these suggestions, and first used the terms "biological control" and "suppressive effect" in plant pathology. Using 40 bacteria, 24 fungi, and three actinomycetes isolated from soil or roots and culms of wheat, they individually inoculated them in pots of sterilized soil along with the take-all fungus, *Gaeumannomyces graminis* var. *tritici*. In another series, the sterile filtrate from a liquid culture of each antagonist was used instead of the organism. The pathogenicity of *Gaeumannomyces* was suppressed to an infection rating of 0–10% by inoculation with any of six fungi and 15 bacteria, and to a rating of 10–20% by six fungi, eight bacteria, and one actinomycete. The filtrates from one fungus, one actinomycete, and six bacteria resulted in a rating of 0–10%, and that from one actinomycete and one bacterium, a 10–20% rating. The filtrates were less active than the microorganisms. Sanford and Broadfoot concluded that "the pathogenicity of *O. graminis* may be . . . at times, controlled by the association with it of a number of soil inhabiting bacteria, fungi, and actinomycetes."

Henry (1931) showed that nontreated Edmonton black loam had an inhibitory effect on *Helminthosporium sativum* and *Fusarium roseum* 'Graminearum' compared with the same soil sterilized. Sterilized soil gave 47.6 and 54.7% diseased plants when inoculated with *H. sativum* and Graminearum, respectively, but in nonsterilized soil the figures were 7.5 and 5.4%. Adding a trace of nonsterilized to sterilized, pathogen-inoculated soil in pots reduced the disease to 7.8% with *H. sativum*, and a one-to-one dilution of sterile in nonsterile soil gave 5.5%. Sterilized soil plus four fungi, two actinomycetes, and two bacteria isolated from this soil, when inoculated with *H. sativum*, gave 2.8% diseased plants in pot tests. The microorganisms multiplied in 24 days so that a trace of soil inoculum added to sterilized soil was almost as suppressive as nonsterilized soil. This apparently was the first demonstration that a total antagonistic soil microflora could be transferred to produce a pathogen-suppressive soil.

Henry (1932) further explored microorganism interactions as influenced by soil temperature. He held sterilized and nonsterilized black loam soil infested with *G. graminis* var. *tritici* in pots at temperatures ranging from 10 to 27°C. The infection rating of the seedling blight, foot rot, and root rot phases of take-all was maximal at 18°C in nonsterilized soil and decreased as the temperature was increased from 18 to 27°C. In sterilized soil, disease was uniformly severe at 18, 23, and 27°C. The significantly greater severity of take-all at 23 and 27°C in sterilized than in nonsterilized soil was attributed to the activity of resident soil microorganisms (Figure 2.1).

C. J. King (1923) showed that the cotton root-rot fungus (*Phymatotrichum omnivorum*) in alfalfa fields tended to die out in the center of infected areas, the first published example of decline of a disease with continued monoculture of the host plant. Alfalfa or cowpeas replanted in the centers of the

Figure 2.1. Relative amount of blighting of Marquis wheat seedlings in sterilized and nonsterilized soil infested with *Gaeumannomyces graminis* var. *tritici* and kept at the indicated temperature. Pots left to right: check; two pots of non-sterilized soil; two pots of sterilized soil. (Reprinted with permission from Henry, 1932.)

spots remained healthy; the largest spots showed the greatest "tendency toward immunity" from reinvasion. King et al (1934) first showed that heavy application of organic matter controlled *P. omnivorum* on cotton. This was confirmed by F. E. Clark (1942), who related the rate of breakdown of organic matter to the rate of death and decomposition of sclerotia of the pathogen. Garrett (*in* Holton et al, 1959) referred to *P. omnivorum* as "undoubtedly the most studied and best understood root disease fungus in the world today," a comment now more applicable to the take-all fungus, *Gaeumannomyces graminis* var. *tritici* (Asher and Shipton, 1981). H. Fellows and C. H. Ficke (1934) and M. D. Glynne (1935) reported that take-all declined with wheat monoculture in Kansas and Woburn, England, respectively, so that the disease was relatively unimportant by the fourth year of wheat culture.

The changing attitude of the times was shown in the 1930 presidential address of H. S. Fawcett (1931) to the American Phytopathological Society entitled "The Importance of Investigations on the Effects of Known Mixtures of Microorganisms."

A series of classical papers by R. Weindling appeared from 1931 to 1941 on the antagonism of *Trichoderma viride* to soilborne pathogens. This common soil saprophyte was shown to parasitize *Rhizoctonia solani*. Acidification of citrus seedbeds to pH 4.0 with aluminum sulfate favored the inhibition of *R. solani* when *Trichoderma* was introduced but not when it was absent. The purified gliotoxin from the antagonist was lethal to *R. solani* when diluted to one in 300,000. Webster and Lomas (1964) have since questioned whether the gliotoxin reported by Weindling and Emerson (1936), and the viridin reported by Brian and McGowan (1945) actually may have been from *Gliocladium virens* (*Hypocrea gelatinosa*) instead of *T. viride* (*H. rufa*). *Gliocladium* spp. are less common in soil than *Trichoderma* spp., and *Trichoderma* spp. apparently produce little or no antibiotic. Gliotoxin and viridin are stable under acid but not under alkaline conditions.

It was, and still is, difficult to demonstrate the presence of antibiotics in soil because they are produced in such small quantities in situ, probably in microsites, and are sorbed on clay particles, decomposed rapidly by other microorganisms, or both. E. Grossbard (1948, 1952) demonstrated the presence of clavacin (patulin) in autoclaved soil following the addition of 3.5% glucose, 5% sterilized wheat straw, and inoculum of *Penicillium patulum*. The amended soil was leached with water after one week and, when the leachate was tested on agar culture, it proved antibiotic to three bacteria. When leached after six weeks, no inhibitory activity was detected. Similar results were obtained with *Aspergillus clavatus* and *A. terreus*. Production of antibiotics was confirmed by D. G. Hessayon in 1951 for *T. roseum* in autoclaved nonsupplemented soil; by K. F. Gregory, O. N. Allen, A. J. Riker, and W. H. Peterson in 1952 for *P. patulum* in nonsterile supplemented soil; and by D. Gottlieb and P. Siminoff in 1952 for *Streptomyces venezuelae* in sterilized nonsupplemented soil (demonstrated with paper chromatography).

J. M. Wright, in seven papers from 1952 to 1957, showed that *Trichoderma viride* (*Gliocladium?*) produced 10.0 μg of gliotoxin per gram of autoclaved nonsupplemented soil, none in nontreated nonsupplemented soil, 320 μg per gram of autoclaved soil supplemented with wheatmeal, 2.4 μg per gram of autoclaved nonsupplemented soil reinfested with a 1/60 dilution of normal soil, and 320 μg per gram of autoclaved, supplemented, and reinfested soil (Wright, 1956). Noninoculated pea seeds sown in nonsterilized soil containing *T. viride* contained gliotoxin in the seed coats, and inoculation of mustard seed with *T. viride* diminished damping-off caused by *Pythium* sp. Wright also showed that *Penicillium nigricans* produced griseofulvin in the same soil sterilized and supplemented but that the antibiotic was rapidly degraded by a *Pseudomonas* sp. In her studies, the antibiotics were extracted with ether or ethanol and were assayed by testing chromatography-paper strips on cultures of *Botrytis allii* or *Bacillus subtilis*. The antibiotics were thought to be formed in bits of organic matter, and other microorganisms were considered to affect their production by inhibiting the growth or changing the metabolism of the producer or by destroying the antibiotic as it was formed.

Despite these studies, there still was skepticism by 1963 "that antibiotics produced under natural conditions are capable of exerting a significant and generalized, rather than restricted and generally inconsequential, effect" (Pramer, *in* Baker and Snyder, 1965). However, recent work has more firmly established the natural occurrence and importance of antibiotics in soil. G. W. Bruehl et al (1969) showed that *Cephalosporium gramineum* produces a wide-spectrum antibiotic in wheat straw that enables it to retain possession of that substrate for two to three years, whereas nonantibiotic-producing mutants in straw are overrun by saprophytes in a few months. This was the first use of genetically marked (mutant) strains to demonstrate a role of an antibiotic in the ecology of a soil microorganism. J. W. Kloepper and M. N. Schroth (1981c) showed that inoculation of seeds with *Pseudomonas* spp. that produce wide-spectrum inhibitors increased plant growth, whereas the inhibitor-negative mutants were ineffective. Both the inhibitor-producing strain and inhibitor-negative mutants were shown, by use of other genetic markers, to colonize the plant roots, but only the inhibitor-producers displaced rhizosphere microorganisms pathogenic to roots. This was the first demonstration, again with genetically marked strains, that ability to produce inhibitors in the rhizosphere is a key attribute of an antagonist introduced to displace one or more root pathogens.

H. H. McKinney (1929) showed that inoculation of plants with one mosaic virus could protect the plants against another virus. "When tobacco plants . . . with one of the writer's light-green mosaics were reinoculated with the . . . virus of yellow mosaic, no change in symptoms occurred. The plants continued to produce typical light-green mosaic after three reinoculations with the virus of yellow mosaic." This demonstration was verified by T. H. Thung in 1931 and R. N. Salaman in 1933. This effect, variously referred to as acquired immunity, interference, cross protection, and antagonism, is biological control.

MYCOPARASITES

Among the early descriptions of fungi there were reports on mycoparasites. Micheli (1729), Bolton (1788–1791), and Bulliard (1791–1812), for example, described *Nyctalis* spp. as parasitic on *Lactarius, Russula*, and *Collybia* spp. By 1890, records of mycoparasites were sufficiently numerous that Zopf (1890) published a list of them. However, there was little mention of these parasites as significant in controlling plant pathogens (K. F. Baker, 1980) until Weindling (1932) and Weindling and Fawcett (1936) studied parasitism of *Trichoderma viride (Gliocladium?)* on *Rhizoctonia solani* in relation to control of seedling damping-off. Drechsler began in 1933 a series of classical studies on fungi parasitic on Phycomycetes and nematodes. In our first book (Baker and Cook, 1974) few significant examples of biological control by mycoparasitism could be cited (Chapter 8), but soon thereafter the subject began to receive the attention it merits.

SUPPRESSIVE SOILS

O. A. Reinking and M. M. Manns (1933) studied the soils of Central America "resistant" and "nonresistant" to Panama disease of banana (caused by *Fusarium oxysporum* f. sp. *cubense*). The disease was most damaging on sandy soils, where a planting succumbed in 10 years or less, and least damaging on clay soils, where a planting survived for 20 years or more. The pathogen, which appeared able to live saprophytically, could be isolated from nonresistant soils 10 years after bananas were grown but could not be so isolated from resistant soils. For many years, the United Fruit Company depended on recognition of resistant soils and the use of field flooding for control of the disease.

A similar relationship to soil type was studied by J. C. Walker and W. C. Snyder (1933) in Wisconsin. Pea wilt (caused by *F. oxysporum* f. sp. *pisi*) was severe in sandy loam but not in Wisconsin red clay soil. Plants grown in the conducive sandy loam that had been transported in containers to a red clay area developed severe disease, but those in the nearby clay soil did not. In glasshouse pot tests, three successive sowings of peas gave 60% disease in noninfested sandy loam mixed 10 to 1 with infested sandy loam, but clay so handled resulted in only 7.5% disease.

These two studies, and others reviewed by Toussoun (*in* Bruehl, 1975), should have established the value of suppressive soils in control of plant pathogens. However, the rapid development of resistant crop cultivars after about 1920 (by 1930, one fourth of U.S. agricultural land was said to be planted to disease-resistant crops [Coons, 1937]) relieved the economic pressure. Biological control achieved with antagonistic microorganisms was "put on hold" until the early 1960s, when interest was revived because of concern about environmental pollution, development of pesticide resistance, lack of adequate or reliable resistance in crops to many important pathogens, and the trend toward more intensive farming with less crop rotation.

SOME LANDMARK EVENTS OF THE PAST FIFTY YEARS

Clearly, by 1934 research had shown that both resident and introduced antagonists could diminish plant disease; antagonists could produce antibiotics; some root pathogens were unable to infect the host or even to survive in soil without a nutrient base; the biological balance could be manipulated by altering the soil organic matter content, temperature, or pH; some soils were naturally suppressive to a soilborne pathogen or could become so following the occurrence of disease caused by that pathogen; the total microflora of resistant (suppressive) soil could be transferred to a nonresistant (conducive) soil, making it resistant; and inoculation of a plant with a mild virus could protect that plant against a more damaging virus. Some of the key words (e.g., antibiosis, antagonist, suppressive effect, immunity, biological control, and protective inoculation) had been introduced, and some of the important principles of the subject had been elucidated.

Since details of more recent investigations essentially are the subject matter of this book, it is only necessary here to indicate some of the landmarks in the further development and application of biological control.

R. Leach (1937, 1939) showed that *Armillaria mellea* could be controlled in tea plantations in Nyasaland by ringbarking the phloem of forest trees one year before felling them for clearing. This severing of phloem caused depletion of carbohydrate in the roots and their invasion by harmless saprophytic fungi, rendering the roots unavailable to *A. mellea*. Twenty of 24 forest trees not ringbarked developed armillaria root rot, but only one (which was questionable) of 12 ringbarked trees developed the disease. Leach also showed that if tea prunings were left on the surface of the ground for one month before burial, they were invaded by saprophytes that prevented their invasion by *A. mellea* after burial.

Key Contributors to Biological Control of Plant Pathogens

W. C. Snyder 1904–1980

A. S. Costa 1912–

D. E. Bliss 1903–1951

J. D. Menzies 1915–

F. E. Clark 1910–

C. Leben 1917–

J. Rishbeth 1918–

T. Kommedahl 1920–

J. H. Grente 1921–

R. J. Cook and G. W. Bruehl (1968) showed that the stubble-mulch or trashy-fallow method of tillage, in addition to conserving summer fallow water in eastern Washington, also allowed the stubble to become colonized rapidly by saprophytes that prevent subsequent invasion by *Fusarium roseum* 'Culmorum' when it was plowed under.

D. E. Bliss (1951) studied the control of *A. mellea* in citrus orchards by soil fumigation with a sublethal dosage of carbon disulfide, a method first reported by W. T. Horne (1914). The pathogen was not killed by fumigation in infected roots in sterilized soil, but was killed in soil in which *Trichoderma viride* was present. The mycelium of the pathogen was "weakened" and rendered more susceptible to destruction and displacement by *T. viride*, an effect later shown (Ohr and Munnecke, 1974) to result from inhibition of production of an antibiotic by the mycelium of *A. mellea* that protects it from *T. viride*. This apparently was the first commercial use of a chemical-biological control combination.

R. A. Fox (1977; *in* Baker and Snyder, 1965; *in* Toussoun et al, 1970) developed an effective integrated control program for the white, brown, and red root diseases of plantation rubber trees caused respectively, by *Rigidoporus* (*Fomes*) *lignosus*, *Phellinus* (*Fomes*) *noxius*, and *Ganoderma philippii* (*pseudoferreum*). Old jungle or rubber trees are poisoned to accelerate their death and permit rapid colonization by saprophytes before felling. A mixed cover of creeping legumes is later sown between the newly planted rows of rubber trees, leaving a clean-culture strip 1 m wide on each side of the trees. This cover provides conditions favorable for: 1) growth of the pathogens and invasion of the legume roots before rubber tree roots are large enough to provide a food base for infection; 2) accelerated decay of stumps and other infested material, with proliferation of pathogen fructifications and rhizomorphs further depleting the essential food base; and 3) development of actinomycetes and other antagonists that inhibit and lyse the rhizomorphs. Fox referred to these effects as "decoy, decay, and deter" and "biological control with limited manual assistance." This is one of the most ingenious and effective controls of root diseases of tropical plantation crops by cultural practices. This work was based on the previously mentioned studies of K. Bancroft (1912) with *R. lignosus*.

K. F. Baker and C. M. Olson (1960) devised a thermally selective treatment of soil to eliminate plant pathogens while leaving a portion of the established non-pathogenic microflora to luxuriate and provide competition for any pathogen that might be introduced. Such treatment with aerated steam at 60°C for 30 minutes is now used commercially in glasshouses. The method also provides a useful, flexible, and experimental means to selectively separate different segments of the antagonistic microflora because of their differential sensitivity to moist heat. This permits tentative identification of an antagonistic microbiota (Baker and Cook, 1974).

Common Scab of Potato

Biocontrol of the pathogen responsible for this disease illustrates the use of 1) crop monoculture to alter pathogen-antagonist balance; 2) maintenance of moist soil during early tuber formation to favor bacterial antagonists; 3) use of green soybean plants turned under before potatoes are planted, to favor antibiotic production by *Bacillus subtilis*, which inhibits multiplication by the pathogen.

Pathogen: *Streptomyces scabies* (Eubacteria, Actinomycetales).

Hosts: Potato; infrequent on beet, turnip, radish.

Disease: Occurs as irregular, concentric, scabby layers of cork on the surface of tubers; the center is not powdery as with *Spongospora subterranea*. This is largely a cosmetic disease.

Life Cycle: Slender, nonmotile, segmented filaments in soil break up into minute gonidia that may become windborne in dust. Infection is through stomata, young unsuberized lenticels, or fresh wounds; proliferation of corky tissue results. The pathogen is common in soil (especially dry soil) and is favored by organic matter (especially manure); it is introduced into soil with infected seed pieces or infested manure (having survived passage through animals).

Environment: The pathogen is most active in dry, neutral to alkaline soils, and disease may be suppressed by maintaining soil moisture levels below (wetter than) -0.4 bar or soil pH values below 5.2–5.5. Liming the soil increases scab severity. Optimum soil temperature for scab development is 23–25° C, with little scab developing at 11° C or above 30° C.

Biological Control: Disease is decreased by 1) long-term potato monoculture (Chapter 4), which apparently favors the pathogen at first but eventually favors antagonists; 2) maintaining moist soil during early tuber formation so that bacterial antagonists on the tubers remain active until all lenticels are suberized; 3) turning under green soybean plants in irrigated soil before planting (*Bacillus subtilis* forms antibiotic[s] in the soybean residue that are inhibitory to multiplication of the pathogen). The maintenance of moist soil perhaps accounts, in part, for the low incidence of scab on some 182,000 ha of irrigated potatoes in states in the Pacific Northwest.

References: Labruyére (1971), Lapwood and Hering (1970), Lewis (1970), Menzies (1959), Weinhold et al (1964).

P. Broadbent et al (1971) and Broadbent and K. F. Baker (1974a, 1974b) showed that soil in an avocado grove in Queensland, Australia, was suppressive to *Phytophthora cinnamomi*. Although the pathogen was present, root rot was controlled. The successful treatment developed by the grower, G. Ashburner, was to plant continuous cover crops and to apply abundant poultry manure to maintain a high soil organic matter content nearly like that of the surrounding undisturbed rain forest, and to apply sufficient dolomitic limestone to keep the pH about neutral. This treatment is now widely used in avocado groves in eastern Australia. The soil became conducive when steamed (100°C for 30 minutes) but remained suppressive when treated with aerated steam (60°C for 30 minutes), owing to survival of antagonistic actinomycetes and spore-forming bacteria. This was the first commercial use of organic matter to maintain the natural suppressiveness of field soil, and the first use of selective thermal treatments to select out the specific antagonist types active in a suppressive field soil.

Following the laboratory tests of Henry (1931), J. D. Menzies (1959) showed in pot tests that the antagonistic microbiota of a soil suppressive to potato common scab also could be transferred to a nontreated conducive soil, making it suppressive. He suggested that "a disease-controlling equilibrium which may take many years to appear naturally, may be induced in one season by appropriate treatments." P. J. Shipton, R. J. Cook, and J. W. Sitton (1973) made the first successful field transfer of a suppressive biota from a field suppressive to *Gaeumannomyces graminis* var. *tritici* to a field that had been treated with methyl bromide. Suppressive soil from a field in central Washington where wheat had been grown in monoculture for 12 years, when rototilled to a 12–15 cm depth (0.5% w/w) into a fumigated field in western Washington to which the pathogen was introduced, gave some suppression of take-all the first year and nearly equaled the suppressive site in the second and subsequent years. Five other eastern Washington soils introduced in the same way, as well as nontreated native soil from the western Washington site, had no effect on take-all in either the first or second crop. This method for increasing the quantity of suppressive soil still has not been commercially exploited.

M. Gerlagh (1968) increased the antagonism of soil to *Gaeumannomyces graminis* var. *tritici* by growing four successive wheat crops per year in pots in a glasshouse. He also showed that the antagonists in suppressive soil were destroyed by steaming at 60°C for 30 minutes. His work provided the first conclusive evidence that suppressiveness of soil causing take-all decline is microbiological in nature. This was confirmed by Z. D. Vojinović (1972, 1973), who showed that the rhizosphere microflora was changed by four successive wheat plantings, the rhizosphere soil from each planting being added to the soil in the next planting. Gram negative, nonspore-forming, motile rods were abundant after the fourth wheat crop. The rods were difficult to culture, highly specific, and antagonistic to *G. graminis* var. *tritici*. They were killed by heat of 40–60°C. P. J. Shipton et al (1973) also confirmed that suppressive-soil from Washington was sensitive to aerated steam at 60°C for 30 minutes.

Y. Henis, A. Ghaffar, and R. Baker (1978, 1979b) used a similar technique to raise the level of suppressiveness of soil to *Rhizoctonia solani* by five successive weekly radish plantings with the pathogen added to acid soil. *Trichoderma harzianum* increased during this period. Only two of five soils so treated became suppressive, indicating some specificity of soil. I. Chet and R. Baker (1981) found an acid soil in Bogotá, Colombia, that was naturally suppressive to *R. solani*. *Trichoderma hamatum*, the antagonist, could be transferred to conducive Colorado soil, making it suppressive. The *Trichoderma* produced no antibiotics, but it produced three enzymes that enabled it to digest mycelium of *R. solani* and *Pythium* spp.

G. C. Papavizas and J. A. Lewis (1981) induced, by ultraviolet irradiation, new biotypes of *T. harzianum*, some of which gave enhanced biocontrol of rhizoctonia damping-off and white rot of onion, greater tolerance to benomyl, and superior survival in soil. This may be the first selection of more efficient antagonists from among induced mutants for improved biological control.

H. T. Tribe (1957) showed that *Coniothyrium minitans* parasitized sclerotia of *Sclerotinia trifoliorum* in loam or clay field soil, killing 85–99% of them in 11 weeks. This perhaps is the first field demonstration of the effectiveness of mycoparasites of sclerotia in biological control.

John Rishbeth, in classical work from 1950 to 1970, showed that inoculation of freshly cut pine stumps with the weak pathogen *Peniophora gigantea* would prevent invasion by *Heterobasidion* (*Fomes*) *annosum* and its spread to nearby standing trees. This method is now used on more than 62,000 ha of pine forests in England. Rishbeth devised a dry tablet inoculum and a liquid spore suspension that greatly expedited use of the method. Oidia of *P. gigantea* also have been added to lubricating oil used for chain saws, inoculating the stump surface as it is cut (Artman, 1972). This was the first time an introduced antagonist was used for commercial control of a pathogen of aerial plant parts, and *P. gigantea* was the first antagonist registered by the U.S. Environmental Protection Agency for biological control of a plant pathogen in the United States.

N. J. Fokkema (1968) showed that resident antagonists would colonize pollen grains and other debris on aerial plant parts as a food base and would decrease infection by pathogens arriving later but that inoculation with pathogens and pollen grains (food base) together increased infection. J. P. Blakeman and A. K. Fraser (1971), and Blakeman and I. D. S. Brodie (1977) showed that *Sporobolomyces* sp. and *Pseudomonas* spp. on

Key Contributors to Biological Control of Plant Pathogens

R. A. Fox 1924–

G. C. Papavizas 1922–

A. Kerr 1926–

R. R. Baker 1924–

J. Webster 1925–

M. V. Carter 1926–

H. T. Tribe 1927–

J. Louvet 1928–

J. L. Ricard 1926–

leaf surfaces competed with germinating conidia of *Botrytis cinerea* in water drops. These studies clarified microorganism interactions on the phylloplane.

The destructive chestnut blight disease (caused by *Endothia parasitica*) was observed by A. Biraghi in 1951 to be diminishing in Italy, with spontaneous healing of cankers. This was studied in France by J. Grente and S. Sauret after 1969 and in the United States at the Connecticut Agricultural Experiment Station by P. R. Day and associates after 1975. Grente and associates introduced the concept of transmissible hypovirulence, which Day et al (1977) found to be associated with double-stranded RNA determinants. A hypovirulent strain infected with one or more double-stranded RNA entities was found to anastomose with compatible virulent strains in cankers, weakening the pathogen and enabling the canker to heal. It is estimated that the disease in southern Europe will be controlled naturally in 60 years, and Grente (Chelminsky, 1979) expects that inoculating every 10th chestnut tree will bring about control in about 10 years. This is the first commercial field use of hypovirulence in biocontrol of a plant pathogen.

A. S. Costa and G. W. Müller (1980) demonstrated commercial cross protection of citrus trees against tristeza virus in Brazil by prior inoculation with attenuated strains of the virus.

Inoculation of seeds with antagonists probably developed from the long-used inoculation of legume seeds with *Rhizobium* spp. R. K. S. Wood and M. Tveit (1955) found that Brazilian oat seed naturally contaminated with *Chaetomium globosum* and *C. cochlioides*, when planted, provided protection against *Helminthosporium victoriae*. They also found that when oat seeds heavily infested with *Fusarium nivale* were sown in the field along with oat-straw cultures of *C. cochlioides*, fair disease control was obtained. I.-P. Chang and T. Kommedahl (1968), and Kommedahl and I. C. Mew (1975) found that corn seed, inoculated with *C. globosum* or *Bacillus subtilis* and sown in fields that had a moderate inoculum density of *F. roseum* 'Graminearum', controlled seedling blight about as well as captan or thiram seed treatment.

Allen Kerr and associates showed in a landmark series of papers after 1972 that *Agrobacterium radiobacter* K84 applied to seed of fruit trees before sowing, or to nursery plants at the time of root pruning, provided excellent field control of crown gall. Numbers of the antagonist had to be at least equal to those of the pathogen, *Agrobacterium radiobacter* pv. *tumefaciens*. This treatment has proved effective over much of the world (Moore, *in* Schippers and Gams, 1979) and is the most widely used biocontrol of plant pathogens by an introduced antagonist. The antagonist produces a bacteriocin that inhibits the pathogen.

M. V. Carter and T. V. Price (1974) devised a method for controlling *Eutypa armeniacae* on apricot trees in South Australia, by inoculating fresh pruning wounds with a spore suspension of *Fusarium lateritium* plus 0.3 ppm

Key Contributors to Biological Control of Plant Pathogens

R. Mankau 1928–

M. N. Schroth 1933–

D. H. Marx 1936–

J. P. Blakeman 1939–

B. R. Kerry 1948–

I. Chet 1939–

M. Gerlagh 1940–

N. K. Fokkema 1940–

J. W. Kloepper 1954–

of benomyl. The benomyl inhibited the *Eutypa* until the *Fusarium*, which is insensitive to benomyl, was established to provide long-term protection against infection by *Eutypa*. The inoculum is now applied with specially designed pruning shears in commercial orchards (Carter, 1983). This example, hopefully, has introduced a new era of combined chemical and biological control.

R. I. Papendick and R. J. Cook (1974) showed that keeping a high water potential in wheat plants allows the natural resistance of wheat to function against the weak parasite *F. roseum* 'Culmorum', and controls the foot rot caused by this pathogen in Washington. The water potential was kept high by a combination of practices that make more water available to the crop (e.g., creating a dust mulch after summer rains, chisel plowing in the fall to increase water infiltration), that reduce the rate at which the crop uses the limited water supply (e.g., by seeding later and by using less nitrogen to keep the leaf area small), and by the use of water-efficient cultivars (Cook, 1980). The disease control achieved by these practices involved management of plant water potentials at a level relatively more favorable to the host than to the pathogen. Bacteria antagonistic to *Fusarium* were also involved, because they become inactive at a matric water potential of −5 bars[1] and below (Cook and Papendick, 1970). Biocontrol is thus favored by conserving moist soil longer into the growing season. This, and decreasing water stress on the wheat plants, provides effective control of fusarium foot rot in the Pacific Northwest of the United States.

It has been known since at least 1930 that exudates from plant roots determine which microorganisms occur in the rhizosphere (Rovira, *in* Baker and Snyder, 1965). However, T. G. Atkinson, R. I. Larson, and J. L. Neal, Jr. in a series of papers from 1970 to 1974 (Neal et al, 1973) were apparently the first to show that "the genotype of the host governs the magnitude and composition of bacterial populations in the rhizosphere with surprising specificity . . . presumably through its control of the quantity or quality of root exudates, or both." Unfortunately, this important work has not been continued in relation to biocontrol, although it adds a promising new dimension to breeding for disease resistance and illustrates the interrelationships of plant genotype and biocontrol. This type of work is being continued, however, in relation to nitrogen-fixing bacteria (Rennie, 1981).

S. Y. Yin and associates (1957, 1965) in the People's Republic of China, selected a *Streptomyces* sp. from among 4,000 isolates of actinomycetes from

---

[1] The recommended Standard International unit for water potential is pascal. However, in this book, we use the unit bar because essentially all plant pathology literature of the past 15 years — since the concept of water potential was introduced — has used bar, and to convert all values to pascals when citing them would be confusing. 1 bar = 10,000 pascals.

roots of cotton and alfalfa on the basis of its in vitro antibiosis to *Rhizoctonia solani* and *Verticillium albo-atrum*. This culture, strain 5406, has been used on 6 million hectares of cotton over the past 30 years, giving increased crop growth (Chapter 4).

Studies in the Soviet Union, England, and Australia had found by 1970 that inoculation of seeds or vegetative propagules with random soil bacteria before planting (bioenhancement of plant growth; bacterization) sometimes gave increased growth of crop plants, but the studies were plagued with variability. P. Broadbent et al (1971, 1977) and P. R. Merriman et al (1974; *in* Bruehl, 1975) selected bacteria and *Streptomyces* spp. for testing in steamed glasshouse soil and in field tests, on the basis of their broad-spectrum inhibition of several plant-pathogenic fungi and bacteria. The most inhibitory microorganisms gave the greatest increase in plant growth and produced little or no indoleacetic acid, gibberellin, phosphate solubilization, or nitrogen fixation. These findings prompted Broadbent et al (1977) to suggest that the increased plant growth resulted from the destruction of phytotoxins produced by the rhizosphere microbiota or from inhibition of nonparasitic pathogenic microorganisms in the rhizosphere that inhibited plant growth (Suslow and Schroth, 1982a). J. W. Kloepper and M. N. Schroth showed clearly in a series of papers after 1978 that effective *Pseudomonas* strains that occur in the rhizosphere can be found readily by selecting for inhibitory effect and that "deleterious rhizobacteria" occur in lower populations on roots colonized by the selected inhibitory pseudomonads. Noninhibitory mutants of effective inhibitory bacteria (Kloepper and Schroth, 1981c), as well as effective isolates grown under gnotobiotic conditions, failed to give increased plant growth (Kloepper and Schroth, 1981b). Siderophores and other substances produced by the introduced bacteria were thought to inhibit the deleterious rhizobacteria. Kloepper and Schroth concluded that "the ability to produce antibiotics in vitro was directly related to the capacity of rhizobacteria to significantly increase plant growth. However, antibiosis was not related to the root-colonizing capacity of these bacteria. . . . " *Bacterization thus seems to be a unique type of biocontrol (Chapter 11), offering a biological means of increasing crop yields without increasing energy or land demands or environmental pollution.* This type of investigation expands the horizon of plant pathology to include deleterious microorganisms, subclinical pathogens, and quasipathogens, in addition to the traditional concern with pathogens that penetrate the host tissue.

D. H. Marx (1972; *in* Bruehl, 1975) demonstrated that ectomycorrhizal fungi increased the resistance of pine roots to *Phytophthora cinnamomi* by: 1) using the available nutrients on the root surface, thereby decreasing those available to the pathogen; 2) the sheathing fungus mantle providing a mechanical barrier to penetration of cortical cells by the pathogen; 3) reinforcing the support of the rhizosphere to an antagonistic microflora; 4) inducing production of volatile and nonvolatile inhibitors by root cortical cells that limit pathogen in-

Cereal Cyst Nematode

Biocontrol of this nematode exemplifies the effective use of resident hyperparasites of eggs and cysts and is accomplished by monoculture of cereals susceptible to the nematode. The population density of the hyperparasites is dependent on the population of the nematodes.

Pathogen: *Heterodera avenae* (Nemata, Heteroderidae).

Hosts: At least 20 species in *Avena, Bromus, Festuca, Hordeum, Lolium, Phalaris, Secale,* and *Triticum*. Resistance exists in some cereals.

Diseases: Seedlings show symptoms when the fourth leaf emerges with a red tip. Roots are short, shallow, with multiple branches, bunchy in appearance, and have small galls. Fields later show areas of sparse, stunted, pale yellow-green plants.

Life Cycle: Lemon-shaped females are embedded in host tissue. Eggs usually are retained in the female, which is converted to a brown, tough cyst. Males are formed. Eggs, released as the cyst wall decomposes, hatch under stimulus of host exudates. Larvae penetrate very young seedlings, usually at root tips, and induce formation of giant cells in host tissues. Larvae molt four times. The disease is important on cereals in England, Continental Europe, Canada, and Australia and was recently found in Oregon in the United States.

Environment: *Heterodera avenae* is favored by moist soil in April and May. Severe damage depends on soils having adequate moisture and being suitably warm for hatching and infection by the nematode at a time when the host is in the seedling or young-plant stage of growth and hence most vulnerable.

---

fections; and 5) producing antibiotics. *Leucopaxillus cerealis* var. *piceina*, for example, produces a polyacetylene antibiotic (diatretyne nitrile) that inhibits germination of zoospores of the pathogen at 50–70 ppb and kills them at 3 ppm. Mycorrhizal pine roots did not exhibit decay of feeder root tips in soil infested with *P. cinnamomi*. Marx stated that "mature trees and seedlings with significant quantities of ectomycorrhizae growing in soils containing feeder root pathogens would have very little susceptible (nonmycorrhizal) root tissue exposed to attack." He suggested that the antibiotic might be translocated or diffused into nonmycorrhizal roots, protecting them. Methods of mass production of mycorrhizal fungi for field inoculation have been devised (Chapters 8 and 9).

K. M. Old and associates showed after 1967 that amoebae perforated and killed hyaline or pigmented mycelium, conidia, chlamydospores, and sporangiospores of many, but not all, fungi, as well as nematodes and their eggs (Old,

Biological Control: *Heterodera avenae* failed to increase on susceptible cereals grown continuously and intensively in England, Germany, and Denmark. Application of formaldehyde (3,000 liters per hectare) resulted in increased *H. avenae* populations, but only in suppressive soils. This led to the discovery that *Nematophthora gynophila, Verticillium chlamydosporium, Catenaria auxiliaris,* and an unidentified fungus were providing biocontrol; the first two hyperparasites were the most important. Females were destroyed in less than seven days at 13°C by the obligate parasite *N. gynophila;* zoospores encountered and infected female nematodes, and cyst and egg numbers were reduced by 95–97%. *Verticillium chlamydosporium,* an unspecialized parasite, sometimes attacks females but primarily attacks eggs. Egg infection by *V. chlamydosporium* sometimes exceeds 30%. The nematode population at these stages is concentrated as eggs in cysts and is maximally vulnerable to these natural enemies. Low summer rainfall resulted in less parasitism by *N. gynophila,* which requires moist soil for infection and for zoospore movement through soil pores. In England, the cereal cyst nematode is kept below the economic threshold of crop damage (<10 eggs per gram of soil) by these parasites. There are 200–400 spores of *N. gynophila* per gram of soil (50–100 million/m^2 of plow depth) when such effective nematode control is maintained. Several other *Heterodera* spp., but not *Globodera rostochiensis,* are also attacked by *N. gynophila.* Oospores of *N. gynophila* survive in soil for at least two years.

    In Australia, the wheat cultivar Festiguay acts like a trap plant and, when used in the rotation, lowers the population of *H. avenae.*

References: Kerry and Crump (1977, 1980), Kerry et al (1980), Kerry (*in* Papavizas, 1981), Rovira and Simon (1982), Thorne (1961), Williams (1969).

1977; Old and Darbyshire, 1978; Old and Patrick, 1976). This confirmed the earlier reports of C. Drechsler (1936, 1937) that *Geococcus vulgaris* "applies its mouth flush to the oospore wall, calks the zone of contact with a yellow secretion . . . and gradually perforates the delimited portion of spore wall, probably by some sort of digestive action." The potential of these amoebae in biological control is still unknown.

Parasites of phytophagous nematodes have been known and studied for many years, but only recently have some been found to be effective in reducing nematode populations in the field. R. Mankau and associates have been studying them intensively since the early 1960s. Mankau (1980) found that *Bacillus penetrans* effectively invaded *Meloidogyne* spp., reducing galling of tomatoes in glasshouse experiments. G. R. Stirling, M. V. McKenry, and R. Mankau (1979) found a fungus egg-parasite, *Dactylella oviparasitica,* that gave good field control of root-knot nematode in California. B. R. Kerry, D. H. Crump, and L. A. Mullen (1980) found two fungi, *Nematophthora gynophila* and *Verticillium chlamydosporium*, that attacked females and eggs, respectively, of the cereal cyst nematode (*Heterodera avenae*) and gave effective control in England in fields intensively cropped to cereals. As the nematode population increases with consecutive wheat crops, the hyperparasite also increases and thereby keeps the number of nematode cysts below an economic threshold. This is one of the few examples of density dependency in biological control of plant pathogens (Chapter 3). Treatment of infested soil with dilute formalin decreased the fungi but not the nematodes, thereby resulting in greater nematode damage to the wheat and demonstrating the effectiveness of the biocontrol.

CONFERENCES AND SYMPOSIA ON BIOLOGICAL CONTROL

The western states of the United States have, for the past three decades, been one of the most active centers for investigations of subterranean plant pathogens and of biological control of soilborne plant pathogens. The first annual Pacific Coast Conference on the Control of Soil Fungi was organized in 1953 by W. C. Snyder, W. A. Kreutzer, K. F. Baker, G. A. Zentmyer, and L. D. Leach. These conferences have continued to the present and have provided both a potent stimulus to research and informal opportunities to exchange ideas with fellow investigators.

Interest in biological control had reached such a level in the early 1960s that this group arranged an international symposium to summarize the available information and lay the groundwork for future studies. The National Academy of Sciences – National Reseach Council sponsored such a symposium at the University of California in Berkeley in April 1963, with 310 investigators attending from 24 countries. The proceedings were published as *Ecology of Soil-Borne Plant Pathogens, Prelude to Biological Control* (Baker and Snyder, 1965). Similar symposia have been held at five-year intervals since then.

As outlined by S. D. Garrett (*in* Toussoun et al, 1970), the 1963 symposium inspired the decision reached at the Tenth International Botanical Congress in Edinburgh in 1964 to initiate an International Congress of Plant Pathology, to be held at five-year intervals. The last four symposia on soilborne plant pathogens have been organized as part of the first four of these Congresses.

Western Regional Project W-38, on the Nature of the Influence of Crop Residues on Fungus-Induced Root Diseases, began in the western United States in 1955 and met annually through 1969 (Cook and Watson, 1969). Western Regional Coordinating Committee No. 12 (WRCC-12), on Management of the Biological Balance of Soil to Achieve Root Health for Efficient Crop

Production, began in 1973. It continued until 1976, when it became Western Regional Project W-147, on the Use of Soil Factors and Soil-Crop Interactions to Suppress Diseases Caused by Soilborne Plant Pathogens. W-147 has continued and meets annually. Thus, two organizations in the western United States have been meeting annually to present and discuss research on soilborne plant pathogens; the subject quite literally has been in the air there for the past three decades.

The second international symposium was held in London in July 1968 and was published as *Root Diseases and Soil-Borne Pathogens* (Toussoun et al, 1970). The third, held in Minneapolis, Minnesota, in September 1973, was published as *Biology and Control of Soil-Borne Plant Pathogens* (Bruehl, 1975). The fourth was held in München, Federal Republic of Germany, in August 1978, and was published as *Soil-Borne Plant Pathogens* (Schippers and Gams, 1979). The fifth is to be held in Melbourne, Australia, in August 1983, and the proceedings will be published. Other important symposia on the subject include:

Symposium, 63rd Annual Meeting, American Phytopathological Society, Philadelphia, Pennsylvania, in 1971; published as "Biological Control of Soil-Borne Pathogens – Mission Impossible?" *in* Soil Biology and Biochemistry 5:707–737, 1971.

International Symposium on Biological Control of Root Pathogens, Burnley, Victoria, Australia, in 1972; not published.

International symposium, Lausanne, Switzerland, in 1973; published in 1973 as *Perspectives de Lutte Biologique Contre les Champignons Parasites des Plantes Cultivees et de Pourritures des Tissues Ligneus.*

International symposium in the First Intersectional Congress of the International Association of Microbiological Societies, Tokyo, in 1974; published *in* Soil Biology and Biochemistry 8:269–283, 1976.

International symposium, "Interactions Between Microorganisms," at a meeting of the Federation of British Plant Pathologists, the British Mycological Society, and the Society for General Microbiology (Ecology Group), London, England, 1976, published *in* Annals of Applied Biology 89:89–114.

International symposium on Biological Control of Plant Pathogens, Rydalmere, New South Wales, Australia in 1977; published in 1977 as *Seminar Papers*, Annual Conference, Australian Nurserymen's Association.

Sections on biological control in IX International Congress of Plant Protection, Washington, D. C., in August 1979; published *in* Proceedings of Symposia of IX International Congress of Plant Protection (Kommedahl, 1981) and Abstracts of Papers.

International symposium, Symposia in Agricultural Research No. V, Beltsville, Maryland, in May 1980; published as *Biological Control in Crop Production* (Papavizas, 1981).

International symposium on Microbial Antagonism — The Potential for Biological Control, Second International Microbial Ecology Symposium, Warwick, England in September 1980; published *in Contemporary Microbial Ecology* (Ellwood et al, 1980) and Abstracts of Papers.

Discussion session, Pacific Division, American Phytopathological Society, Davis, California, in June 1980; not published.

International symposium on Suppressive Soils, 73rd Annual Meeting, American Phytopathological Society, New Orleans, Louisiana, in 1981; published as *Suppressive Soils and Plant Disease* (Schneider, 1982).

National Science Foundation workshop on Biological Control in Plant Pathology, Tucson, Arizona, December 1981; not published.

National Interdisciplinary Biological Control Conference, Las Vegas, Nevada, February 1983; not published.

International symposia on "Biological Control of Soilborne Plant Pathogens" and "Biological Control of Pathogens of Ornamental Plants," 75th Annual Meeting and Diamond Jubilee, American Phytopathological Society, Ames, Iowa, 1983, to be published as abstracts in Phytopathology.

The Commonwealth Agricultural Bureaux of England in 1980 started a journal, "Biocontrol News and Information," that includes some work on biocontrol of plant pathogens.

Thus, at least 19 important symposia, meetings, or workshops have been concerned with biocontrol of plant pathogens in the last 21 years, 15 of them international and 14 with published proceedings. Biological control clearly is taking its place beside other control procedures in plant pathology.

COMPONENTS OF BIOLOGICAL CONTROL

*Ideas or concepts and propositions are patterns for action
in the experimental sciences, and determine not only the questions
asked, but also the answers anticipated, and, unless one is
always on guard, the answers obtained.*
— G. K. K. LINK, 1932

Biological control is used in both plant pathology and entomology to mean control of one organism by another. However, the concept itself has developed along quite different lines within the two disciplines. A distinction between the two disciplines is apparent even in the earliest work. Thus, whereas entomologists introduced a specific parasitic insect from one country to control an insect pest in another, the first work in plant pathology emphasized management of resident antagonistic soil microorganisms through such practices as soil organic amendments, crop rotation, or burying of the residue of diseased plants. The recognized mechanisms of biological control of plant pathogens were also different from those of insects, namely, antibiosis first (Sanford, 1926), competition second (Millard and Taylor, 1927), and parasitism third (Weindling, 1932).

These differences continue to be reflected in the definitions used in the two sciences. Entomologists define biological control as "the action of parasites, predators, or pathogens in maintaining another organism's population density at a lower average than would occur in their absence" (DeBach, 1964). Plant pathologists have emphasized not only the reduction of the population (inoculum) density, but also biological protection of plant surfaces and biological control inside the host plant. The latter includes the host as a biological system acting alone (resistance) or in concert with other organisms that are antagonistic to the pathogen after infection or that induce host-plant resistance to the pathogen.

In entomology the tendency has been to recognize host plant resistance, behavior-modifying chemicals, and mass release of sterile males as "biological methods of control", "biological forms of control", and "biological pest suppression" (Jones and Solomon, 1974; Lundholm and Stackerud, 1980), all considered to be distinct from classical biological control. From our perspective as plant pathologists, we consider as unnecessary such a restrictive usage of the term biological control and the introduction of alternative terms and phrases that mean biological control.

Our thesis in this book is that biological control is the use by man of any organism for pathogen control. We see biological control as a continuum that ranges from reduction in longevity or virulence of the pathogen (such as may be accomplished by predatory fauna, microorganisms, or virus-like agents) to reduction of disease-producing activities of the pathogen by the host acting alone. Biological control microorganisms are seen to act directly through antagonism of the pathogen, or indirectly by providing stress or other conditions necessary for lethal expression of a latent virus in the pathogen. Biological control agents may induce host plant resistance, or the host plant itself may be the agent of biological control. The pathogen, host, and antagonists are all components in biological control.

The biological control of *Agrobacterium radiobacter* pv. *tumefaciens* by *A. radiobacter* strain K84 illustrates why the concept of biological control of plant pathogens must not be restrictive. Biological control by strain K84 is apparently an example of *biological protection of the infection court* on the host plant or at the surface of the susceptible host cell. The protection results from inhibition or exclusion of the pathogen by the antagonist. The antagonist is actually a close relative of the pathogen and carries a plasmid that codes for production of agrocin 84, a bacteriocin (Figure 3.1).

Biological control of crown gall also occurs when cells of an avirulent strain become attached at receptor sites on the host cell, in which case the sites are no longer available to the pathogen. The receptor sites may be occupied by dead as well as living cells of *A. radiobacter*, or by a mutant strain no longer able to produce agrocin 84 (Cooksey and Moore, 1982a), indicating that this effect is mainly one of protection, interference, or physical blockage rather than induced resistance. Such biological control exemplifies *cross protection*, achieved with an avirulent strain acting as a competitor of the pathogen inside tissues of the host plant.

If, as may develop in the future, genes for bacteriocin production are spliced into the host genome, thereby producing a host with the ability to manufacture agrocin 84 and protect itself, this would be biological control achieved by *host resistance*. The prospects for such gene splicing are not beyond the realm of possibility, since the induction of crown gall itself results from this mechanism, namely DNA from the tumor-inducing plasmid (T DNA) of the pathogen associated with (probably inserted into) the plant nuclear genome. A genetic change resulting in host resistance achieved by conventional plant breeding would also be biological control. Clearly, the pathogen, host, and antagonist, as interacting biological systems, all are components of biological control.

Crown Gall

Biocontrol of the crown gall pathogen exemplifies the use of an avirulent, bacteriocin-producing strain of a bacterium to control a related virulent strain. Bare-root nursery stock or seed is inoculated with the bacteriocin-producing K84 strain to achieve a population higher than that of the pathogen and to prevent transfer of the tumor-inducing (Ti) plasmid (T DNA) from the pathogen to the host.

Pathogen: *Agrobacterium radiobacter* pv. *tumefaciens* (Eubacteria, Rhizobiaceae).

Hosts: Wide range of woody and herbaceous plants in 93 families of Angiospermae.

Disease: Galls of various sizes are formed on crowns, roots, or stems; they may kill the plant or reduce its growth or yield. Galls may be decayed by secondary invaders. Infected nursery stock is unmarketable.

Life Cycle: Bacteria present in soil and rhizosphere invade the susceptible host through wounds caused when nursery stock is dug bare-root for transplanting. Bacterial cells attach to exposed wounded host cells, and DNA from their T DNA becomes associated with (probably incorporated into) the host nuclear genome. Host tissue becomes tumorigenic and, among other changes, begins to synthesize octopine or nopaline, unusual amino acids used by the pathogen as a source of carbon and nitrogen.

Environment: The pathogen is favored by moist soil of neutral pH and moderate temperature.

Biological Control: Transplants are inoculated before replanting with a suspension of cells of *A. radiobacter* pv. *radiobacter* strain K84 to give an inoculum density greater than that of the pathogen. Seeds also may be inoculated before planting. The avirulent K84 lacks the Ti plasmid but has another plasmid that codes genetically for production of a bacteriocin, agrocin, to which K84 is insensitive. Infection sites are occupied by K84, protecting them from infection by the pathogen. Because strains lacking the plasmids for bacteriocin production give little or no control, bacteriocin production appears to be an active control mechanism. However, some control mechanism, perhaps the prior occupation of infection sites, operates in addition to agrocin production; a mutant of K84 that did not produce agrocin prevented gall development when inoculated 24 hours before the pathogen, and K84 so applied prevented crown gall when the plant had been inoculated with an agrocin-resistant strain. K84 is effective against biotypes 1 and 2 of the pathogen but ineffective against biotype 3 (pathogenic on grape) and possibly other biotypes. The method is preventive rather than curative.

References: Cooksey and Moore (1982a, 1982b), Kerr (1980), Moore and Cooksey (1981), Moore (*in* Schippers and Gams, 1979).

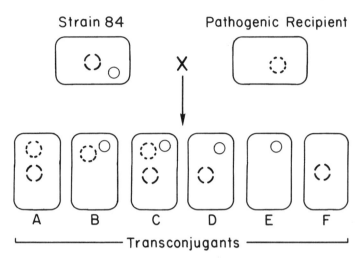

Figure 3.1 Diagrammatic representation of a cross between strain 84 of *Agrobacterium radiobacter* (nonpathogen) and *A. radiobacter* pv. *tumefaciens* (pathogenic recipient). Chromosomes are not shown. Of the two plasmids in strain 84, one codes for the production of agrocin 84 (solid line) and the other for catabolism of nopaline and for conjugation (broken line). The one plasmid in the pathogen (dotted line) codes for pathogenicity and for sensitivity to agrocin 84. The cross results in six plasmid transconjugants, of which B and C combine pathogenicity with resistance to agrocin 84. (Reprinted with permission from Kerr, 1980.)

DEFINITION OF BIOLOGICAL CONTROL

Earlier, we (Baker and Cook, 1974) defined biological control as "the reduction of inoculum density or disease-producing activities of a pathogen or parasite in its active or dormant state, by one or more organisms, accomplished naturally or through manipulation of the environment, host, or antagonist, or by mass introduction of one or more antagonists." This definition is here shortened to: *Biological control is the reduction of the amount of inoculum or disease-producing activity of a pathogen accomplished by or through one or more organisms other than man.* "Disease-producing activity" involves growth, infectivity, aggressiveness, virulence, and other qualities of the pathogen, or processes that determine infection, symptom development, and reproduction. The organisms include 1) avirulent or hypovirulent individuals or populations within the pathogenic species itself, 2) the host plant manipulated genetically, by cultural practices, or with microorganisms toward greater or more effective resistance to the pathogen, and 3) antagonists of the pathogen, defined as microorganisms that interfere with the survival or disease-producing activities of the pathogen.

Biological control may be accomplished through: cultural practices (habitat management) that create an environment favorable to antagonists, host plant resistance, or both; through plant breeding to improve resistance to the pathogen or suitability of the host plant to the activities of antagonists; through the mass introduction of antagonists, nonpathogenic strains, or other beneficial organisms or agents.

THE PATHOGEN AS A COMPONENT
OF BIOLOGICAL CONTROL

Pathogens include several types of agents: pathogenic fungi, bacteria (prokaryotes, including actinomycetes, rickettsialike bacteria, mycoplasmas, and spiroplasmas), nematodes, seed plants, algae, viruses, and viroids. Plant pathogens produce plant disease, defined as the destruction of living plant tissue or the impairment of plant function such that the life of the plant is shortened or its ability to grow normally, compete, or reproduce is diminished.

A fundamental but often overlooked concept is that pathogen and disease are not synonymous. Disease is a process that begins when the pathogen makes contact with the host (inoculation). Infection is the process by which the pathogen colonizes host tissue and establishes a nutritional relationship with the host plant. Symptoms and damage follow infection, and the disease usually ends with the pathogen reproducing on or in the injured or dead host tissue. A pathogen may be classified in the semipopular category "pest," but disease, being a process, is not a pest. To illustrate the distinction another way, a fly is a pest, but referring to a disease such as tuberculosis as a pest would be ridiculous.

This book is concerned with the nature and practice of biological control of plant pathogens during any part of their life cycle, including after infection or during reproduction. Although the focus is on the pathogen, obviously the purpose of biological control of pathogens is to suppress the economic disease they cause.

Plant pathogens do not form a natural grouping in the way that insects or weeds do. Consider the diversity: seed plants (mistletoe), fauna (nematodes), microorganisms (fungi, bacteria, and algae), and complex replicating molecules (viruses and viroids). In general, the concepts of biological control used in entomology apply to nematodes, and the concepts for biological control of weeds probably apply to biological control of parasitic seed plants. However, neither set of concepts is easily applied to biological control of fungi, bacteria, viruses, or viroids. *Biological control of plant-pathogenic fungi and bacteria may be accomplished through the destruction of existing inoculum, exclusion from the host, or the suppression or displacement of the pathogen after infection. Biological control of viruses, viroids, and certain prokaryotes is limited mainly, if not entirely, to systems operative within the plant host or vector.*

Biological control of plants (weeds) and biological control of plant pathogens are opposites and are very different subjects; the former attempts to favor a pathogen at the expense of the plant (Charudattan and Walker, 1982) and the latter to suppress the pathogen so as to favor the plant. This book deals only with biological control of plant pathogens.

The physical relationship of a pathogen to its host during pathogenesis may be either as an epiphyte (on the surface of the plant), an endophyte (inside the plant tissues), or both. Certain strains of *Erwinia herbicola* (Lindow et al, 1976; 1978a) and *Pseudomonas syringae* (Lindow et al, 1975b; 1977) initiate ice-crystal formation in supercooled water (water at temperatures between−2 and−5°C) on plants; if this occurs on the leaves of tomatoes, corn, or other frost-sensitive plants, the tissues are damaged and the responsible bacteria are then ideally positioned to colonize the dead cells (Lindow et al, 1975a). These bacteria exemplify pathogens that initiate plant damage as epiphytes. Certain toxicogenic or deleterious bacteria that occur on the roots of some plants (Suslow and Schroth, 1982a) also fall into this category. Viruses, viroids, and some prokaryotes (e.g., rickettsialike bacteria, mycoplasmas, and spiroplasmas) exist entirely within their host or vector and are rarely, if ever, exposed to the outside environment; such pathogens have no natural enemies except other plant pathogens or possibly secondary invaders of diseased tissue.

Most fungus pathogens have both epiphytic and endophytic phases. *Taphrina deformans* can survive on leaves as an epiphyte but produces disease as an endophyte. Powdery mildew fungi have simultaneous epiphytic and endophytic stages, the former as superficial mycelium and conidiophores and the latter as haustoria. Vascular-wilt fungi have a very brief epiphytic existence, namely, the period of prepenetration growth (if any) between germination of the propagule and penetration of the host root, but thereafter they exist entirely as endophytes within the root cortex and vascular tissue. In general, **the more internal the pathogen during the host-pathogen interaction, the less vulnerable the pathogen to control by antagonists.**

Plant pathogens may be categorized on the basis of the kind of nutritional relationship they maintain with the host. Plant-pathogenic bacteria in their residency or epiphytic phase on leaves and roots (Leben, 1981) exist mainly if not entirely in an apparent state of *commensalism* with the plant; they obtain nutrients (as leaf or root exudates) from the plant but cause no harm to it. However, given the right conditions for pathogenesis, the bacteria kill and destroy the host tissues through action of toxins and enzymes and then multiply in the dead tissue.

Pathogens that obtain their nutrients from a plant by killing the tissue in advance of colonization are *necrotrophic parasites* (necrotrophs) and include many important plant-pathogenic fungi and bacteria. Pathogens that obtain their nutrients from living cells at a cost to those cells but without killing them are *biotrophic parasites* (biotrophs). The mode of parasitism by plant viruses is entirely biotrophic, with no known period of necrotrophy; the infection cycle is completed before the death of the host cell. The mode of parasitism of plants by *Sclerotium rolfsii* is necrotrophic with no known period of biotrophy. Many parasites use both methods of parasitism, first biotrophy then necrotrophy; the duration of each varies according to the parasite, host, antagonists, and abiotic environment. Vanderplank (1978) suggested that the presence or absence of a living nucleus in the host cell affords a valid distinction between biotrophy and necrotrophy; if the host cell involved in the host-parasite interaction has a living nucleus, the parasitism is biotrophic, if not it is necrotrophic. In general, **biological control of strict biotrophs must occur on or within the living host or through the vector (when involved), whereas biological control of necrotrophs may or may not require the presence of the host.**

Many microorganisms may establish as epiphytes and endophytes (e.g., species of *Alternaria* and *Cladosporium* in leaves, and *Fusarium* spp. on roots) but exist in a largely commensal relationship with the host until the tissues become senescent. These early occupants are then ideally positioned to establish themselves more thoroughly in the senescent or dead tissue in advance of competing microorganisms. However, some of these pioneer plant colonists are suspected of accelerating the onset of plant senescence (Dickinson, *in* Jenkyn and Plumb, 1981) with obvious benefits to the fungus but at a cost to the host. Others commence aggressive colonization of the tissues at different stages of aging or in response to drought stress (e.g., low water potential) of the host. Such fungi are weak parasites but can be potent pathogens. **Biological control of pathogens that are favored by weakened, senescent, or stressed host tissue can be accomplished by methods that reduce disease-producing activity, for example by development of cultivars or use of cultural practices that delay or prevent stresses on the host.**

Pathogens may be subject to antagonism at any time during their life cycle, whether dormant or active, during their saprophytic or parasitic existence. Even biotrophic parasitism may prove inadequate as a means of escaping biological control, since the damage biotrophs cause to plant tissue opens the way for colonization by secondary organisms. In some cases, the secondary colonists add to the plant damage, as in the case of biota that decompose root-knot nematode or club-root galls. However, secondaries also have the potential to retard the spread of pathogens in host tissue, to suppress their reproduction, and even to displace pathogens in host tissue. An example is *Fusarium roseum* on pustules of snapdragon rust (*Puccinia antirrhini*).

Under humid conditions, the *Fusarium* commonly invades the open rust pustules. At 10–16°C, only the pustules are invaded, but at 21–32°C, the fungus also advances into healthy host tissue, decays leaves, and girdles stems. The *Fusarium* does not infect nonrusted host tissue but is able to drastically diminish inoculum production by the rust fungus (Dimock and Baker, 1951). Because isolates of the *Fusarium* vary in aggressiveness and virulence, it should be possible to select strains for inoculation whose invasion is restricted to pustules.

The longer the time between infection and sporulation, the more vulnerable the pathogen to displacement by nonpathogens (secondaries). *Scytalidium uredinicola, Sphaerellopsis (Darluca) filum*, and *Tuberculina maxima* invade perennial rusts such as those caused by *Cronartium* spp. on trees and suppress sporulation by the pathogen. In contrast, rusts such as those caused by *Puccinia* spp. on Gramineae may be susceptible to antagonism by colonists of the pustules, but the short time of only 7–10 days between infection and sporulation presents less opportunity for such biological control.

The specialized metabolism of pathogens, which enables them to invade their host and reduce their environmental vulnerability, can be used to advantage in controlling them. Such pathogens generally are more sensitive to unfavorable abiotic factors than are nonpathogens (strict saprophytes). In soil, such organisms can be eliminated by selective treatments with heat or chemicals that do not kill the saprophytes. Within the host tissues, highly specialized biotrophs may be restrained somewhat by environmental stresses on the host (e.g., temperature or water potential suboptimal for host cells); in contrast, less specialized or weak parasites generally become more aggressive and cause more damage with environmental stress on their host (Baker and Cook, 1974). Treatments or practices that alleviate environmental stress on the host assist biological control of these weak parasites by permitting expression of the intrinsic resistance of the nonstressed host.

Many pathogens continue to live on host tissues after the plant is dead. Such microorganisms are then saprophytes and obtain their nutrients and energy through *saprotrophy*. Necrotrophy and saprotrophy nutritionally are the same; apparently, the main distinction is that if the moribund tissues are still part of a living plant, obtaining nutrients from them is necrotrophy, but if the whole plant is dead, the continuing nutrition of the microbial inhabitants is saprotrophy. This distinction can be useful, but confusion may arise, as with fusarium and verticillium wilts. The pathogens responsible for these diseases invade wounds or root tips of the host plants through the differentiating tissues and then become established in the xylem. Nonpathogenic strains of *Verticillium* and *Fusarium oxysporum* may also colonize the cortical tissues and perhaps enter the stele of these same plants, but only the pathogens grow or are carried upward in the plant xylem. However, since the xylem vessels are nonliving tissue, the invader obtaining nutrients from the xylem fluid is marginally parasitic. The plant dies because of increased resistance to water flow (plugging of the xylem; Chapter 6). The pathogen then colonizes the dead stem tissues much as a strict saprophyte would do, but it does so from the inside out rather than, as epiphytes do, from the outside in. Since the entire plant is dead by this time, the continuing nutritional relationship is saprotrophy. *Fusarium oxysporum, V. dahliae*, and other vascular invaders might best be considered as saprophytes that are also pathogens.

Of the great variety of microorganisms that come into contact with plants, only a small proportion are capable of establishing a pathogenic relationship with them. Moreover, although pathogens are generally believed to have coevolved with one or possibly several host species, pathogens important on modern cultivated plants are not necessarily the same pathogens important on those plants or their progenitors in the natural habitat. An epidemic depends on the coincidence of available virulent pathogen inoculum, susceptible host tissue, and a favorable environment. The selection pressure presented by an agroecosystem provides different opportunities for the pathogen than does that of the natural ecosystem, including the presence of plants never encountered by the pathogen in its natural habitat. The life cycle of a pathogen may also be different in an agroecosystem than in a natural ecosystem, because the selection pressure is different. Gibbs and Harrison (1976) point out that apple mosaic virus is important because cloning and vegetative propagation are used to produce the trees; how the virus spreads in a natural situation or even if it is important there is unknown. The so-called gene-for-gene diseases likewise may be peculiar to agriculture, where the pathogen population is presented with an exceptionally rapid succession of hosts with monogenic resistance, in contrast to the natural situation where monogenic, oligogenic, and polygenic resistances are intermixed and the proportion of each changes more slowly (Browning et al, *in* Horsfall and Cowling, 1977). Biological balance achieved in an agroecosystem likewise may bear little or no resemblance to that achieved over millenia in a natural ecosystem; however, the principles of biological balance are the same.

Wilson (*in* Jones and Solomon, 1974) takes exception to the concept that biological control of insects attempts to restore the natural balance by duplicating conditions in the native home of the pest, e.g., by importation of natural enemies of the pest. He cites several examples of biological control of an insect pest achieved by agents that "had no connection with the natural balance of the native habitat." Brader (1980) goes one step further by suggesting that biological control of insects should involve "the deliberate manipulation of naturally occurring biological control elements" in contrast to the more traditional introduction of exotic parasites and predators. Plant pathologists have used this approach all along but are also now attempting to obtain biocontrol with introduced antagonists, including exotic microorganisms. Both approaches are needed.

Inoculum is the biomass of the pathogen (Mitchell, *in* Schippers and Gams, 1979), i.e., the living mass of the pathogen in the form of cells, spores, sclerotia, mycelium, or other propagative units available for infection. Most measurements of the amount of inoculum attempt to quantify *inoculum density*, i.e., the number of discrete propagative units of the pathogens per unit weight or volume of soil, volume of air, or surface area of the host. Such measurements are most useful if made when the pathogen is in a state of dormancy or quiescence; densities so obtained provide a measure of the "seed bank" of the pathogen. However, the amount of inoculum of a pathogen is a dynamic quality and changes over time (Bouhot, *in* Schippers and Gams, 1979) in response to energy flow from the host or nonhost plants, plant debris, or other sources. The increase in pathogen biomass should be a function of the available nutrients and of energy for growth and should follow a normal growth curve, with a lag phase, an exponential growth phase, and finally a stationary or probably declining phase if no contact is made with a host.

The biomass of some fungus pathogens may increase considerably before infection, i.e., from the time the propagule first breaks dormancy until infection has occurred (Figure 3.2). As an example, the biomass of *Fusarium solani* f. sp. *phaseoli* consists of dormant aging chlamydospores in the absence of an external energy supply but expands rapidly into numerous large colonies of mycelia (thalli) in the presence of bean roots or hypocotyls (Cook and Snyder, 1965). These thalli on the surface of the host are a source of hyphal tips that penetrate the host. The greater and more favorable the nutrient supply from the host or available in the soil solution, the greater and more vigorous the prepenetration growth (increase in biomass) of the *Fusarium* and hence the greater the amount of infection (Toussoun et al, 1960). *Gaeumannomyces graminis* var. *tritici* is another example; the amount of infection by this fungus relates closely to the size of the fragment of infested host residue (food base) and the availability of essential nutrients in the rhizosphere to support prepenetration growth (Chapter 7). Many pathogenic *Pythium* spp. are highly successful saprophytic colonists of fresh plant debris and may undergo considerable increase in biomass through saprophytism in response to a dead substrate as well as to a living host (Garrett, 1956). A biomass consisting of a few aging oospores is thus transformed into a much greater biomass consisting of numerous new oospores or actively growing hyphae available for infection of a susceptible host. The period of dormant survival in the absence of the host or other energy supply is the period of declining biomass. **Biological control of pathogen inoculum is obtained through the use of antagonists, the host plant, a trap plant, or other biological agent to retard or prevent increases in**

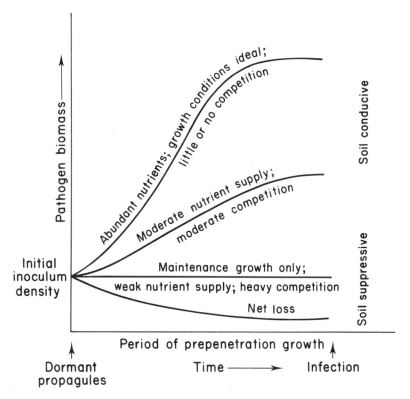

Figure 3.2. Some hypothetical examples of possible changes in pathogen biomass during prepenetration growth near a host root in suppressive and conducive soils.

biomass of the pathogen or preferably to contribute to decreases in pathogen biomass.

Infection capacity (Mitchell, *in* Schippers and Gams, 1979), or amount of infection by a given species or genotype of the pathogen, is related to the amount of pathogen biomass actually in contact with the surface of the host to be infected. Garrett (1970) recognized this in his expanded definition of *inoculum potential:* "the energy for growth of a parasite available for infection of a host at the surface of the host organ to be infected per unit of the host surface."

Energy, as implied in this definition, will be manifested as a greater biomass of the pathogen available at the surface of the host to be infected because of greater prepenetration growth. For example, large conidia of *Fusarium roseum* produced on a high-nutrient medium caused more damage to carnations than did an equal density of smaller conidia produced on a low-nutrient medium (Phillips, 1965), and large propagules of *Rhizoctonia solani* were more infective to bean seedlings than small propagules (Henis and Ben-Yephet, 1970). The larger or better nourished propagules may be said to have relatively more energy for infection of the host, i.e., more inoculum potential (Baker, *in* Horsfall and Cowling, 1978). However, this greater energy is likely to be manifested as more hyphae growing from the propagule to reach the host surface, or more prepenetration growth on the host surface and hence more pathogen biomass on the surface of the host to be infected. The infection capacity will also be a function of the physiological status of a given unit of pathogen biomass and the biotic and abiotic environment of the infection court. Garrett (1970) gives the following three factors as determinants of inoculum potential: 1) the cross-sectional area of the fungus in contact with the unit area of the host surface; 2) the relative vigor of the fungus hyphae attempting to invade the host (determined by the nutritional status of the protoplasm in apical regions of the hyphae); 3) the collective effect of environmental conditions that vary from optimal to completely inhibitory, and determine the actual or realized energy of growth from a possible maximum down to zero.

Virulence is the relative capacity of a given isolate of a pathogen to produce severe disease, and aggressiveness is a measure of the rate of growth or reproduction of the isolate, or other attributes that confer ability to produce a given size lesion in less time or produce more inoculum for infection of other tissues. Although virulence and aggressiveness are separate and important attributes of the pathogen (Nelson, *in* Nelson, 1973), biological control aimed at these traits generally involves the same agents and tactics. Mycelium of *Endothia parasitica* infected with one or more dsRNA determinants is less virulent than is mycelium of the same clone (genotype) of the pathogen but without the determinants (Dodds and Day, *in* Lemke, 1979). Such an infection might also reduce aggressiveness, as suggested for viruslike infection in mycelium of *Gaeumannomyces graminis* var. *tritici* (Ferault et al, 1979).

The frequency of virulent and/or aggressive individuals can also be managed biologically at the population level of the pathogen. An example is the use in England of two, three, or more barley cultivars, each having different genes for resistance to powdery mildew (caused by *Erysiphe graminis* f. sp. *hordei*) but sown as a mixture (Wolfe and Barrett, 1980). A race of *E. graminis* f. sp. *hordei* with virulence for one component of the mixture is either avirulent on the other two components or must compete with the race having virulence for that component only. Because of entrapment of inoculum by the resistant plants in the mixture, possible competitive effects between races, and possible induced resistance by inoculum making contact with resistant leaves, no one race is likely to become dominant in the field. The use of a cultivar mixture is therefore a means to obtain biological control of the frequency of either virulent or aggressive pathotypes in the population.

The pathogen may have the potential to bring about its own biological control, as when individuals within the population carry determinants for destruction of themselves or others. Hypovirulence in strains of the chestnut blight fungus (*Endothia parasitica*) caused by one or more dsRNA determinants is one example. Biological control of crown gall by agrocin 84 also fits in this category; the determinant for destruction of the pathogen is a plasmid for bacteriocin production carried by avirulent strains related to the pathogen.

Plant pathogens with insect vectors may bring about their own biological control by weakening the vector and thus reducing its fitness or life span within the insect population. Black-streak dwarf virus of rice, a rhabdovirus, apparently shortens the life of its planthopper vector. This disease was severe in southern China in 1970–1972, then disappeared abruptly (Li Debao, Zhejiang Agricultural University, Hangzhou, Zhejiang, People's Republic of China, *personal communication*). Disappearance of the disease coincided with reduced efficiency of transmission by the vector population, suggesting that the more efficient vectors had been replaced by biotypes of the vector that were less able to acquire the virus and were therefore healthier and more aggressive.

The pathogen becomes directly involved in biological control when it initiates a host defense (incompatible reaction) that limits or terminates its activity. An objective of biological control is to increase the chances for incompatibility between host and pathogen. This has been accomplished conventionally by plant breeding; it can also be accomplished with avirulent strains that somehow initiate or accelerate the host defense response or protect the host against virulent strains. Either way, biological control comes into play.

ANTAGONISTS AS COMPONENTS OF BIOLOGICAL CONTROL

An antagonist, broadly defined, is an opponent or adversary. Within biology, the term refers to a member of an interaction that interferes with another member, ranging from interference among molecules (e.g., growth factors) to that among higher plants (e.g., inhibition of one plant by another). In biological control of plant pathogens, *antagonists are biological agents with the potential to interfere in the life processes of plant pathogens*. Antagonists include virtually all classes of organisms: fungi, bacteria, nematodes, protozoa, viruses, viroids, and seed plants (e.g., trap plants).

Antagonists are the equivalent of "natural enemies" used in entomology. However, natural enemies (parasites, predators, and pathogens) of insects usually exhibit some form of specificity for their prey, in contrast to many antagonists (except viruses and viroids used for cross protection), which affect pathogens coincidentally or fortuitously. An aggressive root-colonizing bacterium with inhibitory activity against a wide array of microorganisms may protect the roots against damage from pathogens (Kloepper and Schroth, 1981c), thereby providing biological control. However, the effect on any particular pathogen of that plant is probably coincidental. The concept of "density dependency," (i.e., that the population of the natural enemy depends on and fluctuates with the population of its target pest) may apply to some antagonist-pathogen combinations, but this remains to be verified, since, in general, density-dependency has been studied very little where plant pathogens and their antagonists are concerned. One difficulty in the direct application of the concept to microorganisms is the matter of what constitutes an individual. Motile microorganisms that seek or are somehow attracted to their prey are the most likely to exhibit a density-dependent population in the classical sense. However, nonmotile antagonists may grow through soil from one pathogen propagule to another (e.g., *Sporidesmium sclerotivorum* on sclerotia of *Sclerotinia sclerotiorum*), increasing their own biomass but not necessarily their population expressed as number of individuals. Such a density relationship may be termed *biomass dependency*.

Antagonists that have a parasitic relationship with their microbial host presumably will have populations or total biomass dependent on the density or biomass of that host, although few measurements have been made to determine whether such relationships exist. The bacterial parasite *Bdellovibrio bacteriovorus* has been shown to exhibit a classical density dependency relationship (Figure 3.3) with a host bacterium, *Erwinia*, in liquid culture (Stolp and Starr, 1963) and presumably is also dependent in soil on density of bacterial hosts. However, many parasites of microorganisms in soil have more than one means to maintain their population or total biomass. The fungus parasite of eggs of *Meloidogyne* spp., *Dactylella oviparasitica*, also survives as a saprophyte on dead roots in the absence of nematode eggs (Stirling et al, 1979) and can therefore persist and even multiply through this nutritional relationship. It is not, therefore, density-dependent in the classical sense.

Mild strains of a plant virus may be classed as antagonists if, when introduced into a plant, they cross-protect that plant against a severe strain and thereby give biological control. As pointed out above, the only antagonists of plant viruses within living tissues may be other viruses capable of surviving and multiplying in those same tissues. The successful control of citrus tristeza virus in Australia (Fraser et al, 1968) and Brazil (Costa and Müller, 1980) indicates the potential for this approach to biological control. Moreover, the antagonist need not be a complete virus. A small satellitelike self-replicating RNA molecule "encapsidated with and dependent upon, but not part of the viral genome" of cucumber mosaic virus (CMV) (Kaper and Waterworth, 1977) causes increased virulence of CMV in tomato but reduces disease symptoms caused by CMV in squash, tabasco pepper, and Bantam sweet corn (Waterworth et al, 1979). The agent, referred to as CARNA 5

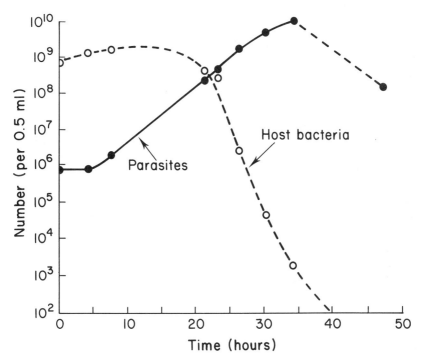

Figure 3.3 The change in cell density of a host bacterium (*Erwinia* sp.) relative to the cell density of the bacterial parasite, *Bdellovibrio bacteriovorus*, in a liquid culture, illustrating density dependency between the parasite and its host. (Reprinted with permission from Stolp and Starr, 1963.)

(CMA-associated RNA 5), regulates disease expression; when disease expression is suppressed, CARNA 5 is acting as a biocontrol agent – an antagonist of CMV.

Viruses or viruslike agents of plant pathogens (e.g., dsRNA mycoviruses) are, in effect, *hyperpathogens*, i.e., pathogens of pathogens. Such agents are contained within the pathogen itself, have the potential for biological control of the pathogen, and may also be classified as antagonists. Besides the use of dsRNA mycoviruslike agents as hyperpathogens to control a fungus pathogen directly, such infections might also be used to create an antagonist of a pathogen. The potential for this latter approach is exemplified by the development of killer wine yeasts *(Saccharomyces cerevisiae)* having dsRNAs that cause secretion of a molecule containing killer protein. S. Hara, of the National Research Institute of Brewing, Tokyo, Japan, has developed such killer strains to eliminate natural contaminating yeasts in wine making (Anonymous, 1982). Perhaps a similar approach could be used to develop a superior antagonist.

Antagonism is actively expressed opposition and includes antibiosis, competition, and parasitism and predation (Park, *in* Parkinson and Waid, 1960). *Antibiosis* is the inhibition or destruction of one organism by a metabolic product of another. *Competition* is "the endeavor of two or more organisms to gain the measure each wants from the supply of . . . a substrate, in the specific form and under the specific conditions in which that substrate is presented . . . when that supply is not sufficient for both" (Clark, *in* Baker and Snyder, 1965). Competition among microorganisms is mostly for food (carbohydrates, nitrogen, and growth factors) but also may be for space (e.g., receptor sites on cells) or oxygen. *Parasitism and predation* of one microorganism by another are fairly common.

Mycoparasites are fungus parasites of fungi and, like parasites of higher plants, may be separated into necrotrophs and biotrophs (Barnett and Binder, 1973). A necrotrophic mycoparasite kills its host, sometimes without infecting it, and then uses the nutrients released from the dead hyphae. Methods of killing the host hyphae are not understood, but may include release of toxic substances, cell wall-degrading enzymes, or other effects. A biotrophic mycoparasite obtains its nutrients directly from the living cells of its fungus host, either by growing in intimate contact with the host or by penetrating and growing inside the host cell (Barnett and Binder, 1973). Biotrophic mycoparasites tend to obtain nutrients while causing little harm to their fungus host, at least during the early stages of mycoparasitism. This may explain why most examples of biological control involving mycoparasitism are of the necrotrophic rather than the biotrophic type.

An antagonist may use more than one form of antagonism, and the action of some antagonists may fit under more than one mechanism. Some *Gliocladium* spp. cause death and dissolution of their host hyphae by secreting one or more antibiotics; these antagonists, coiled around the host hyphae, are then able to grow on the dead cell contents. The antagonism begins with a form of antibiosis but is usually classified as necrotrophic parasitism. Antibiosis and competition also overlap, as when a microorganism in a mixed population obtains a nutrient in short supply by first secreting a metabolite, sometimes no more than its own staling products, to discourage competitors. A microorganism may produce a siderophore that is "antibiotic" in the sense that the compound produced inhibits growth of another organism nearby. In fact, the compound allows the producer organism to compete for iron, and the inhibitory effect on neighboring microorganisms results from their starvation for iron. A fungus antagonist that grows in close association with a host hypha, stealing nutrients as they are made available from the substrate by enzymes from the host, is effectively a mycoparasite, but the form of antagonism is competition. An antagonist using more than one form of antagonism in sequence or concurrently may present a problem in semantics, but this does not lessen the value of the concepts themselves and only confirms the limits of terminology intended to describe the

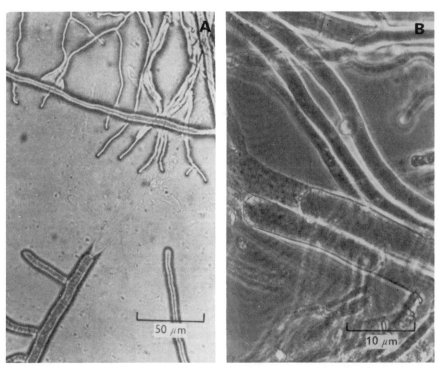

Figure 3.4 Hyphal interference by *Coprinus heptemerus* (hyphae of smaller diameter) against *Ascobolus crenulatus* (hyphae of larger diameter). A) The interference effect is limited to the *Ascobolus* cell contacted by hyphae of *C. heptemerus*. B) The septa separating the affected *Ascobolus* cells from those not affected bulge into the former. (Reprinted with permission from Ikediugwu and Webster, 1970a.)

multiplicity of ways by which one organism may restrict the activities of another.

Ikediugwu and Webster (1970a, 1970b) and Ikediugwu et al (1970) have described a mechanism of antagonism, which they refer to as *hyphal interference*. *Coprinus heptemerus* hyphae, when in contact with hyphae of *Pilobolus crystallinus* or *Ascobolus crenulatus* (Figure 3.4), and *Peniophora gigantea* when in contact with hyphae of *Heterobasidion annosum* (Figure 3.5) cause a drastic alteration of permeability of their cells. The affected cells, having lost semipermeability of their membranes, also lose turgor; death follows. The effect occurs even when the producer and sensitive hyphae are separated by a membrane. Hyphae only mildly affected show cessation of growth of the apex and extensive branching, as observed when hyphae placed in a hypertonic solution lose turgor. This ability of one organism to affect the selective permeability of cells of another organism is advantageous mainly in competition; growth of the competitor is limited, leaving more substrate for

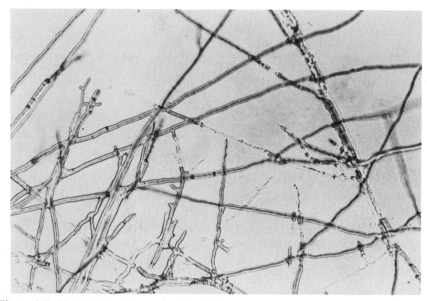

Figure 3.5. Hyphal interference by *Peniophora gigantea* (growing from left to right) against *Heterobasidion annosum* (growing upwards), showing zone of contact. The contacted cells of *H. annosum* are killed and are transparent or granular. (Reprinted with permission from Ikediugwu et al, 1970.)

the producer species. Conceivably, hyphal interference or a similar mechanism is one method by which a necrotrophic mycoparasite inhibits and then kills its fungus host cell. Hyphal interference has wide implications for biological control; for example, in situations where the desire is to obtain displacement of a pathogen in a lesion by secondary colonists, or in crop residue by saprophytes (Chapter 5).

Lysis is a general term for the destruction, disintegration, or decomposition of biological materials (Lamanna et al, 1973). Lysis may result from antibiosis (as when a bacteriocin causes internal disruption and ultimate disintegration of the bacterial cell), from competition (which causes starvation and ultimately lysis of the cell), from necrotrophic parasitism, or from enzymatic digestion of the host cell walls by a predator, resulting in spillage of the cell contents into the environment of the antagonist. These different mechanisms that lead to lysis are of two basic types: *endolysis* (internal dissolution of the cell protoplast without prior or concomitant dissolution of the wall) and *exolysis;* (dissolution of the cell wall and membrane, followed by spillage of the cell contents) (Baker and Cook, 1974). Endolysis includes autolysis, which is often difficult to distinguish from lysis caused by antibiotic action. Exolysis includes such effects as digestion of the hyphal walls by necrotrophic mycoparasites; and the perforation of cell walls, such as caused by vampyrellid amoebae (Old and Patrick, *in* Schippers and Gams, 1979). *Pathogens are subject to the effects of antagonists in soil during dormancy, in crop residue during saprophytic survival, on the surfaces of plant tissues during prepenetration activities, or even inside plant tissues during pathogenesis.*

Antagonism during dormant survival can lead to *biological destruction* of the pathogen inoculum and includes the familiar forms of predation, perforation, and parasitism of the propagules. Antagonism during dormant survival of the pathogen also includes suppression of propagule germination through competition for carbon, nitrogen, iron, oxygen, or other essential materials by microorganisms. Antagonism in crop refuse during saprophytic survival involves the action of saprophytes adapted to life in that residue and having the potential to destroy or displace the pathogen or inhabit the residue with the pathogen.

Antagonism on the plant surface involves mainly competition and antibiosis but also may involve mycoparasitism and predation. Such antagonism can provide *biological protection of plant surfaces* through exclusion, displacement, or inhibition of the pathogen on roots, leaves, or other plant parts. Antagonists that protect plant surfaces are herein called *phytosanitizing microorganisms.* Those that protect roots are *rhizosanitizing,* and those that protect leaves are *phyllosanitizing.* These terms are intended to denote microorganisms that improve plant health and yield by protecting the plant from pathogens, as contrasted with microorganisms that promote plant growth through liberation of growth factors or by providing nutrients in the rhizosphere.

The living plant supports many nonpathogenic or weakly pathogenic microorganisms as epiphytes and endophytes. These microorganisms, like many pathogens during certain stages of their life cycle, probably exist in a commensal or weakly parasitic relationship with the plant, but the consistency of types gives a clear indication that many are specialized in their roles as plant inhabitants and also as plant protectants. These microorganisms, by their superficial but early establishment on or in the tissues, have an advantage over competitors for the tissues when the plant dies. Such organisms are the important antagonists of pathogens on and within plants because of their ability to act as competitors for nutrients on the plant surface or as secondary invaders of lesions, and perhaps because of their ability to displace the pathogen, suppress its reproduction, or induce resistance to it. Some of the most effective antagonists of plant-pathogenic fungi are strains or species related to the pathogen, that are adapted ecologically to the same plant tissues occupied by the pathogen but are themselves nonpathogenic. **Any practice that favors success of the general microbiota or specialized groups of microorganisms over the pathogen, wherever the pathogen resides, is a means to achieve biological control of that pathogen.**

THE HOST PLANT AS A COMPONENT
OF BIOLOGICAL CONTROL

The host plant is a participant in virtually any biological control aimed at suppression of the disease-producing activities of the pathogen, as well as in many biological controls aimed at regulation of the amount of pathogen inoculum. Biological control accomplished by the host acting directly includes the suppression or termination of pathogenesis or reproduction of the pathogen by one or more mechanisms of host resistance. The host becomes indirectly involved in biological control when it provides the site for action of one or more antagonists of the target pathogen.

Host resistance may result from relatively specific biophysical or biochemical characters of the host that are effective against a specific pathotype or group of related pathotypes of the pathogen. Resistance of this type usually involves one or several genes that can be transferred through plant breeding. This kind of resistance ranges from a highly specific immune-type and often monogenic resistance, effective against a specific pathotype or virulence gene (e.g., vertical resistance as defined by Van der Plank, [1968]) to a less complete but more broadly effective and usually polygenic type of resistance, effective against all pathotypes of the species (e.g., horizontal resistance as defined by Van der Plank, [1968]). Many terms have been proposed to describe the monogenic, oligogenic, and polygenic forms of host-plant resistance (Browning et al, *in* Horsfall and Cowling, 1977).

Plants also have *general resistance* to pathogens; "resistance . . . that cannot yet be defined genetically, but which is controlled by the collective action of multiple physiological factors associated with active host metabolism and growth" (Baker and Cook, 1974). General resistance, as used here, is not the same as the genetic "general resistance" used by Browning et al *(in* Horsfall and Cowling, 1977) as the counterpart of "specific resistance"; it is the resistance of plants to strict saprophytes and weak parasites. The classification of a microorganism as a strict saprophyte or weak parasite is the reciprocal of describing the plant as being resistant to that microorganism. A more accurate phrase may be to say that such plants are not fully susceptible as long as the tissues are functioning normally. Such tissues become more susceptible when injured by freezing, heat, drought, improper nutrition, or reduced supply of oxygen. Plants become susceptible to colonization by strict saprophytes only when dead, but may become susceptible to aggressive colonization by weak parasites when near death (when severely stressed). General resistance in a plant can be enhanced by mycorrhizae, which help prevent deficiencies of phosphorus and possibly other nutrients, or by good crop husbandry (fertilization, improved water status of the plant, or other treatments that improve plant vigor or prevent the occurrence of plant stress). A wheat cultivar may have vertical resistance to a specific race of *Puccinia striiformis*, and other cultivars may have horizontal resistance to all races of *P. striiformis*. In addition, wheat has general resistance to foot rot caused by *Fusarium roseum* 'Culmorum', becoming susceptible only when tissues are severely stressed by very low (dry) water potential (Cook, 1973). Plant breeders select for vertical and horizontal resistance by exposing potential cultivars to populations of the pathogen. They can select for general resistance to a weak parasite that is favored by stress by exposing cultivars to the adverse environment (e.g., water stress) and picking the lines best adapted to the environment and therefore least prone to the stress.

In a review of our earlier book, Hirst (1974) asked, "What is biological control?" He expressed a prevalent view that plant breeding was unacceptable as part of biological control. Most would agree that suppression of the disease-producing activities of a pathogen by antagonists acting on or within the host tissue is biological control, as with *Phialophora graminicola* or other avirulent fungi related to *Gaeumannomyces graminis* var. *tritici* that superficially colonize the root cortex of wheat and thereby limit the disease-producing activity of *G. graminis* var. *tritici* (Deacon, 1976a). Most would agree further that, whether the host is an active participant or merely provides the battleground for pathogen and antagonists, the result is biological control. The divergence of opinion concerns mainly whether host-resistance mechanisms managed through plant breeding are any less biological control than resistance to a pathogen achieved with a "third party" agent. Biologically, these two situations are part of an inseparable continuum, and we therefore consider that both are biological control. As Luptin (*in* Jones and Solomon, 1974) stated, "Changing the genotype of the target organism (host plant) is entirely biological."

Some forms of plant resistance to pathogens can be used in biological control much as an antagonist might be used. Examples are trap plants such as *Crotalaria spectabilis*, which the root-knot nematode infects but in which it is unable to reproduce, and "inhibitory plants" such as *Tagetes* spp., which liberate terthienyls toxic to nematodes and some fungi (Baker and Cook, 1974).

A great variety of terms and expressions have been used to describe the phenomenon whereby a plant disease caused by one microorganism is lessened if the plant is inoculated first with a different microorganism. These include interference, cross protection, induced immunity, induced resistance, and defenses triggered by previous invaders (Horsfall and Cowling, 1980). Cross protection originated in virology to describe cases where infection of a cell by one virus lessened the likelihood that the cell would be damaged by a second virus, usually a related strain (Hamilton, *in* Horsfall and Cowling, 1980). The term has since been used in reference to protection involving one fungus against another. Cross protection is probably very common; one pathogen will rarely invade tissues already invaded by another pathogen, whether of the same or a totally different strain or species. By analogy, cross protection is like one warrior protecting his chief against another warrior—but if both are cannibals, the chief may be eaten either way. Induced resistance has a more specific meaning; it implies that the warrior has enabled the chief to defend himself. Most examples of diminished disease obtained following inoculation with a nonpathogen of that host do not reveal whether true induced resistance is involved. When the mechanism is not known, we will use the more general term cross protection. Induced resistance will be used only where this mechanism of biological control has been shown to occur.

Competition, antibiosis, or hyperparasitism is just as possible between cohabitants of living tissues as between those of dead substrates. Thus, while induced resistance to the pathogen may be involved when the antagonist occurs inside the tissues, direct antagonism of the pathogen cannot be ruled out simply on the basis that the habitat is living tissue.

By serving as a food base for pathogens, the host plant as residue continues to be involved in biological control even after its death. A residue with a low C:N ratio and a high proportion of readily available sugars or amino acids (e.g., roots and some leaves) theoretically will result in more intense activity and hence probably more antagonism among the associated microbiota than will a residue with a high C:N ratio and a low proportion of readily available sugars (e.g., wood or straw high in cellulose and lignin). This may explain why pathogens successful as inhabitants of crop residue are mainly those that live in wood (e.g., *Armillaria mellea*) or cereal straw (e.g., *Gaeumannomyces graminis, Cephalosporium gramineum, Pseudocercosporella herpotrichoides*, and others). Plant residues of different C:N ratios may be used to intensify competition between soil microorganisms and a pathogen, as with the addition of a green-manure crop to control phymatotrichum root rot of cotton (Streets, 1969), or of mature barley straw to control fusarium root rot of bean (Snyder et al, 1959). However, care must be exercised in the case of pathogens that are also effective saprophytic colonists of added fresh plant residue or other organic materials (e.g., *Pythium* spp., *Rhizoctonia solani, Sclerotium rolfsii*), which they use to increase their biomass and energy for attack of the next crop. Either the displacement of pathogens already in residue by nonpathogens, or preemption of pathogens by nonpathogens as saprophytic colonists of residue is subject to manipulation by cultural practices and has potential in biological control.

The host plant also becomes involved in biological control by its influence on the physical environment around and even within a given tissue. The temperature within a leaf may be higher than ambient during the day because of absorption of incoming radiant energy, and lower than ambient at night because of a net loss of radiant energy. The actual leaf temperature also is partly a function of the size, shape, and angle of the leaf and therefore is under the control of the host. Within host tissue, pathogens will encounter a variable or possibly a limiting supply of oxygen or a high partial pressure of carbon dioxide, ethylene, or other gases that could influence its growth and metabolism, perhaps even its ability to compete with secondary invaders. The water potential can also vary within a tissue according to the supply of soil water and rate of transpiration. Wheat stem tissues develop water potentials down to −40 to −50 bars under conditions of severe water stress; such water potentials are well below the ideal for normal growth of wheat but are even less suitable for the growth of the take-all fungus, *Gaeumannomyces graminis* var. *tritici* (Cook, *in* Asher and Shipton, 1981). The balance between host and pathogen is thus in favor of the host and take-all is suppressed. Conceivably, strictly saprophytic fungi able to grow at −40 to −50 bars (e.g., certain *Fusarium* spp. [Inglis, 1982]) may begin to displace *G. graminis* var. *tritici* in such plant tissues. The constraint on a pathogen that results from an adverse environment tied to features of the host is no different conceptually than the constraint imposed by a microorganism that causes an adverse pH or a limiting supply of oxygen; all are biological control operating with help from an organism.

ROLE OF THE ABIOTIC ENVIRONMENT IN BIOLOGICAL CONTROL

The abiotic environment includes temperature, water potential, radiation, pH, surface charges, partial pressures of gases, ions and elements, and energy-containing carbon compounds. Each of these factors and their subcomponents vary greatly over time and space, and each interacts to some extent with every other factor. The abiotic environment must, therefore, be recognized as dynamic, heterogeneous, and a complex of gradients. Most measurements of the environment, on the other hand, are understandably crude, particularly as they may apply to an infection court or other habitat of a pathogen or an antagonist.

The abiotic environment is indirectly involved in biological control. The direct inhibition or death of a pathogen caused by heat, desiccation, freezing, anaerobiosis, ultraviolet radiation, low or high pH, or ion toxicity is part of the broader *environmental control* but does not qualify as biological control unless 1) the unfavorable abiotic environment is produced by the host or an antagonist, or 2) the control results from antagonism or host resistance made possible by the environment. For example, certain wood-inhabiting pathogens of trees in New Zealand die within three to four weeks if the wood containing these pathogens is buried in flooded nonsterile soil but survive in the wood buried in flooded sterile soil (Taylor and Guy, 1981). A flooded nonsterile soil will quickly become anaerobic because microorganisms in the soil consume the oxygen faster than it can be supplied through the water-filled soil pores. Oxygen consumption in the sterile soil will be limited to that used by mycelium of the wood-inhabiting pathogens and may not exceed the rate of supply. Death probably results from lack of oxygen in the nonsterile soil. Such death is nevertheless biological control since the oxygen-deficient environment was created by action of the soil microbiota. Such death from oxygen deficiency is similar to death from starvation caused by the competitive action of the soil microbiota. *Observations on the influence of environment on a pathogen or disease must consider the possibility that the effect may be mediated through one or more biological systems.*

Water, like thermal energy, flows from areas of high to low energy status. The flow of water through the soil-plant-air continuum exemplifies this principle in that the water potential is highest in soil (e.g., field capacity is approximately −0.3 bar), lower in roots, lower still in leaves, and lowest in the air around the leaves. Microorganisms live in equilibrium with the water potential of their surrounding environment. When inside plant tissue, the microorganism will be at the same water potential as that tissue. Most soil bacteria cease multiplication at matric potentials below (drier than) −5 bars and at osmotic potentials below −15 to −25 bars. Of the limited number of Phycomycetes and Basidiomycetes studied, most cease growth at about −30 to −40 bars osmotic potential (Griffin, 1977; Cook and Duniway, *in* Parr et al, 1981). In contrast, most Ascomycetes and their imperfect stages have been found to grow at matric potentials down to −70 to −80 bars or even drier and at osmotic potentials down to −90 to −120 bars (Cook and Duniway, *in* Parr et al, 1981). There is considerable diversity, however; *Gaeumannomyces graminis* var. *tritici* does not grow below −40 to −50 bars osmotic or matric potential, in contrast to some *Penicillium* and *Aspergillus* spp., which grow at osmotic potentials down to −250 to −350 bars. The optimal osmotic potential may be −1 to −5 bars for bacteria, Phycomycetes, and Basidiomycetes and −5 to −25 bars for many Ascomycetes.

The dynamics of water potential encountered by microorganisms in nature can be illustrated by using a hypothetical mycelium of a fungus in a fragment of crop residue. If the soil containing the residue is at −1.0 bar (barely moist enough to support germination of a seed) then the residue (and hence the hyphae) will also be near or at −1.0 bar. If the hyphae have a hypothetical osmotic potential of −15 bars, then their turgor would be about +14 bars (the net is −1 bar as required for equilibrium). Suppose that after a period of evaporation, the soil (and hence residue and mycelium) dries to −15 bars (equilibrium relative humidity in the pore space is about 99%). Through osmoregulation, the hyphal tips must develop osmotic potentials of −29 bars in order to maintain the same +14 bars turgor. Osmoregulation is an energy-requiring process and may be accomplished internally by conversion of constitutive osmotica or by uptake of ions from the external solution (Chapters 5 and 7). Some fungi will be more adept at this adjustment than others. A more likely situation is that osmoregulation will not be adequate and the hyphae will have less turgor in the drier substrate. However, a complete loss of turgor pressure obviously terminates hyphal growth.

If the hyphae manage to grow into the root of a host, they will then equilibrate with the water potential of the root tissue, and if they enter the xylem, they will equilibrate with water potential there. The water potential of the tissues may be near the optimum for the pathogen, in which case severe disease is likely to result, especially if the prevailing water potential is less than optimal for the host (Cook, 1973). On the other hand, if the water potential of an infected area drops to a level near or below the minimum for the pathogen, secondary invaders with the ability to make good growth at this dry tissue water potential have a good chance to displace the pathogen and give biological control.

The flow of water from the environment to a microbial cell, or the reverse, is sufficiently rapid that gradients rarely if ever develop around microbial cells. In other words, microorganisms probably do not use or give off water at a rate sufficient to lower or raise the water potential (or equilibrium relative humidity) of the niche occupied. On the other hand, microorganisms may affect the temperature of their environment by their metabolic activity. An example is the elevated temperature of compost piles. The natural pasteurization (elimination of heat-sensitive pathogens) that results from composting is a form of biological control.

Phytophthora root rot of pineapple caused by *P. cinnamomi* can be controlled in Queensland, Australia, by application of elemental sulfur to the soil (Figure 1.2). This treatment lowers the soil pH to 3.8 or below, a condition suitable for growth of pineapple but inhibitory to reproduction by *P. cinnamomi* (Pegg, 1977b). The reduced soil pH results from production of sulfuric acid by sulfur-oxidizing bacteria, *Thiobacillus* spp. This is an example of biological control in which the antagonist, *Thiobacillus*, produces an acid pH unfavorable for the pathogen. This is also an example in which the activity of the antagonist is independent of the pathogen population; indeed, the effect is coincidental and the fact that *P. cinnamomi* becomes inhibited can be attributed simply to "bad luck" for the pathogen but "good luck" in terms of potential control. *Trichoderma* spp. are favored by the reduced soil pH and also play a role in the biological control of *P. cinnamomi* by sulfur applications.

The application of lime to raise soil pH favors take-all (Cook, *in* Asher and Shipton, 1981). Similarly, ammonium nitrogen fertilizer lowers the rhizosphere pH and suppresses take-all, in contrast to nitrate nitrogen, which elevates the rhizosphere pH and favors take-all (Smiley and Cook, 1973). Reis et al (1982) found that deficiencies of zinc, copper, and possibly other trace nutrients increase the intrinsic susceptibility (probably lower the general resistance) of wheat tissues to take-all. Furthermore, trace nutrient uptake by wheat was markedly less at soil pH values above 6.5–7.0. Practices that increase the availability of trace nutrients to wheat, either by lowering soil pH or by direct application, improve biological control of take-all, acting through the mechanisms of increased general resistance of the host to the pathogen (Chapter 6).

BIOLOGICAL CONTROL RELATED TO OTHER CONTROLS

Biological control obviously is not chemical control, defined as pathogen control achieved by application of synthetic or natural (extracted) chemicals. On the other hand, biological control may come into play following the application of a chemical; for example, the displacement of *Armillaria mellea* in citrus wood by *Trichoderma viride* following application of carbon disulfide (Bliss, 1951). Overlapping of boundaries is inevitable in any system of classification. It would seem simple enough to accept such examples as either chemical control, biological control, or biological control induced by a chemical treatment.

Many pathogens are controlled by cultural practices such as adjustment of the planting date, plowing under of crop refuse, or pruning of diseased branches. Cultural control is a broad term that includes most disease control practices involving soil or crop husbandry. Adjustment of planting date of wheat so that the period of anthesis does not coincide with the period of peak inoculum production by *Fusarium* (Cook, *in* Nelson et al, 1981) provides an escape mechanism from wheat scab and is biological control acting through the host. The adjustment of planting date to expose the wheat seedling to lower soil temperature and thereby take advantage of the general resistance in wheat to this same *Fusarium* (Dickson, 1923) is also biological control acting through the host. Both systems are cultural controls. Although biological control can be achieved with cultural practices, not all cultural controls should be considered as examples of biological control. Thus, the use of a dry fallow to kill cells of *Pseudomonas solanacearum* by desiccation (Sequiera, 1958) is a cultural control (direct effect of environment) but probably not a biological control. On the other hand, the use of a dry fallow to predispose sclerotia of *Sclerotium rolfsii* to decay caused by soil microorganisms (Smith, 1972) would be biological control. Cultural practices achieve disease control by any of several practical means, of which biological control is among the most important (Chapter 10). However, as pointed out earlier (Baker and Cook, 1974), such practices work, and from the practical standpoint, whether the mechanism is escape, direct effect of environment, eradication, or biological control is not important.

A stable productive agroecosystem with effective biological control of plant pathogens can be equated with a natural ecosystem in biological balance.

A balanced ecosystem is characterized by diversity of organisms, each adapted to the prevailing and cyclic environment, each a source of nutrients for one or more other organisms, yet each with the ability to endure or escape its natural enemies. A disease outbreak can commonly be traced to some ecological shock causing biological imbalance. Disease itself is an ecological force and will eventually restore balance within the ecosystem. Conventional agriculture contributes to biological imbalance by replacing biological diversity with a single plant genotype, often in monoculture; by placing crop plants in an environment to which they are poorly adapted; by exposing the crop to inoculum of pathogens without benefit of normal endurance or escape mechanisms; and by creating biological voids with tillage, pesticides, and other practices. The transfer of genetic resistance from wild progenitors to domestic cultivars, the breeding of better adapted cultivars, the greater use of mycorrhizae, the shifting of planting dates, the use of practices to increase soil organic matter, the greater use of crop rotation, the use of multilines and variety mixtures, the stabilization of cropping practices, and the introduction of antagonists of pathogens — all help restore biological balance and hence are means to biological control.

As a general guideline, the concept of biological control of plant pathogens as developed and applied in this book is any control achieved through a living system. Man is the only exclusion from our definition, since not to do so would be to include quarantines, production of pathogen-free stock, and even application of chemicals. While the concept is broad, we hold that a broad concept is in the best interest of biological control now and especially for the future. Excessive and unnecessary compartmentalization of this subject can only be counterproductive. The subject must be treated as a whole.

4

APPROACHES TO BIOLOGICAL CONTROL

*Advance is made by the study of cases which cannot
be embraced by a general principle, by the possession
of an eye to detect exceptions and of a mind
willing to examine them instead of putting
them aside because they are not in
harmony with preconceived ideas.*
—SIR RICHARD GREGORY, 1916

In the first experiments on biological control of plant pathogens, antagonistic microorganisms were added directly to soil in an effort to exclude or eliminate pathogen inoculum from soil or keep it in a state of suppression. A later approach used antagonists to protect the infection court, as exemplified by Rishbeth's (*in* Bruehl, 1975) control of *Heterobasidion annosum* by *Peniophora gigantea* applied to freshly cut stumps. The objective was to prevent the pathogen from colonizing the host; any effect on the population of the pathogen was of secondary importance. Still another approach has emerged, namely, cross protection and induced resistance aimed at biological control after host penetration. The lines of demarcation between these examples are blurred and even overlapping. Nevertheless, these three approaches—biological control of inoculum, biological protection of plant surfaces, and cross protection/induced resistance—are convenient divisions in the continuum and are therefore used as the basis of discussion in this chapter.

BIOLOGICAL CONTROL OF INOCULUM

Biological control of inoculum includes: 1) destruction of the propagative units (propagules) or biomass of the pathogen by hyperparasites, hyperpathogens, or predators; 2) prevention of inoculum formation; 3) weakening or displacement of the pathogen in infested residue (the food base) by antagonists; and 4) reduction of vigor or virulence of the pathogen by agents such as mycoviruses or hypovirulence determinants.

Biological Destruction of Dormant Propagules

Virtually all pathogens that spend any part of their life cycle in soil out-side the protection of their host are subject to attack by predators and hy-perparasites. Protozoa were reported to diminish the soilborne population of the cabbage pathogen, *Xanthomonas campestris*, by five orders of mag-nitude, thereby increasing their own numbers 100-fold (Habte and Alexander, 1975). Two species of Collembola (*Proisotoma minuta* and *Omychiurus en-carpatas*) destructively altered the inoculum density of *Rhizoctonia solani* in nonsterile soil under greenhouse conditions, resulting in less damping-off of cotton and greater plant growth (Curl, 1979). Vampyrellid amoebae per-forate and kill the propagules of many soil fungi (Figure 4.1), including con-idia of *Cochliobolus sativus* (Old, 1977) and chlamydospores of *Thielaviop-sis basicola* (Anderson and Patrick, 1978), with mortality rates commonly above 50%. Many different soil microorganisms, including phycomycetes, chytridiomycetes, hyphomycetes, actinomycetes, and bacteria were observed to parasitize oospores of *Phytophthora megasperma* var. *sojae, P. cactorum, Pythium* sp., and *Aphanomyces euteiches*, with 60–80% parasitism not un-common in soil (Sneh et al, 1977). Pratt (1978) observed in studies of oospore germination of *Sclerospora sorghi* in soil that "false" germination (i.e., growth of hyphae of hyperparasites from the oospore) was more common than true germination. *Trichoderma* and *Gliocladium* spp., *Coniothyrium minitans, Sporidesmium sclerotivorum*, and other fungi may cause high rates of mor-tality of sclerotia of phytopathogenic fungi in soil (Ayers and Adams, *in* Papavizas, 1981; Papavizas and Lumsden, 1980). Nematode-trapping fungi abound in soil and have been a source of fascination to soil mycologists for decades.

The remarkable amount of natural parasitism and predation of plant-pathogenic inoculum by antagonists in soil documents Darwin's "struggle for life." As Darwin said, "Everything is born to eat and to be eaten" (Stone, 1980).

Inoculum will eventually die whether exposed to antagonists or not. The difference between a septic (e.g., natural soil) and an aseptic (e.g., sterile soil) environment is that in general, the death rate is greater or biomass of the pathogen available for infection decreases more rapidly in natural than in sterile soil. In natural soil, this higher death rate results from the sum of many forms of biological stress on the propagules, including parasitism and predation, un-timely germination followed by starvation and lysis (Papavizas and Lumsden, 1980), and more rapid expenditure of endogenous reserves that results from the various sublethal stresses imposed by the associated microbiota. Most soil-borne pathogen populations, denied a susceptible host by crop rotation, will eventually subside below a threshold amount of inoculum needed to produce an economic level of disease. An objective of biological control is to somehow hasten the death of the propagules so that a susceptible crop can be grown in

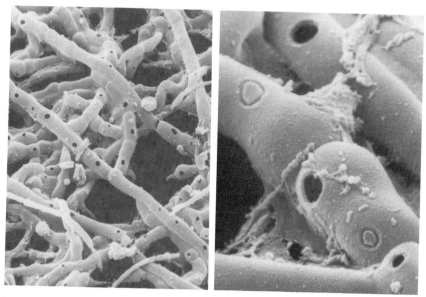

Figure 4.1. Photomicrograph of the annular depressions and perforations of the pigmented hyphae of *Gaeumannomyces graminis* var. *tritici* caused by vampyrellid amoebae. (Reprinted with permission from Homma et al, 1979.)

the field more frequently, or even on an annual basis, without economic loss from disease. The incorporation of organic materials (e.g., barnyard manure, alfalfa meal or hay [Menzies and Gilbert, 1967], or composted materials) into the soil can accelerate the death of propagules of some pathogens (Baker, *in* Pimintel, 1981). Many early workers—e.g., Sanford (1926), Hildebrand and West (1941), and King et al (1931)—observed the controlling effect of organic amendments on soilborne plant pathogens, but Clark (1942) apparently was the first to demonstrate that organic amendments can reduce the population density of a pathogen in soil. His work showed that sclerotia of *Phymatotrichum omnivorum* either decayed, or germinated and then decayed, in response to the amendments. Since Clark's work, germination-lysis (i.e., stimulation of germination by nutrients from decomposing organic material, followed by lysis of the germ tubes or hyphae through action of soil microbiota) has been shown to occur for the propagules of many soilborne pathogens in response to the nutrients supplied by organic amendments (Papavizas and Lumsden, 1980).

Among all the time-tested practices for elimination of the resting propagules of soilborne pathogens, probably none is more effective or quicker than flooding (Chapter 10). In the People's Republic of China, cotton and rice are rotated as the summer crops to control *Fusarium oxysporum* f. sp. *vasinfectum*. The procedure, referred to as the "wet-dry method," involves growing paddy rice at least once for every two crops of cotton. The combination of large quantities of organic manures (for fertilizer) and several months of flooding creates an environment that is intolerable for fungus pathogens of plants. Only those pathogens on the soil surface or able to float (e.g., sclerotia of *Rhizoctonia solani* responsible for rice sheath blight) are likely to survive.

The United Fruit Company used flood-fallowing in an effort to eradicate *F. oxysporum* f. sp. *cubense* in banana fields (Stover, *in* Holton et al, 1959). Unfortunately, recolonization of the drained soil by the pathogen was often rapid, and in some cases the incidence of fusarium wilt was greater where soils were previously flooded for four to six months than where never flood-fallowed. There was also evidence that soils unsuited to survival of the pathogen became more favorable for the pathogen after flooding. Recolonization apparently occurred from propagules of *F. oxysporum* f. sp. *cubense* that survived on the soil surface or were reintroduced on banana propagules. Algae may help provide oxygen for survival and even growth of aerobic organisms at the soil-water interface of a flooded field.

The problem of recolonization encountered by the United Fruit Company is similar to that encountered when fumigation is used to control pathogens (Chapter 7). The treatment may eliminate 99.9% of the inoculum, but even one surviving propagule per thousand propagules of initial inoculum density can be too many, especially since the treatment also shocks or destroys other components of the soil microbiota, leaving the soil conducive to rapid recolonization by the pathogen once the soil is drained or the fumigant dissipated and a susceptible crop planted. The "wet-dry method" used to control fusarium wilt of cotton in the People's Republic of China is similar to fumigating the soil for every two crops; the problem of recolonization is largely solved in this case by simply repeating the treatment every other year.

Solar heating beneath a polyethylene tarp (Katan et al, 1976) eliminates microsclerotia of *Verticillium dahliae* (Ashworth, 1979; Pullman et al, 1981a), sclerotia of *Sclerotium rolfsii* (Grinstein et al, 1979; Katan, 1980), and the propagules of *Pythium* spp., *Thielaviopsis basicola*, and *Rhizoctonia solani* (Pullman et al, 1981a). The treatment applied in unshaded areas of a four-year-old pistachio orchard (Figure 4.2) in southern California reduced the inoculum density of *V. dahliae* to trace amounts down to a depth of 60 cm after six weeks, with no serious effect on the trees themselves (Ashworth et al, 1982). The treatment works in part because of the prolonged elevated soil temperature achieved by entrapment of solar radiation (Pullman et al, 1981b). However, heat alone may not account for all propagule mortality, especially at greater depths, where propagules die even though soil temperature changes little or not at all (Katan, 1980). Some of the same effects achieved in soil beneath a layer of water may be operative, namely reduced gas exchange and hence effects associated with anaerobiosis (Chapter 7). Tarping holds considerable promise

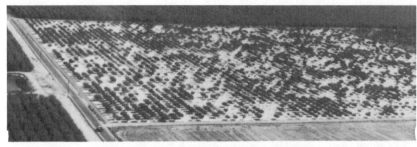

Figure 4.2. Tarping of the soil surface in pistachio groves with sheets of polyethylene for control of verticillium wilt of pistachio trees. Top, an aerial view of a grove fully tarped. Bottom, rows of pistachio trees with tarp. (Reprinted with permission from Ashworth et al, 1983.)

for elimination of inoculum of soilborne pathogens in areas with adequate sunshine.

To this point, our examples of biological destruction of propagules of soil-borne plant pathogens have been limited to those achieved with cultural practices or soil treatments. In each case, no one specific antagonist can be credited with the destruction; rather, control is achieved through multiple effects involving many different microorganisms, often with the propagules predisposed to their effects. All of these examples involve management of resident antagonists, but biological control of pathogen inoculum also may be achieved with introduced antagonists (Chapter 8). Commercial application is still very limited, but the progress is encouraging, particularly with the hyperparasites of sclerotia of *Sclerotium rolfsii*, *Sclerotinia* spp., and other sclerotia-producing pathogens.

Wells et al (1972) were among the first to demonstrate the potential for biological control of sclerotia in the field by application of a hyperparasite. They used *Trichoderma harzianum* isolated originally from rotting sclerotia of *S. rolfsii* and applied to a field where *S. rolfsii* had been a problem on tomatoes the previous year. The field was planted to tomatoes again, and the antagonist, grown on ground annual ryegrass seed, was applied one to three times to the soil surface as a band over each row of tomato plants. The earliest application was when plants were 4 cm tall. The percentage of disease-free plants at harvest ranged from 91.4 for a single treatment applied early, to 99.5 for three applications; only 21.9% of plants in the nontreated rows were healthy. Backman and Rodriguez-Kabana (1975) developed a mass-delivery system whereby *T. harzianum*, grown in diatomaceous earth impregnated with molasses as an energy source for the pathogen, was used to control *S. rolfsii* on peanuts. The incidence of disease was reduced and yields were increased significantly in tests over a three-year period. Commercial preparations of *Trichoderma* spp. are now available in several countries (Papavizas et al, *in* Papavizas, 1981). Such preparations presumably would be used in the United States if treatments were economical.

Several hyperparasites have been tested in the field for efficacy against sclerotia of *Sclerotinia* spp. (Ayers and Adams, *in* Papavizas, 1981). Turner and Tribe (1975) obtained good control of sclerotia of *S. trifoliorum* in England using a preparation of *Coniothyrium minitans* broadcast on the soil surface. *Coniothyrium minitans* has also controlled *S. sclerotiorum* in the field in Australia (Trutmann et al, 1980) and *Sclerotium cepivorum* in England (Ahmed and Tribe, 1977). Huang (1977) showed that *C. minitans* readily parasitizes and kills the sclerotia of *S. sclerotiorum* shortly after they form on the roots and within the stems of wilted sunflower. Huang (1980) subsequently obtained significant biological control of *S. sclerotiorum* on sunflower in Manitoba by applying a preparation of *C. minitans* in the seed furrow at the time of planting. This suggests that for row crops, it may be necessary to destroy sclerotia only within the row and ignore sclerotia that, if germinated, would not reach the plant anyway. In Maryland, *Sporidesmium sclerotivorum* introduced into field plots heavily infested with *S. sclerotiorum* destructively parasitized the dormant sclerotia of this pathogen. The incidence of lettuce drop caused by *S. sclerotiorum* on romaine lettuce the following spring and summer was significantly less in treated than in nontreated plots (Ayers and Adams, *in* Papavizas, 1981).

Sclerotia are generally more susceptible to attack by antagonists after germination (Coley-Smith, *in* Schippers and Gams, 1979). Perhaps part of the benefit of applying hyperparasites of sclerotia at the time of planting is that the sclerotia are stimulated to germinate by the host and are then more susceptible to antagonists. However, the evidence for *C. minitans* and *S. sclerotivorum* leaves little doubt that these hyperparasites destroy sclerotia, whether or not the sclerotia have germinated.

While the tendency is to apply hyperparasites of pathogen propagules just before sowing the crop, it may be better to apply them when the propagules are forming, or at least before the propagules are disseminated into the soil. By this means, the antagonist is present in highest concentration when the inoculum density of the pathogen is highest; this should maximize the number of initial "hits." Continued infections and destruction of pathogen propagules by a hyperparasite in soil might be expected to follow a pattern of infections in foci (Van der Plank, 1975), with the amount of secondary spread of the hyperparasite among propagules being limited or nonexistent owing to the great distance between propagules. Pathogen propagules tend to become ever more uniformly distributed in soil over time, which increases their chance for contacting a host root but also maximizes their chances to escape an antagonist growing or spreading from other infected propagules. A model of McCoy and Powelson (1974) for the relationship between propagule density and distance between the nearest propagules uniformly distributed in soil indicates distances of 0.8 and 1.1 mm for 2,000 and 1,000 propagules per gram, respectively, but a much greater distance as the population drops below 1,000/g. Applying the antagonist just as the inoculum is formed or released might help circumvent this advantage of escape otherwise gained by the pathogen when distributed more uniformly and sparsely in the soil.

It may be possible to establish antagonists with the inoculum at the source, e.g., on or within the host plant, so that each inoculum unit of the pathogen is released into an environment already infested (or infected) by an antagonist. Merriman (1976) observed that sclerotia of *Sclerotinia sclerotiorum* taken directly from lettuce plants in the field decayed faster than sclerotia taken from pure culture. The natural sclerotia entered the soil already harboring a variety of potentially antagonistic fungi, including *Trichoderma hamatum, T. koningii, T. viride,* and *Coniothyrium minitans* (Trutmann et al, 1980). In contrast, sclerotia from pure culture entered the soil free of these colonists. Merriman et al (1979) showed further that sclerotia taken from the surface of lettuce plants yielded three times more fungi when surface sterilized and plated out than sclerotia taken from inside the host plant. Similarly, with *S. sclerotiorum* on wilted sunflower plants in Canada, by the end of the growing season 59, 76, and 29% of the sclerotia on the root surface, inside the root, and inside the stem, respectively, were killed by *C. minitans.* Sclerotia inside the stem were the most likely to escape the antagonist, but even many of these were infected (Huang, 1977). In tests where *C. minitans* was introduced, field-grown sclerotia decayed sooner than culture-grown sclerotia (Trutmann et al, 1980), indicating the advantage to biocontrol of prior establishment of *C. minitans* and other antagonists in the sclerotia before their burial. *A strategy designed to permit maximum colonization of the sclerotia by antagonists on and inside the host before the sclerotia become mixed with the soil may present the best chance for adequate destruction of the sclerotia after their burial in the soil.*

Prevention of Inoculum Formation

For many plant pathogens, the destruction of individual propagules by a one-on-one or even a mass-action system involving antagonists will be too inefficient or impractical. A more efficient way may be to prevent the inoculum from forming. The logic of this approach for cyst nematode control has been pointed out by Wilcox and Tribe (1974), who state that trapping by fungi is too inefficient and that a better way is to use parasites of females and cysts. Sayre (1980) suggested that this principle applies to nematodes in general. In at least three examples, resident hyperparasites have been shown to control nematodes and to have potential as introduced antagonists or as resident antagonists managed by cultural practices. The examples are parasitism of females and cysts of *Heterodera avenae* by *Nematophthora gynophila* and possibly other hyperparasites, which help to maintain safe populations of the nematode under intensive cereal culture in England (Kerry et al, 1980); parasitism of females and eggs of *Meloidogyne* spp. by *Dactylella oviparasitica*, which provides biological control of this nematode in California orchards of peaches on susceptible Lovell rootstock (Stirling et al, 1979); and the parasitism of *Meloidogyne* spp. and other nematodes by *Bacillus penetrans* in many soils (Mankau and Prasad, 1977). In all three examples, the parasites or pathogens of the nematodes are effective against the source of inoculum — females, cysts, or eggs — rather than against the larvae responsible for plant infection.

Prevention of inoculum formation may be the best method of biological control of inoculum of aerial pathogens, particularly those pathogens having inoculum of the "compound interest" type (Van der Plank, 1963). The antagonists in these cases will probably include microorganisms that colonize lesions on the host plant and then suppress sporulation through antibiosis, direct parasitism, starvation, or possibly hyphal interference (Chapter 6). The antagonist may be a hyperparasite of the target pathogen or an aggressive saprophyte that is adapted to the diseased tissue and has the ability to displace the pathogen in plants in time to prevent or retard sporulation by the pathogen.

Biological control of pathogen reproduction may play a role in the natural suppression of rust diseases of trees. A survey in 1965 of 48 nonsprayed stands of white pine in the U.S. Pacific Northwest revealed that 62% of all lethal-type trunk cankers caused by *Cronartium ribicola* were inactivated and that aecial production on them was reduced to 5.4% of the potential (Kimmey, 1969). *Tuberculina maxima* was thought to be the principal cause of the inactivation. *Tuberculina maxima* invades tissue infected by rust mycelia but dies when the rust fungus dies. In western Canada, more than 80% of rust galls caused by *Endocronartium harknessii* were infected by *Scytalidium uredinicola*. This antagonist destroys hyphae of the rust fungus in wood tissue to a depth of 300 μm below the sori and causes hyphal distortions at even greater depths (Tsuneda et al, 1980). Death of the rust fungus is thought to result from toxin or from lytic enzymes produced by *S. uredinicola*. It would be interesting to know whether the mechanism of hyphal interference reported by Ikediugwu and Webster (1970a, 1970b) also plays a role in the displacement of the rust fungus by *S. uredinicola*.

Scytalidium uredinicola and *T. maxima* are also hyperparasites of galls caused by *Cronartium quercuum* f. sp. *fusiforme* on oak in the southeastern United States; *T. maxima* reduces the frequency of oak galls that produce aecia, and *S. uredinicola* reduces production as well as germinability of aeciospores (Kuhlman, 1981a, 1981b). *Scytalidium uredinicola* reduced aeciospore production by 72% (Figure 4.3). In another study, Kuhlman and Matthews (1976) observed through surveys in Florida that about 90% of the uredial sori of *C. strobilinum* on oak were parasitized by *Sphaerellopsis (Darluca) filum* and that this antagonist also became established on uredia and telia of *C. quercuum* f. sp. *fusiforme* on six oak species in North Carolina. In all of these examples, the value of the antagonist may be in reducing the disease from potentially epidemic status to a largely endemic status, by suppression of secondary inoculum formation.

More work is needed on ways to foster the activity and especially the efficiency of antagonists of rusts on perennial hosts, if they are to be used more effectively. Avoiding practices that interfere with the activity of these parasites may be as important as the deliberate use of treatments to favor them. Thus, failure of the antibiotics phytoactin and cycloheximide to control white pine blister rust in the Pacific Northwest was thought actually to have been favorable to the rust, rather than simply ineffective, because the antibiotics were toxic to *T. maxima* (Wicker, 1968). In addition to choosing practices favorable to or not unfavorable to antagonists, considerations should be given to development of technology for their mass introduction. The rust diseases of trees seem suited for this kind of biological control.

Lophodermium seditiosum causes an important needle-cast disease of pines in nurseries, young plantations, and commercial plantings of Christmas trees in Great Britain and North America. The same fungus, as a saprophyte, may colonize needles killed prematurely by thinning, snowbreak, fire, animals, or other kinds of damage to the supporting branches. This "trash" is an

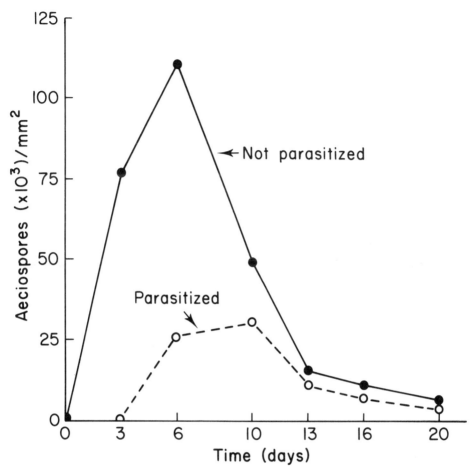

Figure 4.3. Average aeciospore production by *Cronartium quercuum* f. sp. *fusiforme* per square millimeter of gall surface on *Pinus taeda* in nonparasitized areas (solid line) and in areas parasitized by *Scytalidium uredinicola* (dashed line) following the appearance of unruptured aecia. (Reprinted with permission from Kuhlman, 1981a.)

important source of the ascospore inoculum responsible for the infections of the new or current-year needles of the young pines (Minter, 1980). Trash needles are also the habitat for the related *L. conigenum*, a strong saprophyte but a weak parasite not important as a cause of needle cast on new needles (Minter and Millar, 1980). A survey in Scotland and England revealed that *L. seditiosum* occurred mainly where *L. conigenum* was absent (Minter, 1980), i.e., in small isolated areas where pine was not native or had become extinct and was then reintroduced. In contrast, *L. conigenum* was dominant on trash needles in native stands. Minter (1980) proposed that *L. seditiosum*, being common on cone scales often mixed with seed, was introduced into the new plantings, whereas *L. conigenum*, having no such advantage for transportation, was not introduced. The introduced fungus was thus able to invade a habitat normally occupied by *L. conigenum*. Minter (1980) has proposed introducing *L. conigenum* where it is currently absent, to displace or preempt *L. seditiosum* in trash needles and thereby limit this important source of inoculum. This is an example of how a species related to the pathogen, adapted to the same ecological niches as the pathogen, but nonpathogenic itself, may be used to obtain biological control (Chapter 3).

Another example of biological control achieved by suppression of inoculum production of the pathogen occurs with oak wilt (caused by *Ceratocystis fagacearum*). In the United States, oak wilt has not spread rapidly in the southern part of its range (Kentucky, Tennessee, Arkansas, and Oklahoma), compared with its spread in more northerly states (Pennsylvania, Maryland, and West Virginia) (Tainter and Gubler, 1973). Temperatures in the southern states are somewhat limiting to the oak-wilt fungus, but Tainter and Gubler (1973) presented evidence that competition from *Hypoxylon atropunctatum* is also involved in the slow progress of wilt in this area. This secondary colonist of trees is favored by the same high temperatures that inhibit *C. fagacearum* and is thought to be the "primary determinant of survival of *C. fagacearum* in the South, especially in Arkansas." The antagonist as an aggressive competitor effectively displaces the pathogen in trees with oak wilt, thereby suppressing inoculum production and spread by the pathogen. A different antagonist of *C. fagacearum* apparently provides natural biological control of oak wilt in Minnesota. Gibbs (1980) demonstrated that the naturally occurring but nonpathogenic *C. piceae*, if introduced into fresh wounds 24 hours in advance of *C. fagacearum*, prevented infection by the oak-wilt fungus. Besides being adapted to grow in the fresh wood that serves as the infection court for *C. fagacearum, C. piceae* was also observed to overgrow the sporulation mats of *C. fagacearum*, thereby suppressing inoculum formation or distribution.

Many clones of saprophytic *Fusarium roseum*, by virtue of their aggressiveness as secondary colonists of lesions, may have potential for biological control of aerial pathogens if applied on the diseased plant so as to prevent inoculum production. For example, *F. roseum* 'Semitectum' invades both diseased tissue and reproductive structures of plant pathogens, including the leafy proliferations of the so-called "green-ear" caused by *Sclerospora graminicola* on pearl millet, where it parasitizes the oospores and makes them nonviable (Rao and Pavgi, 1976), and cercospora leaf spots of mulberry, phalsa, and nightshade, where it kills hyphae, conidiophores, and conidia of *Cercospora* (Rathaiah and Pavgi, 1973). *Fusarium roseum* 'Sambucinum' was "the most virulent hyperparasite of ergot from a worldwide collection" tested by Mower et al (1975). *Fusarium* spores applied as an aerosol to ergot in the honeydew stage on field-grown rye under sprinkler irrigation in the field gave an estimated 95% control. This *Fusarium* also contributed to the breakdown of ergotamine to psychotropically inert substances. The closely related *F. heterosporum* (*roseum* 'Heterosporum'), inoculated into rye and male-sterile barley florets within 12 days after inoculation with *Claviceps purpurea*, did not prevent infection but prevented sclerotium formation of the ergot fungus (Cunfer, 1975). The *Fusarium* used the honeydew as an energy source, infected sclerotia of *C. purpurea*, and caused their decay while still in the host-plant spikelet or shortly after they reached the soil (Mower et al, 1975). These fusaria are nonpathogenic on plants, growing there on a variety of organic substrates and forming abundant conidia; they seem reasonably well suited for mass production and application to aerial parts of plants.

Pathogens that increase their biomass through saprophytic colonization of dead plant tissues (crop residue) can be controlled biologically by prior establishment of nonpathogens in the residue. This approach makes use of the advantage held by the primary colonists of a substrate over secondary colonists (Bruehl, *in* Bruehl, 1975). Leach (1939) was among the first to knowingly apply this principle to obtain biological control of a plant pathogen. Working in Africa, he showed that tea bush prunings, left on the soil surface for a while to permit their colonization by airborne saprophytes, were then protected from colonization by *Armillaria mellea* when subsequently buried. The result was less *Armillaria* inoculum and hence less root rot of tea. In the U.S. Pacific Northwest, wheat stubble is left above ground for some time after harvest to reduce wind erosion of the soil; the practice also results in thorough colonization of the straw by airborne saprophytes, which then prevent colonization by *F. roseum* 'Culmorum' when the straws finally come into more intimate contact with the soil (Cook, 1970; Cook and Bruehl, 1968). Unfortunately, if the stubble is already colonized by Culmorum (because of earlier pathogenesis), leaving it standing above ground permits asexual sporulation by the pathogen; macroconidia formed on the straws are then washed by rains onto the soil, where they convert into chlamydospores.

Pythium and *Phytophthora* spp. are known to multiply in soil in response to added fresh organic material and to then cause greater damage as pathogens. Kaiser and Horner (1980) observed increased damage to lentils from *Pythium* in Iran when the soil was fertilized with fresh cow manure. In Hawaii, fresh papaya residue incorporated into the soil becomes heavily colonized by *Pythium aphanidermatum* and *Phytophthora parasitica*, which then destroy the next planting of young papaya seedlings (Trujillo and Hine, 1965). These fungi, especially *Pythium* spp., are quick to respond saprophytically to a fresh nutrient supply but are unlikely to respond to or colonize the substrate if it is already colonized by other fungi (Barton, *in* Parkinson and Waid, 1960). Fresh papaya residue, lawn clippings, cow manure, or virgin substrates should be composted or somehow allowed to become thoroughly colonized and the readily available nutrients used by saprophytes before being mixed into soil. Such prior colonization by saprophytes will reduce the utility of the substrate to pathogens such as *Pythium* or *Phytophthora*. Such composting also may reduce the amount of phytotoxic compounds that otherwise may accumulate during early stages of residue decomposition and predispose the host to *Pythium* (Toussoun and Patrick, *in* Baker and Snyder, 1965).

In England, oospores produced in leaves of apple through saprophytic colonization after leaf fall apparently make up a major portion of the inoculum of *Phytophthora syringae* responsible for storage rot of apple and pear, and for attack of the bark of apple trees at the soil level (Harris, 1979). An estimated 125 g of dry apple leaves occurs per square meter of soil surface, resulting in about 10^8 oospores per square meter. Preemption of this pathogen by prior colonists of the leaves would require the establishment or encouragement of antagonists very early, perhaps as epiphytes or endophytes of the leaves even before leaf fall.

Soils in avocado orchards on Mt. Tamborine in Queensland, Australia managed by the Ashburner system (large quantities of organic materials, poultry manure, and dolomitic limestone incorporated on a regular basis) are suppressive to *P. cinnamomi* because of effects on formation and longevity of sporangia (Broadbent and Baker, *in* Bruehl, 1975). The suppression is expressed as formation of fewer sporangia or as lysis and abortion of those formed (Figure 7.3). Leachate from two suppressive soils contained 290–774 zoospores per milliliter, in contrast to leachate from two conducive soils, which contained over 4,000 zoospores per milliliter. Suppressiveness was eliminated and healthy sporangia formed in profusion in suppressive soil steamed at 100°C for 30 minutes and then reinfested, but not in soil treated with aerated steam at only 60°C for 30 minutes. The factor eliminated by the higher temperature either is suppressive to bacteria that otherwise would stimulate sporangium formation, destroys the stimulatory material, or acts directly against the sporangia. The organic treatment is thus thought to control *P. cinnamomi* through suppression of reproduction by the pathogen. Prevention of sporangium formation is also implicated in the control of *P. cinnamomi* root rot and heart rot of pineapple in Queensland following application of elemental sulfur (Pegg, 1977b). Oxidation of sulfur by *Thiobacillus* spp. produces sulfuric acid, which lowers the soil pH to 3.8 or less. This low pH may decrease the activity of the bacteria necessary to stimulate zoosporangium formation, and of those that convert ammonium to nitrate nitrogen, as well as increase the ionic concentration in the soil and greatly stimulate *Trichoderma* spp. For disease control, Pegg (1977b) recommends 1,200 kg of elemental sulfur per hectare for soils of low buffering capacity.

Still another approach to inoculum control is used in Western Australia (Malajczuk, *in* Erwin et al, 1983; *in* Schippers and Gams, 1979), where *P. cinnamomi* causes a devastating dieback of the jarrah (*Eucalyptus marginata*). A hot fire is used to destroy the understory of proteaceous species (e.g., *Banksia grandis*, susceptible to *P. cinnamomi*) and promote instead an understory of leguminous species (e.g., *Acacia pulchella*, resistant and inhibitory to *P. cinnamomi*). The net result is less inoculum of the pathogen available to infect the jarrah (Chapter 7).

Prevention of inoculum formation through elimination or management of alternate host plants has long been used to control insect-vectored pathogens, many of which multiply in symptomless hosts such as weed species. The recent development of ELISA and other sensitive methods for detecting pathogens is beginning to reveal that alternate and symptomless hosts are exceedingly common in nature, particularly for viruses, viroids, and certain prokaryotic plant pathogens. Runia and Peters (1980) reported for The Netherlands that, although only chrysanthemum and cucumber were affected by viroid agents under commercial conditions, 18 of 53 agricultural or horticultural plant species studied were susceptible to cucumber pale fruit viroid and 13 were susceptible to chrysanthemum stunt viroid. Twenty-five of the 53 test plants were susceptible to citrus exocortis viroid. Alternate and symptomless hosts or potential hosts of viroids may be much more common than presently recognized. The wall-less *Spiroplasma citri*, responsible for citrus stubborn disease, can be carried within the bodies of at least 11 different insect species (Rana et al, 1975), including the beet leafhopper, and may be transmitted to clover and other plants in addition to citrus. The ratoon stunt bacterium can multiply in bermuda grass as well as in sugarcane, where it causes ratoon stunting (Davis et al, 1980). The Pierce's-disease bacterium of grape has many host species and is spread from plant to plant by sharpshooter leafhoppers. Clearly, we have only scratched the surface in our understanding and use of this traditional yet extremely important approach to biological control of plant pathogens — elimination of inoculum formation from alternate and often symptomless hosts (Chapter 6).

The use of trap plants is yet another example. For example, when *Crotalaria spectabilis* is used as a cover crop, its roots become infected by the female larvae of *Meloidogyne* spp., but giant cells do not form in roots; the nematodes starve and reproduction is prevented (Baker and Cook, 1974). In Brazil, *C. spectabilis* was as effective as either of two common nematicides in reducing populations of *M. incognita*, and recolonization of the soil by the nematodes was much slower where the trap plants had been used (Huang et al, 1981).

Armillaria Root Rot

Biocontrol of the pathogen responsible for this disease exemplifies the use of resident antagonists to displace or preempt a pathogen in crop residue. It is accomplished by: 1) chemical or thermal treatment of soil containing the pathogen in host residue, thereby reducing the pathogen's antibiotic defense mechanism and allowing the antagonist to displace the pathogen in the residue; 2) ringbarking trees before felling, exposing prunings to aerial saprophytes before burial (the possession principle), and painting stumps with ammonium sulfamate to kill surface cells and encourage saprophytic colonists, all of which favor establishment of saprophytes in host tissue in advance of the pathogen.

Pathogen: *Armillaria mellea* (Basidiomycotina, Hymenomycetes). In the Southern Hemisphere, the pathogen is called *Armillariella elegans.*

Hosts: Extremely wide host range, including broad-leaved and coniferous forest and shade trees, fruit trees, and shrubby and herbaceous ornamentals. Greatest damage is perhaps to fruit trees.

Disease: Infected trees decline in vigor, growth is retarded, leaves or needles may fall, and death ensues when crown or roots are infected. The crown or roots, when exposed, display a characteristic fan-shaped, feltlike white mycelium between the bark and wood in areas of decayed bark. Black rootlike strands of hyphae (rhizomorphs) 1–2 mm in diameter grow over the surface of diseased roots and into the soil. In later stages, honey-colored mushrooms appear around the plant base. The pathogen spreads slowly outward from tree to tree. White mycelia may penetrate into rhizomes, forming feltlike mats along the medullary rays. Diseased trees in a forest commonly center around infected stumps.

Life Cycle: Survival (at least six years) is principally as mycelium in infected roots, dead or alive. From these food bases, rhizomorphs spread through the soil and make contact with roots, which they infect either through wounds or directly. In Zambia and Zimbabwe, rhizomorphs are not formed, owing to toxic material(s) in the soil; spread of *Armillaria* is then by root contact.

No conidial stage is formed, and airborne basidiospores are thought to be of minor importance in the spread of the pathogen. The sporophores, formed in the fall, are ephemeral. The caps are honey-colored, 5–30 cm in diameter, and borne on stipes 8–25 cm long. Basidia are borne on the gills on the underside of the cap; the spores are white in mass. The basidiospores may infect freshly cut stumps not yet colonized by other fungi; these colonized stumps may serve as foci for localized spread by rhizomorphs to surrounding trees.

Environment: The pathogen is active in warm, moist soil and inhibited in cool or dry soil. Thus, native oaks in California are commonly infected but rarely injured in the dry, warm summers and cool, wet winters. However, if the soil is irrigated in the summer, the oaks quickly succumb. The disease-favoring effect of warm, moist soil acts, in part, on the host (predisposition). The optimum temperature for root growth of peach, casuarina, pepper tree, apricot, and geranium is 10–17°C and that for root damage by *A. mellea* is 15–25°C; for citrus and rose, the figures are 17–31°C and 10–18°C, respectively. Host stress from defoliation by gypsy moth, leaf rollers, stem-boring insects, drought, or aphid injury also increases severity of *Armillaria* attack on trees. Rhizomorphs of the pathogen continue to elongate while covered with a water film, but without the water film, more oxygen reaches the outer cells of the rhizomorph, pigmentation occurs, and growth is checked. These rhizomorphs must be connected to a food base such as an infected root or stump in order to grow and to infect. Nutrients are translocated from the food base to the rhizomorph tip but not the reverse.

Biological Control: By 1914, carbon disulfide fumigation of infested soil in California was known to control *A. mellea* in orchards. By 1951, it was found that a high dosage of carbon disulfide killed the pathogen in two to three days, whereas moderate dosages killed it only after 30–50 days in field soil and not at all in sterilized soil. *Trichoderma viride*, the active antagonist, was thought to attack *Armillaria* only when it was "weakened" by the fumigation. Methyl bromide had an effect similar to that of carbon disulfide; benomyl controlled *Trichoderma* but not *Armillaria*. Heating soil to 33°C for seven days with aerated steam also weakened the pathogen, and 36°C for seven days or 43°C for two hours killed it; since living citrus and peach roots were not injured by such temperatures, heat treatment of planted trees is a potential measure for valuable specimen trees. The various stress treatments (heat, chemicals, drying) and weakening of *Armillaria* were found in 1976 to result in less production by *Armillaria* of the antibiotics that protect it against *Trichoderma*. Perhaps such exposure also reduced growth of and increased exudation from *Armillaria* mycelia. In broad-leaf forests, treatment of freshly cut stumps with 40% aqueous solution of ammonium sulfamate favors growth of saprophytes that preempt colonization by *Armillaria*, an example of integrated control.

Ringbarking of forest trees before felling for tea plantations caused reduction of root carbohydrate reserves, favored colonization of the roots by saprophytic soil fungi, and prevented expansion of localized *Armillaria* infections on roots. Leaving tea prunings on the soil surface for a time before burying them permitted their colonization by saprophytes, which then prevented *Armillaria* invasion after their burial. These two treatments illustrate use of the possession principle.

References: Bliss (1946, 1951), Heald (1933), Munnecke et al (1976), Ohr and Munnecke (1974), Raabe (1962), Rishbeth (1979).

Displacement of the Pathogen in Crop Residue

This approach to reduction of inoculum applies to pathogens that use crop residue as a food base or springboard for infection. This includes many important root pathogens (e.g., *Heterobasidion annosum, Armillaria mellea,* and *Gaeumannomyces graminis* var. *tritici*) that use the residue of their host as a substrate and that are unable to form specialized survival structures of their own. It also includes many important aerial pathogens (e.g., *Septoria, Venturia,* and *Helminthosporium* spp.) that overwinter (or oversummer) in residue and then form fruiting structures at the appropriate time for dissemination to the aerial parts of susceptible hosts. In all cases, the pathogen becomes established in the residue initially through pathogenic activities, while the substrate is still functional tissue on the host. The pathogens usually remain confined to the lesion areas, with disease-free areas becoming colonized by numerous saprophytes. When nutrients in the residue are exhausted, the pathogen dies.

Since pathogens living in dead host-plant residue were the pioneer colonists of that residue, attempting to displace them becomes a problem of how to unseat a resident species that is highly adapted to the niche occupied. Displacement of a pioneer colonist by a secondary colonist occurs mainly when 1) the pioneer is finished with the substrate, 2) the pioneer becomes dormant or semidormant because of adverse environment, while the competing colonist remains active, or 3) the secondary colonist can kill the pioneer through antibiosis, parasitism, hyphal interference, or some other mechanism.

As pointed out in the previous section, very few antagonists have been identified that can displace a pathogen within infected tissue of a living plant, i.e., while pathogenesis is underway. Those identified (e.g., *Sphaerellopsis [Darluca] filum, Tuberculina maxima, Scytalidium uredinicola,* and certain strains of *Fusarium roseum*) seem specialized in this role. On the other hand, pathogens can be displaced during their saprophytic existence in residue by a wider array of potentially antagonistic species, many not particularly specialized in their role. Some adverse environmental condition must prevail to weaken, predispose, or cause dormancy of the pioneer colonist, thereby permitting secondary colonists to overrun and displace it.

Two classical biological control systems for soilborne plant pathogens can be explained on the basis of accelerated displacement of the pathogen in crop residue. One is the Chamberlain system for controlling take-all under intensive spring barley culture in England (Garrett and Buddin, 1947), and the other is the use of carbon disulfide to control *Armillaria mellea* in orchards replanted to citrus or other fruit trees in southern California (Bliss, 1951). Neither of these systems was widely adopted in areas beyond those in which they were developed. The Chamberlain system may not have been used other than by the farmer who developed it, Mr. M. P. Chamberlain in Oxfordshire. This fact is less important than the fact that the respective systems worked.

The Chamberlain system involves sowing a mixture of trefoil (*Medicago lupulina*) and Italian ryegrass (*Lolium multifolium*) in the spring at the time of sowing barley, or as a second sowing as much as one month later (about April). The trefoil and ryegrass remain small during early growth of the spring barley and do not compete with the barley. After harvest, the trefoil and ryegrass make good growth and take up the available nitrogen in the top layer of soil containing the barley residue infested with *G. graminis* var. *tritici*. This hastens starvation of *G. graminis* var. *tritici*, which needs the external nitrogen to maintain its metabolism in the residue (Garrett, 1940). The trefoil-ryegrass mixture is plowed under in late autumn or early winter; it decomposes and releases N for the next spring barley crop. Garrett and Buddin (1947) concluded that this system provides maximum nitrogen when the crop is growing, which allows the host to outgrow the pathogen, and minimum nitrogen when the pathogen is attempting to survive saprophytically.

The system of controlling *A. mellea* in citrus wood involves removal of all large roots and then the injection of carbon disulfide into the soil containing the remaining infested root fragments. The rate applied is sublethal to *A. mellea* in the wood, but nevertheless weakens the pathogen while at the same time favoring activity of *Trichoderma* spp., which displace *A. mellea* in the wood.

A similar biological control was developed in Malaysia to control *Rigidoporus (Fomes) lignosus, Phellinus (Fomes) noxius*, and *Ganoderma philippi (pseudoferreum)*, causal agents of white, brown, and red root diseases of rubber, respectively (Fox, *in* Baker and Snyder, 1965). The system involves preventing saprophytic establishment of the pathogens in residue, displacing the pathogens already established in residue, and accelerating the exhaustion of the food base of the pathogens (Chapter 2). In Oregon, urea applied to forest soils results in displacement of *Phellinus (Poria) weirii* by *Trichoderma* spp. in root wood of alder within eight weeks (Nelson, 1975), thereby revealing a specific management practice for control of this residue-inhabiting pathogen.

Complete displacement and death of the pathogen may not be necessary in all cases, as revealed by the results of Lin and Cook (1979) on the mechanism by which root rot of lentil caused by *F. roseum* 'Avenaceum' is suppressed by soils in eastern Washington. The pathogen, in a food base such as colonized lentil stems (a natural source of inoculum) or autoclaved oat grains (an artificial source of inoculum), caused small brown root lesions but not seedling blight in the natural soil. In contrast, use of the same sources and amount of inoculum resulted in nearly 100% pre- and post-emergence seedling blight of the lentils in soil that had been previously

Figure 4.4. Lentil seedlings growing in Palouse silt loam to which *Fusarium roseum* 'Avenaceum' was added. Before adding the pathogen, the soil was not treated (suppressive to the pathogen), treated with moist heat for 30 minutes at the indicated temperatures, or treated at the indicated temperatures and then amended with 1% (w/w) of nontreated soil. (Reprinted with permission from Lin and Cook, 1979.)

treated with aerated steam at 55°C for 30 minutes. The soil treated at 50°C for 30 minutes was still suppressive (Figure 4.4). Inoculum fragments buried in suppressive soil for 24 hours, then transferred to heat-treated (55°C for 30 minutes) soil no longer produced seedling blight, but inoculum buried in heat-treated soil and transferred to heat-treated soil did. Further study revealed that the fast-growing sugar fungi, most notably *Trichoderma viride, Mucor hiemalis*, and *M. pulmbulus*, colonized the residue occupied by Avenaceum. The pathogen was still alive in the food base, usually in the most central portions, but was a prisoner in the residue and now shared the nutrients with aggressive saprophytes. Outbreaks of the disease occurred when lentils were direct-drilled (sown without prior seedbed preparations) into bluegrass sod killed by herbicide. The occurrence of disease in this particular management system suggests that perhaps *F. roseum* 'Avenaceum' is more successful as an occupant of bluegrass residue than of lentil stems or oat grains. W. Hunter and G. W. Bruehl (unpublished) demonstrated that bluegrass thatch in unburned grass fields in eastern Washington is commonly occupied by this *Fusarium*.

Soilborne pathogens with the ability to form specialized survival structures (e.g., oospores, chlamydospores, or microsclerotia) have an advantage because of their ability to convert the host residue into their own specialized storage organs (Bruehl, 1976). These storage organs are like personal bank vaults. Secondary colonists are left with the spoils. Hancock and Benhan

(1980) showed for *Verticillium dahliae* that the population of microsclerotia in buried cotton stems does not drop as the stem decays, i.e., the fungus can withstand the effects of the decay process. On the other hand, microsclerotia of this fungus contained within old cotton stems remain viable longer than do microsclerotia free in the soil (Evans et al, 1966), presumably because the stem tissues in which they are embedded provide protection from antagonists in soil. Nevertheless, decay of the infested residue is advantageous to the pathogen, by allowing the propagules to be distributed more uniformly in soil, thus increasing their chance for contact with new host roots. There is no information on whether a fungus such as *V. dahliae* that depends for survival on specialized structures is also effective in active possession of host remains. Pathogens able to form resting propagules and to survive passively are possibly less capable in active possession (as defined by Bruehl, *in* Bruehl, 1975) of a food base than those entirely dependent for survival on active possession.

Any attempt to introduce one or more antagonists to displace a pathogen in crop residue should be done early in the disease cycle, perhaps even before onset of pathogenesis. In this way, the antagonist becomes established as a cohabitant with the pathogen in the diseased tissues. This may partly explain why fluorescent pseudomonads are suppressive to wheat take-all; members of this bacterial group colonize the root and inhibit take-all, but also become established in the root lesions, where they are then ideally situated for continued inhibition and possibly even displacement of *Gaeumannomyces graminis* var. *tritici* (Wilkinson et al, 1982). Strains are being selected for the ability to control take-all by protection of the root when introduced on seed (next section, and Chapters 7 and 8), but the possibility of benefits extending to inhibitory effects during saprophytic survival between host crops cannot be ignored. This approach also may have applications to biological control of aerial pathogens that survive between hosts in diseased crop residue on the soil surface and later produce inoculum to infect new aerial growth of the host. An antagonist applied to the developing lesion, if selected for aggressiveness as a secondary colonist and as a saprophyte under the prevailing soil conditions, might displace the pathogen before the sowing of the next crop or prevent it from responding to the next crop.

Biological control by destruction of the pathogen in crop residue can be accomplished through cultural practices (e.g., those developed by Chamberlain for take-all, by Bliss for armillaria root rot of citrus and other fruit trees in California, or by Fox and associates for the root diseases of rubber trees in Malaysia). Another example is the Dutch *rigolen* process, whereby infested crop residue or other sources of pathogen inoculum are buried by deep plowing (75–90 cm) and allowed to decompose before they are brought back to the soil surface (Drayton, 1929). Thorough fragmentation and complete burial of tobacco crop residue to eliminate mycelium of *Peronospora tabacina* is used to help control blue mold in the United States, Australia, and probably elsewhere (Lucas, 1975).

Modern agriculture presents many choices that affect the control of residue-inhabiting pathogens, including time and depth of burial, no burial (no-tillage), time between sowings, date of sowing, flooding, and fertilizer management (Chapter 10). The best method for controlling residue-inhabiting pathogens is crop rotation, which can be highly effective against pathogens dependent on host residue for survival. A single year with no wheat or barley planted in the field is considered sufficient to prevent buildup of *G. graminis* var. *tritici*, provided grass weed hosts are controlled. There is a wide variety of choices and combinations of cultural practices for control of residue-inhabiting pathogens. We have hardly begun to reveal through research the effects of temperature, water potential, oxygen supply, and other factors that influence active possession of residue. Even more important, but also more difficult, we need information from field experiments properly designed to compare the longevity or activity of pathogens in crop residue under different types of management (Cook et al, *in* Oschwald, 1978), to supplement information obtained under controlled conditions. This is particularly important in view of recent trends toward reduced tillage and shorter or no crop rotation. The pathogen is particularly vulnerable to antagonism from the associated microbiota during its saprophytic existence in residue. Our task is to find ways to exploit this vulnerability.

Manipulation of Factors Inherent in the Pathogen

This approach involves the reduction of vigor, aggressiveness, fitness, pathogenicity, virulence, or other attribute of the pathogen essential either to its saprophytic or parasitic activities, accomplished through factors inherent (or carried) in the pathogen itself. Examples include altered sex ratio in nematodes, production of self-limiting inhibitors (e.g., killer proteins), and hypovirulence in fungi caused by mycoviruses or other determinants. The potential use of this new research area is still relatively unexplored. Nevertheless, limited work to date suggests that the factors inherent in the pathogen itself have potential in biological control, perhaps comparable to the use of hormones and sterile males for insect control.

Numerous records show that under conditions of deficient nutrition, populations of nematodes of *Meloidogyne* spp. may become largely male and hence unable to induce root knot (Baker and Cook, 1974; Triantaphyllou, 1973). Crowding or starvation may produce hormonal disturbances in this nematode as in other animals. True sex reversal apparently occurs before or during the third larval stage. There is also evidence that different sex ratios in *Globodera rostochiensis, Heterodera schachtii,* and *H. avenae* are influenced by environment, nutrition, or other factors. Sex reversal also occurs in tropical fish, even after sexual maturity (Warner, 1982). We repeat our earlier suggestion (Baker and Cook, 1974) that this relatively unexplored opportunity for biological control deserves more attention.

Biraghi (1951) reported that cankers caused by *Endothia parasitica* on chestnut trees in Italy were healing. Working in nearby southern France, Grente and Sauret (1969a, 1969b) demonstrated that isolates of the pathogen from healing cankers were less virulent than those from nonhealing cankers and showed that when hyphae of a virulent isolate anastomosed with hyphae of an isolate from a healing canker, the culture was no longer virulent. They coined the term "hypovirulent." Although the proof is not unequivocal, little doubt remains that hypovirulence in this fungus is the pathological result of infection by one or more dsRNA determinants (Day et al, 1977).

In southern France, a program is underway whereby every 10th chestnut tree in each grove is treated with a mixture of hypovirulent strains. Because of vegetative incompatibility, spread of the hypovirulent strains is limited in the United States, and it has been necessary to treat each canker with a mixture of hypovirulent strains representing 10 different compatibility groups. Nevertheless, many hundreds of trees have now been treated by this method in the eastern United States, with nearly 100% success. "Field results have been sufficiently successful that there is little doubt that a given canker on American chestnut can be cured with selected hypovirulent strains" (Day and Dodds, *in* Lemke, 1979). The difficulty remaining in the United States is incompatibility between healthy and hypovirulent strains and the lack of means for natural spread. More work is needed, including work with vectors of hypovirulent strains, to achieve biocontrol of this disease in natural stands of American chestnut (Chapter 5).

Rhizoctonia solani is also a candidate for biological control by transmissible hypovirulence associated with dsRNA mycoviruslike agents (Castanho and Butler, 1978a, 1978b; Castanho et al, 1978). Introduction of a mixture of normal and hypovirulent inoculum of *R. solani* into soil in greenhouse pots resulted in significantly less damping-off of sugar beet than resulted from the same quantity of inoculum composed of normal *R. solani* without hypovirulent fungus. The hypovirulence determinant was transmissible only to the cured culture of the same isolate. Vegetative incompatibility among pathogenic isolates might pose a problem even more severe than that encountered with *E. parasitica*.

Hypovirulent *R. solani* survives poorly in soil and would, therefore, require reintroduction with virtually every crop. However, the more rapid death of hypovirulent *R. solani* than of healthy mycelium raises a new question, namely, whether so-called latent dsRNA mycoviruses affect either the longevity of infected strains or their ability to withstand environmental stresses. If so, this manifestation of dsRNA infection may have potential in biological control in addition to reduced virulence. Lemke (*in* Lemke, 1979) speculates that fungi have coevolved with their viruses and that such viruses are latent because those not latent eliminated their host from the population. The possibility exists, however, that "symptomless" carrier strains might be adversely affected by their viruses if subjected to the right environmental stress. If these conditions were known, they could be imposed and biological control would come into play.

Work in France (Lemaire et al, 1971) demonstrated the occurrence of hypovirulent or weakly virulent strains in populations of *G. graminis* var. *tritici* associated with mycovirus infections. Subsequent studies have confirmed that mycovirus infections are common in the mycelium of this fungus (Rawlinson and Buck, *in* Asher and Shipton, 1981) but probably are not responsible for take-all decline (Asher, 1980; Cook and Naiki, 1982). Nevertheless, mycovirus infections have been associated with loss of aggressiveness and virulence (Ferault et al, 1979) and deserve attention as potential agents for biological control of the take-all pathogen. In France, a hypovirulent strain was introduced on wheat seed with the first wheat crop after a break (rotation) crop (i.e., before take-all was a problem in the field), the objective being to limit the occurrence of take-all in the second and subsequent crops (J. Lemaire, personal communication). Approximately 100 ha of wheat were sown with such treated seed in 1980. Biological control has been reported for up to four years in some fields following introduction of a hypovirulent strain (Tivoli et al, 1974). Unfortunately, studies of "hypovirulence" in *G. graminis* var. *tritici* are complicated by the apparent loss of pathogenicity in strains of this fungus during maintenance in culture, possibly caused by change in a heterokaryotic condition of the fungus (Naiki and Cook, 1983a, 1983b). Care must be taken to distinguish between true hypovirulence caused by dsRNA mycovirus infections, and loss of pathogenicity caused by other genetic phenomena in the fungus thallus (Chapter 5).

The killer reaction in *Ustilago maydis* is genetically controlled by a cytoplasmic determinant, apparently dsRNA mycovirus infections (Koltin and Levine, *in* Molitoris et al, 1979). The inhibitory substances are proteins, and each killer strain is insensitive to its own protein but inhibited by the protein of another killer strain. However, smut was not restricted in corn plants inoculated with sexually compatible monokaryons, one a strain with the cytoplasmic killer determinant and the other a sensitive strain, where cell fusion and heterokaryon transfer occurred (Day and Dodds, *in* Lemke, 1979). The killer reaction is apparently ineffective in the host. Day and Dodds (*in* Lemke, 1979) speculate that the most promising prospect for use of the killer reaction in smut control "would be to confer on the cereal host the property of producing killer protein," perhaps by insertion of the structural gene for production of killer protein into the plant genone using a DNA vector system of genetic engineering.

BIOLOGICAL PROTECTION AGAINST INFECTION

This approach to biological control involves the establishment of antagonistic microorganisms in or around the infection court so as to provide biological protection of the host against the pathogen. This approach also may involve some form of direct inhibition (or limitation of prepenetration growth) of the pathogen by antagonists; the host provides the site of pathogen-antagonist interaction but is not itself actively involved. Many examples of biological control by resident antagonists fall into this category, including control ascribed to organic treatments and to suppressive soil phenomena and characterized by lack of propagule germination or prepenetration growth in the rhizosphere (Chapter 7). Some of the most effective biological controls achieved to date with introduced antagonists are of this type. This section is limited mainly to ways in which introduced antagonists can be used to obtain biological protection of plant surfaces. More detail on the use of introduced antagonists is given in Chapter 8, and biological protection of the plant achieved with resident antagonists and cultural practices are treated in detail in Chapters 7 and 10, respectively.

Protection of Planting Material

A potentially very important approach to biological control, unique to plant pathology, is the inoculation of seeds, seedpieces, transplants, or other planting material with a biological protectant before planting. The most widely used biological control with an introduced antagonist is of this type, namely, the method of A. Kerr and associates (Kerr, 1980) for controlling crown gall of stone fruit trees and ornamental plants, using the avirulent, bacteriocin-producing *Agrobacterium radiobacter* strain K84 (Chapters 3 and 8). This work illustrates at least four principles with implications for future biological control systems.

1. Antibiosis can provide a weapon for biological control. Early concern that antibiotic production has no practical use in biological control can be dismissed on the basis of the success achieved against crown gall and of other work cited below. Strains unable to produce agrocin 84 generally are less effective against the crown gall pathogen than those that can produce it.

2. An antibiotic-producing strain used for biological control must also be an aggressive colonist of the plant surface or infection court. The original strain K84 is an effective root colonist as well as producer of agrocin 84, whereas new strains developed through transconjugation to produce agrocin 84 have given variable or ineffective biological control, because they are less effective as root colonists (Ellis et al, 1979). Isolates from

areas other than Australia have not proved as effective as K84 (Moore, *in* Schipper and Gams, 1979), perhaps because they are less aggressive.

3. Biological control agents can be developed through genetic manipulations of microorganisms. Once the necessary attributes of an effective antagonist are known (e.g., antibiotic production and aggressiveness as a root colonist) and once tests for these characters are available, it should be possible with techniques already available to genetically engineer the microorganisms to produce a desired antagonist.

4. Biological control obtained with an antibiotic as specific as agrocin 84 also presents the risk of the pathogen population developing resistance or insensitivity to the antibiotic. Three biotypes of *A. radiobacter* pv. *tumefaciens* have been recognized to date, and only biotypes 1 and 2 are sensitive to agrocin 84 (Kerr and Panagopoulos, 1977). Biotype 3 occurs on grapevines and is not controlled by strain K84. Certain strains in California are also not controlled by K84 (Schroth and Moller, 1976), and resistant mutants have been shown to occur within populations otherwise sensitive to agrocin 84 (Cooksey and Moore, 1982b; Süle and Kado, 1980). On the other hand, resistant strains have not presented a problem in Australia, where the control has been in use for several years, apparently because the predominant biotype there is sensitive to K84.

One of the earliest demonstrations of the potential for seedling protection using an antagonist applied as a seed treatment is that of Tveit and Moore (1954), in which *Chaetomium globosum* was applied on oat seeds to protect the seedlings against *Helminthosporium victoriae*. The isolate of *C. globosum* was a common natural inhabitant of oat grains from Brazil. The findings support the concept that effective antagonists are best found by looking where a plant disease is expected to occur but does not.

The most complete and sustained studies on biological seed treatments for seed and seedling protection have been by Kommedahl and associates, conducted over the past 10–15 years and recently summarized (Kommedahl and Windels, *in* Papavizas, 1981). This work has revealed that *Bacillus subtilis*, *Trichoderma* spp., *Penicillium* spp., and *C. globosum* are as effective as thiram or captan for protection of corn, peas, and soybeans. The work reveals further that isolates differ in effectiveness and that through selection, more powerful strains can be found. Work with *P. oxalicum* applied on seeds of peas produced no evidence that the antagonist grew into the rhizosphere from the seed or that it was active in the rhizosphere (Windels and Kommedahl, 1982); recovery of the antagonist from the rhizosphere was thought to result from passive movement of dormant conidia downward with growing roots or with percolating water. Perhaps strains of the antagonists isolated initially from the rhizosphere would then be more successful as rhizosphere colonists. The improvement in stands resulting from biological seed treatment apparently results from protection against seed decay and preemergence seedling blight and not from control of root disease or postemergence seedling blight. Nevertheless, protection during seed germination and seedling emergence is often the most critical to stand establishment and vigor. Moreover, the common chemical seed treatments for peas, corn, and most other crops likewise give protection mainly against preemergence seed decay and generally not against postemergence seedling blights or root rots. Biological protection, being self-perpetuating, is likely to last longer than chemicals in soil.

The soilborne pathogens responsible for preemergence seed decay are stimulated to grow by nutrients (exudates) that leak from the seed during swelling and sprouting. Germinating pea seeds, for example, release sugar (mainly sucrose) that stimulates *Pythium ultimum* to grow and invade the seed (Flentje and Saksena, 1964; Kerr, 1964). A fast-growing saprophyte present as a coating on the seed at the time of planting, and having the ability to use the seed exudates and dominate the spermosphere (region or soil surrounding the seed), would have a high potential to protect the germinating seed simply by preempting the energy supply used by the pathogen to increase its biomass for infection. This may be the basis of the protection afforded pea seeds by *P. oxalicum*. This antagonist applied as a coating of spores on pea seeds was observed to form a hyphal network and to sporulate on the germinating seeds in soil (Windels, 1981), presumably by use of the seed exudates.

Trichoderma hamatum applied as a coating of spores on the seed was shown by Harman et al (1980, 1981) to protect peas and radish against seed decay caused by *Pythium* spp. and *Rhizoctonia solani*, respectively. This antagonist was isolated originally from a soil from Bogotá, Colombia, that was naturally highly suppressive to *R. solani* (Chet and Baker, 1981), owing at least in part to antagonism by *T. hamatum* (Liu and Baker, 1980). The fungus was shown to produce cellulase, chitinase, and β-(1-3) glucanase and to be able to degrade both the cellulose in the walls of *Pythium* spp. and the chitin and glucans in the hyphal walls of *R. solani*. *Trichoderma* spp. are widely recognized for their aggressiveness as saprophytic colonists of fresh substrates, and part of the mechanism by which *T. hamatum* protects pea and radish seed against decay fungi probably involves preemption of the seed exudates at the seed surface. However, the preferred antagonist not only should be a competitor of the pathogen for host exudates but also should have the ability to inhibit or destroy the pathogen through antibiosis or parasitism. *Trichoderma hamatum* has at least one of these additional attributes (Chet and Baker, 1981). The fact that this particularly effective strain of *Trichoderma* was obtained from a *Rhizoctonia*-suppressive soil is further support for the suggestion (Baker and Cook, 1979) that *effective antagonists can be found by looking where the target pathogen is expected to be causing disease but is not.*

Transplants, bulbs, and corms can also receive biological protection; as in the biological treatment of seeds, the antagonist is applied to the surfaces where the protection is needed. Carnations grown from cuttings under glasshouse conditions are subject to decay by *Fusarium*, which invades through the cut surface. Aldrich and Baker (1970) showed that dipping the freshly cut surfaces into a suspension of cells of *Bacillus subtilis* protected the cuttings against *Fusarium*. Michael and Nelson (1972) controlled the pathogen for up to four months in propagation beds by dipping the unrooted carnation cuttings into a cell suspension of a pseudomonad isolated from soil. Magie (1980) demonstrated that inoculation of freshly harvested gladiolus corms with *F. moniliforme* 'Subglutinans' protected them as well as did a benomyl dip against the yellows and corm rot pathogen, *F. oxysporum* f. sp. *gladioli*. Treatment of corms with certain fungicides, by reducing the activity of the nonpathogens more than that of the pathogen, was observed to increase rather than decrease corm rot. Whether this protection results from antagonists in the infection court or induced resistance is uncertain. Magie (1980) has proposed that corms be inoculated with a combination of antagonists or that a benomyl-resistant isolate of *F. moniliforme* 'Subglutinans' (or other aggressive nonpathogenic colonist of corms) be used to inoculate the corms. Magie considered the use of biological control on corms to be "practical and inexpensive."

In Israel, the introduction of *Trichoderma harzianum* into strawberry nursery beds after soil fumigation significantly protected the plants from black root rot caused by *R. solani*, not only in the nursery where it was applied but also on the plants transferred to commercial production fields (Elad et al, 1981a).

The biological barrier providing plant protection need not be a known specific or specialized colonist of the plant surface. Ko (1971, 1982) reported a simple method of biological control for papaya root rot (caused by *Phytophthora parasitica*) in Hawaii. Papaya stands are established in Hawaii as transplants, usually introduced into soil where papaya was grown previously. Roots of the young transplants are rapidly invaded by *P. parasitica*, which produces severe root rot and often plant death, referred to as the "papaya replant problem." Because seedlings are susceptible to *P. parasitica* only when young, plants transplanted into holes filled with virgin soil (not infested with *P. parasitica*) are protected during the critical period of greatest susceptibility (Ko, 1971, 1982). The small "islands" of virgin soil present a biological barrier to *P. parasitica*. The method is used on the island of Hawaii near Hilo.

The ectomycorrhizae formed by *Leucopaxillus cerealis* var. *piceina* protect pine roots against infection by *P. cinnamomi* and possibly other root pathogens. The mechanism of protection involves a combination of complete dominance of the root surface and production of an antibiotic, diatretyne nitrile. The antibiotic inhibits germination of zoospores of *P. cinnamomi* at 50–70 ppb and kills zoospores at 2 ppm (Marx, *in* Bruehl, 1975). The protection of pine roots by *L. cerealis* var. *piceina* exemplifies biological control by protection of a natural plant surface otherwise highly attractive to a destructive plant pathogen. Krupa and Nyland (1972) and Krupa et al (1973) found that the ectomycorrhizal fungi *Boletus variegatus, Pisolithus tinctorius,* and *Cenococcum graniforme* produced volatile terpenes in pine roots that were inhibitory to *Heterobasidion annosum* and *Phytophthora cinnamomi.*

Protection of Roots with Biological Seed Treatments

For most root pathogens, and probably leaf pathogens as well, successful biological protection of the plant surface depends on the ability of the antagonist to spread over that surface and even from one plant part to another. This attribute is essential, for example, if root or shoot protection is to be achieved with antagonists introduced on seeds or other planting material.

Advances in the past 10–15 years give strong evidence that root protection against pathogens is feasible with antagonists applied on seeds. Moreover, the plant-growth responses achieved with such biological seed treatments are similar to the kinds of growth responses reported to occur with "seed bacterization" and sometimes attributed (Brown, 1974) to plant hormones or other growth-promoting substances produced by the microorganism on the seed or root (Chapter 11). Some of the plant-growth responses are similar to that achieved when root pathogens are eliminated by soil fumigation (Cook, *in* Bezdicek and Power, 1983). Because root pathogens infect roots in response to stimulation by root exudates, and because the limited amount of exudates released from roots is a source of food for many rhizosphere microorganisms in addition to pathogens (Chapter 6), any aggressive, root-colonizing microorganism, given the advantage of being the first to colonize the root, as may occur with seed inoculation, has the potential to preempt the nutrient supply of the pathogen and hence protect the root. The antagonist need only limit the supply of one essential nutrient (e.g., nitrogen, iron, or carbon) to be effective. Beneficial microorganisms that are strong competitors for one or more nutrients on the root surface and that are also able to inhibit pathogens directly by production of antibiotic should provide the best or most consistent root protection and hence produce the greatest plant-growth responses. Plant-growth responses are to be expected when roots and especially rootlets are maintained in the state of health necessary for uptake of nutrients and synthesis of growth factors for the tops (Cook, *in* Bezdicek and Power, 1983).

Broadbent et al (1971, 1977) conducted extensive studies in a New South Wales commercial nursery to determine the feasibility of increasing the growth rate of bedding-plant seedlings with biological seed treatments. Their approach differed from those of earlier studies with "seed bacterization" in that the isolates tested were screened first on agar plates for antibiotic activity against nine different plant-pathogenic fungi. One isolate, *Bacillus subtilis* A13 obtained from the surface of mycelia of *Sclerotium rolfsii*, exhibited antibiotic activity against all nine test fungi. This same organism was particularly effective in improving the stand and increasing seedling top weight of portulaca, delphinium, and snapdragon. In addition, when applied at 10^6-10^7 cells per seed of barley that was planted in a field with the "bare-patch" disease (caused by *Rhizoctonia solani, Pythium* spp., and *Fusarium* spp.) it resulted in increased tillering, up to 9% increase in yield, and the plants matured about two weeks earlier than plants from nontreated seed. In other field tests, treatment of oat seeds with *B. subtilis* A13 increased grain yield by 40% in one trial but had no effect on wheat or barley in two other trials (Merriman et al, 1974). A 48% yield increase occurred with carrots grown from seed pelleted with the bacterium. Increased plant growth also occurred in response to seed treatments with other isolates of *B. subtilis* and also with certain isolates of *Streptomyces griseus* selected originally on the basis of in vitro antibiotic activity to *R. solani* (Merriman et al, *in* Bruehl, 1975). The plant-growth response was at least partly from protection against root pathogens (e.g., *R. solani*).

Broadbent et al (1977) reported plant-growth responses with plants in steam-treated (60 or 100°C) soil inoculated with selected bacteria. This result might be interpreted to indicate that the introduced bacteria had an effect in addition to that of pathogen control. However, the steamed soil would be rapidly recolonized by airborne microorganisms (Schippers and Schermer, 1966; Watson, pages 96–97 *in* Baker and Cook, 1974). The possibility exists that some of these invaders would have been deleterious to the plants had the plants not been protected by the test bacteria.

Burr et al (1978) and Kloepper et al (1980b) obtained 10–15% increased growth and yield of potatoes by treating the seedpieces with strains of *Pseudomonas fluorescens* and *P. putida* isolated originally from the surfaces of freshly dug tubers or from roots of potato or celery. In another study, strains of fluorescent *Pseudomonas* spp. isolated orginally from the rhizosphere-rhizoplane of field-grown beets produced significant 13% average increases in root weight and total sucrose for sugar beets when used as seed treatments in field trials over three years in Idaho and California (Suslow and Schroth, 1982b). The effective strains used by Burr et al (1978) and Kloepper et al (1980b) were screened initially for in vitro inhibitory activity against *Erwinia carotovora* pv. *carotovora* and several other test microorganisms, and those used by Suslow and Schroth (1982b) on sugar beets exhibited in vitro inhibition of *Rhizoctonia solani, Pythium ultimum, P. aphanidermatum,*

P. debaryanum, Erwinia carotovora pv. *carotovora* and pv. *atroseptica*, *Pseudomonas marginalis* pv. *marginalis, and P. syringae* pv. *syringae*, pv. *phaseolicola*, and pv. *tomato*. Inhibitory ability was essential in the response of potatoes, because noninhibitory mutant strains still colonized the roots but no longer increased plant growth (Kloepper and Schroth, 1981c). Effective strains genetically marked with resistance to rifampicin and nalidixic acid were shown to colonize the roots of potato and other plants in populations of 10^4–10^5 per centimeter of root when introduced as seedpiece inoculation. Effective strains also had a generally suppressive effect on rhizosphere and rhizoplane microorganisms in general, as indicated by 50–75% fewer rhizoplane fungi and 25–30% fewer Gram-positive bacteria. Specific pathogens or quasipathogens (toxicogenic or subclinical pathogens) were among the microorganisms displaced by the introduced strains. No plant-growth response occurred with the bacterial seed treatments when the plants were grown in sterile soil (Kloepper and Schroth, 1981b; Suslow and Schroth, 1982a), further indicating that the response resulted from protection against root diseases (Schroth and Hancock, 1982).

Root-colonizing fluorescent *Pseudomonas* strains inhibitory to *Gaeumannomyces graminis* var. *tritici* were found by Weller and Cook (1982) to occur at significantly higher populations on roots of wheat grown in suppressive soil (soil from fields where take-all had occurred at one time but had declined) than on roots of wheat grown in conducive soil (Figure 7.1). Strains isolated from the roots and selected for strong inhibitory activity against the pathogen in vitro, when introduced as a seed treatment at 10^8–10^9 cells per seed, gave significant biological control of take-all and 15–20% increased yield in field trials at Pullman (Figure 4.5) and elsewhere in Washington. No evidence was found for increased growth or yield of wheat with treated seeds planted in soil in the absence of the take-all pathogen. A strain with in vitro inhibitory activity against *G. graminis* var. *tritici* was also suppressive to the pathogen in vivo, but a mutant strain with no in vitro inhibitory activity against the pathogen was ineffective in vivo (D. M. Weller, unpublished). A mutant strain marked genetically with resistance to rifampicin and nalidixic acid, when introduced as a seed treatment, made up about 50% of the total bacterial population and more than 90% of the total population of pseudomonads on wheat roots one month after planting in natural soil in the field at Pullman (D. M. Weller, unpublished). These results provide further documentation that microorganisms with ability to colonize and protect roots can be introduced on seeds. They further show the importance not only of ability as a root colonist, but also ability to inhibit since, in general, strains negative for inhibitory activity against *G. graminis* in vitro gave little or no control. On the other hand, inhibitory activity alone is not sufficient; the organism must also be an aggressive root colonist.

Figure 4.5. Winter wheat growing in a plot at Pullman, showing biological control of take-all by a fluorescent *Pseudomonas* sp. introduced on the seed. The plot area had been treated with methyl bromide to eliminate naturally occurring root pathogens of wheat and to make the soil conducive to take-all. The healthy plants in the three rows on the left received no inoculum of *Gaeumannomyces graminis* var. *tritici* and the seeds were not treated. The take-all pathogen was introduced in the other six rows, but plants in the center three rows were grown from seed treated with cells of bacteria inhibitory to *G. graminis* var. *tritici*. (Reprinted with permission from Weller and Cook, 1983.)

Some of the strains effective against *G. graminis* var. *tritici* match the characters of known nonpathogenic pseudomonads, but others produce a hypersensitive reaction in tobacco, typical of plant-pathogenic pseudomonads. *Pseudomonas cepacia*, a weak pathogen of onion and inhibitory to *Fusarium oxysporum* f. sp. *cepae* in vitro, provided excellent control of the *Fusarium* in naturally infested soil when introduced on the onion seeds (Kawamoto and Lorbeer, 1976). The bacterium introduced on the seed was recovered from the roots and probably is well adapted to onion roots. That they have certain traits of pathogens may account for the success of some of the pseudomonads as root colonists. The possibility that they may provide biological control through cross protection or even induced resistance cannot be ruled out. Obviously, it will be necessary to verify lack of virulence to the host before proceeding with any antagonist as an agent of biological control.

Biological seed treatments are less consistently effective than chemical treatments. Much of the inconsistency undoubtedly relates to the early success of the antagonist as a root colonist; if conditions are unsuitable for rapid and early root colonization, the antagonist itself may be permanently excluded from the developing root system by indigenous rhizosphere colonists. Another reason may relate to the *need* for root protection. P. Geels and B. Schippers (personal communication) found in Holland that root-colonizing pseudomonads applied to potato seedpieces gave no growth response in field soils where potatoes had been grown in rotation with wheat for the past 18 years but gave a strong growth and yield response in adjacent soils where potatoes had been grown continuously (without rotation) for the previous 18 years. Yields had declined in the continuous-potato plots, but the biological seed treatment elevated potato tuber production to a level not significantly different from that in the rotation plots. Thus, greatest plant-growth responses to root protection obtained with biological seed treatments might be expected to occur where damage from root pathogens is greatest.

Howell and Stipanovic (1979, 1980) obtained improved emergence of cotton seedlings by treating the seeds with strains of fluorescent pseudomonads isolated originally from the rhizosphere of cotton. They found evidence of specificity, however, with one strain providing control of seed decay caused by *R. solani* but not that caused by *Pythium* spp. and another strain having the opposite effect. The two strains produced the antibiotics pyrrolnitrin and pyroluteorin, respectively. Treatment of cotton seeds with the antibiotics alone also gave improved emergence in soil infested with *R. solani* and *P. ultimum*, respectively. Whether these pseudomonads also colonized and protected the emerging radicle of the cotton seedlings was not determined, but the fact that the purified antibiotics were about as effective as treatments with whole cells indicates that, as with treatments tested by Kommedahl et al (1981), the main benefit is against preemergence seed decay. Nevertheless, many fluorescent pseudomonads are aggressive root colonists, and the potential exists for use of such strains to protect roots as well as germinating seeds against *Pythium* spp. and *R. solani*.

A success story apparently unnoticed until recently is that of S. Y. Yin and associates (1957, 1965) in the People's Republic of China. Their *Streptomyces* strain 5406 has been used on more than six million hectares of cotton during the past 30 years (Cook, *in* Papavizas, 1981). Strain 5406 was selected from among more than 4,000 isolates of actinomycetes, half originally obtained from roots of alfalfa and the other half originally obtained from roots of cotton. All were screened initially in vitro for antibiotic production against *R. solani* and *V. albo-atrum*. Isolate 5406 was selected because of strong in vitro antibiotic activity. The response of cotton and also wheat to 5406 can be striking in fields where soilborne pathogens limit stand establishment and plant vigor (S. Y. Yin, personal communication).

Biological Protection of Foliage and Flowers

The natural protection provided to green leaves by the cosmopolitan epiphytic and endophytic microflora has become evident from many studies, including

Heterobasidion Root Rot of Pine

Biocontrol of the pathogen responsible for this disease illustrates the use of an introduced antagonist to protect a freshly cut plant surface against infection. Freshly cut, nearly sterile stump surfaces are inoculated with a relatively weak pathogen that inhibits, by hyphal interference and other antagonistic effects, the invasion of the stumps and the interconnected trees by *Heterobasidion annosum*.

Pathogen: *Heterobasidion* (*Fomes*) *annosum* (Basidiomycotina, Polyporaceae).

Hosts: Conifers in the Northern Hemisphere, especially pine, spruce, and larch.

Disease: The fungus breaks down cellulose and lignin in infected roots, causing extensive decay that may lead to wind-throw, decreased growth, heart rot, and death. The characteristic fructifications form at or near the soil level at the base of the tree or stump in less than one year. In first plantings of a forest, the disease appears only after stumps have been created. The disease is minimal in natural stands and maximal in intensively managed forests or plantations of pure stands, where pathogen spread belowground is expedited by extensive root grafting.

Life Cycle: The fungus is essentially root-inhabiting; the mycelia have very limited ability to grow through or survive in soil but can survive many years in stumps and logs. Airborne spores are released from sporophores and may directly infect wounds in trees, but root infection from free mycelium or spores in soil apparently is unimportant. The usual spread is by mycelia growing from stump roots into living roots in contact with or grafted to the stump roots. Stumps are colonized either at the cut surface by spores or from previously infected roots.

Environment: Infection by *H. annosum* is favored by moist conditions and the presence of freshly cut stumps.

studies of brown leaf spot of rice caused by *Helminthosporium oryzae* (Akai and Kuramoto, 1968), leaf spot of rye caused by *H. sativum* (Fokkema, 1971, 1973), fire blight of apple and pear caused by *Erwina amylovora* (Riggle and Klos, 1972), and brown spot of tobacco caused by *Alternaria* sp. (Spurr, *in* Papavizas, 1981). In all these cases, and others that could be cited, disease has been less on leaves containing the normal leaf microflora than on leaves where the microflora was simplified or removed by prior treatment (e.g., a fungicide sprayed on the foliage). The exact mechanism(s) by which the epiphytic microflora provide natural protection of leaves is still a matter of speculation, but prior colonization or consumption of nutrients on the leaf surface clearly is important (Blakeman, *in* Blakeman, 1981; Chapter 6). Antibiosis and hyperparasitism may also play a role in the outcome between antagonist and pathogen, as is suspected

Biological Control: *Peniophora gigantea*, a prolific fungus with limited ability to cause a white rot of wood, is an excellent antagonist of *H. annosum*. As a weak pathogen, it does not attack healthy trees. It produces sporophores the first year after stump inoculation, but immediate inoculation of the freshly cut stump surface is necessary to ensure colonization and dominance of the stump by *P. gigantea*, because natural spore production is greatly decreased by both high and low temperatures. Strains vary in growth rate in different trees and, to select the best strains, tests must be made in the field. *Peniophora gigantea* can displace *H. annosum* in some trees but not in resinous roots. Asexual oidia of *P. gigantea* are produced in culture and used for inoculum. The oidia were formerly distributed as dehydrated tablets that survived four months at 20° C or less. Each tablet dissolved in water treated about 100 trees. The fungus is now distributed in fluid suspension in sachets of 1 ml containing 5×10^6 viable oidia. This is diluted in 5 liters of water plus 5 g of dye and applied from a plastic bottle with an attached brush. A single sachet will treat 100 stumps and costs 0.6–1.2 pence per stump (about the same as chemical treatment). These treatments have been used in England since 1963; by 1973, 30 forests (62,000 ha) had been treated. The method is used to a lesser extent in Finland and the United States. The oidia have been added to lubricating oil used for chain saws in the United States to inoculate the stump surface as it is cut. *Peniophora gigantea* colonizes stumps of some species (Douglas-fir, spruce) very slowly; application of oidia in 5% ammonium sulfamate enhances infection in these cases by killing host cells on the surface of the stump. The antagonism of *P. gigantea* to *H. annosum* apparently is by preemption (establishment before *H. annosum*). Antagonism is also by hyphal interference, which may explain the ability of *P. gigantea* to displace *H. annosum* in some cases.

References: Artman (1972), Rishbeth (1963; *in* Bruehl, 1975; 1979).

to occur in some antagonist-pathogen interactions with belowground plant parts.

Fokkema (1978) obtained partial control of both *Septoria nodorum* and *Cochliobolus sativus* on leaves of wheat, using a spray of two yeasts, *Sporobolomyces roseum* and *Cryptococcus laurentii* var. *flavescens*. He and his associates (Fokkema et al, 1979) attempted to manipulate the resident population of yeasts by adding nutrients (2% sucrose plus 0.5 or 1% yeast extract) and antibiotics (penicillin and streptomycin), but the population was increased only when the leaves were sprayed with cells of the yeasts themselves. These findings support the premise that antagonistic microorganisms can be applied for protection of foliar surfaces. Best results will be obtained with strains that are selected or developed for adaptation to the site (e.g., isolated from the site or from a similar kind of plant surface initially), that are highly aggressive (competitive) as colonists of the site to be protected, and that preferably are also capable of antibiosis and/or parasitism of the pathogen.

Swinburne et al (1975) obtained protection of apple leaf scars against *Nectria galligena* by spraying the trees after leaf fall with two antibiotic-producing strains of *B. subtilis*. Both strains could be recovered from leaf scars during the dormant season as well as in the spring, suggesting that the antibiotic strains grew during the winter. Antagonistic bacteria also have been used successfully against fire blight caused by *Erwinia amylovora*. A nonpathogenic *Erwinia* sprayed onto apple blossoms as a cell suspension protected against fire blight to about the same degree as streptomycin (Beer et al, 1980), and in California, strains of psuedomonads with ability to colonize pear blossoms were about as effective as streptomycin sprays against fire blight (Thomson et al, 1976). In Norway, Tronsmo and Dennis (1977) applied aerial sprays of conidia of *Trichoderma viride* and *T. polysporum* to strawberries, starting at early flowering; the protection thus obtained against subsequent rot of the fruit during storage was equivalent to that provided by dichlorofluanid.

Plant-pathogenic bacteria are highly successful as colonists of aerial parts, for which the only known possible source of inoculum is the seed at sowing. These bacteria are now recognized as having an epiphytic or residency phase (Leben, 1965a, 1981) on leaves and roots of their host plant, during which they cause no symptoms. If plant-pathogenic bacteria introduced on seeds at sowing can establish and multiply epiphytically on the emerging shoot and leaves in apparently high populations, then certain nonpathogenic bacteria should be able to do likewise. Indeed, some have proposed that the cosmopolitan and standard array of epiphytic yeasts and bacteria on leaves of green plants are carried on seeds from site to site and establish initially on the emerging shoot. Lindow (1982) showed that some strains *E. herbicola* introduced on potato seedpieces at planting not only became established on the potato foliage but also partially displaced *P. syringae* responsible for ice-nucleation. The introduced strains were well suited as leaf epiphytes but were negative for ice-nucleation; displacement of *P. syringae* that was ice-nucleation active resulted in less frost damage to full-grown potato plants. Thus, the prospects also exist for use of biological seed treatments to protect the tops against pathogens.

Inoculation of Pruning Wounds with Antagonists

The first practical use of an antagonist established in an infection court to protect against a pathogen was that of *Peniophora gigantea* (Figure 4.6) to control *Heterobasidion (Fomes) annosum*, a poor saprophytic competitor.

Figure 4.6. *Peniophora gigantea* on a pine stump, typical of the growth eventually made by this antagonist when applied to freshly cut stumps for biocontrol of *Heterobasidion annosum*. (Photo courtesy of J. Rishbeth.)

The control was developed by Rishbeth (1963, *in* Jones and Solomon, 1974, *in* Bruehl, 1975). The method involves inoculation of freshly-cut pine stumps with a suspension of oidia of *P. gigantea*. Earlier attempts to establish *Trichoderma viride* on the cut surface by first killing the wood with ammonium sulfamate had shown some success, especially if followed by *P. gigantea*, an effective competitor (Rishbeth, *in* Jones and Solomon, 1974). *Peniophora gigantea* is a weak pathogen, which enables it to establish in the living wood of the freshly cut stump, yet it produces no disease, as does *H. annosum*. The antagonist also limits the growth of *H. annosum* by the mechanism of hyphal interference, whereby the affected hyphae lose turgor and cease to grow. The method was initially developed for control of root rot of pine in East Anglian plantations in England, where it had been used on 62,000 ha by 1973 (Greig, 1976). *Peniophora gigantea* has subsequently been used with success in several other countries and was the first agent for biological control of a plant pathogen to receive registration by the U.S. Environmental Protection Agency for use in the United States.

    The principle of early and rapid establishment of a competitor or weak pathogen in wounds or similar infection courts has been applied successfully by Pottle et al (1977), and Smith et al (1979), using *Trichoderma harzianum* to control wood-rotting Hymenomycetes that otherwise invade summer wounds of red maple (*Acer rubrum*). Carter (1971, 1983) applied conidia of *Fusarium lateritium* to pruning wounds of apricot in Australia to protect against infection by *Eutypa armeniacae*. The antagonist was shown to give best control when used in combination with benomyl, to which it is relatively insensitive. Ricard and Bollen (1968) established the potent antibiotic producer *Scytalidium lignicola* as a primary colonist of Douglas-fir poles to protect the poles against decay by *Poria carbonica*. Grosclaude et al (1973) used *T. viride* applied to pruning wounds of plum trees to protect against infection by *Chondrostereum purpureum*, the cause of silver leaf disease. Ricard (1977) coined the term "immunizing commensals" for organisms having ability to grow in and protect plant tissue against harmful organisms without themselves causing harm. These and other examples of biological protection of wounds are discussed further in Chapter 8.

CROSS PROTECTION AND INDUCED RESISTANCE

Cross protection and induced resistance are types of biological control that take place mainly inside the plant, another approach unique to the control of plant pathogens. Examples of cross protection and induced resistance are well known in plant pathology (Kuć, 1982), going back to the early work of McKinney in 1929 on plant viruses. The first review of the subject and associated problems was by Chester (1933).

The previous chapter attempted to clarify the concepts of cross protection and induced resistance by distinguishing between instances in which the first organism in the plant acts directly (as an antagonist) against the second (challenger) organism and those in which the first organism initiates (induces) a process whereby the host plant itself inhibits the challenger. The former (direct antagonism) could include any one or a combination of antibiosis, competition for sites or nutrients, hyphal interference, or parasitism of the second organism by the first within the host tissue. The latter is indirect and could include any combination of the forms of active self-defense of plants against microorganisms as long as it is stimulated by the first organism. The distinction between direct antagonism and induced host-plant resistance is clear in only a few cases. Some possible mechanisms are discussed in Chapter 6.

From the practical standpoint, the use of cross protection and induced resistance implies that the inducer strain is, itself, nonpathogenic or only mildly pathogenic. As a rule, but with exceptions (Baker et al, 1978), the organism used to achieve such biological control is related to the pathogen for which control is desired, or the organism is pathogenic on similar tissues of some other crop. This is to be expected; related strains (or pathogens of similar tissue) are most likely to compete for the same materials available from a tissue, like any two organisms adapted to the same ecological niche. Related organisms (or those pathogenic in a similar tissue) are also the most likely to trigger the same defense reaction in the plant. The likelihood that the protecting strain or species will be a parasite or even a mild pathogen is also to be expected in view of the job to be done; it must establish in healthy host tisue not already occupied by the target pathogen, which requires at least some ability as a parasite. A strict saprophyte may exist as an endophyte in healthy plant tissue in a commensal relationship with the host (Chapter 3), and through this position and as a prior colonist, it may provide protection against a pathogen of those tissues. However, most induced resistance specifically and cross protection generally will probably require that the pioneer colonist attempt to establish more than a superficial or commensal relationship with the host.

The risks with this approach to biological control have been mentioned (Chapter 1). Even so-called mild or weak strains of a pathogen may have a negative effect on plant vigor or productivity, or an organism mildly pathogenic on one host may be acutely pathogenic on another grown in the same agroecosystem. Although a mild strain of a virus may protect against related strains, it can sometimes interact synergistically with unrelated strains to produce severe symptoms in a host. The CARNA 5 described by Kapper and Waterworth (1977) and Waterworth et al (1979) is an example; it gives attenuation of cucumber mosaic virus (CMV) in some hosts but results in more severe expression of disease when combined with CMV in tomato. Because of examples such as this, and a reluctance among plant pathologists to intentionally spread or inoculate crop plants with pathogens, this approach has found only limited use in commercial agriculture. On the other hand, the lack of use to date does not mean the lack of potential for biological control.

Four examples are given here to illustrate the ways by which pathogens or parasites can be used to enhance host-plant resistance, give cross protection, or both. These are: 1) inoculation of the plant with a mild strain of a virus, which then protects the plant against damage from a more virulent strain; 2) cross protection (including induced resistance) for control of fungus pathogens; 3) use of multilines or mixtures of the host to favor a race mixture of the pathogen and thereby decrease inoculum effectiveness, increase the chances for interrace antagonism, and possibly obtain induced host resistance; and 4) the use of mycorrhizae to improve plant vigor and hence the general resistance of plants to pathogens.

Tristeza Disease of Citrus

Biocontrol of the pathogen responsible for this disease illustrates effective use of cross protection, accomplished by inoculation of the host with a mild strain of a virus before its exposure to a virulent one.

Pathogen: Closterovirus, with highly flexuous filamentous particles.

Hosts: Restricted to Rutaceae. Citrus varieties budded on intolerant sour orange (*Citrus aurantium*) rootstocks are severely affected; those on tolerant sweet orange (*C. sinensis*), mandarin (*C. reticulata*), *C. trifoliata*, Rangpur lime (*C. limonia*), or Troyer citrange are largely unaffected by the disease. *Aeglopsis, Afraegle,* and *Pamburus* are also susceptible.

Disease: The virus was introduced from Africa in the 1920s and almost wiped out the citrus industry in Argentina, Brazil, Uruguay, and Java by the 1940s. It became very important in California after 1940. Symptoms include quick decline of citrus on sour orange rootstock, stem pitting on grapefruit, dieback of lime, and decline of mandarin and Pera orange. Sweet orange on sour orange rootstock suddenly wilts, declines, and dies, or it develops overgrowth at the bud union. The disease is worldwide in tropical and semitropical areas.

Life Cycle: The virus is spread in the field by its most efficient vector, citrus oriental aphid (*Toxoptera citricidus*), and less efficiently by *Aphis gossypii, A. spiraecola,* and *T. aurantii.* The virus may be acquired by the vector in seconds and transmitted in seconds; it may be retained 24 hours. The virus is circulative, styletborne, and nonpropagative. It is also transmitted by tissue union

Virus Control by Cross Protection

One of the most successful commercial uses of cross protection for virus control is the inoculation of citrus with a mild strain of citrus tristeza virus (CTV) to control severe strains in Brazil (Costa and Müller, 1980). CTV is a phloem-limited closterovirus spread by aphids, the citrus oriental aphid (*Toxopter citricidus*) being particularly efficient. CTV spread rapidly in Argentina, Brazil, and Uruguay following its introduction from Africa in the 1920s. Budding commercial citrus types onto tristeza-tolerant root stocks gave only partial control, and breeding for resistance was considered to have even less chance for success. The search for mild strains for cross protection began in the early 1960s; trees having good growth and general appearance were sought in orchards or areas where the disease occurred and where escapes were unlikely. This is another example of biological control found by looking where the disease is absent for no apparent reason. Some 45 mild strains of CTV naturally present in Brazil were recovered

(grafting and budding) and experimentally by dodder union but not by sap or seed. Particles exist in large numbers in phloem cells but not elsewhere. Destruction of phloem in sour orange causes quick decline of grafted trees.

Biological Control: In 1951, mild strains in natural tristeza virus complexes in Brazil were found to protect against severe strains. Since the tristeza-tolerant rootstocks did not fully control the disease there on West Indian lime, grapefruit (*C. paradisi*), or sweet orange, cross protection was studied. Buds were taken from a large number of healthy trees in uniformly and severely injured orchards (an illustration of the principle of seeking antagonists where disease is expected but not found), and budded on tolerant and intolerant rootstocks that became naturally infected with tristeza. Some of these infected but apparently healthy trees, when inoculated with severe virus isolates, remained healthy. Strain 45 of the virus was selected and widely used for cross protection. Protection was less for tissue-union inoculation than for natural aphid spread, perhaps because aphids transmit the mild strain. The studies, begun in 1951, led to commercial testing by 1968; by 1977, five million trees were cross-protected, and by 1980, eight million were cross-protected in Brazil. This illustrates the value of persistent long-term studies of biocontrol by cross protection. Grapefruit infected with mild stem pitting is similarly protected from severe tristeza in Australia.

References: Bennett and Costa (1949), C.M.I. Descriptions of Plant Viruses 33, Costa and Müller (1980), Fraser et al (1968), Grant and Costa (1951).

by this approach. Of these, three showed potential for satisfactory cross protection against severe CTV in Pera sweet orange, two in Galego lime, and one in Ruby Red grapefruit. A satisfactory result meant that the plants grew well, showed little or no effect of CTV (including none of the stem-pitting reaction to CTV), and produced good yields. CTV-protected budwood on tristeza-tolerant root stock was made available to growers in 1968. Within 10 years, five million cross-protected citrus trees had been planted, and no evidence has been found thus far of breakdown in protection.

Mild strains from a given type of citrus apparently give the best protection if reintroduced into the same citrus type. Thus, the best isolates for Pera sweet orange were those collected from Pera sweet orange; they produced a more severe reaction if used in Galego lime. It was also necessary to avoid the presence of other pathogens (e.g., exocortis viroid) in the propagative material. Some "breakdown" in protection (less than 1%) is considered the consequence of virus blending when cross-protected budwood is budded onto a root stock naturally infected with the regular CTV complex (Costa and Müller, 1980).

A slightly different situation has been reported from Israel, where CTV was introduced in the period 1928–1937 but did not spread until recently, owing to an absence of aphid vectors (Bar-Joseph, 1978). Although *Aphis gossypii* is present, it was not a vector of the strains of CTV in Israel until recently. Bar-Joseph (1978) believes that a mutant strain of CTV transmissible by *A. gossypii* has developed from within the original CTV complex of strains in Israel. This mutant strain is now spreading, whereas the original strains, although severe, have not spread. Bar-Joseph suggests further that, over the past 30 years, the original CTV strains have provided cross protection against any multiplication of mutant strains that may have occurred. However, as these trees have aged, a progressively larger amount of tissue in them has been left unprotected, thereby providing niches for multiplication of a mutant strain transmissible by *A. gossypii*. A single mutant successfully acquired by *A. gossypii* would be enough. Because efficient vectors are already present in Brazil and presumably move ample amounts of virulent virus into each orchard annually, the possibility exists that, after another 10–20 years, the protected trees in Brazil will succumb to CTV. The solution to this problem might be to replace the trees at an age slightly younger than normal replacement age.

Cross protection against tobacco mosaic virus (TMV) of glasshouse tomatoes is used commercially in Canada, England, Holland, Japan, and many other countries. Its use in Europe was greatly expanded in 1972–1973 with the development of an artificial (nitrous oxide-induced) symptomless mutant of TMV identified as MII-16 (Rast, 1972). The strain has been used successfully and extensively in England for protection of glasshouse tomatoes, and most failures have been traced to escapes (unprotected plants) because of improper inoculation technique (Fletcher and Rowe, 1975). On the other hand, the use of MII-16 in England resulted in a marked shift toward a higher frequency of strain 1 of tomato TMV, in contrast to the earlier situation in which strain 0 tended to be dominant (Fletcher and Butler, 1975). An increase in the incidence of strain 1 occurred once before, when cultivars having the Tm-1 gene for TMV resistance were widely used in the industry. Removal of cultivars with the Tm-1 gene brought a return of strain O as the dominant type. The mutant MII-16 was from strain 1 originally, and Fletcher and Butler (1975) suggest that the return to dominance of strain 1 (as much as 94% in some nurseries where cross protection had been used since 1973) may relate to contaminant virulent strains in the inoculum, mutations back to virulence, or selection for a new strain 1 for which cross protection was ineffective. These workers propose that a mild mutant strain coming from strain 0 but combined with the use of cultivars with the Tm-1 gene would present a more secure situation.

Control of Fungus Pathogens by Cross Protection

Although no example of commercial application exists for cross protection achieved by introduction of specific avirulent fungi, several examples of naturally occurring biological control by this mechanism are known or are suspected to occur. Moreover, experimental work in progress suggests that in the future, biological control by inoculation of the plant with one or more avirulent fungi will be possible.

Wheat grown in rotation with grasses (for pasture or hay) and susceptible to *Gaeumannomyces graminis* var. *tritici* can be severely damaged by take-all already in the first crop after grass because the grass roots provide a source of inoculum of the pathogen (Sprague, 1950). In England, where wheat-grass rotations are common, take-all is surprisingly uncommon in the first year of wheat but increases with subsequent wheat crops (Deacon, 1973a, 1974a, 1976a). The evidence is strong that, although the grass roots maintain a population of *G. graminis* var. *tritici*, these roots also support a large population of *Phialophora graminicola*, avirulent on wheat roots but able to colonize the outer layer of wheat root cortical tissues (Deacon, 1974b; Scott, 1970; Speakman and Lewis, 1978). Given the advantage of being the first colonist, *P. graminicola* protects wheat roots against invasion by *G. graminis* var. *tritici*. *Phialophora graminicola* produces no root necrosis and does not colonize the stele (Deacon, 1974a, 1974b). *Gaeumannomyces graminis* var. *tritici* is able to grow longitudinally inside the stele but does not do so in regions where the surrounding cortex is colonized by *P. graminicola* (Speakman and Lewis, 1978). The avirulent fungus apparently modifies the root in a way that even tissues not actually occupied (e.g., the stele) become less susceptible to the pathogen.

The avirulent *Phialophora* occurs naturally throughout Britain on Italian and perennial ryegrass, timothy, meadow foxtail, lawn and turf grasses, and others (Deacon, 1973a). However, although high populations carry over and protect the first wheat crop, such populations do not persist to provide the same level of protection of subsequent wheat crops. Cultivation and other practices associated with wheat growing are somehow less favorable to *P. graminicola* than the practices used when grasses are grown. With a decline in the population of *P. graminicola*, a niche is opened for *G. graminis* var. *tritici* and take-all becomes important. The eventual decline of take-all with long-term wheat monoculture is another pheomenon, and avirulent fungi related to *G. graminis* var. *tritici* apparently are not involved (Deacon, 1976a). *Phialophora graminicola* also may be a natural protectant of turf in England against the patch disease caused by *G. graminis* var. *avenae* (Deacon, 1973b).

In Australia, the avirulent or weakly virulent *G. graminis* var. *graminis* is a common colonist of roots of many grasses and, like *P. graminicola* in England, declines in frequency with cultivation and cropping of the field to wheat. Also like *P. graminicola, G. graminis* var. *graminis* colonizes the outer layer of cortical tissues of wheat roots and, if established first, protects the roots of wheat against either *G. graminis* var. *tritici* or *G. graminis* var. *avenae* (Wong, 1975).

Turf grass (*Agrostis stolonifera*) used for golf greens and some lawns is frequently established on soils fumigated with methyl bromide to eliminate weed seeds and pathogens. *Gaeumannomyces graminis* var. *avenae* is among the first pathogens to reestablish in these new plantings, and the resultant patch disease can be severe. Wong and Siviour (1979) showed that any of four avirulent fungi *(G. graminis* var. *graminis*, a *Phialophora* sp. [probably *P. radicicola*], *P. graminicola*, and an isolate of *G. graminis* var. *tritici* that had lost virulence), when introduced into fumigated soil in advance of planting the turf grass, protected the turf against infection by *G. graminis* var. *avenae*.

Avirulent fungi related to *G. graminis* var. *tritici* are generally lacking in wheat-field soils in Australia. Wong and Southwell (1980) demonstrated that take-all can be suppressed in the field by introducing either *G. graminis* var. *graminis* or *P. radicicola* in colonized oat grains with the wheat seed at the time of sowing. Protection was best where the amount of in-oculum of the take-all fungus was lowest. One problem in the use of *G. graminis* var. *graminis* is that it does not grow as well as the take-all fungus at lower temperatures. Wong (1980) selected a low-temperature isolate of *G. graminis* var. *graminis* and thereby obtained more com-plete biological control of take-all in tests conducted at the lower tempera-ture.

The work with avirulent fungi related to *G. graminis* var. *tritici* and var. *avenae* leaves little doubt that some benefit is already derived from them through altered cultural practices. The work of Wong and Siviour (1979) and Wong and Southwell (1980) indicates that a potential exists for introduc-tion of these fungi in some situations if technology for mass production is developed.

Many taxonomic groups of fungi exist in nature as a mixture of virulent and avirulent species and subspecies. Melouk and Horner (1975) isolated an avirulent *Verticillium nigrescens* from roots of mint showing mild symptoms of verticillium wilt caused by *V. dahliae*. Inoculation of roots of either spearmint or peppermint with *V. nigrescens* two or more days before inocula-tion with *V. dahliae* greatly reduced the incidence of wilt, but less or no reduction in wilt occurred if inoculation was at the same time or two or more days after inoculation with *V. dahliae*. The role of natural protec-tion by *V. nigrescens* in the field is unknown but may be significant. Soil from the Chateaurenard region of France suppressive to fusarium wilt of muskmelon has a high population of saprophytic (nonpathogenic) clones of *F. oxysporum* (Alabouvette et al, *in* Schippers and Gams, 1979). The non-pathogenic members of *F. oxysporum* are well-known to colonize the cortical tissues of roots of many plants but to cause no evidence of disease. These findings raise the possibility that root protection through prior colonization by nonpathogens related to the pathogen may be partly responsible for the general absence of fusarium wilt in this soil. Introduction of nonpathogenic clones of *F. oxysporum* into steamed soil produced a wilt-suppressive soil, but the effect was more pronounced in some soils than others (Chapter 7). This approach to biological control may have potential for wilt control of glasshouse crops where soil mixes are prepared as rooting media or where seedlings are produced for transplanting, provided the proper strains are used.

A number of investigators have demonstrated induced resistance or cross protection between pathogens of *F. oxysporum*, achieved by prior inoculation of the plant with a forma specialis (or race) pathogenic on another plant species (or cultivar) (Baker and Cook, 1974). Again, this approach to biological control may have the greatest application with plants produced under glasshouse conditions or normally handled as transplants. The seedlings could be produced in soil infested with the nonpathogenic race or forma specialis or roots of the seedlings could be dip-inoculated with a spore suspension of the nonpathogenic strain before being transplanted into the field soil. Wymore and Baker (1982) showed for tomato that cross protection against *F. oxysporum* f. sp. *lycopersici* by *F. oxysporum* f. sp. *dianthi* required an inoculum density of the cross-protection agent equal to or higher than that of the tomato pathogen. However, protection was effective only if the biocontrol agent was applied a few days in advance of the pathogen and protection lasted only three to four weeks. Wymore and Baker (1982) suggest that since protection is only temporary, inoculation with avirulent strains of *F. oxysporum* may not be a practical control of fusarium wilt of tomato.

On the other hand, the potential for cross protection as an approach to biological control may depend on obtaining the right nonpathogen for inoculation. K. Ogawa and H. Komada (personal communication; Chapter 8) obtained full-season protection of sweet potatoes in the field in Japan by inoculation of cuttings with a nonpathogenic strain of *F. oxysporum* originally obtained from the stems of sweet potato. The *Pseudomonas* spp. effective as antagonists on roots of wheat (Weller and Cook, 1983), potatoes (Burr et al, 1978), and onions (Kawamoto and Lorbeer, 1976) were obtained originally from the roots of these same respective plant species. Similarly, in the work of Lindow (1982), the strains of *Erwinia herbicola* effective as leaf colonists for protection against *Pseudomonas syringae* responsible for ice-nucleation were originally from leaves. **An antagonist sought for use in soil, on a plant surface, or within a plant tissue is most likely to be adapted to the niche where needed if it is obtained from that niche originally.**

The protection of carnation cuttings against stem rot caused by *F. roseum* 'Avenaceum' by inoculation of the cuttings with a nonpathogen (e.g., *F. roseum*

'Gibbosum') results at least in part from induced resistance in the cut stems (Baker et al, 1978). The cut surface of a carnation stem becomes resistant to Avenaceum naturally within 96 hours, but inoculation with a nonpathogen achieved the same resistance in 24 hours. Direct forms of antagonism between the agent and the pathogen (i.e., antibiosis, parasitism, or competition for nutrients) were ruled out as mechanisms of protection. Evidence was presented instead for induction of a phytoalexin type of response in the host by the first organism, which inhibited the pathogen.

Cross protection against fungus pathogens is usually a localized phenomenon. However, systemic protection occurs in some interactions, as exemplified by the protection of cucurbits against *Colletotrichum lagenarium* by inoculation of the first true leaf (cotyledon) with *C. lagenarium* (Caruso and Kuć, 1977a; Kuć et al, 1975). Cucumber, watermelon, and muskmelon all were more resistant to infection by *C. lagenarium* on the second or subsequent leaves, starting about 96 hours after the first leaf was inoculated with the same pathogen. The increased resistance was evident as fewer and smaller lesions and lasted for four to five weeks. A second ("booster") inoculation three weeks after the first extended the protection for cucumber into the fruiting period (Kuć and Richmond, 1977). Protection was proportional to the number of spores applied to the first leaf, but even a single lesion on the first leaf conferred significant protection to subsequent leaves. Significant protection of cucumber and watermelon was achieved in field trials by application of spores to the first true leaf of the plants about seven weeks before transplanting them in the field (Caruso and Kuć, 1977b).

Increased resistance to *C. lagenarium* in cucurbits can also be induced systemically by inoculation of the first leaf with pathogens other than *C. lagenarium*, including *Pseudomonas syringae* pv. *lachrymans* and tobacco necrosis virus (Jenns and Kuć, 1979). Increased resistance becomes detectable beginning with the first evidence of lesion development on the inoculated leaf. The induced resistance is not, therefore, so specific as to require inoculation by a race of *C. lagenarium*. The use of this kind of biological control has potential with a crop such as cucurbits, which is transplanted and can therefore be inoculated in one environment (e.g., in the bedding glasshouse) before being moved to another (e.g., the field). The fact that several pathogens will give the effect is also advantageous; a pathogen having low potential for spread in the field would be the probable choice.

Use of Multilines and Variety Mixtures

The vulnerability of a genetically uniform crop to pathogens has been revealed many times in the past, at great cost to farmers and to society. An alternative is genetic diversity, not only from field to field within a geographic area, but also from plant to plant within each field. A multiline is "a mixture of isolines that differ by only one or a few genes for reaction to a pathogen" (Browning and Frey, *in* Jenkyn and Plumb, 1981). Cultivar mixtures are mixtures of two, three, or more cultivars differing in genetic backgrounds as well as having different genes for resistance to the pathogen (Wolfe and Barrett, 1980). Cultivar mixtures can be prepared by blending seed of existing cultivars, whereas multilines require special effort in plant breeding. Multilines are preferred where a high degree of agronomic uniformity is required, but cultivar mixtures provide the greatest diversity and flexibility. With either approach, the components must be sufficiently similar in plant characteristics to provide uniform maturity and end-use quality.

Multilines and cultivar mixtures have been used as approaches to control of pathogens populations that are host-specific and in which each new source of host resistance provides directional selection for virulent types. Such populations in nature are mixtures of genetically different pathotypes (races), with some components of the mixture occurring at a very low frequency depending on previous selection pressures. Each pathotype in the population carries a gene or combination of genes necessary for virulence on specific genotypes of the host. The phenotype is the degree of compatibility (or incompatibility) evident from the host-pathogen interaction. The parasitic cycle of a compatible interaction is complete when the parasite produces abundant spores, which then become disseminated to fresh tissue to begin the cycle again. An incompatible interaction terminates during the early stages of attempted parasitism, and few or no spores are produced to perpetuate that pathotype. Thus, a host population consisting of a single genotype provides selection pressure in favor of pathotypes having genes for compatibility on that host. This pathotype generally remains the dominant type in the population so long as that host cultivar is grown.

A mixture of host genotypes favors multiplication of a mixture of pathotypes. Each pathotype in the mixture will have the necessary gene or genes to produce a compatible phenotype on one or more of the host genotypes in the mixture. On the other hand, each pathotype is also likely to lack the gene or genes necessary to produce a compatible phenotype on one or more other components in the host mixture. Thus, a cultivar mixture creates a situation in which a portion of the pathogen inoculum is likely to be avirulent or weakly virulent on a significant portion of the host population. By intercepting inoculum, resistant plants surrounding a susceptible plant in the mixture become barriers to the movement of any given race to that susceptible plant. Most inoculum produced on any given host tends only to move to adjacent plants. Interception of this inoculum by resistant plants is one of the benefits of a mixture. In addition, avirulent inoculum on resistant plants in the mixture provides an opportunity for pathogen interactions, including possible induced resistance to races otherwise virulent on that plant. Three-component mixtures of spring barley cultivars for control of powdery mildew (caused by *Erysiphe graminis* f. sp. *hordei*) in England generally exhibit less than half as much mildew as the mean of the components grown alone (Wolfe et al, *in* Jenkyn and Plumb, 1981). About 15% of the disease control has been estimated to result from induced resistance. A mixture of wheat cultivars has also been shown to retard epidemics of *Septoria nodorum* and *Rhynchosporium secalis* in Wales (Jeger et al, *in* Jenkyn and Plumb, 1981) and is under consideration as a means to control *Heterodera glycines* on soybean in the southeastern United States (Riggs et al, 1980).

A variety mixture provides selection pressure for a superrace — one capable of attacking all components of the host mixture. However, this superrace on any given host must compete with the other virulent pathotypes having genes for virulence on that host, some of which are probably better adapted to that host. For example, in a mixture of ABC of the host, superrace abc must compete with pathotypes a, ab, and ac on host A, pathotypes b, ab, and bc on host B, and pathotypes c, ac, and bc on host C. Van der Plank (1968) makes the case that stabilizing selection is for individuals having no more genes for virulence than necessary. These individuals are sometimes less aggressive or less competitive than those with exactly the right complement of virulence genes to produce a compatible phenotype. The chances of selecting a superrace are also decreased by making regular and strategic changes in the host mixture (Wolfe et al, *in* Jenkyn and Plumb, 1981; Wolfe and Barrett, 1980).

Actual trials with mixtures have demonstrated less disease and more grain yield than in genetically uniform crops. In 1978, 37 different three-component spring barley cultivars were grown beside the same cultivars in pure stands at 26 sites in England and Scotland. Comparisons of the two groups revealed 9% higher average yield for the cultivar mixture where powdery mildew was considered important and 3% higher yield where the disease was considered unimportant. Cultivar mixtures are now commonly used in England to control powdery mildew of barley. In addition, oat multilines have been used in Iowa since 1968 as part of a program to control crown rust caused by *Puccinia coronata* var. *avenae* (Browning and Frey, *in* Jenkyn and Plumb, 1981), two multiline cultivars of wheat were released by the Rockefeller Foundation program in Colombia for control of stripe and stem rust (Browning and Frey, *in* Jenkyn and Plumb, 1981), and a 10-component multiline ('Crew') was recently released for growers' use in the Pacific Northwest to control stripe rust of wheat (R. E. Allen, personal communication).

Vesicular-Arbuscular Mycorrhizal Fungi

Considerable attention has been devoted recently to the question of whether vesicular-arbuscular (VA) mycorrhizal fungi, which increase the uptake of phosphorus and other nutrients by plants, also affect susceptibility of plants to disease. If such effects occur, whether expressed as greater or less root damage, then management of these symbionts becomes a means to achieve biological control of the pathogen(s) affected. Moreover, VA mycorrhizal hyphae growing from the roots of one plant may interconnect with the roots of neighboring plants of the same and different species, thereby serving in the transfer of nutrients between plants (Chiariello et al, 1981; Whittington and Read, 1982). Perhaps the biological control benefits of VA mycorrhizae are also transferred by interconnections between plants.

Experiments under controlled conditions indicate that in general (but with some notable exceptions), roots with VA mycorrhizae are less damaged by pathogenic nematodes and fungi than are nonmycorrhizal roots (Hussey and Roncadori, 1982; Schönbeck and Dehne, 1979). *Glomus fasciculatum* established first on roots of tomato seedlings significantly reduced the number and size of galls caused by *Meloidogyne incognita* and *M. javanica* when the seedlings were transplanted into a nematode-infested soil (Bagyaraj et al, 1979). *Glomus macrocarpum* on roots of soybean resulted in a larger root system and fewer galls caused by *M. incognita*, although the effect was more evident with some soybean cultivars than others (Kellam and Schenck, 1980).

Among root-pathogenic fungi, *Thielaviopsis basicola* caused less damage to mycorrhizal roots (*Glomus mosseae*) than to roots lacking mycorrhizae, an effect shown for cotton (Schönbeck and Dehne, 1977), tobacco (Baltruschat and Schönbeck, 1975), and possibly alfalfa (Baltruschat and Schönbeck, 1975). Phytophthora root rots of plants (Baltruschat and Schönbeck, 1975) may be more or less severe depending on the plant species and pathogen. Ross (1972) observed that 88% of mycorrhizal soybean plants developed severe stem rot caused by *P. megasperma* var. *sojae*, and 33% of the plants died, while only 17% of nonmycorrhizal plants developed severe stem rot and none died. Similarly with avocado infected by *P. cinnamomi*, mycorrhizal roots were more severely damaged than were nonmycorrhizal roots (Davis et al, 1978). In contrast, mycorrhizal citrus roots showed significantly less damage from *P. parasitica* than nonmycorrhizal roots in low phosphorus soil but no difference in soil with high phosphorus (Davis and Menge, 1980). Bärtschi et al (1981) obtained highly effective biological control of *P. cinnamomi* root rot of the woody ornamental, *Chamaecyparis lawsoniana* 'Ellwoodi' by inoculation of the roots with spores of a mixed population of VA mycorrhizal fungi six months before inoculation with *P. cinnamomi*. Inoculation two months in advance was ineffective, as was inoculation with *G. mosseae* as much as eight months in advance. The evidence suggests that a mixture increased the chance for obtaining the right symbiont-host combination but that establishment of good mycorrhizae was also necessary for control. Graham et al (1982) showed that *Glomus* spp. sometimes colonized the cortex of roots without forming extensive external mycelium in the soil; such isolates did not enhance plant growth.

The tops of plants with mycorrhizal roots are likely to be more susceptible to damage by pathogens (Schönbeck and Dehne, 1979). Diseases caused by *Helminthosporium sativum* and *Erysiphe graminis* f. sp. *hordei* on barley, *Colletotrichum lindemuthianum* and *Uromyces phaseoli* on French bean, *E. cichoracearum* on cucumber, *Botrytis cinerea* on lettuce, and TMV on tobacco all were more severe on mycorrhizal than on nonmycorrhizal plants. Thus, any attempt at biological control of one disease (e.g., a root disease) by introduction or management of VA mycorrhizal fungi must include precautions against a possible increase in another disease.

At least part of the influence of VA mycorrhizae on damage from plant pathogens results from the improved nutrition (mainly phosphorus) of mycorrhizal plants. Adding phosphorus was about as effective as inoculation with *Glomas fasciculatum* in reducing the damage to citrus roots caused by *P. parasitica* in low-phosphorus soil (Davis and Menge, 1980). The incidence of verticillium wilt caused by *Verticillium albo-atrum* in cotton in a phosphorus-deficient soil was increased by *G. fasciculatum* in the cotton roots, and adding phosphorus in the absence of the mycorrhizal fungus also increased the incidence of wilt (Davis et al, 1979). Take-all caused by *Gaeumannomyces graminis* var. *tritici* is exacerbated by deficiency of phosphorus, and infection of wheat roots by *G. mossae* was suppressive to the disease in a phosphorus-deficient soil (Graham et al, 1982). The improved growth rate and increased size of the tops resulting from improved nutrition may also help explain why so many foliar pathogens cause more damage on mycorrhizal plants; conditions or treatments that enhance the growth rate of the plant and especially those that increase the size of the tops or total leaf canopy will generally favor foliar pathogens, particularly those having a biotrophic-type of parasitic relationship with the plant.

On the other hand, simply adding phosphorus is not always adequate, even in phosphorus-deficient soils; Plenchette et al (1981) demonstrated with apple seedlings transplanted from pots into a phosphous-deficient natural soil in a field that introduction of VA mycorrhizal fungi produced significantly better growth of the trees than application of phosphorus at a rate of 112 kg/ha (100 lb/A). Inoculation with mycorrhizal fungi was necessary in this case in spite of naturally occurring VA mycorrhizal fungi and the benefits could not be duplicated with phosphorus, although the main benefit to the trees of the introduced fungi was apparently improved nutrition. Native VA mycorrhizae may be ineffective because no external mycelium are formed in soil (Graham et al, 1982).

Some of the benefits of these fungi in disease control may also involve root protection other than relief from nutritional stress. Prior establishment of VA mycorrhizal fungi in roots limited infection by *Olpidium brassicae* in a localized way (Schönbeck, *in* Schippers and Gams, 1979). A similar alteration or reduced attractiveness was proposed to account for less infection of tomato roots by *F. oxysporum* f. sp. *lycopersici* (Dehne and Schönbeck, 1975).

The obvious approach with mycorrhizal fungi should be to encourage their presence on plant roots wherever needed to help the plant obtain phosphorus or other nutrients, or wherever benefits can be demonstrated but not duplicated by simply adding fertilizer. If plants with VA mycorrhizal associations are more susceptible to certain diseases (e.g., verticillium wilt, rust, mildew), this is not justification for elimination of the mycorrhizae. Our approach should be to provide conditions for vigorous, healthy plant growth; such plants will thereby have a higher level of general resistance to pathogens favored by a stressed host, and for those diseases that may increase in severity, some other method of disease control should be found.

5

THE PATHOGEN IN BIOLOGICAL CONTROL

Pathogenic fungi . . . provide for the future in many ways. . . .
pioneer colonists . . . possess the residue more surely by hoarding . . .
than by . . . contesting the remains with all challengers. . . .
Active possession relies upon slow utilization of the substrate
and upon persistent defense of it against competitors.
—G. W. BRUEHL, 1975

Survival is determined by how well energy
is conserved . . . and by how well the organism can
protect that energy against antagonist
elements in the microbial community.
—J. E. MITCHELL, 1979

Plant pathogens, like other living things, have existed throughout their evolutionary history with the problem of obtaining, protecting, and storing food in the presence of aggressive competitors. Those in existence today are the best that nature has produced, a self-evident demonstration of survival of the fittest. This chapter emphasizes the energy management of the pathogen and how it beats the competition—that is, why it has been successful.

Plant pathogens usually are viewed as aggressors, and mechanisms of defense in the disease process have been considered mainly from the perspective of the host. However, pathogens also must be defensive, even during pathogenesis. They must defend themselves against displacement by the many associated microorganisms on plant surfaces and even within plant tissue. Every aspect of pathogenesis must be considered in the broader context of not only how the pathogen obtains but also how it protects its food and energy (Bruehl, 1976). In Chapter 6 of our earlier book (Baker and Cook, 1974), we discussed the mechanisms of defense used by soilborne pathogens after death of the host plant. The present chapter examines the active as well as the dormant phases of pathogen life cycles from an ecological point of view, in order to better understand the strengths and weaknesses of the pathogens in relation to biological control.

FOOD AND ENERGY MANAGEMENT BY PARASITES DURING PARASITISM

Plant leaves and roots support a mosaic of interacting microbial communities, yet pathogens appear to establish themselves within these tissues in nearly pure culture. Secondary invaders that have potential for biological control may not begin to colonize lesions to any significant extent until after reproduction by the pathogen has begun. The older portions of a lesion commonly are the first to be colonized by secondaries, but the opposite may occur. With rice blast caused by *Piricularia oryzae*, the pathogen is concentrated in the center of the lesion and the margins become colonized by *Nigrospora, Cladosporium*, and *Fusarium* spp. (Suzuki, *in* Horsfall and Cowling, 1980). As pointed out in Chapter 4, few microorganisms have the ability to invade lesions and displace pathogens. How do plant pathogens protect themselves against secondary colonists during pathogenesis? The ideas set forth by Bruehl (*in* Bruehl, 1975) that plant pathogens exclude their competitors by food management and by creating an environment less favorable for the competitors than for themselves probably apply during parasitism as well as during saprophytism.

Pathogens use the following strategies to protect their energy supply during pathogenesis: 1) They allow no excess accumulation of readily available nutrients, which may attract competition in or on host tissue, 2) they cause the infection site or lesion area to become polluted, so as to discourage competitors, 3) they modify the physical environment of the tissue so as to limit growth of competitors, and 4) they convert the nutrients into "personalized packages" as rapidly as possible.

Preventing Accumulation of Excess Nutrients

One of the earliest host responses during a compatible host-pathogen interaction is an increase in permeability of the host plasma membrane (Thatcher, 1939; Wheeler, 1975). Nutrients from these leaky cells diffuse into the cell wall and probably the intercellular spaces (i.e., from the plant symplast into the apoplast), where they are more available to the pathogen — the organism responsible for the membrane change. Because nonpathogens and pathogens may coexist in the infection court and because the nonpathogens are likely to be aggressive saprophytes (scavengers) endowed, like the pathogen, with enzymes to break down carbohydrates, proteins, and lipids that may become available, it becomes important for the pioneer (the parasite) to cause no more release of nutrients than it can absorb almost immediately. A flood of nutrients from the host symplast into the intercellular spaces would only attract competition. It is, therefore, not surprising that during initial parasitism by strict biotrophs, host cell membranes become leaky but do not deteriorate and rupture until later in the disease process. This allows a slow and controlled supply of food to the parasite. Meanwhile, the parasite thoroughly colonizes the tissue so that, as the cells die, possession of the substrate is assured, at least until the parasite can reproduce.

Consider the downy mildew fungus of lettuce, *Bremia lactucae* (Ingram et al, *in* Friend and Threlfall 1976; Sargent et al, 1973). Spores germinate on the surface of lettuce cotyledons in response to substances from the leaf that reverse a water-soluble germination inhibitor contained within the spore. An appressorium forms at the end of a short germ tube and, from this, an infection peg emerges to penetrate the cuticle and the underlying host cell wall. The enzymes needed to digest these host barriers apparently are contained in the fungus wall. Ingram et al (*in* Friend and Threlfall, 1976) cite two advantages of enzymes bound by the fungus wall: less host tissue is damaged than would occur by production of enzymes in solution, and the method is more economical for the fungus. A third reason can be added: digestion of host material is limited to the immediate area of the absorbing hyphae, thereby causing minimal damage to the host and reducing the risk of excess nutrients, which could invite unwanted company. Of special interest is the fact that upon formation of primary and secondary vesicles within the invaginated cell protoplast, *B. lactucae* seals off the outer germ tube and appressorium by forming a callose plug in the infection peg. In addition to stopping water loss from the host, the plug presumably prevents leakage of nutrients to the leaf surface. In short, *B. lactucae* is not only a good manager of its food supply, it even covers its tracks.

The ability to change the permeability of living membranes toward greater leakiness may be a fundamental distinction between parasites and strict saprophytes. Saprophytes scavenge for nutrients on plant surfaces and within wounds and natural openings of plants. Saprophytes as endophytes also may live on the nutrients that become available naturally within the intercellular spaces (e.g., as exudates) but probably are unable to accelerate the flow of nutrients across a plant cell plasma membrane, into their environment. The ability to accelerate the flow of nutrients from the symplast is a feature of parasites, although many also get nutrients from substances naturally available in the apoplast (Hancock and Huisman, 1981).

A saprophyte, to be useful as a competitor of a pathogen during pathogenesis, must be an aggressive colonist highly adapted to the physical environment presented by the particular plant tissues (in their normal state or as they are modified by the pathogen) and preferably dependent for growth on the nutrients needed by the pathogen. A likely candidate for this role would be an organism closely related to the pathogen but nonpathogenic or only weakly pathogenic itself. This may account for the common effectiveness of nonpathogens as antagonists of pathogens to which they are related. The ability to produce an antibiotic inhibitory to the pathogen and/or to parasitize the pathogen as well as to compete with it for nutrients would be even better traits of microorganisms antagonistic to pathogens in lesions. Still another mechanism, potentially very effective, would be hyphal interference, which is responsible for the inhibition of one fungus by another in dung (Ikediugwu and Webster, 1970a, 1970b) and wood (Ikediugwu et al, 1970). By this mechanism, one fungus (the antagonist) causes hyphae of a nearby fungus to lose their selective permeability and turgor. Hyphal interference is like "turning the tables" on the parasite; while the parasite causes the host cell to lose selective permeability, the antagonist causes the parasite's cells to undergo a similar change, leading to its demise.

The mildew and rust fungi probably do not have any significant mechanism of defense against secondary colonists other than very careful food manage- ment. Bushnell (1970) found for stem rust on wheat (caused by *Puccinia graminis*) that increased respiration of host cells was negligble beyond the hyphal tips of the pathogen. If increased respiration is an indication of increased membrane leakiness, owing to the greater work requirement and hence greater energy need within the cell (Daly, *in* Heitefuss and Williams, 1976; Chapter 6), then the results indicate that wheat leaf cells do not become leaky beyond the hyphal tips of the stem rust fungus. Similar results were obtained for powdery mildew fungi (Bushnell, 1970). The haustoria formed by these fungi are probably as much defensive as offensive; they provide a maximum absorptive area for obtaining nutrients from the host cells, yet leave only a minimal portal for nutrients to escape into intercellular spaces where antagonists may reside. That host cells infected by rust and powdery mildew fungi remain alive for many days after infection also serves as a defense for these parasites; a live cell with an intact plasma membrane means a more controlled flow of nutrients, which will benefit the parasite but probably not its natural enemies.

Biotrophic parasites are well known to cause both short- and long-distance movement of host carbon toward the site of infection. Lesions are strong sinks for photosynthate. The mobilization of host carbon, which the parasite then uses and stores in its spores, is a major cause of economic damage by parasites on crop plants (Daly, *in* Heitefuss and Williams, 1976). Carbon transport can be explained by flow along a concentration gradient, where the lesion has the lowest concentration, owing to consumption and storage (as spores) by the parasite and also to lessened photosynthesis because the cells are damaged. Transport is highly regulated and compartmentalized, which probably is to the advantage of the pathogen over the secondaries. This is another reason for suggesting that **to suppress, displace, or preempt a pathogen in plants, an antagonist must be specialized as an aggressive colonist of the modified host tissue, must be able to use most of the nutrients needed by the pathogen, and preferably should have some means to inhibit growth of the pathogen, although it lacks the** pathogen's ability to initiate the flow or transport of food from the host.

Horsfall and Cowling (1979) refer to the "hungry parasite." To this we add "the even hungrier saprophyte." Given enough energy to increase its own biomass, a saprophyte may displace the primary pathogen, like a hungry dog that, when thrown a morsel, leaps on its master's lap and eats the whole meal. The challenge for biological control is to find ways to upset the delicate balance of food management by the pathogen in a way that permits secondaries to colonize the lesion earlier or more rapidly. Such colonization may not save that particular plant or plant part but would suppress disease progress and probably inoculum production by the pathogen, thereby retarding or preventing an epidemic.

Polluting the Infection Site or Lesion

Necrotrophic parasites kill host cells relatively quickly during pathogenesis and then live largely if not entirely as saprophytes in the dead tissues (Chapter 3). Such parasites seem highly vulnerable to competition from nonpathogens that are aggressive saprophytes, yet somehow they manage to dominate the lesion area, at least for a while. The ability to produce or promote the accumulation of antimicrobial substances of a kind and concentration more limiting to secondaries than to themselves would be a significant competitive advantage for the pathogen. Such selective antimicrobial substances may be products of the pathogen's own metabolism, or the pathogen may induce the host to produce substances that confer this advantage.

Pollution of the lesion to the selective advantage of the pathogen may be one role of phytoalexins. Müller and Borger (1940) proposed that phytoalexins are compounds produced by the host as an active defense against pathogens. Subsequent work has shown that phytoalexins — e.g., pisatin produced in tissues of pea (Hadwiger and Schwochau, 1971) and phaseollin produced in tissues of bean (Rahe and Arnold, 1975) — are stress metabolites formed by the damaged plant tissues and are not produced specifically in response to a particular infection. Ellingboe (1981) argued against phytoalexins as a host resistance mechanism on the basis of genetic evidence. Vanderplank (1978) proposed that phytoalexins may provide a form of defense for plants against secondary rather than primary infections, i.e., phytoalexins formed in the course of cell death protect the tissue against invasion by secondaries but not against the primary pathogen. As an extension of this proposal, perhaps phytoalexins actually act to the advantage of the pathogen because of their selectively inhibitory effect on secondaries. Phytoalexins are known to be more toxic to nonpathogens than to pathogens of the plant from which they are isolated (Cruickshank, in Horsfall and Cowling, 1980). It would seem advantageous to the pathogen to trigger just enough phytoalexin formation to limit its competitors but not itself. This hypothesis does not rule out the possibility that in some plants, the pathogen triggers too much phytoalexin and also becomes inhibited.

Pathogens of birdsfoot-trefoil must be adapted to high concentrations of cyanide (Millar and Higgins, 1970). The tissues contain cyanogenic glucosides and alinamarin (and lotaustralin?), which, when the tissues are wounded, are hydrolyzed by β-glucosidase into glucose and an aglycone, the latter then hydrolyzing to HCN. Cyanide production damages the tissue, which then becomes substrate for necrotrophs. *Stemphylium loti* (a pathogen on leaves of trefoil) had greater tolerance to HCN than did eight nonpathogens tested (Millar and Higgins, 1970). Millar and Hemphill (1978) showed further that *S. loti* produces its own β-glucosidase, indicating that the selection pressure on the fungus has been for the ability to trigger production of HCN as well as to be insensitive to it. The advantages to *S. loti* of both enzyme production and HCN tolerance would be twofold — it would have a mechanism of pathogenesis and a mechanism of defense, both achieved with a single process.

Pathotoxins of the selective type are commonly responsible for disease symptoms and probably do not serve the producer organism in any way other than as a mechanism of offense to the host (Wheeler, 1975). In contrast, certain nonselective pathotoxins may serve not only in offense to the host, but also in defense against competitor microorganisms. A candidate in this regard is syringomycin; this toxin is inhibitory to a broad spectrum of microorganisms in addition to being phytotoxic to host tissues (Sinden et al, 1971). Another candidate is fusaric acid, produced by pathotypes of *Fusarium oxysporum* and once thought to play a role in the production of wilt. Wilting is now readily explained on the basis of increased resistance to water flow in the plant (MacHardy and Beckman, *in* Nelson et al, 1981; Van Alfen and Allard-Turner, 1979; Van Alfen and McMillan, 1982). Yet selection pressure on certain pathotypes of *F. oxysporum* has resulted in their ability to produce fusaric acid. Perhaps fusaric acid is a weapon of defense used by the pathogen to discourage competitors during saprophytic colonization of the dead host. **The more the kinds and the greater the numbers of microorganisms unable to grow in a host tissue, the greater the ecological advantage to those that can grow and reproduce in such tissues.**

Modifying the Physical Environment of the Host

The pathogen may gain an advantage over secondaries by modifying the physical environment of the host tissues. A healthy corn leaf will develop leaf water potentials down to −20 to −25 bars by midday, but tissues killed by toxin of *Helminthosporium maydis* Race T tend to equilibrate with the atmosphere (if the air relative humidity is 90%, the dead leaf tissues will probably dry to an equilibrium water potential of about −120 bars at 25°C). Epiphytic bacteria will cease growth at −10 to −15 bars (probably much wetter). *Helminthosporium maydis* probably can grow at water potentials near or below −100 bars (Cook and Duniway, *in* Parr et al, 1981). A low substrate water potential is, therefore, to the ecological advantage of a pathogen such as *H. maydis*. Antagonists of such a pathogen must also be able to grow at very low substrate water potentials if they are to become fully established as cohabitants with the pathogen in the leaf.

Phytopathogenic bacteria cause water-soaking, and the intercellular spaces, where the bacteria are located, commonly become filled with polysaccharides and other components of slime. The diffusion of oxygen through such tissues will be restricted compared with that in healthy tissues. Phytopathogenic bacteria, being microaerophilic, or even facultatively anaerobic, have a selective advantage over fungi in such tissues. Bacteria similarly adapted to low oxygen status, nonpathogenic themselves but aggressive competitors or capable of producing antibiotics that are effective against the pathogen, probably have the highest potential use as antagonists of phytopathogenic bacteria in host plant tissues.

Sclerotium rolfsii produces oxalic acid, which lowers the pH of host tissue (Bateman and Beer, 1965). Antagonists in this case obviously must be adapted to acid conditions to have any chance as competitors of *S. rolfsii* in the damaged tissue. *Trichoderma* spp., which are adapted to low pH, have a high potential as biological control agents for this pathogen.

A means to encourage the establishment of secondaries as cohabitants with necrotrophic parasites in lesions during or near the end of pathogenesis would be important to biological control. This would not only retard disease progress or suppress the production of secondary inoculum but would also limit subsequent survival of the pathogen in the crop residue and suppress production of primary inoculum, perhaps even a year later. Thus, an antagonist adapted to corn leaf tissues killed by *H. maydis* might not save that particular leaf but could affect reproduction or survival of the pathogen at a later date. A pathogen in full possession of crop residue as the pioneer colonist is difficult to displace with saprophytic colonists. Establishment of saprophytes as cohabitants of the residue during the early stages of pathogenesis might help overcome this difficulty (Chapter 4).

Converting Food into New Propagules

Strict biotrophs, such as rust and powdery mildew fungi, undergo reproductive activity during the disease cycle and have little or no continuing need for the host cells after they are dead. These biotrophs use a form of *passive possession*, a scenario described by Bruehl (*in* Bruehl, 1975): "invade thoroughly, digest extensively, store the surplus food in resting structures, and then abandon the fragile, exhausted host remains to scavengers." **The longer the time between infection and sporulation, or the longer the period of sporulation, the greater the opportunity for secondary invaders to suppress sporulation.**

The perennial rusts of trees (*Cronartium ribicola, C. quercuum* f. sp. *fusiforme,* and *C. strobilinum*) persist for years in sporulating galls. As pointed out in Chapter 4, *Tuberculina maxima, Scytalidium uredinicola,* and *Sphaerellopsis (Darluca) filum* may invade these galls and gradually destroy or displace the rust fungus. These particular antagonists are apparently unique among the vast array of potential secondary colonists that surely make contact with the galls on these trees at one time or another. It would be interesting to know what mechanisms of exclusion or discouragement operate against the many potential colonists of perennial rust galls and how these barriers are overcome by *T. maxima, Scytalidium uredinicola,* and *Sphaerellopsis filum.* Such information might provide important clues to traits needed by the ideal antagonist.

Many necrotrophic parasites, including biotrophs that complete their life cycles as necrotrophs, rapidly compartmentalize food from the host into forms unavailable to competitors, such as oospores, chlamydospores, and sclerotia. Any reduction in the energy supply caused by competition from secondary colonists should be reflected in fewer or smaller propagules produced by the pathogen, making it less efficient in infection (Henis and Ben-Yephet, 1970). In addition, the opportunity exists at this stage of the pathogen cycle for establishment of hyperparasites on the pathogen propagules. As suggested earlier (Chapter 4), *hyperparasites of pathogen propagules may be most efficient if established at the source so that they travel with the propagule, rather than being dispersed so that they must then contact the propagule.*

Verticillium dahliae and the pathotypes of *Fusarium oxysporum* are particularly efficient and thorough in the way they colonize their respective host plants and then package the host plant tissues into propagules. These fungi invade the xylem of their respective hosts through root tips and are then carried as microconidia upward with the xylem stream or may grow up the stem. The xylem is ideal for rapid and thorough movement within a plant, but it is a poor place to gather and store food. *Verticillium dahliae* has been shown to erode xylem vessel walls; possibly some nutrients are also obtained from xylem parenchyma. More important, these fungi, being in the xylem, are ideally positioned to colonize the entire stem, petiole, or leaf when the plant dies. The only important competition is on the outside of the dead plant. *Verticillium dahliae* and *F. oxysporum* can grow saprophytically at temperatures up to 35°C and at water potentials down to −120 bars

(Manandhar and Bruehl, 1973), the kind of environment likely to exist in tissues of a dead plant (Chapter 6) still standing erect in the field. This saprophytic ability provides an important competitive edge over the epiphytic fungus flora, especially since the inside of the dead plant presents a far more stable and suitable environment for fungus growth than does the outside. The best method of biological control aimed at preventing propagule formation by *Verticillium* and *Fusarium* wilt fungi in stems of their hosts may be early and thorough burial of the plant residue before colonization and propagule formation begin, thereby increasing the chances for preemption of these internal plant inhabitants by saprophytic colonists from the soil. Ioannou et al (1977c) proposed an even more effective treatment: bury the infected plant residues before microsclerotia have formed, and immediately flood the soil to shut off the oxygen supply the pathogen needs to form microsclerotia.

MYCOVIRUS INFECTIONS AND HYPOVIRULENCE

Viruses in fungi (mycoviruses) are mostly isometric particles 25–50 nm in diameter that contain a segment of double-stranded RNA (dsRNA) as their genome. An exception is the viruslike particles in *Endothia parasitica*, which contain dsRNA but are somewhat pleomorphic and club-shaped rather than isometric (Dodds and Day, *in* Lemke, 1979). Another exception has been found in the cultivated mushroom, namely, bacilliform particles measuring 19 × 50 nm (Van Zaayen, *in* Lemke, 1979). Mycoviruses with a single-stranded DNA genome have been found, but not in phytopathogenic fungi (Bozarth, *in* Molitoris et al, 1979). Mycoviruses or viruslike particles have been found in all major groups of fungi, including many plant pathogens (Hollings, 1982). Multiple infections, i.e., infection by more than one mycovirus in the same thallus, are common.

The majority of mycoviruses are latent in their host, causing no recognizable physiological or morphological change. On the other hand, virus infections may cause yield reductions in cultivated mushrooms (Van Zaayen, *in* Lemke, 1979) and they cause or have been associated with reduced virulence in *E. parasitica, Rhizoctonia solani*, and *Gaeumannomyces graminis* var. *tritici* (Day and Dodds, *in* Lemke, 1979). Both the aggressive and nonaggressive strains of *Ceratocystis ulmi* are infected with dsRNA determinants, but the nonaggressive strain carries more dsRNA determinants (multiple infections) (Pusey and Wilson, 1982). Mycovirus infections are transmitted by hyphal anastomosis or through spores that may anastomose upon germination, although transmission of mycoviruslike agents in *G. graminis* var. *tritici* and *E. parasitica* through ascospores is uncommon (Rawlinson and Buck, *in* Asher and Shipton, 1981). The transmissibility of mycoviruslike agents and their pathological effects on some plant-pathogenic fungi indicate their potential in biological control. Even a partial reduction in virulence, aggressiveness, or competitive ability of the pathogen would be significant if colonization of the lesion or residue and displacement by saprophytes were then favored.

Hypovirulence in *Endothia parasitica* and its Role in Biocontrol

J. Grente used the term "hypovirulent" to describe the strains of *E. parasitica* obtained from healing cankers (Figure 5.1) on chestnut trees in Italy and southern France. Day and Dodds (*in* Lemke, 1979) distinguish between hypovirulence, a transmissible trait, and the loss of virulence caused by mutations, which is not transmissible. They limit usage of hypovirulence to the transmissible trait, and this same limitation to the term is used here.

Grente and Sauret (1969b) proposed that a transmissible cytoplasmic factor was responsible for hypovirulence in strains of *E. parasitica*, and this was subsequently demonstrated by Berthelay-Sauret (1973) and also by Van Alfen et al (1975). Day et al (1977) presented the first evidence that the factor was one or more dsRNA mycoviruslike determinants.

Initial studies using French hypovirulent strains to control blight cankers caused by *E. parasitica* on American chestnut were unsuccessful, because vegetative incompatibility factors were a barrier to hyphal anastomosis (Anagnostakis, 1982; Van Alfen, 1982). Van Alfen et al (1975) produced an American hypovirulent strain under laboratory conditions by pairings with a French hypovirulent strain; this "French-derived American hypovirulent strain" (Day and Dodds, *in* Lemke, 1979) controlled blight when inoculated into cankers. Transfer of hypovirulence occurs mainly between strains of the same vegetative compatibility group but occurs in about 20% of pairings between strains of different vegetative compatibility groups, depending on the strains (Anagnostakis and Day, 1979). The presence of dsRNA infections has been suggested to override vegetative compatibility in some cases, or possibly some transmission occurs before complete expression of an incompatible interaction (Anagnostakis and Day, 1979). Conversion following anastomosis between compatible strains occurs in about two days.

Large American chestnut trees that have survived for many years with persistent *E. parasitica* infections were found to be chronically infected with hypovirulent strains (Jaynes and Elliston, 1982; Griffin et al, *in* MacDonald et al, 1979). Hypovirulent strains of *E. parasitica* must be sufficiently competitive to survive and reproduce in nature, and they must remain hypovirulent. Hypovirulence is a quantitative trait; isolates range from totally avirulent to nearly as virulent as a noninfected strain. The ideal

Chestnut Blight

Biocontrol of the pathogen responsible for this disease is an example of the use of hypovirulence. It is accomplished by introducing into the canker a hypovirulent strain of the same vegetative compatibility group as that of the virulent strain responsible for the canker. One or more dsRNA viruslike agents contained in the hypovirulent strain are transmitted to the virulent strain when their hyphae anastomose. The virulent strain becomes hypovirulent, and the canker heals (Figure 5.1).

Pathogen: *Endothia parasitica* (Ascomycotina, Pyrenomycetes, Sphaeriales).

Hosts: The American (*Castanea dentata*), European (*C. sativa*), Chinese (*C. mollissima*), and Japanese chestnut (*C. crenata*) are the main hosts, but all *Castanea* species are susceptible, as are *Quercus virginiana, Q. stellata*, and possibly other oaks.

Disease: Cankers form in the bark of twigs, branches, and the trunk, causing death of distal parts; sprouts may grow up from the stump, but these are also soon cankered. Probably the most devastating plant disease known, it has almost extinguished the American chestnut. The pathogen was introduced into the United States from Asia by 1904, into Europe by 1924, and into Italy by 1938. However, the disease is no longer a problem on cultivated chestnuts in Italy, due to the natural spread of hypovirulence. Natural hypovirulence also occurs in Michigan, Tennessee, Virginia, and possibly elsewhere in the United States.

Life Cycle: The pathogen infects only through wounds, the septate mycelium from spores invading the bark, cambium, and sometimes the outer sapwood. After a month, the pycnidial pustules form in stromata and small slime spores are produced copiously in tendrils, each of which may contain about 10^8 spores. These spores are spread by rain, insects, and birds. Pycnidial pustules are converted to perithecial stromata, in which asci produce and eject windborne ascospores. Cankers remain active into the second season, advancing about 16 cm per year in the north, perhaps twice that rate in the south. The pathogen may

hypovirulent strain must be vigorous but unable to invade cambium tissue of the tree. There are several different dsRNA components, and Jaynes and Elliston (1980) showed that strains with "superinfections" (i.e., infection of strains already infected) were consistently less virulent than the original hypovirulent strain. Such strains also were less competitive, and some no longer produced spores. A typical treatment involves the simultaneous inoculation of a minimum of 10 hypovirulent strains of different vegetative compatibility groups into the canker to increase the chances of transmission to the strain responsible for the canker. If "superinfections" develop, the chances

grow saprophytically on the bark of oak, maple, pecan, and sumac. Ascospores are the principal means of pathogen spread.

Environment: Formation and release of both pycnidiospores and ascospores depend on rainy conditions. Ascospores are not discharged at 3.7° C or colder, only a few are discharged up to 12.1° C, and discharge is maximal at 20–26.7° C.

Biological Control: A. Biraghi observed in Italy, near the site of the first appearance of the disease there, that by 1950 many cankers were healing, the disease was spreading unusually slowly, and mycelium of the pathogen did not penetrate as deeply into the bark. J. Grente found that the pathogen had become "hypovirulent." Hypovirulent strains produce markedly fewer pycnidia and ascospores, but some conidia and most or all ascospores give rise to virulent isolates. Hypovirulence was found by P. Day and associates to be associated with dsRNA viruslike genetic determinants in the cytoplasm. The dsRNA is packaged within membranes of vesicles, an atypical condition for viruses. These determinants are transmitted to healthy isolates by hyphal anastomosis; transmissibility is thus affected by the anastomosis compatibility grouping of the pathogen. In France, only one anastomosis group may be found in an area, and in Italy, only nine anastomosis groups are known. In the United States, 10 groups may occur in one tree; 77 compatibility groups are known. The desirability of field inoculation with a mixture of isolates of different anastomosis groups is under study. Natural spread of hypovirulence is prevalent in Italy and France but not in the United States, perhaps due to the greater number of anastomosis groups present here. Insect vectors of hypovirulent strains are not known but are being sought. Because few conidia or ascospores are formed by diseased isolates, and because not all conidia carry the viruslike agent, a vector may prove essential to spread of hypovirulence in the field. In France, distribution of hypovirulent inocula to at least 10 places per hectare nearly eliminated active cankers within 10 years; 18,000 ha will be so treated over four years at a cost of $1.3 million.

References: Anagnostakis (1982), Grente and Sauret (1969), Heald (1933), MacDonald et al (1979), Van Alfen (1982).

for natural spread in the forest may be reduced. It becomes important, therefore, to select the proper combination of strains for introduction into the canker.

Anagnostakis (1981) showed that the failure of dsRNA to persist in some strains (Anagnostakis and Day, 1979) was not a barrier to the use of hypovirulence since, with other strains, the dsRNA components persisted through subculturing and were consistently transferred. Thus, while much work remains, biological control of *E. parasitica* exemplifies the potential use of mycoviruslike agents. This success story is a model that should capture

Figure 5.1. Two American chestnut sprouts artificially infected about one year earlier with a virulent (normal) strain of *Endothia parasitica*. The sprout on the left received no further treatment; it became girdled and died. The canker on the sprout on the right was treated with a curative (hypovirulent) strain of *E. parasitica* six weeks after infection; the canker ceased to develop further and the wound has begun to close. (Photos courtesy of R.A. Jaynes.)

our imagination as a new and innovative method to achieve biological control.

Pathological Effects of Hypovirulence
in *Rhizoctonia solani*

The pathological effects of dsRNA mycovirus infections of *R. solani* are expressed as loss of dark pigmentation in the mycelium, slower growth rate, and production of fewer sclerotia (Castanho and Butler, 1978a). Healthy cultures could be recovered at a low frequency by hyphal tip isolation. Transmission does not occur between isolates of different anastomosis groups, but within the same anastomosis group the agent transmitted to a healthy culture was shown to move quickly throughout the culture. Three different segments of dsRNA having molecular weights of 2.2, 1.5, and 1.1×10^6 occurred in the infected strain 189a (Castanho et al, 1978). The cured strain (189HTS) added to soil caused 23% preemergence and 79% postemergence damping-off of sugar beet seedlings, but when both 189HTS and the hypovirulent 189a were added to soil, no significant damping-off occurred, suggesting biological control (Castanho and Butler, 1978b). The limitations to use of hypovirulence for biological control of *R. solani* are similar to those with *E. parasitica*, namely, lack of transmission between incompatible strains and reduced ability of the strain to survive if severely diseased. Nevertheless, these barriers are being overcome with *E. parasitica* and should not deter efforts to use hypovirulence in *R. solani* for biological control of this important pathogen.

Hypovirulence in *Gaeumannomyces graminis* var. *tritici*

The epidemiological significance of dsRNA infections of the wheat take-all fungus is less certain than for *E. parasitica* and *R. solani*, in spite of the fact that studies of mycovirus infections in this fungus preceded work with the other two fungi. Lapierre et al (1970) and Lemaire et al (1970), working in France, were the first to suggest a possible role of viruslike infections in reduced virulence of *G. graminis* var. *tritici*. Two hypovirulent isolates of the pathogen were shown to contain viruslike particles about 29 nm in diameter (later revised to 35 nm [Lapierre, 1973]). The infected isolates came from two different monoculture wheat fields where take-all decline had occurred.

As summarized by Rawlinson and Buck (*in* Asher and Shipton, 1982), two types of isolates were recognized by the French workers. The first type included abnormal, virus-infected, hypovirulent isolates that came from plants from fields where take-all decline had occurred. These caused lesions that were limited to roots, rarely formed perithecia, tended to sector, lacked diurnal zonation in culture, and were difficult to maintain on agar. The second type included normal isolates without virus, which were strongly pathogenic, produced perithecia readily on plants or agar, did not sector in culture, and formed concentric zones of growth in response to light.

The inference that take-all decline may result from hypovirulence in the pathogen population prompted field experiments in France on the use of such strains to obtain biological control of take-all. The introduction of a hypovirulent isolate of *G. graminis* var. *tritici* reduced take-all on wheat in field plots by 46% (Brun et al, 1976), increased yield by 10–30% (Lemaire et al, 1975, 1976), and was reported to suppress disease in the field for up to four years. These successes notwithstanding, the role of mycovirus infection and/or hypovirulence in the epidemiology of take-all is still uncertain.

Rawlinson et al (1973) confirmed the existence of viruslike particles in isolates of *G. graminis* var. *tritici* from England but found no relationship between mycovirus infection and take-all decline, virulence of the isolates, ability of the isolates to form perithecia, or difficulty of maintaining the isolates on agar. Apparently healthy mycelium contained some particles measuring 27 nm and others measuring 35 nm in diameter. A third particle size of 40 nm was subsequently found (Rawlinson and Buck, *in* Asher and Shipton, 1981). Mycovirus-infected strains were also found in a field in France cropped to wheat for only the second year after a break, and surveys (Rawlinson, 1975) in England revealed that 65% of 300 isolates of *G. graminis* var. *tritici* contained virus particles. Frick and Lister (1978) showed by serology that 20 of 22 isolates from Indiana contained dsRNA and 19 contained virus particles. Their work further revealed the existence of serologically unrelated viruses in *G. graminis* var. *tritici*, including several dsRNA components, with different isolates harboring different single or combination dsRNA components. Virus infections also occur in *G. graminis* var. *avenae, G. graminis* var. *graminis*, and *Phialophora graminicola* (Rawlinson and Buck, *in* Asher and Shipton, 1981). The evidence for occurrence of mycoviruses in *G. graminis* is therefore unequivocal, but it also indicates that mycoviruses in *G. graminis* are latent or unexpressed, much as reported for viruses in other fungi.

Asher (1980), working in England, and Cook and Naiki (1982), in the Pacific Northwest of the United States, compared virulence of isolates of *G. graminis* var. *tritici* representing populations in fields of short-term and long-term wheat cultivation. In both studies, the virulence differences between the respective populations were insignificant and could not account for the lack of take-all in the fields cropped many years to wheat. In the Northwest, 63% of 731 plants randomly selected from seven fields that had undergone take-all decline yielded *G. graminis* var. *tritici* upon isolation in the laboratory, and at least 90% of the isolates were highly virulent when tested in a rooting medium conducive to take-all. Moreover, the range of variation in virulence was the same for the populations from long- and short-term wheat culture, based on analysis of single-ascospore cultures (Cook and Naiki, 1982).

A new interpretation of the seemingly conflicting observations on hypovirulence in *G. graminis* var. *tritici* has been offered by Naiki and Cook (1983a, 1983b). Following a report by Romanos et al (1980) that some isolates of *G. graminis* var. *tritici* produce a fungus inhibitor on potato dextrose agar at pH 4.0, Naiki and Cook (1983b) found highly significant negative correlations (r values of −0.70 to −0.95, depending on the culture collection) between pathogenicity of the isolate to wheat seedlings in a vermiculite rooting medium and ability of the same isolates to produce inhibitor on potato dextrose agar at pH 4.0. More significantly, whereas 90–99% of the isolates were pathogenic (and produced little or no inhibitor) when fresh from nature, the isolates gradually became nonpathogenic (and produced more inhibitor) after successive transfers or storage on agar media. Analyses of single-ascospore and single-cell cultures obtained from the same parents before and after they had become nonpathogenic (and had become producers of inhibitor) indicated that 1) the fungus is heterokaryotic with respect to nuclei having the genetic information that results in production of inhibitor, and 2) the frequency of nuclei having the genetic information that results in production of inhibitor is thought to be low in the wild type (possibly only 10%) but increases several-fold (possibly to 90% or more) with the selection pressure provided in agar culture.

Naiki and Cook (1983b) suggest that the ability to produce a small amount of inhibitor is advantageous to the fungus in nature; the amount produced would not inhibit the producer but would possibly inhibit potential competitors in the same way *Cephalosporium gramineum* benefits from antibiotic production during its survival in wheat straw (Bruehl et al, 1969). The correlation between apparent loss of pathogenicity and ability to produce inhibitor was thought to be an artifact of the pathogenicity test, which involved the use of agar disks (or other artificial substrate) as the source of inoculum of the pathogen. These disks contained inhibitor that possibly prevented the expression of virulence by retarding or preventing growth of the pathogen. Naiki and Cook (1983b) proposed that isolates produce the proper amount of inhibitor in nature but become overproducers of inhibitor during their maintenance on agar, possibly because producer cells in the heterokaryotic mycelium have a selective advantage under this condition. Such isolates then inhibit themselves.

The "hypovirulent" strains of *G. graminis* var. *tritici* used for biological control of take-all in France apparently had become weakly virulent or avirulent after maintenance in culture (Lemaire, personal communication). Conceivably these cultures have become overproducers of inhibitor. Lemaire et al (1979a,

1979b) reported that the physiology of the wheat plant is altered by seed inoculation with a "hypovirulent" culture. Perhaps the Q factor, as the inhibitor was termed (Romanos et al, 1980), becomes systemic in the wheat seedling and provides chemical protection against infection by virulent *G. graminis* var. *tritici.*

Romanos et al (1980) found no consistent relationship between occurrence of one or more dsRNA mycovirus infections in the mycelium of *G. graminis* var. *tritici* and ability of the isolate to produce inhibitor. On the other hand, Ferault et al (1979) observed that cultures that had no detectable mycovirus when fresh from nature had at least two mycoviruses (based on particle size) after 17 months in culture; the appearance of the mycovirus infections coincided with loss of "aggressivity" of the culture. Perhaps cultures become overproducers of the inhibitor by the mechanism proposed by Naiki and Cook (1983b), which then provides the opportunity (e.g., stresses) for expression of previously latent viruses.

A great deal more research is needed on the genetics, nature, and role of inhibitor production in the life cycle of *G. graminis* var. *tritici.* Perhaps a way can be found to apply selection pressure in nature in favor of overproduction of the inhibitor. Alternatively, it may be possible in the future through genetic engineering to transfer genes for production of inhibitor from the fungus into a root-colonizing microorganism nonpathogenic to wheat. It might even be possible in the future to splice the genetic material for production of inhibitor into the wheat genome, thereby producing a resistant variety.

MAINTENANCE AND DEFENSE IN THE ABSENCE OF A SUSCEPT—THE BEAR MARKET AFTER A BULL MARKET

Financiers refer to ascending and descending stock markets as "bull" markets and "bear" markets, respectively. During a bull market, the mood is to buy, not sell. Investments pour in and capital increases; investors are confident, bold, and sometimes even reckless. During a bear market, investors are conservative; the mood is to sell, not buy. Investments are withdrawn, capital decreases, and investors prepare for harder times.

The wise investor knows that a bull market will be followed by a bear market and consequently uses the bull market to prepare for the harder times ahead. The wise investor also recognizes the beginning of the next bull market and strives to take advantage of the opportunity quickly. The greatest fear of an investor is a "crash market." A great many checks and balances have evolved to allow moderate fluctuations between bull and bear markets but to prevent a crash market.

Plant pathogens cycle through the equivalent of bull markets and bear markets. The appearance of a suitable suscept signals the beginning of a bull market, when capital will increase. The capital in this analogy is biomass in the form of mycelium, cells, spores, or other units of the pathogen. The death of the susceptible host signifies the beginning of a bear market, in which withdrawals eventually will exceed deposits and capital will decrease.

A wise investor uses several methods to prepare for and survive a bear market, including: 1) diversification, i.e., making many small investments provided there is at least some return from each one; 2) developing a large reserve; and 3) tightening up on expenditures. Plant pathogens use equivalent methods for maintenance and defense in the absence of their susceptible host, including: 1) diversification, i.e., parasitizing more than one host or a single host in more than one kind of environment or multiplying in or on symptomless hosts; 2) taking possession of a large food base and then living saprophytically on the host tissues claimed earlier through pathogenesis; and 3) entering into dormancy, where expenditures are held to a minimum. Knowledge of the method(s) of pathogen survival in the absence of a regular susceptible host is necessary to achieve biological control of the pathogen.

Diversification as a Method of Resource Management

Diversification may be the best method for ensuring against financial difficulties because of a bear market. Diversification provides a steady income that will grow with the economy, in contrast to fixed savings that will probably lose value over time. It is not surprising that most successful plant pathogens are diversified by one means or another.

Diversification can occur in many ways. *Pythium ultimum* is diversified by its ability to invade the roots of virtually any angiosperm or gymnosperm that may pass by (Sewell, *in* Thresh, 1981). *Sclerotinia sclerotiorum* is diversified by its ability to invade below-ground plant parts following growth as mycelium from sclerotia (Huang and Hoes, 1980) or aerial plant parts by ascospores released from sporocarps formed from the sclerotia (Coley-Smith, *in* Schippers and Gams, 1979). *Fusarium roseum* 'Culmorum' is diversified by its ability to attack the aerial parts of cereal grains in humid climates, and the roots and lower stem tissues of the same hosts in dry climates (Cook, *in* Nelson et al, 1981). Rice dwarf virus is diversified by its ability to multiply not only in the plant host, but also in its leaf-hopper vector and to pass transovarially to the offspring of the vector (Gibbs and Harrison, 1976). Many viruses, viroids, and bacteria multiply without symptoms in or on host plants other than important susceptible crops.

On the other hand, the ecological niche of each pathogen has limits, no matter how diversified the pathogen may seem. *Pythium ultimum* depends critically on wet and even saturated soil and a temperature unfavorable to the host to cause disease; it must invest at the earliest possible opportunity or forgo the market, and it derives a large but only "simple interest" return on its investment. *Fusarium roseum* 'Culmorum' is active mainly between −15 and −90 bars water potential; the environment may be too dry above ground since −90 bars corresponds to an equilibrium relative humidity of about 94%, and it may be too wet below ground, where soils wetter than the permanent wilting percentage are at water potentials above −15 bars. Rice dwarf virus is limited by the hosts and feeding preferences of its vector; any change in the vector population could spell a crash market for the virus, as may have happened in south China, where the rice dwarf virus disease disappeared abruptly in the mid-1970s (Chapter 4). By understanding the scope and nature of the niche occupied by a pathogen, we can find ways through cultural practices, cultivar modifications, or introduction of antagonists to narrow or fill the niche to the disadvantage of the pathogen (Chapters 8 and 10).

A given pathogen usually is limited in its host range to phylogenetically related plants (e.g., species of Gramineae, Leguminoseae, Cucurbitae, Cruciferae), but some surprises occur. Thus, *Fusarium solani* f. sp. *pisi* was thought to be pathogenically specialized for *Pisum sativum* until Matuo and Snyder (1972) discovered members of this same taxon on branches of mulberry and roots of ginseng in Japan. The Pierce's-disease bacterium, a rickettsialike organism transmitted by sharpshooter leafhoppers, is the cause not only of Pierce's disease of grape, but also of alfalfa dwarf and almond leaf scorch, and has a great many hosts, including many weed species.

Many pathogens are diversified by their ability to grow epiphytically or endophytically on nonsuscepts. In our analogy, this is equivalent to making many small investments for numerous small but important returns. The pathotypes of *Fusarium oxysporum, Verticillium*, and possibly others can grow in the rhizosphere and colonize the cortical tissues of the roots of many plants in addition to their suscepts (Schroth and Hendrix, 1962). Such colonization leads to a few new propagules being formed in place of the aged propagules. Ability to grow epiphytically on different plants is especially important to the success of phytopathogenic bacteria. Leben (1965a, 1981) describes this as the "residency phase," a method of maintenance and perpetuation used by species of *Erwinia, Pseudomonas, Xanthomonas*, and possibly others. The bacteria cause disease only if the proper conditions occur (Hayward, 1974). *Erwinia carotovora* multiplies epiphytically and endophytically on roots and leaves of many nonsuscepts as well as suscepts (Pérombelon and Kelman, 1980), but other phytopathogenic bacteria apparently are more selective. For example, a highly virulent (VI) strain of *P. syringae* causes brown spot of bean but is also adapted to epiphytic survival on hairy vetch (Ercolani et al, 1974).

Of all the attributes of a pathogen, ability to diversify is clearly among the most refractory for biological control. Elimination of alternate or symptomless hosts, one method of biological control, is often impractical where many alternate hosts are involved. The programs of eradication of barberry (the alternate host of *Puccinia graminis* var. *tritici*) to control wheat stem rust and of *Ribes* (the alternate host of *Cronartium ribicola*) to control white pine blister rust have been discontinued, and the effectiveness of these programs has been questioned. Elimination of any source of inoculum, whether on symptomless plants, alternate hosts, weeds, or volunteer host plants, is a cardinal rule for disease control, but experience makes clear that this approach generally is inadequate for pathogens with multiple hosts. Fortunately, biological control of plant pathogens is not limited to approaches aimed only at the sources of inoculum.

Saprophytic Survival

Most parasites of the necrotrophic type are successful largely because of their strong attributes as saprophytes. Indeed, the role of pathogenesis in the life cycles of these organisms may be mainly to establish themselves as the pioneer saprophytes in the tissues and to thereby preempt the competitors. It is hard to imagine a saprophyte more successful than the dominant occupant of a log. *Armillaria mellea* and *Heterobasidion annosum* exemplify such saprophytes. These fungi gain their advantage by their unique additional ability to colonize the log while it is still part of a living tree. Any practice aimed at excluding these fungi from crop or tree residue must begin by preventing them from colonizing the living plant. This is the method Rishbeth used to control root rot of *Pinus* spp. caused by *H. annosum* (Rishbeth, in Bruehl, 1975). *Armillaria mellea* can be preempted as a saprophytic colonist of cut stumps of trees by ringbarking each tree many months before felling, so as to starve the roots and thereby permit their colonization by saprophytes in advance of *A. mellea* (Leach, 1939).

The use of pathogenesis to gain position as the pioneer saprophyte in crop residue is especially well illustrated by *F. roseum* 'Culmorum' on wheat. This fungus occupies at least one and often two and three lower internodes of the culms of diseased plants (Cook, 1968). Simply rotting the crown (tissue where all culms of a given plant are attached) would be sufficient to kill the plant, but by also growing one to three internodes up the culms, the fungus invests for its future as a saprophyte. The fungus is generally unable to colonize straws already occupied by other saprophytes (Cook, 1970; Cook and Bruehl, 1968), but this problem is solved by its ability to establish in the straw while it is still part of a living plant.

Pathogens successful in saprophytic survival in dead plant remains are mainly those that establish in large pieces of fibrous or woody tissue. Such tissues are slow to decompose and may be the easiest to defend against competitors. *Gaeumannomyces graminis* var. *tritici* is more infectious in fragments of wheat crown (basal stem) tissue than in root fragments (H. T. Wilkinson, unpublished). Fragile rootlets or other tissues that decompose quickly do not support a prolonged saprophytic existence for an occupant; the selection pressure on pathogens of such delicate tissues has, therefore, been toward those that store the energy in safe packages, like the oospores of *Pythium ultimum*. That wood and cereal straws are more ideal substrates for prolonged saprophytic existence is suggested by the pathogens that are successful at this method of survival, most of them cereal pathogens (e.g., *Pseudocercosporella herpotrichoides, Cephalosporium gramineum, G. graminis, F. roseum* 'Graminearum') or pathogens of tree roots (e.g., *Heterobasidion annosum, Ganoderma philippii, Rigidoporus lignosus, Phellinus noxius,* and *Armillaria mellea*). A patch of cotton killed by *Phymatotrichum omnivorum* was once traced to a piece of infested pecan wood many feet deep in the soil; the wood was a remnant of a pecan orchard removed from the site 20 years earlier (Streets, 1969).

Saprophytic survival is essentially an extension of necrotrophic parasitism, the only real difference being that the dead host tissue is now crop residue rather than a part of a living plant (Chapter 3). Accordingly, the methods of maintaining control of this food supply in the midst of intense competition differ little from those used during parasitism, and include: thorough establishment in the substrate; solubilization of the substrate at a rate no faster than the rate of use; production of staling products or antibiotics, but only enough to counter the "signals" provided by food escaping from the substrate; and flexibility in response to environmental change (Bruehl, *in* Bruehl, 1975). It is to this last aspect that pathogens appear to be most vulnerable. Using a cultural practice to create an environment that is unfavorably dry, wet, acid, alkaline, hot, or cold for the pathogen, but not equally limiting to other potential colonists of the residue, is probably the best way to displace the pathogen from the residue (Chapter 4).

A few pathogens are able to increase their biomass through strictly saprophytic means. *Sclerotium rolfsii* was reported by Boyle (1961) to grow saprohytically on the leafy plant residue and to thereby attain a larger mass of inoculum and energy to attack peanut plants or other suscepts surrounded by the residue. The "nondirting" method of control (Garren and Duke, 1958) eliminated this important niche for *S. rolfsii* by first covering the residue with soil and then by not bringing soil into contact with the plants during cultivation. *Sclerotium rolfsii* is unable to remain active in nonsterile soil deeper than 1–2 cm. Saprophytic colonization of straw in soil by *F. roseum* 'Culmorum' requires that the straw be bright, unweathered, and not already occupied by other saprophytes (Cook, 1970). Allowing the wheat stubble to stand in the field until weathered and to become colonized by other saprophytes eliminates this niche of Culmorum. *Pythium ultimum*, a successful saprophytic colonist of fresh crop residue, multiplies less in some soils of France (Bouhot and Joannes, *in* Bouhot, 1980) and California (Hancock, 1977; *in* Schippers and Gams, 1979) than in others; the existence of soils suppressive to saprophytic multiplication by *P. ultimum* reveals the potential for closing a niche for this pathogen as well.

Survival as Dormant Propagules in Soil

The pathogen that survives as dormant propagules between suscepts resembles a manager who recognizes that his capital will no longer increase and that survival to the next bull market will be possible only by tightening expenditures. Capital is distributed among numerous safe deposits and loses value each day, but the trade-off is that the capital can be liquidated on very short notice and reinvested at the first opportunity.

Some of the most important plant pathogens use this method of survival. The propagule types include the resting spores of *Plasmodiophora brassicae*; oospores of *Aphanomyces, Pythium*, and *Phytophthora* spp. and many downy mildew fungi; chlamydospores of *Fusarium* spp., *Thielaviopsis basicola*, and *Phytophthora cinnamomi;* pigmented conidia of *Helminthosporium* spp.; microsclerotia of *Verticillium dahliae, Cylindrocladium crotalariae*, and *Pyrenochaeta* spp.; eggs and cysts of *Meloidogyne* and *Heterodera* spp.; and sclerotia of many soilborne fungi. Obviously, selection pressure has been toward this method of survival and resource management by pathogens during dormancy, particularly among soilborne pathogens. Some of the pathogens listed above attack foliage, but all complete part or all of their life cycle in soil between hosts.

Specialized survival structures serve two basic purposes—they provide a means of protection and a means of position. A thick wall or outer rind provides protection from antagonists and from environmental extremes. The dormant state provides protection against overly rapid exhaustion of endogenous reserves. The benefits of position are that the pathogen is so widely and uniformly distributed that some part of its biomass, perhaps only one propagule, will be in the right place at the right time to reinitiate the cycle. The uniform inoculum distribution also increases the chance for a portion of each population to escape antagonists.

Distribution.—Propagules in soil initially may be unevenly distributed, as noted for *Fusarium oxysporum* f. sp. *cubense* (Trujillo and Snyder, 1963) and *Cylindrocladium crotalariae* (Griffin et al, 1981). This results from the initial concentration of propagules near the place of their formation (i.e., on or within decaying residue of the host). Inoculum density may even increase or appear to increase for a brief period immediately after their arrival in soil as a fresh population of propagules from a diseased host, owing to saprophytic multiplication (Baker, *in* Pimentel, 1981) or possibly to fragmentation of multiple-celled propagules into smaller units (Horsfall and Dimond, *in* Baker and Snyder, 1965). As suggested in Chapter 4, the period when propagules are most concentrated in areas near where they formed may be the best time to introduce a hyperparasite to eliminate them, because it will increase the efficiency of secondary spread by the hyperparasite.

With decay over time, and because of tillage and the activities of soil animals, the propagules become more widely and uniformly distributed in soil. This ensures that within a given soil volume, the average distance between propagules will be the maximum possible for that population (McCoy and Powelson, 1974). Baker (*in* Pimentel, 1981) has described an idealized curve for survival of pathogen propagules in soil (Figure 5.2); this includes a brief period of inoculum increase, then a period of rapid decrease in inoculum density, leaving a residual of persistent propagules suited to very long-term survival. The ability of a small portion of the inoculum to persist for a long period may be explained on the basis of different survival capabilities but may also reflect greater safety of isolation achieved by thinning of the population. Hancock (1981) indicates a role of this latter factor in the persistance of a small, stable population of *Pythium ultimum* in soil.

Protection. — Conceivably, the methods by which pathogens elude or discourage antagonists during periods of dormancy are similar to the strategies used during pathogenesis or active saprophytic survival — allow little or no leakage of nutrients to attract antagonists, and foster accumulation of pollutants or development of an environment that discourages antagonists. Propagules with dark pigments in their walls (or rind, in the case of many sclerotia) are generally more resistant to lysis, as shown by Bull (1970a, 1970b) for conidia of *Stemphylium, Alternaria*, and *Helminthosporium* and by Old and Robertson (1970) for pigmented conidia of *H. sativum*. A thick wall also may be a barrier to leakage of nutrients from the propagule, thereby reducing the risk of attracting antagonists. A potential hazard from exudation of sclerotia of *Sclerotium rolfsii* was shown by Smith (1972); when dried and then rewetted in soil, the sclerotia developed cracks in their rinds, leaked nutrients, were colonized by other microorganisms, and rotted. A similar effect has been noted for sclerotia of *Sclerotinia sclerotiorum* when dried and then rewetted (Adams, 1975). Sclerotia of *Rhizoctonia* are protected from microbial attack by the presence of one or more an-

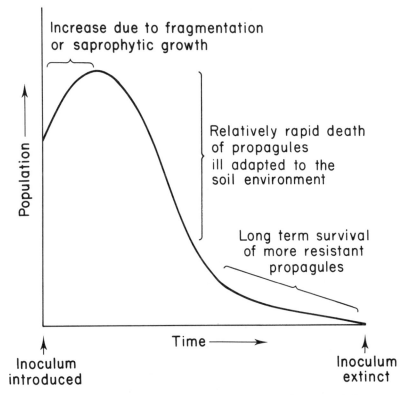

Figure 5.2. An idealized survival curve and the major elements of the curve for propagules of a fungus following their introduction into soil. (Modified from Baker, *in* Pimentel, 1981.)

tibiotics in the sclerotium (Coley-Smith, *in* Schippers and Gams, 1979). Antibiotic production may be a means of counteracting nutrient leakage (Bruehl, *in* Bruehl, 1975) by sclerotium-forming pathogens; this deserves more study.

Accelerating the death rate of dormant propagules in natural soil is probably best accomplished with help from one or more forms of environmental stress. A severe stress may kill propagules directly, but more likely the stress predisposes them to decay by the soil microbiota, much as wetting and drying may predispose sclerotia of *Sclerotium rolfsii* and *Sclerotinia sclerotiorum* to decay by microorganisms. General resistance of plant cells to attack by weak parasites depends on continued active metabolism and function of the cells; conditions that retard or prevent the plant cell from functioning predispose that cell to decay by organisms that otherwise are unable to attack it. Perhaps healthy propagules can also be characterized as having "general resistance" to attack by organisms other than hyperparasites specialized for this role. Conditions that impair cell function might then predispose that cell to decay by the associated microbiota. In the case of sclerotia, the associated microbiota may be superficial inhabitants (commensals) of the rind or other outer layer. In the case of single-celled propagules, the associated microbiota may be contiguous cells or colonies of soil bacteria.

More basic information is needed on how propagules respond to or endure environmental stresses, as a means of showing how to more effectively predispose them to decay in soil. For example, sclerotia of *Phymatotrichum omnivorum* are sensitive to freezing and also to drying. Current views (Grieve and Povey, 1981) hold that freeze-damage results largely from loss of water and hence from osmotic stress in the cell. A cell sensitive to desiccation is likely to be damaged whether the desiccation results from freezing or from drying. By inference, it may be that propagules resistant to freezing (and drying) might have some means to endure or avoid cellular damage from osmotic stress. The microsclerotia of *Verticillium dahliae* survive in cold regions and in dry soils but become sensitive to desiccation (and perhaps also freezing) after they undergo conidiogenic germination (Menzies and Griebel, 1967). Perhaps the germination process reduces the quantity or affects the quality of stored materials in the microsclerotia critical to their osmoregulation or ability to endure osmotic stress. Further investigation of such phenomena might reveal practical ways to increase propagule vulnerability to this or other potential stresses possible in the environment, in keeping with the suggestion (Baker and Cook, 1974) that the stress need not kill the propagule but only weaken it so that the associated microbiota will then destroy it.

INITIATION AND MAINTENANCE OF GROWTH IN RESPONSE TO A SUSCEPT—THE BEGINNING OF A BULL MARKET AFTER A BEAR MARKET

Like money managers, pathogens must recognize the making of a bull market and invest at the earliest opportunity; those that wait or never invest increase the risk of becoming extinct. The successful investor is in the right place at the right time and has the means to liquidate and deliver assets as required. Similarly, the successful plant pathogen has the means to be in the right place at the right time (inoculation) and to invest in its suscept (infection).

Inoculation

Inoculation for the great majority of pathogens is accomplished by chance contact. *Fusarium solani* f. sp. *phaseoli* makes chance contact with the roots and hypocotyls of bean by virtue of its high inoculum density and uniform distribution within the tillage layer of soil (Nash and Snyder, 1962). Roots growing through infested soil cannot avoid growing within striking distance of at least some of these propagules. The method of inoculation used by this fungus is typical for soilborne fungus pathogens; one difference is that those with large propagules (e.g., sclerotia) have the potential to reach the host from greater distances away from the root and to produce economic disease with relatively lower inoculum densities; those with small propagules (e.g., chlamydospores) have a relatively shorter striking distance and require higher inoculum densities to produce economic disease (Baker and Cook, 1974). *Erysiphe graminis* exemplifies a different method for exploiting chance contact, namely, releasing enough spores into the air to ensure at least one contact with the leaf of a compatible host, then rapidly producing new inoculum of the compound interest type to take maximum advantage of the new market. Still another method of chance contact is illustrated by the wheat scab fungus, *Gibberella zeae*. The fungus overwinters in infested crop residue on the soil surface and releases ascosporic inoculum when the wheat or barley flowers the following spring or early summer. The near-perfect timing is a matter of selection pressure; strains that, by chance, released ascospores too early or too late have not survived.

Biological control of the chance contact between host and pathogen is probably best accomplished through the host. Phillips and Wilhelm (1971) demonstrated that the occurrence of less verticillium wilt in the Waukena White breeding line of cotton (*Gossypium hirsutum*) than in the Acala S5-1 cultivar in plots in Tulare County, California, resulted from a different root distribution. The Acala cotton produced most major laterals in the upper 45 cm of soil where most inoculum of the pathogen occurred, in contrast to Waukena White, which produced most laterals at a depth below 45 cm. The difference between the two cultivars was less apparent when they were grown in pots in the greenhouse where roots and inoculum were intimately intermixed. Burke et al (1972) demonstrated reduction of damage from root rot of bean by using deep tillage that broke up the pressure pan and loosened the soil, allowing the roots to penetrate into deeper layers of soil where inoculum density of the pathogen was less (Figure 5.3). Clark (1942) showed that inoculum of the take-all pathogen placed in a small container with soil and several plants destroyed the roots, whereas the same amount of inoculum placed in a large container of soil with only one plant produced

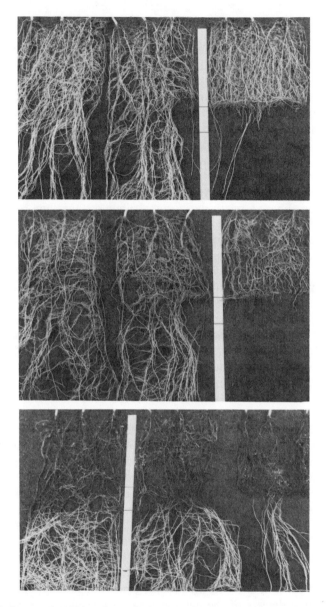

Figure 5.3. Roots of common bean in fumigated soil (top two photos) or soil infested with *Fusarium solani* f. sp. *phaseoli* (bottom photo). The soils were adjusted to either a low (−0.8 bar) (top photo) or high (−0.2 bar) matric water potential (bottom two photos). The bulk density of the central soil layer (indicated by the marker) was, from left to right, 1.2, 1.4, and 1.6 g/cm³. (Reprinted with permission from Miller and Burke, 1974.)

mild disease. Less disease in these examples results, in part, from less inoculation, accomplished largely because a significant proportion of the infection
courts are beyond the striking distance of the pathogen propagules in the
soil.

The inoculation process used by aerial pathogens able to produce secondary
inoculum and to complete several cycles per season (e.g., powdery mildew
on cereals) can be interrupted biologically by use of cultivar mixtures or
multilines. Besides the evidence for biological control by induced resistance,
studies show an effect from the resistant plants catching or "filtering out" the
inoculum that otherwise could spread from one susceptible plant to another if
all plants were susceptible. The epidemiological aspects of powdery mildew
on a mixture of spring barley cultivars is discussed more thoroughly by
Wolfe and associates (Wolfe et al, *in* Jenkyn and Plumb, 1981; Chapter
4).

Changing the planting date or cultivar so as to shift the date of anthesis
out of synchronization with peak ascospore release by *Gibberella zeae* is one
method to control wheat scab (Cook, *in* Nelson et al, 1981). The pathogen
is stimulated in the infection courts of the wheat spikelets by anthers and
pollen, more specifically by choline and betaine in these materials (Strange
and Smith, 1978), and is therefore most damaging when ascospore release
coincides with the presence of anthers and pollen. Workers in the People's
Republic of China have developed barley cultivars that complete flowering
while the head is still enclosed in the boot and therefore are less vulnerable
to scab when the head emerges (Cook, *in* Nelson et al, 1981). Ergot is
considerably more important on male-sterile wheat and barley plants than on
normal fertile plants, yet male-sterile lines are needed to develop hybrid seed of
these cereals. The vulnerability of male-sterile flowers to infection by *Claviceps
purpurea* results from their more open florets, which trap more inoculum.
The rapid outbreaks of ergot in plots of male-sterile lines give some idea of
the amount of failure of inoculation by this fungus on normal male-fertile
lines. Perhaps male-sterile parents can be found with less receptive infection
courts.

Pathogen Response to the Host

Pathogen propagules must have a mechanism whereby germination occurs in
the vicinity of a suitable host and under favorable environmental conditions.
For pathogens of aerial parts of plants, the propagules must be prevented
from germinating before they are dispersed from their place of origin, but
must be released from this inhibition as soon as possible after dispersal. For
pathogens of below-ground plant parts, "dispersal" (i.e., the scattering in soil
at various distances from their place of formation) only means a continuation
of the "waiting game," in which germination must not occur until a root of the
susceptible host arrives.

Initiation of growth by soilborne pathogens. — Most soilborne fungus pathogens depend for propagule germination and prepenetration growth on nutrients
released from plant roots. Some exceptions are pathogens that survive as
sclerotia and germinate by production of conidia (sporogenic germination) or
sporocarps (carpenogenic germination) in response to certain temperatures
(Coley-Smith, *in* Schippers and Gams, 1979). For example, the sclerotia
of *Claviceps purpurea* have a constitutive dormancy that is broken when
the sclerotia are exposed to 0–10°C (Mitchell and Cooke, 1968). Sclerotium
germination by the sporogenic or carpenogenic method usually results in water-
splashed or airborne inoculum timed for contact with aboveground infection
courts. As such, these soilborne pathogens are more like the example of
Gibberella zeae than that of *Fusarium solani* f. sp. *phaseoli* given above.
In cases where sclerotia germinate by production of mycelium (myceliogenic
germination) and roots are infected, root exudates or volatiles usually are involved. The mechanisms that keep the propagules of root-infecting fungi from
germinating precociously, i.e., before the arrival of a suitable host root, are
complex but involve constitutive dormancy factors, requirements for specific
nutrients, and soil fungistasis (Chapter 7).

The nutritional requirements for germination, combined with the metabolic
nonreadiness for germination of some spores (Griffin and Roth, *in* Schippers
and Gams, 1979) and with fungistasis, are usually such that not all propagules
within any given population germinate at once. This helps ensure a continuing
supply of inoculum to offset germinated propagules that fail to infect or
reproduce. The lack of germination by a portion of the propagules may
result from constitutive dormancy factors that disappear with aging; different
propagules may be of different ages and hence differ in the aging time still
left.

An aging or ripening requirement is not the only reason for the tendency
of some propagules in a population to remain dormant. Essentially all chlamydospores of *F. solani* f. sp. *phaseoli* are capable of germination in axenic
culture without nutrients (Griffin, 1970b), but in soil, only 1 or 2% germinate upon addition of water alone, and germination is rarely more than
30–50% in response to an added source of carbon and nitrogen, e.g., glucose
plus ammonium nitrogen or an amino acid (Cook and Schroth, 1965). The
failure of the remaining half or more of the population to germinate results
from competition and other antagonistic effects of the soil microbiota, which
respond quickly to the same carbon and nitrogen sources supplied for the
chlamydospores. Some chlamydospores germinate with a lag phase of only
four hours, but others require more time, e.g., 8, 12, 16, or even 24–30 hours.
In general, depending on the soil, chlamydospores that have not germinated
in 16–20 hours probably will not germinate, since by that time the associated
microbiota has responded to a level of activity that suppresses further germination (Cook and Snyder, 1965). When 1% yeast extract was added with the
usual carbon and nitrogen supply, germination proceeded at the normal rate,

but for only nine hours; chlamydospores not germinated by nine hours did not germinate, although many or even most were probably still alive in the protection of the dormant state. When antibacterial antibiotics were added with the carbon and nitrogen source, germination proceeded for 24–30 hours and reached nearly 90%.

Dormant propagules may be stimulated to germinate in the absence of the host by nutrients from decomposing organic material. Such germination is common among soilborne pathogens that have no special requirements for germination other than a source of carbon and energy, nitrogen, and possibly vitamins. The introduction of relatively large quantities of readily decomposable organic material can stimulate high percentages of propagule germination but, having no host to infect, the germling or developing thallus dies. At least part of the cause of death is starvation; the initial flush of nutrients stimulates the propagule to germinate, but the associated microbiota of the soil is also stimulated to grow and quickly depletes one or more essential factors in the nutrient supply (Chapter 7). Continuing growth and formation of new propagules may be restricted by lack of nitrogen, carbon and energy, or possibly lack of oxygen (and associated low redox potential) in some microsites. The germination-lysis process has been demonstrated for a great variety of soilborne pathogens, including *Phymatotrichum omnivorum, Thielaviopsis basicola, Macrophomina phaseolina*, and *Fusarium solani* (Papavizas and Lumsden, 1980). Unfortunately, there is also a risk with this approach to biological control; if lysis occurs too slowly or inefficiently, the pathogen may form new propagules and increase its population.

Sclerotia of the onion white rot fungus, *Sclerotium cepivorum*, germinate in response to water-soluble alkyl cysteine sulfoxides, which are released only from roots of the host genus, *Allium* (Coley-Smith, *in* Schippers and Gams, 1979). Merriman et al (1981) demonstrated that germination followed by lysis of sclerotia of this pathogen can be achieved by introduction of artificial onion oils into the soil. Once germinated, the sclerotia decay. Sclerotia are commonly more susceptible to decay after germination than before germination (Coley-Smith, *in* Schippers and Gams, 1979).

Prepenetration growth by pathogens in the rhizosphere. — The amount of pathogen inoculum in soil is commonly estimated by counts of the number of propagules. Such estimates are mainly of the "seed bank" (Sewell, *in* Thresh, 1981) of the pathogen. Although such counts are useful and often provide the only quantitative information about inoculum of the pathogen in soil, root disease severity is probably related more closely to the total biomass of the pathogen in contact with the infection court than to the density of the dormant propagules (Chapter 3). Most root-infecting pathogens make at least some growth between germination and penetration (Garrett, 1970); and for some pathogens (e.g., *Sclerotium rolfsii,*

Pythium spp., *Rhizoctonia solani, Fusarium solani* f. spp. *phaseoli* and *pisi*, and *Gaeumannomyces graminis* var. *tritici*), a considerable thallus may develop in the infection court before penetration occurs. Some hypothetical cases of biomass increase (or lack of increase) between propagule germination and penetration are given in Figure 3.2. An obvious objective of biological control is to restrict or prevent increases in pathogen biomass near or on the root and thereby limit the amount of infection (Chapter 3).

Wildermuth et al (*in* Schippers and Gams, 1979) estimated the change in density of runner hyphae of *G. graminis* var. *tritici* on seminal roots with time by counting the number of intercepts between hyphae and two orthogonal lines on an eyepiece graticule and then calculating the mean length of mycelium per unit area of root surface (Figure 5.4). Fifteen days after planting, hyphal density had peaked at about 22 mm/mm² in a fumigated soil but was only 2 mm/mm² in the fumigated soil amended with 1% (w/w) natural soil. By 25 days, the density had declined to 15 mm/mm² in fumigated soil and had peaked at about 7 mm/mm² in fumigated soil amended with natural soil. The increase in density of hyphae followed an S-shaped curve when plotted over time, much as described for increase in propagule numbers of a pathogen (Waggoner, 1977; Baker and Wigetunga, *in* Wicklow and Carroll, 1981), but with the least growth on roots in fumigated soil amended with natural soil. The natural soil used by Wildermuth et al (*in* Schippers and Gams, 1979) was a known suppressive soil from a field that had undergone take-all decline (Cook and Rovira, 1976). It now seems likely that the lack of growth by *G. graminis* var. *tritici* on roots in the presence of this or any other natural soil added to fumigated soil is partly or largely because of greater competition from the soil microorganisms introduced with the natural soil (Wilkinson et al, 1982; Chapter 7).

A large and continuing supply of nutrients and energy as root exudates stimulates propagule germination at greater distances away from the root than without nutrients and results in a greater increase in biomass of the pathogen outside the root and therefore a greater amount of infection. Conversely, a smaller or more limited supply of nutrients from the root stimulates germination of a smaller proportion of propagules, and those that are stimulated are mainly very near or on the root surface.

Soil treatments that enhance biological activity in soil before arrival of the suscept (e.g., a few days before sowing), or that intensify the level of competition for carbon, nitrogen, or other nutrients required by the propagule for germination, can effectively limit propagule germination to those very near to or in actual contact with the nutrient source (e.g., exuding root surface). A well-known example is mature barley straw or other material of high C:N ratio mixed with soil a few days before beans are planted; nitrogen

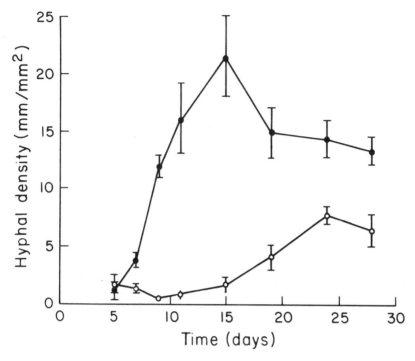

Figure 5.4. Density of runner hyphae of *Gaeumannomyces graminis* var. *tritici* on wheat roots in soil fumigated with methyl bromide and infested with *Gaeumannomyces* alone (solid circles) or with *Gaeumannomyces* plus 1% (w/w) suppressive soil. (Reprinted with permission from Wildermuth and Rovira, 1977.)

is immobilized by action of the soil microbiota, only a few chlamydospores germinate in response to the host, and disease is mild. Baker (*in* Cook and Watson, 1969) refers to this effect as "shrinking the rhizosphere to a rhizoplane" so far as the pathogen is concerned. There is disagreement as to how best to model the rhizosphere and surrounding pathogen-infested soil and how to mathematically treat data on inoculum density and disease incidence (Baker and Drury, 1981; Ferris, 1981; Gilligan, 1979; Grogan et al, 1980; Leonard, 1980; Van der Plank, 1975). Nevertheless, the concept of pathogen propagules infecting from varying distances or only when in contact with the root, depending on availability of nutrients, soil suppressiveness, or other factors is fundamental (Chapter 7) regardless of how it may be modeled.

Aerial pathogens. — For aerial pathogens, spore germination while the spores are still in the vicinity of the parent mycelium (or contained in reproductive structures) is frequently prevented by chemical inhibitors in the parent mycelium or fruiting body. According to Allen (*in* Heitefuss and Williams, 1976), the same or similar chemical inhibitors may be carried with the mature spores of the fungus to the infection court of the next suscept. Germination then depends on dilution and dissipation of these inhibitors. A film of water on the leaf is probably sufficient for this purpose because the inhibitor usually is water-soluble and diffusible.

Germination of the spores of *Colletotrichum musae* is prevented by a water-soluble inhibitor, but spores produced in vitro may lack the inhibitor (Parbery, *in* Blakeman, 1981). Swinburne (*in* Blakeman, 1981) showed that spore germination in *C. musae* is inhibited by iron contained in the spore wall and that spores produced in vitro on an iron-deficient medium germinated freely with no evidence of inhibition. Removal of iron stimulated spore germination and occurred with exposure of the spores to leachates from banana fruit containing anthranilic acid. Conidia metabolize anthranilic acid to 2,3-dihydroxybenzoic acid, which chelates the iron in the walls, thereby releasing the spores from inhibition of germination. Specific iron-binding compounds (siderophores) produced by bacteria on the fruit surface also may be involved in promotion of spore germination by this pathogen. Muirhead and Deveral (1981) further observed that the spores germinate on green fruit but then form dormant appressoria. Anthracnose lesions that developed as the fruit ripened were caused by germination of the dormant appressoria, and not from latent, subcuticular hyphae as previously thought. Subcuticular hyphae established in green fruit either die or fail to resume activity because of host defense mechanisms. The same sequence of prepenetration activities — spore germination, appressorium formation, appressorium dormancy, and appressorium germination — apparently occurs with *C. gloeosporioides* in avocado fruit and possibly with other *Colletotrichum* spp. on their hosts (Parbery, *in* Blakeman, 1981). Muirhead (*in* Blakeman, 1981) suggests that the situation on fruits may also occur on leaves affected by anthracnose.

Clarification of prepenetration events for *C. musae* on banana fruit shows how increased understanding can point the way to possible methods for biological control. Cultivars might be selected on the basis of promoting unusually high iron content in the walls of spores produced on leaves, low concentrations of iron-chelating compounds in the leaf exudates, or both. The fact that the pathogen exists exposed on the surface of the cuticle and not subcuticularly as previously thought, indicates that this pathogen is no more able than other leaf pathogens to escape the antagonistic effects of microorganisms on the plant surface. Some of these microorganisms may compete for nutrients needed for renewed growth of appressoria, or they may compete with bacteria that stimulate spore germination of the pathogen. The stage may be set for new biological approaches to control anthracnose.

Whereas the spores of some aerial pathogens are nutritionally independent and depend on self-contained chemical inhibitors and stimulators for regulation of germination, the spores of others have varying nutritional requirements for germination or nutrients provide an important additional stimulus to germination and prepenetration growth on the leaf surface. Pollen grains scattered on the leaf surface are used by *Helminthosporium sativum* on leaves of rye (Fokkema et al, 1975) and by *Botrytis cinerea* in many hosts (Blakeman, *in* Coley-Smith et al, 1981). However, the preemption of this nutrient source by epiphytic bacteria, yeasts, or nonpathogenic filamentous fungi resident on leaves provides significant natural biological control of these pathogens and has the potential to provide even greater biological control if these populations could be increased (Chapter 6). Conidia of *B. cinerea* also may germinate in response to nutrients initially lost in leakage from the conidia and then resorbed (Brodie and Blakeman, 1975) or in response to nutrients (e.g., amino acids) made available by leaf exudates. When beet leaves were wetted 24 hours before they were inoculated with spores of *B. cinerea*, the natural microflora, having had a head start of 24 hours, preempted the nutrient supply needed by *B. cinerea* for germination (Blakeman and Brodie, 1977). Application of isolates of *Pseudomanas* spp. inhibited germination of conidia of *B. cinerea*, *Phoma betae*, and *Cladosporium herbarum*, on beet leaves and a close relationship was demonstrated between failure of the fungus spores to germinate and uptake of amino acids by these bacteria. Blakeman and Brodie (1977) found that although water films on leaves are important to germination of conidia of *B. cinerea*, either of two *Pseudomonas* strains introduced onto the wet leaves in advance of conidia of *B. cinerea* prevented germination. Uptake of amino acids by the bacteria on the leaf surface was demonstrated.

WATER REQUIREMENTS OF PATHOGENS

Water plays a biochemical role in the growth and survival of microorganisms by its involvement in the metabolic reactions within the cell. It plays a biophysical role through the hydrostatic pressure (turgor) essential to growth and development of microorganisms. Finally, water plays an indirect role in the growth and survival of microorganisms by its effect on diffusion of nutrients and oxygen toward the organism and of waste or inhibitory substances away from the organism (Chapter 7). This section is concerned mainly with the biophysical and indirect roles of water in the growth and survival of pathogens and with how an understanding of these roles is basic to biological control.

Spore germination, the division of bacterial cells, and the elongation of hyphae require a minimum hydrostatic water pressure within the plasma membrane, pressing against the cell wall (Harris, *in* Parr et al, 1981; Robertson,

1958, 1959). This turgor is maintained by osmoregulation so that the osmotic (solution) potential of the cell is less than the water potential of the environment surrounding the cell (e.g., soil, plant tissue, crop residue, agar medium, or other sites). The water potential gradient across the cell membrane results in the movement of water from higher to lower potential until, by the development of turgor, water potential inside the cell is in equilibrium with that outside the cell.

Plant pathogens have characteristic optimal and minimal water potentials for growth, much as they have characteristic optimal and minimal temperatures for growth (Table 5.1). However, the "cardinal" values differ according to whether the available water is under matric or osmotic control and also according to the temperature. Under matric control (e.g., in soil where water potential is mainly a function of adhesion and cohesion), both fungus and bacterial growth seem to be maximal at water potentials near zero (at field capacity or even wetter), provided oxygen is not limiting, and are progressively less with each incremental drop in matric potential down to the lower limit for the organism. The lower limit of matric potential for hyphal growth of *Fusarium roseum* 'Culmorum' is about −90 bars (Cook et al, 1972). The lower limit of matric potential for bacterial activity in soil is about −5 bars (Griffin, 1972). In contrast, the lower limit of osmotic potential for *F. roseum* 'Culmorum' is about −100 to −120 bars and for bacteria, −20 to −25 bars (Chapter 7). *Fusarium roseum* 'Culmorum', as well as most other fungi and bacteria, actually grows faster as the osmotic potential is lowered slightly, to a maximum growth rate (optimal osmotic potential) between −5 and −20 bars, below which growth is progressively more limited down to growth extinction. With *F. roseum* 'Graminearum', the optimal osmotic potential for hyphal growth is about −15 to −20 bars at 20°C and −30 to −50 bars at 35°C (Cook and Christen, 1976).

The stimulation of hyphal growth rate (or rates of multiplication, in the case of some bacteria) by slightly reduced osmotic potentials (but not matric potentials) may relate in part to the availability of exogenous solutes that the cell absorbs and uses to maintain turgor (Griffin, *in* Kozlowski, 1978; Cook and Duniway, *in* Parr et al, 1981). In a matric system, osmoregulation (and hence maintenance of turgor) with decreasing water potential outside the cell must be accomplished mainly or entirely within the cell (Chapter 7). This may be inefficient or even impossible. Thus, macroconidia, ascospores, and chlamydospores of *F. roseum* 'Culmorum' and 'Graminearum' germinate at a uniform 100% in about eight hours on osmotically adjusted (KCl added) water agar over the range of −1 to −20 bars (Sung and Cook, 1981), yet mycelia of these fungi grow only about half as fast at −1 as at −10 to −20 bars. Cook and Duniway (*in* Parr et al, 1981) proposed that spores, being rich in endogenous reserves, can attain adequate turgor for maximal growth rates without an exogenous supply of solutes for osmoregulation. The ability to adjust turgor and maintain growth differs among or-

Table 5.1. Approximate optimal and minimal osmotic potentials for hyphal growth of selected pathogenic fungi, as estimated by measuring rates of colony extension on agar media containing salts, sucrose, or both[a]

| Pathogen | Osmotic potential (bars) | | Reference |
	Optimum	Minimum	
Phycomycetes			
Phytophthora cinnamomi	−10	−40	Adebayo and Harris, 1971
	−15	−45 to −50	Wilson and Griffin, 1975
P. drechsleri		−56	Cother and Griffin, 1975
Pythium ultimum	−5	−25 to −30	
Ascomycetes and Fungi Imperfecti			
Gaeumannomyces graminis	−1 to −2	−45 to +50	Cook et al, 1972
Fusarium roseum 'Culmorum'	−10 to −15	−90 to −100	Cook et al, 1972
F. roseum 'Graminearum'	−15	−90 to −100	Wearing and Burgess, 1979
F. oxysporum f. sp.			
vasinfectum	−10	−115	Manandhar and Bruehl, 1973
Verticillium albo-atrum	−10 to −20	−115	Manandhar and Bruehl, 1973
V. dahliae	−20	−100	Ioannou et al, 1977
Sclerotinia sclerotiorum	−10	−80	Grogan and Abawi, 1975
S. borealis	−10 to −30	−40 to −50	Bruehl and Cunfer, 1971
Fusarium moniliforme	−15	−125 to −140	Wilson and Griffin, 1975
Alternaria tenuis (alternata)	−10	−100	Adebayo and Harris, 1971
Pseudocercosporella			
herpotrichoides	−10 to −15	−100	Bruehl and Manandhar, 1972
Cephalosporium gramineum	−2 to −8	−90 to −100	Bruehl et al, 1972
Fusarium nivale	−1 to −3	−45 to −50	Bruehl and Cunfer, 1971
Basidiomycetes			
Typhula idahoensis	−1 to −2	−30	Bruehl and Cunfer, 1971
T. incarnata	−1 to −2	−30	Bruehl and Cunfer, 1971
Rhizoctonia solani		−45	Dube et al, 1971
Macrophomina phaseolina	−15 to −20	−40	Shokes et al, 1977

[a] Expanded from Cook and Duniway (*in* Parr et al, 1981).

ganisms, and this, with other unique requirements of each organism, creates a dynamic situation of differential growth rates and duration of dormancy among interacting populations. **Antagonists must be active over at least the same range and preferably a wider range of water potentials than the pathogen.**

As a general rule, pathogens that grow maximally at relatively low osmotic potentials (e.g., −10 to −15 bars) and that cease growth only when the water potential is −90 to −100 bars or lower (e.g., *Fusarium* spp. [Cook, *in* Nelson et al, 1981]) cause the most disease under dry soil conditions and in plants with low plant water potentials (Cook and Papendick, 1972; Cook and Duniway, *in* Parr et al, 1981). In contrast, pathogens that grow maximally at relatively high osmotic potentials (e.g., −1 to −10 bars) and cease growth at −50 to −60 bars (e.g., *Pythium* and *Phytophthora* spp., *Rhizoctonia solani, Thielaviopsis basicola, Gaeumannomyces graminis* var. *tritici*) cause the most disease where soils are moist and plants are not stressed for lack of water. Exceptions are the vascular wilt fungi (e.g., pathotypes of *Fusarium oxysporum, Verticillium,* and *Cephalosporium gramineum*), which grow maximally at −10 to −20 bars osmotic potential, cease growth only when the osmotic potential is below about −120 bars, yet produce the most disease where crops are irrigated or soils are wet on a regular basis (Cook and Papendick, 1972). This can be explained by the higher rates of transpiration by the host achieved when soils are kept well watered. This enhances systemic distribution of the pathogen in the host. On the other hand, selection pressure on these fungi obviously has been toward ability to grow at low water potentials, which Cook and Duniway (*in* Parr et al, 1981) suggest is because of the importance of saprophytic colonization to them (i.e., their need to be able to colonize the stems, leaves, or other aerial parts of the wilted plant). *Verticillium dahliae* makes microsclerotia at water potentials down to −60 bars (Ioannou et al, 1977a), as it must be able to do to succeed inside the wilted plant. This is further evidence that vascular pathogens are highly dependent for success on their ability as saprophytes, and any system aimed at their biological control as pathogens should take this into account.

Spores of pathogens of aerial plant parts are commonly observed to require "free water" to germinate. Free water by definition has zero water potential. It seems highly doubtful that such spores cannot obtain adequate water for germination at water potentials less than zero. Indeed, critical studies indicate that even ascospores of *Sclerotinia sclerotiorum* (which require wet leaves to infect) are capable of germination down to −90 bars, which is also the lower limit for hyphal growth by this fungus (Grogan and Abawi, 1975). As pointed out in the previous section, water films are needed to help remove chemical inhibitors of germination contained within some spores (Allen, *in* Heitefuss and Williams, 1976). Water films are also important to supply nutrients in solution to the fungus spore and for diffusion of microbial toxins away from the producer organism. The latter effects may be important to biological control of leaf pathogens by competition from the epiphytic flora.

Propagules usually swell during early stages of germination and before actual sprouting (Yarwood, *in* Kozlowsky, 1978). This swelling results from increasing internal pressure on the spore wall, but the uptake of water may serve more for uptake of nutrients in solution than to meet some minimum turgor pressure for sprouting. The spores of rust fungi may increase in volume only slightly during germination, and those of two species of powdery mildew fungi shrink during germination (Yarwood, *in* Kozlowsky, 1978). These spores are probably nutritionally independent, and hence uptake of nutrients in solution (or by any other means) is unnecessary for germination. The shrinkage of the powdery mildew conidia during germination can only mean that the turgor pressure for the early stages of germ tube extension was maintained adequately within the shrinking spore without additional uptake of water.

Clearly, progress has been made during the past 10–15 years in understanding the water relations of plant pathogens. This progress notwithstanding, much more remains to be done. For example, the optimal and minimal water potentials for growth are known for only a few pathogens (Figure 5.1), yet the optimal, maximal, and minimal temperatures for growth have been known for years for most important pathogens. The differential response to matric and osmotic potentials is significant and deserves much more study. Information on growth rates at different combinations of water potential and temperature has prediction value for the environment likely to favor a microorganism, and can help identify ecotypes within a species of microorganism. Moreover, microorganisms are more responsive to water potential at some stages (e.g., during reproduction) than at other stages (e.g., during vegetative growth) of their life cycle. Future progress on biological control of pathogens through ecological manipulations to favor antagonists, the host, or both will depend on more and better information on the response of organisms to water potential.

6

THE HOST AND BIOLOGICAL CONTROL

One is constantly reminded of the infinite lavishness and fertility of Nature — inexhaustible abundance amid what seems enormous waste. And yet when we look into any of her operations . . . we learn that no particle of her material is wasted or worn out. It is eternally flowing from use to use, beauty to yet higher beauty.
—JOHN MUIR, 1869

The host plant plays both a passive and an active role in the biological control of plant pathogens. In the passive role, it provides a meeting place for the interactions of pathogens, nonpathogens, and antagonists. This passive role begins on the day a new seed comes to rest in the soil and continues throughout the life of the plant, ending only when the residue of the plant has decomposed beyond a stage useful to the pathogen. The active role requires that the host be living and includes the mechanisms of active resistance to pathogens. In the passive role, the host is seen as an environment of absorptive and adsorptive surfaces and of fluctuating temperature, water potential, pH, partial pressures of biologically important gases, and organic and inorganic nutrients. In the active role, the host has mechanisms that recognize and reject, exclude, or somehow respond to microorganisms or their products.

Throughout this book we have attempted to portray pathogens in the ecological sense, as microorganisms that differ from others only in that their specialized ecological niche is the living tissue of plants. Ability of plants to colonize a desert soil, a marine estuary, or the floor of a jungle also calls for specialization, which, to one degree or another, is the rule in biology. Pathogens succeed in the living tissue of a plant because they have the ability to obtain their nourishment and grow in this habitat. However, the requirements for microbial growth are the same everywhere: The organism must be able to osmoregulate or otherwise maintain turgor, there must be a supply of nutrients, and the physicochemical environment must not be restrictive or poisonous. In this respect, the biophysics of the host-pathogen interaction may be as important as the biochemistry of the interaction.

The physical environment provided for microorganisms on and in plants is often not adequately considered in studies of plant-microorganism interactions. Some attention has been given to the physical-chemical environment on plant surfaces, most notably the root surface and more recently the unstirred layer of air (boundary layer) at the leaf surface and the associated phylloplane. Much less attention has been given to the biophysics of the host-pathogen interaction, or to how the physical environment on and inside the plant affects the growth of pathogens and the outcome of the host-pathogen or host-pathogen-antagonist interactions. As pointed out in the previous chapters, *most of the strategies for biological control of plant pathogens depend in one way or another on the manipulation or exploitation of useful biological interactions on or within the host plant itself.* This chapter begins, therefore, with a consideration of the living plant as an environment for growth and reproduction of microorganisms and for biological control.

PLANT WATER POTENTIALS

Plant water potentials determine the ease with which a microorganism in equilibrium with cells or intercellular spaces can obtain water for maintenance of turgor and for metabolism and other cellular functions, including production of enzymes and secondary metabolites (e.g., toxins or antibiotics). The water potential of a plant cell also influences the ability of that cell to respond to or resist a pathogen. Plant water potentials vary with the plant species, the age of the plant, the season, and especially the time of day. The result is a dynamic situation of fluctuating plant water potentials, fluctuating growth rates of the microbial inhabitants (as epiphytes, endophytes, and parasites), and fluctuating degrees of susceptibility of the cells. Plant water potentials can be managed somewhat through plant breeding and by cultural practices, and thus they offer a means to achieve or enhance certain kinds of biological control.

Plant Water Potentials for Pathogen Growth and for Biological Control with Antagonists

The soil-plant-air continuum through which water moves from the soil to the atmosphere can be subdivided into the rhizosphere soil, root hairs, root cortex, root endodermis, root xylem, stem xylem, petiole and leaf xylem, mesophyll layer, substomatal chamber, and leaf boundary layer (Papendick and Campbell, *in* Bruehl, 1975). Resistance to water flow occurs at the soil-root interface, at the root endodermis, in the xylem, in the leaf mesophyll, and at the leaf boundary layer. The water potential tends to change most abruptly (the gradient is the steepest) between roots and soil or between leaves and the atmosphere, remaining fairly uniform from the roots to leaves of a given plant. The gradient between the tops and the root surface is greatest during the day, when stomata are open and the water supply from the roots is least likely to keep pace with transpiration, but there are limits to the gradient beyond which the stomata close. The roots and leaves probably attain the same water potential during the night, when the stomata are closed and the tops equilibrate with the roots. Methods to measure plant water potential are described (Papendick and Campbell, *in* Bruehl, 1975; Barrs, *in* Kozlowski, 1968; Wiebe et al, 1971).

A microorganism, whether in the rhizosphere, within the intercellular spaces of the root or leaf, in the xylem, or within the leaf boundary layer, will be in equilibrium with the water potential of the microenvironment occupied. As the water potential of the environment fluctuates during the day or with the season, so the water potential of a hypha, bacterial cell, or other microbial unit in that environment will also fluctuate. The water potential of a given tissue or cell may be below the optimal water potential for growth of the pathogen (Table 5.1), thereby reducing its ability as a pathogen and increasing its vulnerability to displacement by secondaries not so limited by the prevailing water potential. Turner (1974) recorded leaf-water potentials for maize, sorghum, and tobacco at sunrise as −7, −4, and −6 bars, respectively, but at 0700–0850 hours, leaves of the three crops averaged −18, −22, and −15 bars water potential, respectively. Some midday leaf water potentials measured for different field crop plants during dry periods include: soybeans, −25 to −30 bars (Heatherly et al, 1977); cotton, −18 to −20 bars (Ackerson et al, 1977); sorghum, −25 to −28 bars (Ackerson et al, 1977); wheat, −35 to −40 bars (Papendick and Cook, 1974); and alfalfa, −35 to −45 bars (Cook, unpublished). A microorganism, whether pathogen or antagonist, must be highly responsive and adaptive to remain active in such tissue. In some cases, the plant water potential may drop below the cardinal minimum of the microorganism, in which case its growth will cease. Equally important, **retardation or cessation of growth by the pathogen within the host increases vulnerability of that pathogen to inhibition or displacement by antagonists not so limited.**

The hypothetical situation below illustrates the kinds of changes necessary if a microorganism is to continue uninterrupted growth (i.e., with no loss of turgor) during the diurnal fluctuation in leaf water potential.

	Host Cell Total = Osmotic + Turgor			Pathogen Cell Total = Osmotic + Turgor		
Night	−10 =	−17	+ 7	−10 =	−25	+ 15
Day	−20 =	−24	+ 4	−20 =	−25	+ 5

Note in this particular example that to maintain a uniform 15 bars turgor by day as well as by night (rather than the 5 bars given for day in the chart), the microorganism in the leaf would need to adjust osmotically from −25 bars each night down to −35 bars each day. Such adjustments might be accomplished biochemically within the microbial cell itself, by adjustments in the internal solutes between states or complexes that are more (or less) active osmotically (Chapters 5 and 7). Such adjustments also may be accomplished by exchange of osmotically active substances such as potassium between the microorganism and its environment, e.g., the host tissues. Either method may require expenditure of energy. A more likely happening is that the fluctuations will not be accomplished entirely if at all by osmoregulation; rather, turgor will fluctuate and hence rate of hyphal extension and cell division will fluctuate, probably being greatest when turgor is greatest (by night) and least when turgor is least (by day).

Plants grown in an environment chamber (or glasshouse) differ from field-grown plants in the water potential they develop. Thus, leaf water potentials of wheat plants in the glasshouse never decreased below −25 bars in spite of limited soil water, loss of turgor, and conspicuous symptoms of stress; yet plants of the same age in the field developed leaf water potentials down to −35 to −40 bars before equivalent stress was apparent (Papendick and Campbell, *in* Bruehl, 1975; Cook, 1980). Davies (1977) showed for both soybean and cotton that midday leaf water potentials were −8 to −10 bars in a growth chamber, −10 to −15 bars in the glasshouse, and −10 to −17 bars in the field. In all cases, cotton had slightly lower leaf water potentials than did soybean. This situation suggests that some studies aimed at biological control by management of plant water potential might best be carried out in the field rather than in a glasshouse or growth chamber.

A difference in water potential also may exist between leaves in the top of a canopy and leaves lower in the canopy. Mesophyll cells in leaves at the top of the canopy and in full sunlight commonly have a lower osmotic potential (higher concentration of solutes) than cells in leaves lower in the canopy and partially or totally shaded. The total water potential may be only −1 to −2 bars less in the upper than in the lower leaves (providing the gradient for upward water flow), but the osmotic potential can be −5 to −10 bars lower in the upper leaves, being offset by higher turgor. The total water potential of stem xylem connecting leaves in the upper and lower canopy in this hypothetical example would be about midway between the two. Xylem water potential includes a small (dilute) osmotic component and a large negative turgor (suction) component. A hypothetical example illustrates this point.

	Total	=	Osmotic	+ Turgor
Leaf, upper canopy	− 15	=	− 20	+ 5
Xylem	− 14	=	− 3	− 11
Leaf, lower canopy	− 13	=	− 14	+ 1

Thus, a microorganism within the plant may encounter a similar total water potential but different turgor and osmotic potentials according to its location within the plant.

The leaf boundary layer, being unstirred humid air, presents a significant barrier to the diffusion of vapor from the substomatal chambers to the atmosphere. As such, this layer could be a factor in the development of a vascular wilt. Any factor that retards transpiration of a plant infected with a wilt pathogen will also retard systemic distribution (especially upward movement) of the pathogen in the xylem. Conversely, any factor that increases the rate of transpiration is likely to enhance upward movement of the pathogen, which is carried with the transpirational stream. In this regard, frequent winds or even breezes, by reducing the thickness of the leaf boundary layer, could accelerate transpiration (provided the soil water supply was not limiting) and hence the rate of development of a vascular wilt. Shawish and Baker (1982) observed that the time required for development of wilt was shorter and symptom expression was more severe if tomato, flax, and pea plants inoculated with their respective pathotypes of *Fusarium oxysporium* and incubated in still air in a glasshouse were shaken gently for one minute each day during the incubation period. They suggested that the effect was due to morphological alterations (thigmorphogenesis [Jaffe, 1980]) caused by mechanical stimulation of the plant, and clearly such an effect is possible. Even the noninoculated (check) plants responded to the gentle shaking, as evidenced by their shorter, thicker stems. Nevertheless, reduction of the leaf boundary layer by gentle shaking of the plant on a daily basis might also explain the effect.

The fleshy or storage organs of fruits and vegetables have water potentials near or slightly below those of the growing plant at the time of harvest. Ripe peaches, apricots, raspberries, strawberries, and blueberries all have water potentials in the range of −10 to −20 bars (Figure 6.1). Sweet cherries, on the other hand, have water potentials between −25 and −50 bars (Cook and Papendick, 1978). As might be expected, the water potential of a fruit is related to the sugar content of that fruit. Pie cherries at the hour of picking registered −30 bars osmotic potential and −15 bars total potential (turgor therefore was estimated at 15 bars), but during six days in storage, the turgor deceased until it reached zero and the total and osmotic potentials were both at −30 bars. A microorganism contained within the intercellular spaces of

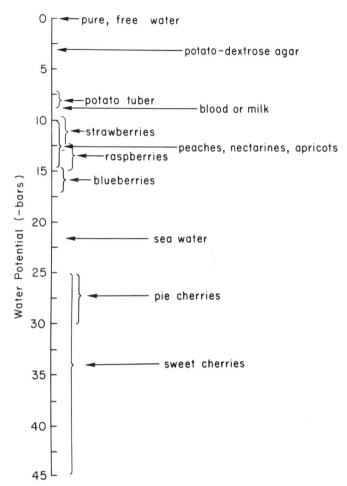

Figure 6.1. Water potentials of some common fruits, potato tubers, and other familiar materials. (Modified from Cook and Papendick, 1978.)

such tissue probably would need to exert twice the energy to obtain water from this fruit on the sixth as on the first day after picking. Most bacteria normally found on plants would be unable to grow in cherry fruits, which may account for the fact that bacterial pathogens are not a problem on cherry fruit.

The water potential of potato tubers was recorded as −6 to −8 bars (Cook and Papendick, 1978), ideal for growth of bacteria. Soft rot caused by *Erwinia carotovora* develops in tubers at −6.5 bars but not in tubers at −7.5 to −8.0 bars water potential (Pérombelon and Kelman, 1980), yet bacterial growth would be only slightly less at −8.0 than at −6.0 bars osmotic potentials (Harris, *in* Parr et al, 1981). The greater amount of soft rot at −6 compared with that at −8 bars is thought to result from the greater turgidity of tuber tissues at −6 bars (Pérombelon and Kelman, 1980). This explanation is supported by Cook and Papendick (1978), who showed that tubers when harvested and fully turgid had a total water potential of −6 bars (including −8 bars osmotic potential and 2 bars turgor potential) but that after drying, turgor was lost and the total potential was then the same as the osmotic potential (−8 bars). Loss of turgor possibly changes the geometric or spatial relationships among the tuber cells in such a way that porosity of the tubers to oxygen is increased. Pérombelon and Kelman (1980) also suggest that turgid cells leak more nutrients than nonturgid cells, which favors growth of the pathogen.

Our discussion thus far has focused on water potential dynamics inside the plant. As might be expected, the water potential of a plant surface is different from that of the underlying tissue and usually fluctuates more widely. The root surface water potential will be somewhere midway between that of the root cortex and the surrounding soil. Papendick and Campbell (*in* Bruehl, 1975) concluded that water potential gradients outward from the root surface are relatively small "until the soil water potential approaches the potential that limits flow." Their model indicates that, as the soil water potential approaches the limit for conductivity, a gradient may extend for several millimeters outward from the root surface. The soil water potential at which a gradient will develop is higher (wetter) as the rate of transpiration is increased. However, in soil drier than the root (as may occur around roots near the soil surface after a period of drying), water probably diffuses from the root into the soil. In this case, the root-surface water potential will be influenced by that of the underlying root tissue and could be suitable for growth of microorganisms even when the surrounding soil water potential would be below the cardinal minimum for the microorganisms.

The water potential of the phylloplane is greatly influenced by that of the underlying leaf tissue. When the leaf surface is not covered with a water film, the vapor pressure of the unstirred boundary layer of air on a leaf approaches that of the substomatal chambers, and provides a microenvironment suitable for microbial growth even when the relative humidity of air beyond the boundary layer is too dry for microbial growth. On the other hand, the phylloplane is more subject than the rhizosphere to drying because of winds and radiation. It seems possible, therefore, that phylloplane microorganims must be more adaptive to fluctuating water potentials and must be able to grow at lower water potentials than their counterparts in the rhizosphere.

It is important to recognize for any parameter of the physical environment that in most or all instances, the success or failure of the organism will be determined not only by the direct effects of the environment on the organism, but more importantly, by the effects of the environment on the ecosystem as a whole. Thus, the water potential on the host surface may be relatively unfavorable for a pathogen of that host but even more unfavorable to antagonists, whereupon a greater ecological niche becomes available to the pathogen and infection occurs. This is partly why *Fusarium roseum* 'Culmorum' causes more disease on wheat in drier soils; the pathogen germinates and germ tubes elongate maximally at soil water potentials near −1 bar matric potential, but antagonism from bacteria is intense at water potentials down to −5 to −10 bars matric potential, leaving −15 bars and drier as the niche most available to the pathogen.

Gaeumannomyces graminis var. *tritici* also grows maximally at matric potentials near −1 bar, where bacterial antagonism is most intense, but it is not endowed like Culmorum with ability to escape the bacteria by growing in drier soils. Its growth rate is severely reduced by −20 bars and prevented at −40 to −45 bars matric potential. This problem for *G. graminis* var. *tritici* may account for the observation (Henry, 1932) that take-all in sterile soil is most severe at 27°C, near the cardinal optimum temperature for *G. graminis* var. *tritici*, and in nonsterile soil between 12 and 16°C, well below the cardinal optimum for the fungus. Warm soil is effectively unavailable as an ecological niche for this fungus because of the greater microbial activity in the soil at 27° than at 12–16°C, the inability of *G. graminis* var. *tritici* to compete under these conditions, and the fact that any limitation of this microbial activity by reduced soil water potentials would also limit *G. graminis* var. *tritici*. The intensity of antagonism encountered by *G. graminis* var. *tritici* at 27°C will not be encountered at 12–16°C, where microbial activity is considerably less than at 27°C. It may be that total microbial activity is retarded proportionately more than the growth rate of *G. graminis* var. *tritici* by this temperature reduction. The lower soil temperature may provide an ecological niche more available to this fungus than that at 27°C. Take-all decline, on the other hand, is thought to result from microorganisms antagonistic to *G. graminis* var. *tritici* at the cooler soil temperatures (Cook and Rovira, 1976). Perhaps the multiplication of these antagonists effectively closes the only ecological niche available to *G. graminis* var. *tritici*.

Influence of Plant Water Potentials on Resistance of Plants to Pathogens

The osmotic, turgor, and total water potentials of plant tissues are just as critical to the functioning of host cells as they are to the function of pathogen or antagonist cells. The following scenario summarizes the sequence of responses in plants undergoing progressively lower or more prolonged exposure to low water potentials. The first response is a decrease in turgor of leaf cells; leaf growth ceases almost immediately (Hsiao, 1973). In sunflower, a turgor of 3.5 bars is needed for blade expansion (Boyer, *in* Kozlowski, 1976). With reduced rates of blade expansion, sugars may accumulate and contribute to cell osmoregulation, thereby helping the leaf to develop a lower osmotic potential and maintain turgor. As turgor approaches zero, abscissic acid increases; stomata then close and reduce further water loss, but this also effectively prevents photosynthesis (Boyer, *in* Kozlowski, 1976). A drop in the supply of assimilates in the leaves is then reflected as less assimilates and energy conducted to the roots. The turgor pressure essential for root penetration into soil is then lost. Cytokinin activity drops in the roots and limits branching in the tops. In tobacco, changes in cytokinin activity commenced in the roots following only 30 minutes of wilting (Itai and Vaadia, 1971). Although the levels of some hormones decrease with water stress, ethylene production increases and is accompanied by accelerated senescence and abscission of lower leaves (Hsiao, 1973; Kozlowski, 1976). The concentration of betaines also increases with plant water stress (Jones and Storey, *in* Paleg and Aspinall, 1982). Some of these processes and biochemical changes tend to help the plant adapt to the shortage of water. McMichael et al (1972) noted for cotton that −8 bars leaf water potential caused abscission of leaves of two-month-old plants, whereas −17 bars was required for abscission of leaves on cotton seedlings. However, with continued shortage of assimilates and energy to the roots, uptake of nutrients by the roots decreases and finally stops, owing to insufficient energy. Respiration is the last process to cease. **With progressive stress, the plant apparently foregoes new growth first, sacrifices old growth second, and maintains function in the youngest growth until the last.**

Low plant water potentials affect pathogens by retarding or preventing biochemical processes essential for expression by the cells of resistance to pathogens (Cook, 1973). Root rots caused by *Phytophthora* species are usually enhanced by wet soil conditions, largely because high soil matric water potentials favor production of sporangia and release and swimming of zoospores of the pathogen (Duniway, 1979). However, drought conditions also may predispose plants to more severe phytophthora root rot, as shown by Duniway (1977) for *P. cryptogea* on safflower. Plants of the *Phytophthora*-susceptible safflower cultivar Nebraska 10, but not those of the resistant cultivar Biggs, developed root rot when watered daily to maintain leaf water potentials between −4 and −6 bars and inoculated with zoospores at three to five weeks of age. In contrast, both cultivars developed root rot when water was withheld two to four days to decrease leaf water potentials to −13 to −17 bars, after which plants were watered and inoculated with zoospores of the pathogen. Water stress somehow caused a failure of the resistance mechanism(s) in Biggs. Nebraska 10 was still more susceptible to root rot than Biggs, with the predisposing stress causing the already susceptible cultivar to become "super susceptible." The actual water potential of the roots during predisposition could not be inferred, but it was presumed to be lower than the −4 to −7 bars soil water potential that occurred around the roots when leaves were drier than −12 bars. *Phytophthora cryptogea* infects the resistant variety without water stress, suggesting that the water stress enhances the development of established infections (Duniway, 1977).

A similar effect of water stress was shown by Blaker and MacDonald (1981) for *P. cinnamomi* on rhododendron. The cultivar Caroline was resistant to the pathogen in the absence of stress, but highly susceptible if stressed in advance of inoculation to leaf water potentials less than −16 bars. In contrast, the cultivar Splendor was susceptible whether or not stressed.

A predisposing effect of low plant water potentials on rate of tissue colonization (lesion development after infection) is also clear from work on European white birch infected with *Botryosphaeria dothidea* (Schoenweiss, 1975). The pathogen was introduced into bore holes in stems of test seedlings and the seedlings then stressed to different extents by withholding water from the roots. Cankers developed only if stem xylem water potentials were −12 bars or lower (drier), with the rate of canker development being proportional to the extent to which water potentials dropped below −12 bars down to about −25 bars, the lowest stem xylem water potential attained in the study. The stress effect was reversible; if seedlings under water stress were watered to elevate xylem water potentials above −12 bars, canker development ceased and callus formation was initiated. Freezing stress (−10 to −30°C) and defoliation stress also predisposed stems to an increased rate of canker development by *B. dothidea* in the stem.

Fusarium foot rot of wheat (caused by *F. roseum* 'Culmorum') became important on winter wheat in the Pacific Northwest in the early 1960s following availability of semidwarf cultivars (Cook, 1980). The pathogen itself is apparently native to Pacific Northwest soils, having been recorded widely by Sprague (1950) as a weak parasite on grasses and cereals throughout the region. Infection of winter wheat occurs in the fall from chlamydospore inoculum confined largely to the top 10 cm of soil (Cook, 1968). Infections occur readily but the pathogen may then remain confined to lesions on roots or cause only mild crown decay. This kind of syndrome was probably the predominant manifestation of the pathogen in wheat in the region in earlier days. Such damage to wheat is chronic and probably causes only a slight loss. More acute disease occurs when the pathogen grows aggressively and destructively through the crown tissues and two or three internodes up the culm. Such an attack almost invariably results in rapid and premature death of the plant, with significant reductions in kernal plumpness and total grain weight. It was this more acute manifestation of the disease that began to appear in markedly higher proportions with the cultivation of semidwarf wheats and that subsequently proved to

Fusarium Root and Foot Rot of Wheat

Biocontrol of the pathogen responsible for this disease illustrates the use of soil and plant management to favor antagonists and to enhance resistance of the host. It is accomplished by: 1) preventing saprophytic increase of the pathogen in wheat straw, by allowing nonpathogenic saprophytes to become established first in the straw; 2) suppressing infection from chlamydospore inoculum, by keeping the soil sufficiently moist to favor antagonistic bacteria; and 3) preventing aggressive attack of the host tissues after infection, by keeping plant water potentials suitably high for expression of host resistance.

Pathogen: *Fusarium roseum* 'Culmorum' (Deuteromycotina, Hyphomycetes, Tuberculariaceae).

Hosts: Wheat, barley, oats, maize, and most other Gramineae, as well as asparagus, beans, clover, carnations, and many other unrelated plants.

Diseases: Virtually any part of a cereal plant is susceptible to attack by this fungus, including spikes and kernels (head-blight and scab), stems (crown and stem rot, i.e., foot rot), roots (common root rot), or seedlings (seedling blight). Plants affected by foot rot die prematurely (usually after heading) when exposed to high evaporative demand (e.g., a hot, dry, windy day). Diseased root, crown, and stem tissues commonly remain intact, become dry-spongy, and develop a chocolate brown color, sometimes with pink or burgundy-colored mycelium evident beneath the leaf sheaths that enclose the stems.

Life Cycle: Soilborne inoculum is mainly chlamydospores in soil or chlamydospores and mycelium in crop residue colonized earlier through pathogenesis while the "residue" was still part of a living plant. Soilborne inoculum is mainly in the tillage layer. Aerial inoculum is not important except in humid climates. Root infections begin with hyphae from chlamydospores or crop debris; they may remain confined to roots or spread into the crown (the area where all stems of a single plant are connected) and one to three internodes

result from predisposing low plant water potentials (Papendick and Cook, 1974).

Fusarium foot rot under Pacific Northwest conditions appeared first where wheat was sown in late August or early September on summerfallow land and where average annual rainfall was only 25–40 cm. Growers were optimistically applying up to 100 kg of nitrogen per hectare in this area in an attempt to reach the high yield potential of these semidwarf cultivars. Wheats receiving these high rates of nitrogen developed a large leaf area, and water loss by transpiration was therefore greater. Plants developed midday leaf water potentials of −35 to −40 bars and even lower (Papendick and

up the stems. The fungus may sporulate on nodes of the dead or dying plant if moisture is suitable. These conidia, washed by rains into the soil, are converted to chlamydospores. Secondary infections are possible aboveground if sporulation occurs sufficiently early in the life of the crop and raindrops splash conidia onto neighboring plants. The fungus persists as a saprophyte in infected tissues after plant death. This fungus can also saprophytically colonize straw, provided the straw is not already thoroughly colonized by other fungi.

Environment: Infections from soilborne inoculum are favored by dry soil (soil water potentials drier than -5 to -10 bars but not drier than -75 to -80 bars matric potential). The pathogen grows maximally at -10 to -20 bars osmotic potential and does not grow below -90 to -100 bars osmotic potential. It is favored most by temperatures between 15 and 25° C. Temperature requirements of Culmorum are intermediate between those of the closely related *F. roseum* 'Graminearum' (found in warmer areas) and *F. roseum* 'Avenaceum' (found in cooler areas). A lesser-known strain (ecotype), *F. roseum* 'Crookwell', is intermediate in many other characters between Culmorum and Graminearum. Soils with acid to neutral pH are more suitable than alkaline soils.

Biological Control: Stubble is left standing after harvest to ensure thorough establishment of airborne saprophytes (e.g., *Cladosporium* and *Alternaria* spp.) before it comes into contact with the *Fusarium* in the soil. Supplemental irrigation of winter wheat in the fall during tillering and early stages of crown root development (when important infections occur) may limit root and crown infections by favoring bacterial antagonism in and near the infection court. The most effective biocontrol is achieved by use of cultural practices (delayed planting date, less nitrogen) that result in less vegetative growth of the plants and hence slower use of the limited water supplies; plants so managed are not stressed and maintain their naturally high resistance to *Fusarium*.

References: Cook (1980), Nelson et al (1981).

Cook, 1974). At 25–30°C, the pathogen grows maximally at osmotic potentials of -15 to -25 bars and ceases growth only if the osmotic potential drops below -90 to -100 bars (Cook and Christen, 1976; Cook et al, 1972). The growth rate at -35 to -40 bars osmotic potential is approximately the same as that at -3 bars osmotic potential. Although the wheat plants are still alive at -35 to -40 bars, the level of general resistance to *Fusarium* probably is not much different than that of a plant near death. Since the concentration of betaines increases in plants under water stress (Jones and Storey, *in* Paleg and Aspinall, 1982) and since betaines are stimulatory to *Fusarium* as a pathogen of wheat (Strange and Smith, 1978), the possibility of betaines stimulating the *Fusarium* in wheat under water stress deserves study. Wheats receiving only moderate rates of nitrogen develop midday leaf water potentials no lower than −30 to −35 bars. Such wheats become infected, but since they are stressed less, the acute manifestation of the disease does not occur.

Cultural practices shown to lessen the chances of plant water stress—and to lower the potential for damage from fusarium foot rot—include the use of less N, later seeding, and wider row spacing (Chapter 10). All of these practices lower the leaf area per unit area of soil surface and therefore the rate of soil water use. Of the three, the most practical and effective is to apply less nitrogen. Often, fields must be seeded early because moisture loss from the seed zone occurs rapidly during August and early September, and a few days delay in seeding can mean no germination until much later when fall rains begin. Row spacing beyond the standard spacing of 40–45 cm is undesirable since weeds are then favored. Some cultivars have been identified as having less tendency to stress. The basis for their higher (commonly 5 bars higher) leaf water potentials is unknown but may relate to deeper rooting, better stomatal control of water loss, less leaf area, or other factors. This system exemplifies a method of biological control effective after the pathogen is established in the plant, operative at least in part through maintenance of general resistance in the host, and accomplished by preventing low water potentials in the plant.

The increase in salt concentrations in the irrigated soils of some areas presents another means by which plants may be stressed and hence predisposed to pathogens. MacDonald (1982) has demonstrated for rooted cuttings of chrysanthemum that exposure of the roots for 24 hours to 0.1 or 0.2 M NaCl and then to zoospores of *Phytophthora cryptogea* resulted in three- to four-fold higher incidence of roots with lesions than occurred in roots inoculated without prior exposure to the salinity stress (Figure 6.2). The roots were returned to solution culture without the salt amendment before inoculation with the pathogen. Since the host but not the pathogen was exposed to the salt, the effect was obviously on the host and not on the pathogen. The number of zoospore cysts attached to roots within one hour after inoculation was directly proportional to the severity of salinity stress, suggesting that stressed roots were more attractive to the zoospores. In addition, lesions developed and spread more rapidly on stressed than nonstressed roots, suggesting that the root tissues were physiologically more susceptible to *P. cryptogea*. The risk of predisposition from salinity stress, either because of direct effects of the salt on the host, or because of possible lower water potentials that may develop in the plant, must be considered along with all other stress factors in the environment if host resistance is to be maximized for biological control purposes.

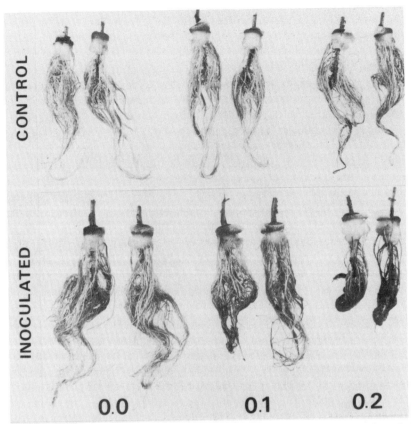

Figure 6.2. Chrysanthemum root systems grown in half-strength Hoagland's solution and given a 24-hour pulse exposure to 0.0, 0.1, or 0.2 M sodium chloride solution. Upper row is noninoculated controls. Lower row is inoculated with 10^6 motile zoospores of *Phytophthora cryptogea* immediately upon relief of the stress. (Reprinted with permission from MacDonald, 1982.)

Reduction in rates of protein synthesis, production of enzymes, and rates of energy production owing to reduced respiration all probably are important to reduction in active resistance mechanisms with plant water stress. Protein synthesis slows very quickly in response to even mild water stress (Hsiao, 1973). With seedlings of maize, polyribosomes began to revert to monoribosomes within 30 minutes after exposure of the seedlings to water stress. The percentage of ribosomes in the polymeric form dropped from 60 to 30 within four hours after the water potential had dropped from −6.5 to only −10 bars. The shift was reversible when the seedlings were again supplied with water. Hsiao (1973) suggested that the early decline of protein synthesis is mainly at the translational level. According to Todd (*in* Kozlowski, 1972), water deficits cause overall decreases in enzyme level, particularly of enzymes involved in synthesis, although those involved in hydrolysis or degradation tend to remain unchanged or to increase. Nitrate reductase is especially sensitive, dropping markedly with wilting of the plant. Phenylalanine ammonia lyase also decreases markedly with water stress (Bardzik et al, 1971). Boyer (1970) found for corn, soybean, and sunflower that respiration dropped linearly as the water potential was lowered from −4 to −16 bars but thereafter remained at 50% of that in the well−watered check, down to −40 bars. Dark respiration was reduced in sunflower only slightly between −4 and −20 bars (Boyer, 1971).

Conceivably any one of the physiological or biochemical changes occurring in the plant during water stress could be important to disease progress, including cessation of leaf or root growth, loss of host cell turgor, changes in the kinds of proteins synthesized, decreased protein synthesis, increase in concentration of betaines, or improper function of enzymes. A particularly important change toward greater susceptibility surely occurs with decreasing rates of respiration and hence supply of adenosine triphosphate within the cell. The respiratory rate of plant cells under water stress may be sufficient for cell maintenance but insufficient to provide defense against pathogens. At some point the plant is functionally near death. Such a plant is then fair game for a weak parasite, which can kill plants that otherwise would recover if relieved of the stress by watering. Weak parasites can therefore be highly important as pathogens. The problem for both plant and microorganisms is compounded by the persistence of the stress for some time after water is provided. Cultural practices or genetic changes in the host designed to reduce plant water stress or help the plant cope with water stress can provide highly effective biological control of these pathogens.

PLANT TEMPERATURE

Studies of temperature effects on disease are usually based on ambient air or soil temperature, rarely on plant temperature. An understanding of the influence of temperature on the activities of pathogens or pathogen-antagonist interactions on plants requires knowledge of temperatures of the plant itself. The differential between the plant and surrounding environment is greatest for the most fully exposed leaves and least between roots and surrounding soil. For example, the internal temperature of sugar beet roots was found by M. Stanghellini (unpublished) in Arizona to be the same as the soil temperature measured 10 cm deep. In contrast to the situation for roots, leaves are exposed to incoming radiation and commonly absorb more energy than they emit during the day and emit more energy than they absorb during the night. Thus, leaf temperatures not only fluctuate

more than root temperatures, they also may fluctuate more than their surrounding air temperatures, being higher than air temperature by day and
lower by night. This situation has obvious implications for growth and
survival of phylloplane microorganisms (Blakeman and Fokkema, 1982).
Plant temperatures also affect host plant susceptibility (or resistance) to
pathogens.

Leaf Temperatures for Growth of Pathogens and for Biological Control by Antagonists

About 99% of the energy absorbed by leaves as incoming radiation is emitted,
mostly as thermal (infrared) radiation but also by conduction, convection,
and evaporative-cooling. Only about 1% of the energy absorbed is stored,
i.e., used in photosynthesis or stored as elevated leaf temperature. Noble
(1974) calculated that storage of only 0.1 calorie per centimeter per minute
(0.5–1% of the total energy flux) would cause a temperature rise of 0.5°C
per minute. Heat moves by conductance through the blade and across the
boundary layer and thereafter by convection between the boundary layer
and turbulent air above. Since air is a poor conductor of heat compared
to water, the thicker the boundary layer at the leaf surface, the greater the
temperature differential between the leaf and the turbulent air. Assuming an air
temperature of 21°C, no wind, and 1.7 calories absorbed per sqaure centimeter
per minute, Noble (1974) calculated that the center area of a large leaf could
develop a boundary layer up to 4 mm thick, in which case leaf temperature
would be 12°C higher than air temperature. The temperature difference for a
boundary layer 1.3 mm thick would be 4°C, and for one 0.24 mm thick, only
1°C.

Wind reduces the thickness of the boundary layer and will therefore enhance
the efficiency of conduction and convection. Since the boundary layer is the
thinnest at the leaf margins, the smaller the leaf (or leaflet), the greater
the rate of heat loss by conduction and convection, and the less likely the
existence of a temperature difference between the leaf and the air. Nearly
half of the energy flux other than thermal radiation is achieved by conduction
and convection, and hence even the slightest change in rate of heat loss by
these means will be rapidly expressed as a temperature change in the leaf.
Heat loss by transpiration and the resultant evaporative cooling accounts
for half of the total energy flux outward from the leaf other than thermal
radiation, and thus a change in rate of transpiration will also markedly affect
leaf temperature. A total cessation of transpiration on a hot day could
produce exposed leaf and stem temperatures of 45°C and greater (Gates,
1980).

Leaf temperature below air temperature at night results when energy as
incoming radiation is greatly reduced, while energy efflux as infrared radiation

continues, especially from leaves exposed to the night sky. As leaf temperature drops below that of the surrounding air, the air begins to move toward the cooler leaf. Not uncommonly, the leaf temperature drops below the dew point of the warmer turbulent air, so that condensation occurs on the leaf surface. This condensation, in turn, produces a heat gain for the leaf. Frost may form on the leaf surface if the heat gain does not warm the leaf above freezing.

Obviously, phylloplane microorganisms must contend with extreme fluctuations in temperature as well as water potential. Indeed, conditions on the surfaces of exposed leaves must be physically limiting to growth and sometimes to survival of microorganisms over a significant period by day, when temperatures may rise above their cardinal maximum or even thermal death point, and by night, when temperatures may drop below their cardinal minimum. Fortunately for biological control, **saprophytic microorganisms tend to have a wider temperature range over which growth is possible than do parasites.** Selection of antagonists for protection of leaf surfaces should be for strains having ability to survive and grow over temperature and water poential ranges greater than those required by the target pathogen.

A dead plant left standing in the field probably develops very high leaf and even stem temperatures during cloudless days since heat loss by transpiration is nil. Significantly, *Verticillium albo-atrum* and *Fusarium oxysporum* f. sp. *lycopersici* can grow at temperatures as high as 35°C, but to do so they also require water potentials below −35 bars; at higher (wetter) water potentials, the cardinal maximum for these fungi is the more familiar 32°C (Manandhar and Bruehl, 1973; Chapter 5). *Pseudocercosporella herpotrichoides* (Bruehl and Manandhar, 1972), and *F. roseum* 'Culmorum' and 'Graminearum' (Cook and Christen, 1976) exhibit a similar reaction. **Plant-pathogenic fungi have the ability to grow at unusually high temperatures when the water potential is low.** Many of these fungi cause plant death just before normal plant maturity, when nighttime air temperatures are often highest and daytime plant temperatures even higher. Such pathogens are thus well-equipped to establish thoroughly in the moribund tissue of their host upon its death. To preempt such colonization with antagonists would require transfer of the dead plant to another environment, perhaps by burial in soil (Ioannou et al, 1977c; Chapter 7).

Plant temperatures can be modified by changes in row spacing, plant density, seeding date, fertilizer rate, or other cultural practices (Chapter 10). Plant temperatures and particularly leaf temperature might also be modified by plant breeding, e.g., selection for different size, shape, or angle of a leaf or of canopy structure. Even a slightly higher leaf temperature on one cultivar compared with that on another could be significant. A small but consistent elevation (or decline) in temperature may seem insignificant when the broad temperature ranges for growth of microorganisms are considered, but it can be highly significant to an interacting population of microorganisms on the phylloplane. Even a slight disadvantage in ability of one organism to compete can effectively eliminate the organism from that ecological niche.

Influence of Plant Temperature on Resistance of Plants to Pathogens

The influence of temperature on the level of general resistance of a plant to damage caused by a pathogen is particularly well illustrated by the greater resistance of corn to *F. roseum* 'Graminearum' at soil temperatures above 24°C and of wheat to the same pathogen at 12°C or less (Dickson, 1923). It is also illustrated by the greater resistance of watermelon to *Pythium ultimum* and *Rhizoctonia solani* at soil temperatures above 20°C and of spinach to the same pathogens below 12°C (Leach, 1947). These early contributions established a principle now widely accepted for general resistance of plants to nonspecialized pathogens, namely, **the greater the plant vigor, as expressed by growth rate, the greater the level of general resistance to weak parasites.** By choosing cultural practices that favor faster emergence and faster plant growth, or by selection of better-adapted cultivars, greater levels of general resistance against pathogens can be achieved.

A very dramatic effect of plant temperature on susceptibility to disease has been shown by M. Stanghellini (personal communication) for root rot of sugar beets caused by *Pythium aphanidermatum*. The pathogen exists in a high inoculum density in rhizosphere soil surrounding the beet roots but causes no disease so long as the internal root temperature is below 27°C. At 27°C and above, on the other hand, the root is rapidly decayed. Apparently, a physiological change occurs within the host at 27°C, permitting its destruction by *P. aphanidermatum*.

The basis for the greater or lesser general resistance with favorable or unfavorable temperature is undoubtedly very complex. Moreover, the lower resistance in warm-temperature crops exposed to a low temperature may involve mechanisms quite different from those in cool-temperature crops exposed to a predisposing warm temperature. The fact that plants exhibit maximum growth rate at the temperature range of greatest general resistance suggests that virtually all metabolic processes essential to maximal cellular activity may be involved.

A direct connection between photosynthesis and general resistance seems unlikely, since, in the studies by Dickson (1923) and Leach (1947), the differential responses were already apparent during seedling emergence, before the photosynthetic machinery was functioning. Moreover, photosynthesis exhibits a broad response to temperature, being 75% of maximum over a range as wide as from 10 to 30°C (Berry and Björkman, 1980). Even so-called cold-adapted plants, such as may be found above 82°north latitude or at elevations of about 3,400 m, show near-maximum photosynthesis over the entire range of 10–30°C, with highest rates at 15–20°C. Plants also exhibit a strong tendency toward "photosynthetic acclimation," i.e., adaptation of the photosynthesis process to the changing temperature so that actual rates ultimately vary relatively little from the maximum possible for the available carbon dioxide and photosynthetically active radiation.

Plant temperature may exert a striking effect on the expression of monogenic resistance of specific cultivars to specialized biotrophic parasites such as rust and mildew fungi and certain viruses. The examples of temperature effect on such resistance are mainly of the type whereby the gene confers resistance at a low temperature but susceptibility at a high temperature (Vanderplank, 1978). A well-known example is the *Sr6* gene in wheat; plants with this gene that are inoculated with race 6 of *Puccinia graminis* f. sp. *tritici* are resistant if incubated at 20°C but susceptible if incubated at 25°C. Temperature also may affect the expression of multigenic resistance. A classic example is the type A and type B resistance of cabbage to fusarium wilt caused by *F. oxysporum* f. sp. *conglutinans*. Plants with type A (single dominant genetic factor) resistance do not develop wilt even at temperatures (28–30°C) highly favorable to the disease. In contrast, plants with type B (several genetic factors acting additively) resistance develop severe wilt in warm soils but are resistant at lower soil tempertures (Walker and Smith, 1930). Vanderplank (1978) cites many examples involving resistance genes in cereals, tomato, cotton, bean, spinach, pea, and tobacco, where the phenotype shifts from resistance to susceptibility when the temperature is raised.

Predisposition by Chilling Injury and Freezing

Cells of warm-temperature plants (e.g., watermelon, maize, sugarcane, tomato, sweet potato, cotton, banana, and bean) undergo a membrane phase change from a flexible liquid-crystalline to a solid-gel structure at temperatures below 10–12°C (Raison, *in* Stumpf, 1980). This phase change is associated with chilling injury in fruit and vegetables adapted to tropical and subtropical areas (Lyons, 1973). As the membrane lipids solidify and contract, cracks and channels are thought to develop in them, thereby leading to increased membrane leakiness (Lyons, 1973). Increased leakage from cells of chilling-sensitive plants following exposure to chilling temperatures has been shown for bean and corn root tips, but did not occur in root tips of cool-temperature plants such as pea or wheat. Increased leakage following chilling has also been shown for cotton roots, cucumber leaves, and fruits of several tropical plants. Mitochondrial membranes also lose their integrity and become nonfunctional in warm-temperature plants exposed for prolonged periods to chilling temperatures. Respirational activity of the cells is thus seriously impaired (Raison, *in* Stumpf and Conn, 1981). In cotton, the adenosine triphosphate level in leaves was more than four times greater in nonchilled than in chilled plants (Raison, *in* Stumpf and Conn, 1981). Glycolysis, on the other hand, is not impaired following chilling, with the result that products of fermentation (ethanol, pyruvate, aldehydes, and acetic acid) accumulate (Raison, *in* Stumpf and Conn, 1981). The effects of chilling are reversible but may be permanent if the low temperature persists for many days.

The combination of greater leakiness of cells, lower oxidative activity (and hence lower energy supply), and greater accumulation of the products of fermentation might account for the greater susceptibility (loss of general resistance) of warm-temperature plants to pathogens such as *Fusarium, Pythium,* and *Rhizoctonia* at lower temperatures. The pathogens themselves are well adapted for growth at temperatures below 10–12°C and even down to 2–3°C and hence will not be limited to the same extent as the host.

With cool-temperature crops (e.g., wheat, spinach, peas), metabolic and especially respiratory activity decreases in cells exposed to warm (predisposing) temperatures, thereby lowering the energy available in these plants for resistance to nonspecialized pathogens whose own metabolic activity is not retarded by a comparable amount.

So-called cold-hardy plants are naturally protected from freeze damage in at least two ways (Burke et al, 1976). In most hardy plants, water moves out of the cell and then freezes extracellularly. The tissues of such plants become highly dehydrated, which is ultimately the mechanism of injury; the greater the tolerance of the tissue to dehydration, the greater its cold-hardiness. Cold-hardiness usually is therefore an equivalent of drought-hardiness. Some woody plant species extremely tolerant of dehydration can survive temperatures down to −196°C.

Another mechanism of hardiness involves supercooling of the water within the protoplasm. Pure water will supercool to −38°C, the point of homogenous nucleation. Normally dust particles (or bacterial cells of *Pseudomonas syringae*) initiate ice formation (ice nucleation) at the so-called freezing temperature of water or slightly below. Plant cells with ability to supercool are thought to undergo compartmentalization of water into finely dispersed tiny volumes of water; by this arrangement, ice nucleation of water in one compartment does not initiate ice nucleation in an adjacent compartment because they are physically separated. Plants or tissues with this mechanism of hardiness are safe down to −38°C. Buds of some woody plants are thought to survive by this mechanism. Ice formation inside the cell destroys membranes and hence the orderly compartmentalization of enzymes; the cell then becomes a medium for strict saprophytes as well as necrotropic parasites when again thawed. Probably the weak parasites poised as epiphytic and endophytic inhabitants of plant tissue require only that the host tissue approach death, at which time they can begin to grow aggressively and possessively into the tissues. Such tissues are stressed but are not killed outright by the cold and could probably recover (Levitt, 1980) except for the advantage gained by the weak parasites that then become important pathogens. Development of cold-hardy cultivars (Hoppe, 1957) or the use of practices that help the plant avoid freezing (Lindow, 1983) are ways to obtain biological control of these pathogens.

OXYGEN IN PLANTS AND IN THE ROOT ZONE

As emphasized throughout this book, any impairment of respiration or growth rate of a plant tissue greatly lowers the level of general resistance of that tissue to pathogens. That flooding and associated oxygen stress predispose roots to disease is best known for diseases caused by phycomycetous phytopathogens, most notably *Pythium, Phytophthora*, and *Aphanomyces* spp. Water-saturation of avocado-orchard soil in Queensland, Australia, temporarily decreases its suppressiveness to *Phythophthora cinnamomi* and increases root rot (Chapter 7). Stolzy et al (1965) showed for citrus root rot caused by *Phytophthora citrophthora* that plants exposed to water-saturated soil for eight hours three times per month had more root decay than plants exposed to water-saturated soil for a few minutes every four hours during the six-month test. Plants exposed from the beginning of the experiment to a constant 15% soil-water content, comparable to that to which they had become accustomed before application of the treatments, showed little root rot unless the oxygen supply was curtailed. Had the plants been reared from the outset under submerged conditions, thereby allowing for root adaptation (Baker and Cook, 1974) and possibly the development of aerenchyma tissue (Russell, 1977), predisposition and the concommitant acute root decay of the citrus seedling might have been less. Lowered oxygen levels in the soil also reduced growth and regeneration of roots.

The concentration of oxygen at any point within a plant or in the soil is a function of 1) the rate of diffusion from the surrounding air to that point, and 2) the rate of oxygen consumption by plant and microbial cells along the path of diffusion. The rate of diffusion is determined by the porosity and hence permeability of the plant tissue or the soil. A plant tissue may be relatively porous, and hence an efficient conductor of oxygen, but entry into the plant may be restricted by water films or natural barriers such as cutin, suberin, or possibly root-cap mucilage. The diffusion rate of oxygen along the path of interconnecting pores in plants or soil is greatly decreased if the pores are filled with water (Chapter 7). Oxygen diffusion is only 1/10,000 as fast through water as through air (Griffin, 1972).

Plant tissues generally are sufficiently porous to allow for rapid diffusion of oxygen and other gases from one tissue to another (Sifton, 1957). Noble (1974) estimates that plant leaves are, by volume, about 50% gas-filled pore space. In addition to the intercellular spaces that appear between turgid swollen cells, large lucunae or aerenchyma tissues may develop and greatly enhance gaseous exchange within and between plant parts. Such aerenchyma tissues are best known in roots of aquatic plants such as rice, water lily, cattail, willow, and cypress. These plants can grow in flooded soil where virtually 100% of the oxygen needs of the roots are met by oxygen diffusion through stems from the tops. Oxygen diffusion from the tops can also be important to root growth of mesophytic plants (Greenwood, in Whittington, 1969; Luxmoore et al, 1970a, 1970b, 1970c).

The oxygen requirement of roots is greatest at the tips, where respiration is greatest. The root tip is the first to cease functioning and to die when soil oxygen is lacking. Root penetration into soil is thus restricted earlier than foliar growth (Greenwood, in Whittington, 1969). Luxmoore et al (1970b) found that at 25°C, respiration was reduced by half at 8% oxygen for roots of maize and at 16% oxygen for roots of rice (air is 20.8% oxygen). The problem of adequate oxygen supplies to the root tips is complicated by the fact that, although respiratory rates are highest in the tips, porosity (and hence the opportunity for internal diffusion of oxygen) is lowest in the root tips, particularly in the meristematic region (Burstrom, in Baker and Snyder, 1965). Porosity of the root tissues was a maximum of 10% for maize and 33% for rice at 6 cm or more back from the root tip, was progressively less between 6 cm and the root tip itself, and was virtually zero in the root tip (Luxmoore et al, 1970b). This problem for the tips is offset somewhat by permeability to external supplies of oxygen, which is greatest at the root tip and progressively less at increasing distances back from the tip, where suberin deposits and other natural barriers are greatest. These results suggest that whereas oxygen needs of older portions of roots (i.e., to within a few millimeters of the tip) are readily served by oxygen diffusion from the tops, the root tip is more dependent (but also more adapted) to oxygen provided through the surrounding soil.

Flooding or even thick water films around root tips increase the dependency of the root tip on oxygen from the leaves (Luxmoore et al, 1970a, 1970b, 1970c). Greenwood (in Whittington, 1969) estimates that oxygen diffusion from the tops can meet the entire demand of the root only to the depth of a few centimeters. Following cessation of respiration at the root tips, the plant responses may include (Russell, 1977): 1) slower uptake of water, with the result that leaf cell turgor is lost and leaves wilt in spite of roots being submerged in water; 2) a decrease in gibberellins and cytokinins and an increase in the ethylene supplied to the tops, with the result that shoots cease growth and leaves begin to senesce; and 3) a continuation of glycolysis for some time, with the result that ethanol and other products of anaerobic fermentation may accumulate and diffuse from the root into the soil. The plant may adapt by producing numerous laterals in the top few centimeters and by greater formation of aerenchyma in the cortex. Oxygen can also diffuse laterally from a portion of the root system that is in drained soil, going through the roots toward another portion of the root system that is in waterlogged soil.

Allen and Newhook (1973) showed that zoospores of *Phytophthora cinnamomi* are chemotactically attracted to the ethanol liberated from roots. Kuan and Erwin (1980) found for phytophthora root rot of alfalfa caused by *P. megasperma* f. sp. *medicaginis* that saturation of soil for one week before inoculation with the pathogen greatly increased root-rot severity; the increase was associated with greater leakiness of roots, more root exudates, and increased chemotactic attraction of zoospores to the roots. They suggested that loss of cell-membrane integrity caused by the ethanol (Kiyosawa, 1975), or possibly resulting from the lack of oxygen, may account for the greater leakiness of alfalfa root cells and hence their greater attractiveness to *Phytophthora* zoospores.

Ethylene is formed and accumulates in soils during periods of anaerobiosis (Smith and Retsall, 1971), with profound effects on plant roots and possibly on disease development. The presence of ethylene accelerates senescence of exposed plant tissues, which increases susceptibility of the tissues to certain weak parasites. Unfortunately, most research to date on the predisposing effects of soil flooding on root disease has not distinguished between effects of low oxygen and those of high ethylene (Chapter 7).

Brief periods of flooding and the associated oxygen stress also exacerbate fusarium root rot of beans (Miller and Burke, 1975). This enhancement of *Fusarium* damage following flooding occurs in six to eight hours in rill-irrigated fields of the Columbia Basin in the Pacific Northwest. Miller and Burke (1975) duplicated the effect by growing beans in plexiglass airtight cells filled with either *Fusarium*-infested or fumigated soil. Nitrogen gas was passed over the soil surface to create an oxygen deficiency in the soil. Roots were rapidly destroyed by *F. solani* f. sp. *phaseoli* in the soil where an oxygen-deficiency was created, but they showed no serious effect of oxygen-deficiency in the fumigated soil. The stress effect did not increase the number of infections, but rather, it increased the rate of destruction of the tissue by the pathogen already in the tissue. Possibly oxygen diffusion from the tops to the infected hypocotyl and roots was sufficient to meet the needs of the pathogen but insufficient to meet the needs of the host tissue under challenge by the pathogen, whereupon severe disease developed.

It is important to recognize in the case of fusarium root rot of bean, indeed for many diseases favored by stress on the host, that the host is susceptible and that lesions develop even in the absence of stress. Stress causes the tissues to become "super susceptible" to the pathogen, much as happens with the susceptible Nebraska 10 cultivar of safflower when exposed to water stress before exposure to *Phytophthora cyptogea* (Duniway, 1977) or with certain Pacific Northwest winter wheat cultivars exposed to low plant water potentials after infection by *F. roseum* 'Culmorum' (Papendick and Cook, 1974). The effect of stress is to tip the balance drastically against the host and in favor of the parasite, so that disease develops more rapidly and severely. Cultural practices or genetic changes in the host aimed at avoidance or greater tolerance of plant stress may not protect the plant against infection, but they can greatly reduce the amount of damage and hence provide significant biological control.

The foregoing discussion is limited mainly to oxygen deficiencies and associated problems for plant health in flooded soil. Obviously, the problem is most acute and easiest to document in this situation, but it is not limited to such extreme circumstances. Oxygen deficiencies can also develop in microsites in soil (Chapter 7) and in plants, with important consequences for microbial activity, host plant resistance, and the host-pathogen-antagonist interaction.

Plant diseases caused by bacteria commonly are characterized by water-soaking of the tissues and by large accumulations of polysaccharide slime. The intercellular spaces of infected leaves and stems are sometimes packed with fluids in which the bacteria reside, spread, and multiply. Stall and Cook (1979) demonstrated for *Xanthomonas vesicatoria* that water-soaking prevented the hypersensitive response in an otherwise incompatible host-pathogen interaction. "Smothering" the host cell, i.e., shutting off its oxygen supply by such barriers as water films or polysaccharide slime, may be one way for the pathogen to "shut down" critical host-response mechanisms. Perhaps this is how phytopathogenic bacteria create an environment within the host to favor themselves while stressing the host.

The relationship between oxygen deficiency inside the tissues and increased susceptibility of the tissues to bacteria is documented best in soft rot of potato tubers (Pérombelon and Kelman, 1980). The amount of soft rot caused by *Erwinia carotovora* var. *atroseptica* increased exponentially with decreasing oxygen concentration. The increased decay with anaerobic conditions in the tubers was attributed to a requirement of the tuber for oxygen to resist the pathogen. The bacteria in pure culture actually grow better under aerobic than anaerobic conditions. An oxygen deficiency in the tubers causes loss of cell-membrane integrity and hence greater leakage of nutrients, favorable to greater multiplication of the pathogen. Tubers covered with a water film and incubated at 22°C became depleted of oxygen within two and one-half hours (Burton and Wigginton, 1970). Less than 10 cells were required to initiate a lesion under anaerobic conditions, compared with 10^{10} cells under aerobic conditions. Temperature is also important, with fewer bacteria required at 30°C than at lower temperatures. According to Pérombelon and Kelman (1980), progress of decay depends on the outcome of an apparent race between multiplication of bacteria and development of tuber resistance.

MINERAL NUTRITION AND IONIC BALANCE

Pathogens most likely to thrive in nutrient-stressed plant tissues are again the nonspecialized types characterized as weak parasites but potentially strong pathogens. Strong biotrophs are likely to be most active (and destructive) when the plant is thriving, i.e., not nutrient deficient (Baker and Cook, 1974). This section considers how mineral nutrition and ionic balance within and near plant tissues are relevant to biological control of pathogens favored by nutrient deficiencies.

Of special significance is the possibility that even a "subclinical" deficiency of some nutrient, e.g., a trace element, may favor some pathogens (Reis et al, 1982). "Subclinical" is used here to suggest that classic deficiency symptoms are not readily apparent in plants in the absence of the pathogen. Reis et al (1982) demonstrated a suppressive effect of zinc on take-all of wheat in eastern Washington in an area where previous fertilizer trials had indicated no need for zinc for wheat. On the other hand, potatoes in this same area responded to applications of zinc, suggesting that the nutrient is near a threshold level, perhaps adequate for wheat with healthy roots but not adequate for potatoes or for wheat with roots damaged by take-all. Phosphorus, trace nutrients, or other elements may not be attainable by a damaged or impaired root system, whereupon the plant with root disease becomes nutrient deficient more rapidly and the disease then progresses more rapidly. Conceivably, as cropping practices become more intensive, nutrient deficiencies and root diseases will become more common and the need for better information and for more corrective practices will increase.

Take-all is favored by mineral deficiences in the host plant (wheat or barley), but some nutrients apparently are more important than others, and different nutrients may help limit or counteract the disease in different ways. The disease develops more rapidly or is more severe in plants deficient in any of phosphorus (Stumbo et al, 1942), potassium (Trolldenier, 1982), magnesium, zinc, copper, manganese, and possibly also iron (Reis et al, 1982). Plants deficient in these critical nutrients also produce fewer secondary roots — new roots that otherwise would tend to compensate for the damaged ones and thus help the plant tolerate the disease (Garrett, 1948). In contrast to deficiencies of these nutrients, deficiencies of calcium or sulfur could not be shown to favor take-all (Reis et al, 1982, 1983). The number of diseased roots and the disease severity index (which takes crown damage into account) were not significantly increased in spite of symptoms of calcium deficiency (e.g., swollen root tips). Withholding these nutrients likewise did not affect the number of secondary roots formed during the course of the study.

Nitrogen deficiency had still another effect on take-all; susceptibility of the tissues to take-all was not different from that of plants well-supplied with nitrogen, but the formation of new roots was less on nitrogen-deficient plants. This latter finding confirmed an earlier conclusion by Garrett (1948), that added nitrogen helps in take-all control by promoting new root development and hence escape or tolerance of the disease but does not improve the intrinsic resistance of the tissues to disease.

Significantly, the deficiencies that favor take-all are of those nutrients important to the host's basal cell maintenance and function. In contrast, those nutrients not favoring take-all when withheld are those that serve more as constituents of plant structure and new plant organic matter. We will consider the latter category first.

Calcium has mainly an apoplastic role in plants, being an important ion in the structure of the cell wall, particularly the middle lamella. Calcium also has a symplastic role, helping to stabilize membrane permeability, membrane structure, cytoplasm structure, and complexing with enzymes (Clarkson and Hanson, 1980). However, the demand of the symplast for calcium is very low, compared with that of the apoplast, and may be easily satisfied even with very low amounts of this ion (Clarkson and Hanson, 1980). Moreover, cell maintenance (the symplastic role) would probably have priority over new growth. Similarly, a severe deficiency of one or both of nitrogen and sulfur would impair basal cell maintenance and function. However, in the work of Reis et al (1982), as in most studies of the effects of macronutrients on take-all, the nutrient supply was limiting but not entirely eliminated. On the assumption that cell maintenance has priority over new cell formation, then limiting (but not totally removing) the supply of nitrogen or sulfur would retard growth, but basal cell maintenance would continue unimpaired. Our proposal for the effects of calcium, nitrogen, and sulfur on take-all is essentially the same for all three: deficiencies in these nutrients may weaken the structure or decrease the rate of formation of new host cells and tissues but probably will not seriously impair the maintenance and function of the existing cells essential to adequate expression of the limited but significant general resistance in wheat to take-all.

Deficiencies of phosphorus, potassium, magnesium, and trace nutrients cause tissues to become supersusceptible to take-all (Figures 6.3 and 6.4). These nutrients have mainly a symplastic role in the function of plant cells. Phosphorus contributes to the buffering capacity of cells, and has a structural role in phospholipids essential to normal membrane structure. extremely important in the capture, storage, and transfer of energy in cells. Magnesium has a structural role in chlorophyll, is required in many enzyme

Figure 6.3. Wheat plants that were grown in a silica sand rooting medium with normal-strength Hoagland's solution (two groups of plants on the right) or the same solution but with the phosphorus concentration reduced by half. *Gaeumannomyces graminis* var. *tritici* was present in the sand used to grow the plants represented by the second and fourth groups, from left. (Photo courtesy of E. Reis.)

reactions, and plays a role in ribosome integrity and the stability of nucleic acids in membranes (Clarkson and Hanson, 1980). Potasssum contributes to osmoregulation in the cell and to electrical balance across the plasma membrane. It also contributes to enzyme activation. Trace nutrients occur as structural chelates or metalloproteins in plant cells, and they participate in many ways in the biochemistry of the cell, most notably in redox reactions. A limiting supply of any of these nutrients will be expressed as an impairment in processes critical not only to plant growth and structure, but also to basal cell maintenance and function, thereby opening the possibility for greater damage from a pathogen such as *Gaeumannomyces graminis* var. *tritici.*

Some take-all control has long been achieved in the field by application of phosphorus, but the results of Reis et al (1982, 1983) indicate that additional disease suppression is possible with application of certain trace nutrients, especially in alkaline soils where trace nutrients are most likely to be lacking. Take-all is favored by soil pH values in the range of 6.5–7.0 and higher (Cook, *in* Asher and Shipton, 1981). Severe take-all has been observed in acid soils, e.g., pH 5.0–6.5, but invariably the addition of lime results in still more disease. Availability of iron decreases 1,000-fold with each unit increase in pH in the range 5.0–8.0. Extractable zinc drops from 6.5 ppm at pH 5.0 to only 0.007 ppb at pH 8.0. Similar reductions in solubility occur with other trace nutrients. Leaf analyses of wheat plants growing in a silica sand rooting medium adjusted to different pH values revealed little difference in uptake over the range of pH 4.5–6.5 but significantly less trace nutrient uptake (especially less copper and zinc) when the pH was elevated from 6.5 to 7.5 or above. The concentration of copper in the leaves was a uniform 18–20 μg/g at pH 4.5, 5.5, and 6.5 but only 3–4 μg/g in leaves when pH of the rooting medium was 7.5 or 8.5 (Reis et al, 1983). Levels of Cu and Zn both were below the sufficiency level reported by Jones (1972) for several crop species.

Take-all is also influenced by the form of nitrogen. Ammonium nitrogen is suppressive to the disease, in contrast to nitrate nitrogen, which may enhance disease, at least under some conditions (Huber et al, 1968; MacNish, 1980; Smiley and Cook, 1973; Trolldenier, 1981). Ammonium influx into the symplast of roots is balanced by an equivalent efflux of cations into the apoplast, in accordance with the mechanics of active and passive transport and maintenance of electrical neutrality across the plasma membrane. A portion of the ions expelled in response to ammonium uptake are hydrogen ions. Influx of nitrate into the symplast of the root is balanced by an efflux of anions, a portion of which are hydroxyl ions. Smiley and Cook (1973) found that the rhizosphere pH for wheat was 5.5 for roots in soil supplied with ammonium N and 7.5 for roots in the same soil supplied with nitrate. They proposed that the suppression of take-all by ammonium nitrogen is largely a pH effect. When the changes in pH were prevented (with acid or base additives), the differences in take-all caused by ammonium and nitrate nitrogen were not evident.

Recently, Sarkar and Wyn Jones (1982) demonstrated in pot experiments with French beans that iron, manganese, and zinc all were less available in the rhizosphere and uptake by the plant was significantly less in soil fertil-

Figure 6.4. Wheat plants with take-all caused by *Gaeumannomyces graminis* var. *tritici*, showing the effect of copper deficiency on disease severity (second group from left). The two groups of plants on the right were provided copper as an amendment to the rooting medium. The first and third groups of plants reading left to right represent the noninoculated checks. Copper supplied through the leaves was as effective as copper supplied through the roots. (Photo courtesy of E. Reis.)

ized with calcium nitrate than in soil fertilized with choline phosphate or ammonium phosphate. This further emphasizes the importance of microsites in disease control; a pH adjustment may not be necessary in the soil generally but may be achieved on the root surfaces exclusively, where the need is greatest. Moreover, the work of Sarkar and Wyn Jones (1982), although done with beans, is evidence that the effect of rhizosphere pH on take-all may involve trace nutrient deficiencies. Obviously this effect will vary with other factors, most notably temperature, oxygen supply, the fluxes of potassium and sodium ions compared with those of hydrogen ions in the maintenance of electrical neutrality, and other factors that affect ion transport and balance.

Hornby and Goring (1972), working with a soil from Rothamsted, could not confirm the results of Smiley and Cook (1973) on the effects of ammonium versus nitrate nitrogen on take-all. The difference could be related to the kinds of ions exchanged for the ammonium or nitrate ions at the root surface. Electrochemical equilibrium can be attained through diffusion by ions in addition to H^+ and OH^-. Verification of an ammonium rhizosphere pH effect would require verification that the rhizosphere pH actually decreased when ammonium nitrogen was applied.

Smiley (1978a, 1978b) produced evidence that fluorescent *Pseudomonas* spp. antagonistic to *Gaeumannomyces graminis* var. *tritici* were significantly more numerous on roots receiving ammonium than on roots supplied with nitrate nitrogen. Increases in populations of fluorescent *Pseudomonas* spp. antagonistic to *Gaeumannomyces graminis* var. *tritici* on the root surface were also implicated by Cook and Rovira (1976) as possibly responsible for the suppressiveness of soil to take-all following take-all decline (Chapter 7). A mechanism to explain the suppression has been suggested by Kloepper et al (1980a), namely that siderophores (powerful natural iron chelaters) produced by root-colonizing fluorescent *Pseudomonas* spp. may remove iron needed by the take-all fungus. A soil suppressive to take-all became conducive if iron was made available, and conducive soil became suppressive with introduction of the siderophore compound. Clearly, progress has been made toward unraveling the complexities of plant nutrition and ionic balance in plants and the rhizosphere as they affect take-all, but much more work is needed to fully explain the many observed phenomena and their interrelations, if any.

In contrast to take-all, fusarium wilts are favored by ammonium nitrogen and are suppressed by nitrate or by liming the soil (Smiley, *in* Bruehl, 1975; Woltz and Jones, *in* Nelson et al, 1981). Moreover, the disease-suppressing effects of lime may be counteracted by ammonium, and the disease-enhancing effects of ammonium may be nullified by lime (Woltz and Jones, *in* Nelson et al, 1981). Thus, the form-of-nitrogen effect on wilts apparently is a pH effect, but opposite to that observed for take-all (Smiley, *in* Bruehl, 1975). Wilt occurs in limed soil if certain combinations of trace nutrients are also applied, e.g., iron, manganese, and zinc (Woltz and Jones, *in* Nelson et al, 1981). This effect is thus also opposite to that on take-all, which is suppressed by application of trace nutrients. Studies with fusarium wilt of tomato indicate that the pathogen has a higher requirement for certain trace nutrients than the host. In Florida, "liming sandy soils to the pH range of 6.0 to 7.5, together with the use of principally nitrate nitrogen have resulted in a significant degree of disease control for . . . aster, chrysanthemum, cucumber, gladiolus, tomato, and watermelon" (Woltz and Jones, in Nelson et al, 1981).

Club root of crucifers (caused by *Plasmodiophora brassicae*) has been decreased in some cases by liming the soil, but results have been variable. Dobson et al (1983) showed that the degree of club root control is proportional to how thoroughly the lime is mixed with the soil. A field soil before application of lime supported 100% seedling infection. Macro soil samples (15 g) from this field were pH 6.0, and micro soil samples (0.5 g) varied between 5.9 and 6.1. A "field mix" of lime reduced the percentage of infected plants only slightly, to 84%, and raised the pH of macro soil samples to 6.4, but the pH range for micro soil samples varied between 5.7 and 6.8. When the lime was sieved first, uniformly applied to the plot area, and then thoroughly mixed with the soil, only 25% infection occurred, and micro soil samples were a uniform pH 6.4–6.65. Thus, even microsites of slightly acid soil can apparently allow for infection of roots by *P. brassicae*.

NUTRITIONAL ENVIRONMENT AND ANTAGONISTIC INTERACTIONS AMONG MICROORGANISMS ON PLANT SURFACES

As in any microbial habitat, the amount of microbial growth (epiphytic and endophytic) on the various plant surfaces is determined by the environment and supply of nutrients. The nutrients occur either as exudates or as debris such as pollen, aphid honey-dew, sloughed cells, or dead microbial biomass (Fokkema et al, 1983). For many pathogens, whether aerial or soilborne, these nutrients become an important source of energy for prepenetration growth. A reduction in the amount of nutrient, whether through a reduced supply or increased competition, has the potential to limit or totally prevent colonization of the plant by a pathogen (Blakeman and Fokkema, 1982). Microorganisms with ability to preempt the nutrient supply of a pathogen on the host plant surface and thereby exclude the pathogen or prevent it from infecting may be thought of as prophylactic in their effect and as phytosanitizing microorganisms. This section discusses the sources of nutrients on roots and aerial parts and the potential for biological control through competitive and other antagonistic effects of nonpathogens on the plant surface.

The outcome in competition between a pathogen and the associated microbiota in the rhizosphere, on the phylloplane, or in other regions where nutrients are available is determined largely by effects of the physical environment on the relative competitive advantage of the pathogen versus that of the other organisms. This can partially explain the frequent observation that temperature, water potential, pH, and other factors optimal for growth of a pathogen in pure culture are different from those most favorable to infection of plants by that pathogen in the field. All factors of the environment combined determine the ecological niche available for the pathogen and antagonist. This section examines plant surfaces as ecological niches for pathogens, with particular attention to the nutritional status and competitive interactions.

Root Exudation, Dynamics of Root Growth, and Colonization of the Rhizosphere and Rhizoplane

Roots growing through soil represent a significant supply of carbon and energy for soil microorganisms, including exudates leaked from the tip region, sloughed cells from older portions of roots, and eventually the entire root as it ages and then dies. Barber and Martin (1976) estimated that, in three-week-old wheat and barley plants, 5–10% of the carbon assimilated in the tops was lost through the roots in sterile soil, with 3–9% of this carbon lost as water-soluble and 17–25% as water-insoluble material. The addition of microorganisms to the sterile soil increased losses of soluble materials from wheat roots up to 12–18%. Soil microorganisms have been shown to increase exudation (Rovira, 1959). Martin (1977) showed that 22–24% of assimilates were lost from wheat roots in nonsterile soil in the first six weeks.

On balance, although the supply of nutrients through roots is significant from the perspective of the plant, all evidence suggests that in most agricultural soils, the supply of carbon and energy for rhizosphere colonists is adequate for only a brief period if at all, and that most of the time microorganisms near roots, as elsewhere in soil, exist in a state of inadequate nutrition. Only 4–10% of the root surface of eight plant species was covered by microbial cells, with the remaining 90–96% being open spaces (Rovira et al, 1974). Probably the nutrient supply was inadequate to support a greater root-surface cover. In addition, the number of microorganisms in the rhizosphere falls off rapidly with distance outward into the soil. The number of bacteria 15–20 μm from the root is only 10% of that in the rhizoplane (Foster and Rovira, *in* Loutit and Miles, 1978). At −0.3 bar matric potential, there were 15 times as many microorganisms on the rhizoplane as at 0.1 mm away from the root and 430 times as many as at 0.8 mm from the root. Most nutrients released from roots apparently are consumed on or very near the root surface. Bacteria tend to be grouped in colonies and concentrated along the junctions of the anticlinal walls of epidermal cells (Rovira, *in* Baker and Snyder, 1965). On *Phalaris* roots, they develop in distinct colonies.

On *Eucalyptus calophylla* roots, there are 50 times more bacteria along these junctions of epidermal cells than in other areas (Bowen and Rovira, 1976). These junctions are the main sites of exudation and seem to be the main avenues of spread of bacteria along roots (Bowen, *in* Ellwood et al, 1980). Illustrations of the root-soil interface are available (Foster et al, 1983).

The evidence that one or more nutrients (most likely carbon or nitrogen) become limiting to microbial growth in the rhizosphere (Benson and Baker, 1970) raises the question of how pathogens obtain adequate nutrients for prepenetration growth in the rhizosphere. Probably in those soils characterized by a large and responsive microbial biomass, the nutrient supply is not sufficient for the necessary pathogen growth, and infection consequently does not occur. Such a soil is commonly pathogen-suppressive (Chapter 7). The suppressiveness in this case results from the total active microbiota acting as a nutrient sink and leaving the pathogen propagules deficient in carbon, nitrogen, or some other essential nutrient. A soil is conducive when the response of the soil microbiota to the root exudates is slower and hence less competitive initially, thereby giving the pathogen propagules more opportunity to respond, grow, and attack the root.

An opportunity for pathogen attack is also provided by root exudates released mainly from the unsuberized advancing root tips into soil where the associated microbiota is in a state of general quiescence. The dormant microorganisms bathed by this fresh supply of nutrient must first pass through a lag phase of growth, during which time the nutrient supply may temporarily exceed consumption rates of the associated microbiota. For this reason, nutrients may diffuse for some distance from the source before being consumed. Log-phase growth may not occur until the root hairs are forming. Stationary-phase growth and successions of colonization can be expected as the root matures. Thus, roots growing through soil provide the equivalent of a moving "window" for pathogens, namely a brief period of adequate nutrient supply in that region of the elongating tip where exudation is greatest and competition is least. The actual size of this window, both spatially and temporally, will vary according to the influence of soil physical conditions on growth rate of the pathogen relative to its influence on competitors. Closing this window with antagonists, resident or introduced, may provide significant biological control.

Roots are coated with deposits of polysaccharide materials of varying thickness, referred to as mucilage (Scott, *in* Baker and Snyder, 1965). A controversial subject in research on roots has been the origin of this mucilage, whether plant or microbial, and whether from one or several sources. It now seems clear that at least for grasses, two types of mucilage occur: 1) a gelatinous material that originates from the root cap (Miki et al, 1980) and that probably serves as a lubricant for penetration of the root tip into soil, and 2) a layer of material over the epidermal cells behind the immediate root tip, previously thought to be mucilage secretions from epidermal cells but actually remnant primary wall material from the outer tangential face of the epidermal cells (Foster, 1981).

Root cap mucilage for grasses is thought to originate near or at the periphery of the root cap and to be moved via Golgi-derived vesicles to the plasma membrane and then released through the cell wall (see Miki et al, 1980 for references). The root cap mucilage of corn and possibly other grasses contains a β-1-4-linked glucose polymer having oligouronic acid side chains and neutral sugars. Pectinlike compounds also may be present in addition to small amounts of cellulose and phenolic groups. This mucilage also contains sloughed root cap cells. Lubrication by root cap mucilage is important to root penetration through soil, especially in drier soils where mechanical impedance is greatest (Hsiao, 1973).

The nature and origin of mucilage overlying the epidermal cells in the root hair zone and older portions of the root were studied by Foster (1981) by application of electron microscopy and histochemistry to cores of soil containing undisturbed roots of wheat, *Paspalum notatum*, and *Pinus radiata* removed directly from fields. In all cases, the outer tangential face of the root epidermal cells consisted of a cuticle, a thickened mucilaginous primary wall, and an inner compact cellulose-rich secondary wall—three distinct layers in all. According to his observations, the primary wall is the only wall during cell elongation; it is multilaminate and includes cellulose microfibrils embedded in a matrix of pectins and hemicelluloses. Root hairs originate from lamellae adjacent to the plasmalemma and penetrate the mucilaginous matrix of the primary wall as they emerge. The secondary wall is laid down beneath and as part of the tangential epidermal cell wall after cell extension has ceased. As the root grows through soil, the cuticle becomes ruptured by mechanical abrasion and by the action of soil microorganisms. Soil particles become forced into the outer primary wall layer and embedded in the mucilaginous materials. According to Foster (1981), the so-called mucilage on the root epidermal cells is actually primary wall material filled with soil particles and cells of microorganisms and is not a distinct mucilaginous layer. Bacteria penetrate the cuticle and divide beneath it. The primary wall is rapidly lysed by microorganisms, some of which secrete capsular material. Slime on older root surfaces is apparently mainly of microbial origin.

There is no evidence that dissolution of the aging primary wall is a pathogenic effect of the rhizoplane colonists. In fact, the contrary seems true, especially with the feeder rootlets of plants, which are naturally transitory, even in pathogen-free soil, as evidenced by an eventual and spontaneous breakdown and collapse of the entire cortex and epidermis with aging of the root. Deacon and Henry (1980) demonstrated that up to 50% of the root cortical cells of seminal roots may be dead, starting behind the root hair zone. In their studies, cortical cells to the endodermis were dead in a sterile as well as a nonsterile

rooting medium. An exception was near the point of emergence of lateral rootlets, where cortical cells remained alive for about 800 μm around the emerging
lateral. Wilhelm and Nelson (*in* Toussoun et al, 1970) showed for rootlets
of strawberry in fumigated soil that, with aging, the entire epidermis and
cortex collapse to form a dense suberized layer surrounding the endodermis.
The vascular tissue enclosed within the endodermis may continue to function
as connecting and conducting tissue between the advancing root tip and the
main supporting root, or the entire rootlet may die and be replaced by a new
rootlet near the point of attachment of the dead rootlet. Rootlet replacement
is a normal healthy process with perennial and possibly all plants and has
been referred to as rootlet exchange (Wilhelm and Nelson, *in* Toussoun et al,
1970).

The older portions of roots with collapsing or disintegrating cortical
tissues are conceivably protected from infection in natural soil by the many
saprophytes likely to be in stationary-phase growth around this portion of
the root. Crossing this barrier of collapsing or disintegrating cortex intermixed with and dominated by competing microorganisms would seem difficult
for most plant pathogens. There is a need to find strains of root-colonizing
microorganisms capable of dominating the younger portions of root tips,
the region most likely to be invaded by pathogens. The pathotypes of
Fusarium oxysporum and *Verticillium* spp. enter the vascular system through
the undifferentiated xylem in the juvenile portion of roots where the endodermis is not yet formed. *Phytophthora cinnamomi* zoospores and also
nematodes accumulate and penetrate in the region of root-tip elongation,
where epidermal walls are thin and exudation is greatest. *Pythium* spp. establish in tissues only 0.1–1.0 mm behind the root cap in wheat (Cook
and Haglund, 1982). *Gaeumannomyces graminis* var. *tritici*, inoculated at
various points along the axis of wheat and barley seminal roots, infected
most readily at the root tips (Deacon and Henry, 1980). Clearly, for many
important root pathogens, the juvenile root tip is the region in need of protection.

Significantly, the *Pseudomonas* strains shown to provide biological control of root pathogens when introduced on seeds or seedpieces (Burr et al,
1978; P. Geel and B. Schippers, personal communication; Kawamoto and
Lorbeer, 1976; Kloepper and Schroth, 1979; Suslow and Schroth, 1982a;
Wadi, 1982; Weller and Cook, 1983; Chapter 4) are aggressive root colonists
and sometimes dominate the population of root-surface microorganisms.
Weller (1982) observed with a marked (antibiotic-resistant) strain applied
on wheat seed sown in October in field plots at Pullman, Washington,
that after one to two months, it not only made up a significant portion
of the total population of aerobic bacteria on the roots (Chapter 4), but
was the dominant bacterium on root tips of plants grown from the treated
seed.

Conceivably, one mechanism of protection by bacteria applied by seed bacterization is their ability to colonize and grow with the root tip, use the exudates, and preempt the pathogen in the niche otherwise most available to it. An organism or mixture of organisms with ability to preempt the pathogen on the juvenile root tip might require special characteristics not found in the majority of rhizosphere colonists: a high rhizosphere competence (Schmidt, 1979), including rapid growth rate, special adaptation to the root tip, and perhaps even ability to obtain nutriment from the root cap mucilage. Perhaps biocontrol strains should be selected for their ability to use remnant primary wall material as it becomes available from the developing root tip (Foster, 1981). Such a trait might add significantly to their ability to dominate and protect root tips.

Most work on the selection of beneficial plant-microorganism combinations for biological control have been directed at finding superior antagonists. The antagonists are then used on the available cultivars. Future research should also be directed at the development of plant cultivars more suitable for the antagonist. Atkinson et al (*in* Bruehl, 1975) have demonstrated that the occurence of a rhizoplane microbiota on wheat antagonistic to *Helminthosporium sativum* is under genetic control of the host. The superior performance of the cotton cultivars developed by Bird (1982), now in extensive field use in Texas, is thought (Bird, 1982) to involve alterations of the microflora on the plant surface in favor of those that protect the plant. This is a relatively unexplored area that deserves more study.

Exudation and Leaching from Aerial Parts

Substances may move from the interior to the surface of aerial plant parts by exudation through the cuticle via ectodesmata or microcapillaries, as well as through stomata, hydathodes, lenticels, and wounds. A portion of these materials may be leached from the plant surface by rain, dew, mist, or fog, a phenomenon noted as early as 1727 by Stephen Hales and 1804 by N. T. de Saussure. Brown (1922) found that the electrical conductivity of water drops on petals increased with time and that this often was directly correlated with increased fungus growth. Many detailed investigations of exudation and leaching were made after radioisotopes came into use in the 1950s.

Exudation and leaching from aerial plant parts involve both active and passive mechanisms, and the materials lost include both waste products and those essential to the plant. Exuded materials include all the common inorganic nutrients (especially potassium, calcium, magnesium, and manganese) as both macro- and microelements, as well as organic materials such as free sugars, pectic substances, sugar alcohols, amino acids, organic acids, growth-promoting substances, vitamins, alkaloids, and phenolic substances (Tukey, *in* Preece and Dickinson, 1971).

Inorganic materials may be leached away in large quantities, e.g., 39 kg of phosphoric acid and 5 kg of calcium oxide from sugar beets per hectare by 18–24 hours of rainfall; 20–30 kg of potassium, 10.5 kg of calcium, and 9 kg of sodium from apple trees per hectare per year; 2–3 kg each of potassium, sodium, and calcium from spruce and pine trees per hectare in 30 days. Those most readily leached are sodium and manganese; those moderately leached are calcium, magnesium, potassium, and strontium; and those difficult to leach are iron, zinc, phosphorus, and chloride. The greatest leakage was of organic substances, especially carbohydrates. Loss of carbohydrate from apple trees may reach 800 kg/ha per year, and up to 6% of the dry weight of young bean plants may be leached in 24 hours (Tukey, *in* Preece and Dickinson, 1971).

Cations for the exchangeable cation pool of leaves move through the cuticle to the surface; little or none is leached directly from within cells or cell walls, but some may move directly from the transpirational stream by diffusion through areas devoid of cuticle or by mass flow through hydathodes or stomata. On the leaf surface, cations are exchanged with hydrogen from the leaching solutions or may combine with carbon dioxide to form carbonates. This accounts for the usual alkaline nature of leachates. The volume of leaching solution has little effect on loss of cations, and dew and dripping fog of long duration may be as effective as brief, heavy rain. The mechanisms of cation exchange and diffusion involved in cation leaching are little affected by changes in temperature, light, and energy levels in the plant. Young, vigorously growing tissues accumulate cations from the exchangeable pool in their cells and cell walls, decreasing cations subject to leaching. In mature tissues, the rate of cellular metabolism and surface leaching are antithetical processes (Tukey, *in* Preece and Dickinson, 1971).

Organic and inorganic compounds leached from aerial plant parts may fall onto the soil or the understory of aerial parts of the same or other plants, where they are metabolized by microorganisms or sometimes absorbed by the plant. A portion of the materials is thus efficiently recycled. This is particularly important for rain forest trees (Broadbent and Baker, 1974a). Materials that do not normally move basipetally in plants (e.g., calcium, strontium, and magnesium) are thus returned to the soil and resorbed by roots.

Plant leachates falling on soil in a rain forest, and perhaps in more temperate areas, may affect soil texture, aeration, fertility, permeability, and base exchange, as well as the soil microbiota. In addition, foliar or root exudates of plants may include substances that have an inhibitory or stimulatory effect on growth of other plants. The inhibitory effect (allelopathy) also may arise from abscised plant parts and volatilized substances.

Exudation from aerial parts, like that from roots, is universal, but some plants lose more than others. Thus, *Ribes nigrum* leaves are more resistant to leaching than are those of apple, pear, plum, and gooseberry; leaves of sugar beet and tomato are more resistant than leaves of wheat or barley (Tukey, *in* Preece and Dickinson, 1971). Leaching of nutrients also occurs from both active and dormant stems and branches, from flower parts, developing seeds, and soft fruit.

In contrast to the exudation from roots, which is greatest from juvenile tissues, loss of materials aboveground increases with the age of the leaves. "Seedling surfaces in general are nutritionally impoverished" (Leben, 1965a). Carbohydrate loss from chrysanthemum and poinsettia increases from initiation of flower buds to full bloom and then decreases during senescence. Blakeman (1972) found a greater quantity of amino acids in leachates from old than from young beet leaves. Tukey (*in* Preece and Dickinson, 1971) reported that "young actively growing [leaf] tissue is relatively immune to loss of mineral nutrients and carbohydrates, whereas . . . tissue approaching senescence is very susceptible to leaching." Loss of potassium from very young leaves may be less than 5% in 24 hours, but as leaves mature the loss may be 80% or more of the total potassium. This is related, at least in part, to the greater resistance to wetting of young than mature leaves. Wetting and drying of leaves increases their wettability and leaching. In addition, leaves injured by microorganisms, insects, adverse climate, nutritional deficiency, wounding, or natural senescence are more apt to lose nutrients than are healthy, vigorous leaves.

Microorganisms on surfaces of aerial plant parts may chemically alter the external layers and thus affect nutrient leakage. Ruinen (1966) found that the common residents, *Aureobasidium pullulans, Cryptococcus laurentii,* and *Rhodotorula glutinis* were lipolytic on leaf surfaces. "In the phyllosphere [phylloplane] the close contact with the cuticle of the organisms makes possible the breakdown of the cutin, and this suggests not only the destruction of already-formed cuticle, but, even worse, the prevention of its formation." *Botrytis cinera* was found by Shishiyama et al (1970) to degrade the cuticle of tomato. Baker et al (1977) found that a common fungus, *Zygophiala jamaicensis* (*Schizothyrium pomi*), was able to destroy the epidermal wax of a wide range of plants. This degradation of surface layers contributes to the greater leakage from old than from young leaves.

Leaf pathogens also affect exudation and phylloplane colonists. The number of yeasts (particularly *Sporobolomyces roseus*) and bacteria on snapdragon leaves increased following infection by *Puccinia antirrhini*, and after the pustules opened, *Erwinia herbicola* increased greatly and became dominant (Collins, 1982). Similarly, the population of phylloplane microorganisms increased on calendula leaves infected with *Entyloma calendulae* (Brady, 1960),

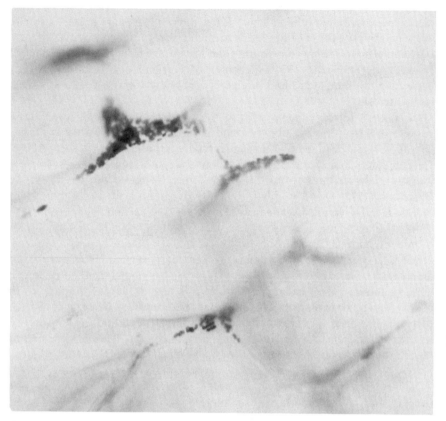

Figure 6.5. Bacteria clustered in depressions along junctures of anticlinal walls of epidermal cells on the surface of a leaf of a cucumber seedling grown under conditions of high humidity. (Reprinted with permission from Leben, 1965b.)

on mint leaves infected with *Puccinia menthae* (Last and Deighton, 1965), and on rose leaves infected with *Diplocarpon rosae* (Hayes, 1982).

Stomata are not the primary pathway of nutrient leakage, as shown by the greater loss from the upper leaf surface, where stomata are fewer, than from the lower leaf surface, where they are more numerous. Leakage on leaf surfaces, as on roots, is greatest in the depressions along the junctures of the anticlinal walls of epidermal cells. Bacteria tend to occur in lines or clusters at these nutrient-rich points (Figure 6.5). Active guttation from hydathodes (Baker et al, 1954; Elleman and Entwistle, 1982), stomata (Bald, 1952; Munnecke and Chandler, 1957) and ectodesmata and trichomes, and secretion from nectaries may also deposit material on the leaf surfaces.

Colonization of Aerial Parts by Microorganisms

Leaves of plants support a characteristic and cosmopolitan flora. At least part of this flora, e.g., members of the bacterial flora and possibly some yeasts, probably are transmitted with seed and other propagative structures. Another significant portion, e.g., members of the dematiaceous fungi, arrive later as airborne spores, probably from fruiting structures on plant residue lying on the soil.

The epiphytic bacterial flora on leaves contain a high proportion of Gram-negative, asporogenous, slime-forming, chromogenic bacteria, the best known of which is the yellow *Erwinia herbicola*. Variants of green-fluorescent *Pseudomonas fluorescens* are widely distributed on plants, and the *P. syringae* group is also common. Certain strains of *E. herbicola* and *P. syringae* on leaf surfaces are ice-nucleation active, and may cause supercooled (-2 to $-5°C$) water there to crystallize, injuring the tissue (Chapter 3). Other genera of bacteria found on leaf surfaces are the Gram-positive *Corynebacterium, Bacillus*, and *Lactobacillus* and the Gram-negative *Aerobacter, Xanthomonas*, and *Flavobacterium* (Blakeman and Fokkema, 1982).

Fungi that commonly occur as epiphytic residents on aerial plant parts include dematiaceous types such as *Epicoccum nigrum, Alternaria* spp., *Cladosporium herbarum*, and others, and the yeastlike fungi such as *Aureobasidium pullulans*, members of the Sporobolomycetaceae (*Sporobolomyces, Tilletiopsis*, and *Itersonilia*) and Cryptococcaceae (*Cryptococcus* and *Torulopsis*), and *Candida* spp. (Blakeman and Fokkema, 1982; Last and Warren, 1972). The suggestion that microorganisms on aerial plant parts tend to be pigmented and thus protected against sunlight is not entirely valid, since many successful epiphytes (yeasts, bacteria, powdery mildew fungi) are hyaline.

A succession of epiphytic microorganisms develops on leaves from the bud to abscission (Dickinson, *in* Dickinson and Preece, 1976) (Figure 6.6). Bacteria are the primary colonizers, then yeasts and yeastlike fungi gradually intrude. Presumably the primary colonizers gain their advantage at least in part because they tend to be seedborne and grow up with the juvenile shoots. Filamentous fungi arrive as airborne spores and become common as the leaf begins to age (Last and Warren, 1972; Ruinen, 1961). This succession also may be due to the ecological interactions of the microorganisms or to aging of the leaves and attendant changes in leaf exudates. Some saprophytes may occupy substomatal chambers as endophytes of healthy leaves and rapidly take over after leaf fall, or perhaps just before leaf fall and therefore before the leaf becomes available to saprophytes in the soil. Dickinson (1981) suggests that these leaf colonists may actually accelerate leaf senescence and thereby contribute to crop loss.

After leaf fall, saprophytic fungi that use sugars, pentosans, and possibly hemicelluloses (e.g., the Mucorales) colonize leaves, followed later by cellulose-

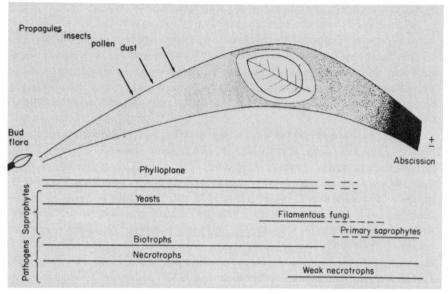

Figure 6.6. Diagrammatic illustration of microbiological activities during the life and death of a leaf. (Reprinted from Dickinson and Preece, 1976, with permission from Microbiology of Aerial Plant Surfaces. Copyright: Academic Press Inc. (London) Ltd.)

decomposing Ascomycetes and Fungi Imperfecti, and finally by the lignin-decomposing Hymenomycetes and Gasteromycetes.

Spread of microorganisms from aerial parts of one plant to those of another may be accomplished by splashing water, by mites and insects, and by airborne spores. The Sporobolomycetaceae forcibly discharge their spores, but the spores of other resident fungi are merely moved by air currents. Gravitational fall is another way by which microorganisms reach the surface of aerial plant parts. The primary colonists may migrate from the seed to the tops by swimming on the plant surface, by continued division on the shoot apex (thus being carried upward as the shoot elongates), by insects and mites, and perhaps by being carried upward inside the plant (Leben, 1965a). Khudyakov (*in* Krasil'nikov, 1958) showed that epiphytic microorganisms migrated from the seed onto shoots and roots of wheat plants in the field. He concluded that epiphytic bacteria on plants were a special ecological group, and he thought they were able to live on roots as well as on aerial parts. The extensive Russian literature on spread of epiphytic bacteria is reviewed by Last and Deighton (1965) and by Leben (1965a). Leben (1963, 1964, 1965b) also found that bacteria migrated from seed of cucumber and tomato onto aerial parts, where they became rather uniformly distributed over the surface. If environmental conditions are suitable for survival and multiplication, it may be as easy to establish an an-

tagonist on leaves as on roots, and the rules of the game probably are
similar.

Debris as a Source of Nutrients on Aerial Parts of Plants

A very important source of nutrients for epiphytic microorganisms (and for
foliar pathogens) is pollen, flower parts, and other debris that fall on aerial
plant parts. One of the earliest observations was that of Newhook (1957),
who noted that establishment of *Cladosporium herbarum* in dead tomato
petals in advance of *Botrytis cinerea* greatly reduced infection by *B. cinerea.*
Warren (*in* Dickinson and Preece, 1976) showed that *Agrostis* pollen on
birch leaves increased the population of yeasts, *Aureobasidium pullulans*, and
Cladosporium spp. from 5.1×10^3 to 260×10^3 per gram of leaf tissue; on linden
leaves the figures were 100×10^3 and 430×10^3. Fokkema (1968, 1971) showed
that the increased number of pollen grains on rye leaves after flowering resulted
in a two- or threefold increase in the number of colonies of *C. herbarum* and
a marked increase in infections by *Helminthosporium sativum* and *Septoria
nodorum* but had no effect on the number of infections by *Puccinia recondita*
f. sp. *recondita. Helminthosporium* and *Septoria* exemplify pathogens that
produce extensive superficial mycelium over the leaves of their host, and this
prepenetration growth was increased by energy provided by pollen, leading
to increased infections. Mycelial growth was about 4,000 μm/mm² in 24
hours on leaf surfaces with pollen, but only 100 μm/mm² on leaves without
pollen. Mycelial density and necrotic area of leaves were positively correlated
($r = 0.85$). Because the rust spores quickly penetrated directly and without
nutrients from surface debris, their infection was little affected by the pollen
and the resultant increase in phylloplane microorganisms. The addition of
phylloplane fungi (*Aureobasidium* and *Cladosporium*) to the plant surface
preempted the energy reserves of pollen grains otherwise used by *Helminthosporium* and *Septoria* and in this way resulted in biological control. Again,
there was no evidence of antagonism other than competition. Saprophytes on
the plant surface seem to act as scavengers, consuming the nutrients of the
pollen.

Warren (1972) found that pollen shed on beet leaves favored increased growth
of saprophytic yeasts, *Cladosporium* sp., and *Aureobasidium pullulans*. Inoculation with spores of *Phoma betae* onto beet plants laden with natural
pollen gave 3–5% infection, and inoculation with spores of *P. betae* plus
additional (rye) pollen gave only 5% infection. However, on plants initially
free of pollen, inoculation with rye pollen plus *P. betae* gave 88% infection. It appears that the normal accumulation of pollen on leaves promoted
a microflora competitive with pathogens, and the addition of more pollen
into this established ecosystem (where the saprophytes capable of growth on

pollen already were present in high numbers) was of little advantage to the pathogen. In contrast, fresh pollen applied simultaneously with spores of the pathogen to pollen-free plants was to the competitive advantage of the pathogen, since saprophytes with ability to compete for the pollen had no head-start advantage. The results of this experiment illustrate a common situation: if a pathogen and a source of nutrients are introduced together into an environment (e.g., soil or the phylloplane), where that particular nutrient source (e.g., pollen) has repeatedly been made available, microorganisms capable of competing for that nutrient source are in an even higher population and even more poised for growth than in sites where the nutrient source had not recently been made available; the pathogen is limited accordingly.

Fokkema and Lorbeer (1974) found that the growth of germ tubes of *Alternaria porri* on onion leaves was reduced by *Aureobasidium pullulans* and *Sporobolomyces roseus* from 1,900 μm per spore to 500 and 600 μm per spore, respectively, and the number of lesions was reduced by 50%. Fokkema and Van der Meulen (1976) showed further that both the superficial growth of *Septoria nodorum* on wheat leaves and leaf necrosis caused by this pathogen were reduced to less than half by *A. pullulans, S. roseus*, and *Cryptococcus laurentii* without pollen. In the presence of rye pollen, neither mycelial growth nor necrosis was reduced by the saprophytes. *Puccinia recondita* f. sp. *recondita* and *Botrytis squamosa* penetrate leaves rapidly without extensive mycelial growth and are thus less affected by epiphytic antagonists competing for the nutrients (Fokkema and Lorbeer, 1974).

Protection of Leaves and Flowers by the Phylloplane Microbiota

As in the rhizosphere, the epiphytic and endophytic microflora of the leaf phylloplane provides important biological buffering against colonization of aerial parts by plant pathogens (Blakeman, 1981; Blakeman and Fokkema, 1982). The effect is particularly important for suppression of leaf pathogens dependent on nutrients on the phylloplane to support their own prepenetration growth (Blakeman, 1978). Such pathogens include *Botrytis cinerea, Cochliobolus sativus, Alternaria alternata*, *Septoria* spp., and others that have a necrotrophic type of parasitic relationship with the host and for which the amount of penetration and leaf damage is related to their total biomass and the energy available for growth in the infection court. **Any preemption of a critical nutrient by nonpathogenic members of the phylloplane flora will probably be reflected in less growth by the necrotrophic parasites and hence less damage to the leaf.**

Blakeman and Fokkema (1982) point out that with few exceptions, the "pathogens . . . sensitive to antagonism normally exhibit some superficial growth over the leaf surface from which penetration of the leaf takes place." Spurr (1979) showed that tobacco leaves immersed for 30 seconds in 70% ethanol to remove the microflora and then rinsed with deionized water developed twice as many leaf spots upon inoculation with *A. alternata*. This agrees generally with observations that application of the cosmopolitan pink and white yeasts, together with *Aureobasidium pullulans, Cladosporium* spp., and others to leaves under glasshouse conditions (where the natural populations of these phyllosphere colonists may be exceptionally low) has resulted in 50% less infection of cereal leaves by *Cochliobolus sativus* and *Septoria nodorum* (Fokkema and DeNooij, 1981) and 50% less infection of onion leaves by *Alternaria porri* (Fokkema and Lorbeer, 1974).

Spores and germlings of rust fungi are subject to inhibition on the leaf surface, but mechanisms other than direct competition for nutrients apparently are involved. McBride (1969) obtained protection of Douglas-fir seedlings against infection by *Melampsora medusae* by applying cells of *Bacillus cereus, B. mycoides*, or an unidentified *Bacillus* sp. under glasshouse conditions. These bacteria were obtained initially from healthy foliage of Douglas-fir. Protection was proportional to the population of bacteria on the seedling needles, but antibiosis was suggested as a possible mechanism because cell-free culture filtrates also had some effect. In another study, *B. cereus* as a cell suspension applied to leaves of leek in a glasshouse study prevented infection by *Puccinia alli* applied 24 hours later (Doherty and Preece, 1978). Spores of the rust fungus did not germinate in the presence of *B. cereus*, even if separated from the bacterium by a cellophane membrane, again suggesting active inhibition of the rust spores. Either *Alternaria* or *Cladosporium* spp. applied to leaves of poplar suppressed spore germination and infection by *Melampsora larici-populina* by penetrating the spores from appressoria and causing the spores and germ tubes to lyse (Omar and Heather, 1979). Spores of these biotrophic parasites germinate without an external supply of nutrients, and thus it is not surprising that the examples of their inhibition on the phylloplane involve probable antibiosis and hyperparasitism but not competition.

Another important form of antagonism of rust and mildew fungi occurs after infection, when certain fungi, e.g., *Sphaerellopsis (Darluca) filum* and *Ampelomyces (Cicinnobolus) quisqualis*, specialized hyperparasites of rust and mildew colonies, suppress inoculum production by these plant parasites (Chapters 4 and 5).

Apparently, a relatively small percentage of epiphytic bacteria produce antibiotics effective against fungal and bacterial pathogens. Of 230 isolates of resident epiphytic bacteria from cucumber leaves, only one was antagonistic to *Colletotrichum lagenarium* (Leben, 1964). Similarly, of 358 isolates of epiphytic bacteria from soybean buds, only four were highly antagonistic to *P. syringae* pv. *glycinea*. On the other hand, *Bacillus* spp. with ability to produce broad-spectrum antibiotics are commonly resident on leaves of many plants, including *B. mycoides* on tobacco leaves inhibitory to *Alternaria alternata* (Spurr, *in* Papavizas, 1981), several *Bacillus* species on the needles of Douglas-fir inhibitory to *Melampsora medusae* (McBride, 1969), and *B. subtilis* on foliage and branches of apple trees inhibitory to *Nectria galligena* (Swinburne, 1973). Certain phylloplane fungi also produce antibiotics (Blakeman and Fokkema, 1982), and their potential to provide biological control should be investigated. Nevertheless, because conditions suitable for growth of any given species or strain of microorganism on most leaves usually is brief, selection on the phylloplane probably has been mostly for ability to respond rapidly over a wide range of physical environments and to quickly convert any readily available nutrients into new propagules. Furthermore, antibiotic production would require suitable sources of carbon and energy. Perhaps the ideal antagonist for introduction onto the phylloplane would be a microorganism selected first as an aggressive and competitive phylloplane colonist but also genetically engineered to produce a desired antibiotic in environments having a suitable energy supply.

The natural protection of leaves provided against pathogens by the buffering capacity of phylloplane microorganisms is clearly demonstrated by cases where upsetting the flora by a chemical spray leads to increased disease (Griffiths, 1981). For example, increased damage from *Botrytis cinerea* on cyclamen plants sprayed with benomyl to control *B. cinerea* was shown by Bollen (1971) to result from a benomyl-resistant strain of the pathogen operating with diminished competition. The benomyl was inhibitory to *Penicillium* spp. that otherwise served as competitors of *B. cinerea* for pollen and other nutrient sources on the leaf surface. Introduction of a benomyl-resistant strain of *Penicillium* restored the balance so that disease severity was again about the same as when no benomyl was applied. More severe outbreaks of *Colletotrichum coffeanum* apparently resulted from use of copper fungicides to control the coffee berry disease, perhaps by increasing spore production after the fungicide no longer persisted (Furtado, 1969). Carter (1971) suggested that fall application of a copper fungicide to control *Clasterosporium carpophilum* on apricot in South Australia intensified dieback (caused by *Eutypa armeniacae*) because it diminished antagonists on the plant surface. Fokkema et al (1975) showed in a two-year study that spraying rye leaves with benomyl resulted in 60% more necrosis from *Cochliobolus sativus* than did spraying with water; the natural microflora of water-sprayed leaves averaged 6,500 propagules per square centimeter and that of benomyl-sprayed leaves averaged only 800. In another study, populations of sensitive yeasts and other fungi were lower for about three weeks on wheat leaves sprayed with either benomyl or difolitan compared with populations on leaves sprayed with water only (Fokkema, *in* Blakeman, 1981). Dimock and Baker (1951) showed that

spraying snapdragon plants with Bordeaux mixture in a humid climate controlled the antagonist *Fusarium roseum*, which infected pustules of *Puccinia
antirrhini*, but had little effect on the rust, thus increasing the severity of
rust.

Chemicals may also be used to increase the epiphytic microflora and augment biological control. Burchill and Cook (*in* Preece and Dickinson, 1971)
showed that spraying trees with 5% urea solution just before leaf fall increased the population of *Cladosporium, Alternaria, Fusarium*, and bacteria
and decreased development of perithecia of *Venturia inaequalis* on fallen
leaves. A pioneer colonist such as *V. inaequalis* usually would have an advantage over secondaries in possession of the tissues, but apparently the urea
served to enhance the competitive advantage of the epiphytic and endophytic
leaf microflora relative to that of *V. inaequalis*. The urea may also have
enhanced decomposition of leaves, affected perithecial development directly,
enhanced microbial antagonism, or increased ingestion of the leaf material
by earthworms — biological control in any case. The leaf microflora shifted
from Gram-positive chromogenic bacteria to Gram-negative nonchromogenic
types.

Most work with potential antagonists applied to leaves has not been extended to the field. Such studies are nevertheless significant because they
show overwhelmingly that biological buffering on the phylloplane is important. Moreover, some progress is being made, and undoubtedly the future will bring an accelerated effort toward greater use of introduced antagonists as prophylactics or phyllosanitizing microorganisms under field conditions.

Spurr (*in* Papapavizas, 1981) used a nonpathogenic isolate of *Alternaria* to
protect tobacco leaves against *A. alternata* under field conditions in North
Carolina; the antagonist probably requires many of the same nutrients for
saprophytic growth as required by the pathogen. Surprisingly, a formulation of *Bacillus thuringiensis*, normally used to control insects, was also
effective against the leaf spot of tobacco caused by *A. alternata*. Tobacco
leaves, being very broad, probably tend to develop an unusually thick boundary layer (unstirred humid air) that favors growth by many bacteria and
could help account for the success of *B. thuringiensis* as an antagonist on
leaves of this crop. Lindow (1979; 1983) obtained some frost protection
of potato leaves against ice-nucleation active (INA) strains of *Erwinia herbicola* and *Pseudomonas syringae* by introduction of an ice-nucleation inactive strain of *E. herbicola* adapted to the niche occupied by INA strains.
The protection apparently resulted from competition and displacement of the
INA strains by the introduced strain and not from inhibition by antibiosis.
Some strains that protected against frost damage were shown to produce
antibiotics inhibitory to INA strains in vitro, but mutant strains negative
for antibiotic production in vitro were nevertheless effective in vivo. Riggle
and Klos (1972) demonstrated that *E. herbicola*, if established first in pear

blossoms, will inhibit the fire blight bacterium *E. amylovora* (Chapter 4), and Beer et al (1980) obtained some success with this and other strains of *Erwinia* against fire blight of apple in New York. *Erwinia herbicola* has also been shown to limit disease caused by *Xanthomonas campestris* pv. *hederae* on ivy and *P. syringae* pv. *phaseolicola* on bean. A white *Erwinia*-like bacterium from cherry leaves, when coinoculated with *P. syringae* pv. *morsprunorum* into leaf scars on cherry, reduced the incidence of infection and the size of resulting cankers (Crosse, *in* Preece and Dickinson, 1971).

The potential clearly exists for manipulation of the phylloplane microflora, but like microorganisms used for biocontrol in other environments, the antagonist obviously must be adapted to the same or similar niche and/or use the same nutrients as the pathogen, and it must have strong attributes as a competitor, producer of antibiotic, or hyperparasite, or have some combination of these traits.

COMPARTMENTALIZATION AND BIOLOGICAL CONTROL OF DECAY IN TREES

The ability of trees to resist decay, a specialized example of the inhibition of pathogens by the host, has been discussed in detail and beautifully illustrated by Shigo and Marx (1977), Shigo (1979, 1982), and Shortle (1979). Wood decay, a natural and necessary process in the overview, starts in a living tree with a wound caused by biotic or abiotic agents. Pioneer invaders (anaerobic, aerobic, and facultative bacteria and nondecay fungi) infect the wound and discolor the wood, to be followed by hymenomycetous decay fungi. Trees do not replace or repair organs but may shed diseased branches or woody roots. The host strategy is to restrict and enclose the pathogen by compartmentalization, in which "a tree gives up a small portion of itself to save the larger whole."

Tree trunks and large branches are highly compartmented by normal tissue structures (Figure 6.7). The ray cells (side wall 3 in Figure 6.7) are discontinuous vertically and radially, and the continuous innermost small cells (produced in fall) in each annual growth ring (inner wall 2) form two-dimensional compartments in nonwounded trees, whether they form heartwood (oaks) or not (maple). The tree reacts to wounding by plugging the vertical vascular system (by tyloses, gum deposits, and pit closures) above and below the wound (vertical wall 1), thus forming a three-dimensional compartment. The cambium then also begins to form a new protective wall separating tissue present at the time of wounding from that formed afterward (wall 4).

The vertical walls (wall 1) are the weakest in defense against decay fungi, and the innermost vertical columns are the weakest of all. Wall 1 resists

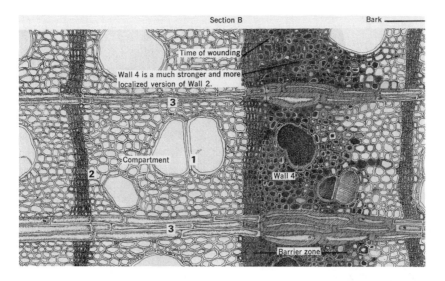

Figure 6.7. Drawing of a section of wood illustrating walls 1 (weakest wall), 2 (inner wall), 3 (ray wall), and 4 (barrier zone and strongest wall) that form in the wood as part of the process of compartmentalization in response to injury. (Reprinted with permission from Shigo and Marx, 1977.)

vertical spread of the pathogen, which may or may not be in the heartwood, depending on the pathogen. The inner walls (wall 2) are the next strongest defense line; they resist inward spread of the pathogen. The ray cell walls (wall 3) are the strongest barrier in the nonwounded tree because of their chemical and anatomical structure; they resist lateral spread of the pathogen. The nonconducting barrier cells (wall 4) are formed by the cambium following injury; they are axial parenchyma cells (plus resin ducts in conifers) in a localized, structurally weak, but protectively strong (because of phenols and resins) defense involving walls 2 and 3. Wall 4, defensively the strongest of the four walls because of its chemical nature, confines the decay to wood present at the time of wounding. As a result of this compartmentalization, each column of decay from each wound is separated from the others. However, the processes require energy, and loss of energy storage in sapwood from repeated wounding and decay weakens the tree and may eventually so decrease wound responses that the tree succumbs. A similar barrier zone also forms in response to vascular pathogens and some leaf-spot and root-rotting fungi (see below).

The ability to activate walls 1, 2, and 3 varies greatly from tree to tree; there is evidence that these features are genetic and that decay-resistant trees may be bred for urban plantings. Certain arboricultural practices complement this natural compartmentalization: dead or injured branches are pruned at the outer side of the branch-bark ridge, not flush cut; barrier wall 4 is never breached by bore holes or by cleaning out decayed wood; and wound dressings are not expected to provide real protection of wounds.

Microorganisms (mostly fungi) antagonistic to wood decay fungi and present as natural inhabitants of bark and wood may complement the natural compartmentalization process. These antagonists are especially beneficial when in fresh wounds in advance of decay fungi; through their competitive effects or perhaps by hyphal interference (Chapters 3 and 5), they provide at least some protection to the tree during the critical period when the wound is healing and compartmentalization is still in the early stages. The direct application of *Fusarium lateritium* to pruning wounds of apricot trees (Carter, 1971) and *Trichoderma* spp. to pruning wounds of plum (Grosclaude et al, 1973) and maple (Pottle et al, 1977) are examples of antagonists introduced to protect wounds. Even the infection of wounds with antagonistic microorganisms such as *Trichoderma* spp. late in the microbial succession will inhibit the decay process, although decay is not stopped. Gibbs (1980) has presented evidence that infection of oak trees in Minnesota by the oak-wilt fungus *Ceratocystis fagacearum* is limited naturally by *C. piceae*, which is adapted to growth in the fresh wood that *C. fagacearum* infects. Wounds on oak are susceptible to infection by *C. fagacearum* for only a few days, after which natural healing is sufficient to limit infection.

Antagonistic fungi have also been shown to protect against heart rot such as may occur as the tree ages or where natural compartmentalization is inadequate to limit the decay. Basham (1973) observed that *Ascocoryne sarcoides* is a widespread colonist of the heartwood of black spruce trees 75 years and older in Ontario, Canada, and that, although causing a negligible color change in the heartwood itself, it greatly limits invasion of the wood and subsequent decay by *Fomes pini*. The frequent and widespread occurrence of *A. sarcoides* in heartwood of mature black spruce was offered as a reason "why black spruce is so much less defective than all other tree species in Ontario." Basham (1973) proposed that *A. sarcoides* be established in young black spruce trees to protect the trees against rot as they age. The fungus was considered an ideal agent of biological control of heart rot because it can confer resistance to decay in the trunk without harming commercial value of the wood.

THE ACTIVE ROLE OF THE HOST IN BIOLOGICAL CONTROL

An understanding of recognition and rejection (or direct inhibition) phenomena in plant-microorganism interactions is important to the study of biological control because it shows how plants actively defend themselves, how resistance responses may be induced, how plants recognize and accept nonpathogens having the potential for cross protection against plant pathogens, and how nonpathogens may be established in leaves or roots to prevent establishment by pathogens. Reviews on this subject, particularly as it pertains to the successful establishment of infections by microorganisms on plants are available in Dazzo (*in* Bitton and Marshall, 1980), Heitefuss and Williams (1976), Horsfall and Cowling (1980), Schmidt (1979), and Vanderplank (1982).

Cell–Cell Communication Between Plants and Microorganisms

Communication, which may ultimately determine compatibility or incompatibility between a plant and a potential pathogen, begins when the cells of the two organisms make physical contact (Sequeira, *in* Horsfall and Cowling, 1980). Phytopathogenic bacteria penetrate plant tissues through wounds, natural openings, or by dissolution of the middle lamella but are generally unable to penetrate the cellulosic walls of plant cells (Kelman, *in* Horsfall and Cowling, 1979). With few exceptions, cell-cell contact between a plant and a pathogenic bacterium is therefore wall-wall contact. Cell-cell contact between a plant and a parasitic fungus of the biotrophic type is commonly fungus wall-host plasmalemma contact. With most parasitic fungi of the necrotrophic type, including weak parasites that depend for pathogenesis on stress of the host, the contact is generally wall-wall until dissolution of the host tissue begins. For saprophytic fungi and bacteria having ability to establish as endophytes in plant tissue, the contact probably remains wall-wall until the plant tissue dies from some other cause, whereupon the saprophyte may begin to digest the dead tissues.

Communication or "recognition" between the contacting surfaces is determined by the informational potential contained in them (Sequeira, *in* Horsfall and Cowling, 1980). This informational potential is thought to reside in the carbohydrates or carbohydrate-containing materials on the respective surfaces, i.e., polysaccharides, lipopolysaccharides, glycoproteins (lectins), glucans, and others. Recognition is determined by the degree of matching or binding between the molecules on one cell surface and those on the other cell surface. Sequeira (*in* Horsfall and Cowling, 1980) states: " . . . surface polysaccharides exhibit a wide array of potentially useful sites for recognition phenomena. In the course of evolution, many organisms have developed mechanisms to recognize each other on the basis of similarities or dissimilarities of surface carbohydrates." The enormous informational storage capacity of these complex molecules is subject to genetic control, and the matching ability of surface-surface carbohydrate molecules or their residues might even be the basis for the gene-for-gene diseases, where each gene conditioning resistance or susceptibility in the host is matched by a complementary gene conditioning avirulence or virulence in the pathogen.

Cells of *Pseudomonas solanacearum* become attached to cells in tobacco by way of a lipopolysaccharide in the bacterial wall and a specific lectin in the plant cell wall (Sequeira and Graham, 1977). Attachment of rhizobia to the root hairs of their legume host is also mediated by host lectins interacting with surface carbohydrates on the *Rhizobium* cell (possibly the extracellular polysaccaride slime), although some data contrary to this explanation have been produced (Schmidt, 1979). Attachment of cells of *Agrobacterium radiobacter* pv. *tumefaciens* to walls of its host is thought to result from binding between lipopolysaccharide in the bacterial cell wall and a specific polysaccharide contained within the pectic substance of the host cell wall (Lippincott and Lippincott, 1969; Lippincott et al, 1977), although recent studies revealed that recognition and attachment occur between host and pathogen cells even if the host cells are wall-less protoplasts (Matthysse et al, 1982). Attachment sets the stage for a more intimate cell-cell interchange that will further decide compatibility. In the case of *A. radiobacter* pv. *tumefaciens*, attachment is a necessary prerequisite to virulence, and the genes for specificity of attachment in the bacterium are carried on the plasmid genome responsible for tumor formation (Whatley et al, 1978).

The possibility exists for recognition between roots of a plant species and root-colonizing nonpathogenic microorganisms. Certain nonpathogenic root colonists tend to occur on roots of the same plant species, suggesting more than a casual relationship. Atkinson et al (*in* Bruehl, 1975) showed further that, on wheat roots, the kinds or groups of microorganisms present are under genetic control of the plant. This occurrence of specific types on the same roots could also be explained by selective stimulation of the microorganisms by nutrients released in exudates. Adhesion or attachment to the root or leaf surface by saprophytic bacteria might not be a limiting factor because bacteria have many mechanisms for attachment even to inert surfaces (Bitton and Marshall, 1980).

Cell Membrane Integrity as a Factor in Success of the Host-Pathogen Interaction

Active resistance mechanisms in plants against invading microorganisms are expressed in many ways, illustrated by the following cases: 1) the pathogen may cease growth or multiplication shortly after penetration, producing no visible symptoms in the plant tissue; 2) a few host cells may die rapidly and exhibit a tanning reaction, typical of the hypersensitive response; or 3) the pathogen may progress for some distance through the tissue before becoming inhibited or even sealed off (compartmentalized) by the host response. In the latter case, although the lesion may be sharply defined (or even jettisoned, as in the case of a shot-hole syndrome), the damage can still be considerable and the reaction is classified as susceptible even though, biochemically or biophysically, a resistance response has occurred. In all cases, the important distinguishing feature is that growth of the pathogen is inhibited, in some cases so quickly that few or no symptoms develop (case 1 above). In other host-pathogen combinations, the host responds more slowly, and lesions of characteristic size develop. In many cases, the inhibition is only partially effective, retarding but not totally preventing growth of the pathogen; in these situations, environmental conditions relatively more favorable to the host than to the pathogen can tip the balance toward more effective inhibition and hence less disease, or those relatively more favorable to the pathogen than to the host can tip the balance more toward disease. In addition, attempted infection by an avirulent strain of a pathogen, or by the pathogen of another plant, can sensitize the plant so that it then responds more rapidly to a pathogen (Kuć, 1982) and is effectively more resistant to the pathogen (induced resistance). Obviously the reason(s) for the growth inhibition of the would-be pathogen inside the tissue of the plant is fundamental to the subject of the active role of the host in biological control.

   We propose that loss of membrane integrity of cells of the host, pathogen, or both is a key factor in the success or failure of a host-pathogen interaction. Consider first what happens to host cell membranes during a host-pathogen interaction. In the so-called compatible host-pathogen interaction, the host cell membrane becomes leaky, and this is followed quickly by a rise in uptake of oxygen by the host cell (Chapter 5). At least part of this increased workload, reflected as increased respiration, presumably results from increased active transport — the plant cell working to keep solutes in, while the tendency is for them to leak out. However, the so-called incompatible host-pathogen interaction, whether involving a fungus, bacterium, or virus, apparently is characterized by an even greater but also more temporary rise in oxygen uptake than is the compatible interaction. Presumably damage to the host cell membrane is even more acute with the incompatible interaction, perhaps so acute that counteraction by host repair and maintenance mechanisms is impossible and death comes quickly. An important question is whether the reverse might also occur, i.e., whether owing to the interchange of materials between the host and pathogen, the pathogen cell becomes leaky and loses turgor. A fungus hypha without turgor would no longer extend (Robertson, 1958, 1959). A bacterial cell without turgor would no longer divide (Harris, in Parr et al, 1981).

   Hypothetically, an ability of the host to modify selective permeability of cell membranes of the pathogen, and the converse, could have any of four effects.

1.   The change in membrane permeability may be nonexistent or insignificant for both host and colonist. The colonist would thus be left to grow on substances available naturally in the apoplast and intercellular spaces of the host, but it would be ideally positioned for rapid proliferation following

any damage to the host (chilling injury, drought stress, herbicide damage, etc.). Such organisms could be endophytes, weak parasites, opportunistic necrotrophs, or strict saprophytes.

2. A change in membrane permeability of the host but not of the colonist may occur. The degree to which host cells (the symplast) give up their contents and become moribund, as balanced against the degree to which they keep their contents and maintain cell integrity and function, would decide the rate and makeup of the food supply for the colonist and hence its rate of growth (pathogenicity) through the tissue (Hancock and Huisman, 1981). This would exemplify the familar susceptible interaction. A situation in which the pathogen was a necrotroph and the host cells supplied abundant food would give the appearance of a highly aggressive pathogen or very susceptible host. The converse, in which the host provided nutrients for the pathogen at a very slow rate, would appear as a nonaggressive pathogen or a moderately resistant host.

3. The membranes of both host and colonist may become totally disrupted, resulting in loss of turgor and possibly death of the cells of both parties. This situation could appear as a hypersensitive response. Capacity to avoid or trigger this situation would be genetically controlled by both host and pathogen and could be specific. Virulent strains of bacteria might avoid or delay such an effect or obtain time for adjustments by their extrapolysaccharide slime coating, which could serve as temporary insulation against physical contact with the host cell. Novacky et al (1973), showed that tobacco leaf tissues could be conditioned so as not to react hypersensitively to live cells of *Pseudomonas pisi*, if exposed first to low inoculum densities of the same bacterium.

4. Membrane permeability of the colonist but not of the host may be altered. This would cause the invader to cease growth while producing no outward symptom, rapid cell death, or other evidence of a host response.

Host Defense by Physical or Chemical Containment of the Pathogen

Some host defense reactions lead to a physical or chemical containment of the pathogen (Beckman, *in* Horsfall and Cowling, 1980) such as the compartmentalization process described for trees in the previous section. The barrier produced may not be totally impenetrable, but more work is required of the pathogen during a period when its energy supply is dwindling and competition from secondary invaders is increasing. Containment can therefore be quite effective. Beckman (*in* Horsfall and Cowling, 1980) describes the effect of containment as "dedifferentiation" and "redifferentiation" of tissues. He gives the examples of the darkened "pits" caused by *Rhizoctonia solani* in potato tubers and the shot-hole disease of *Prunus* spp. Tissues around the infected area, whether storage, palisade, or spongy parenchyma, must first dedifferentiate, then become meristematic (including resumption of cell division and growth), and finally redifferentiate into an abscission layer that forms outside the zone occupied by the pathogen. The outer layers of cells may become infused with phenolics and also may become lignified, thereby presenting an even more formidable barrier to the colonist.

The vascular occlusion response associated with resistance of plants to vascular parasites exemplifies physical containment (Beckman, *in* Horsfall and Cowling 1980; MacHardy and Beckman, *in* Nelson et al, 1981). Microbial cells entering the xylem vessels are sieved free of the vascular fluid by the perforated end walls that occur in the vessels every 2–5 cm in herbacious plants and at 30-cm intervals in banana stems. *Fusarium oxysporum*, trapped on the underside of a perforated wall, grows through the wall within 24 hours. By 72 hours the fungus has formed microconidia above the first end wall and has now become trapped at the next end wall. However, if the variety is resistant, the end wall (which actually consists of primary walls of the connecting vessels, separated by a middle lamella) have already begun to swell around the invading hypha during the first one to two days after fungus-host contact. In addition, materials and the pit membranes in the lateral walls of the vessel swell inward until the entire vessel lumen is plugged for some distance above the trapping site. A protective layer of material is then laid down between the pit membranes in the adjoining parenchyma cells. A process of dedifferentiation of xylem parenchyma then occurs, whereby cells some distance above the trapping site begin to grow much like periderm, producing balloonlike extensions through the pits and into the vessel lumen. These tyloses enlarge, fuse, and totally block the vessel. Finally, a phenolic material becomes infused and then polymerized with the occluding material. The parenchyma cells adjoining the xylem are thus important contributors to host defense against vascular pathogens.

The ability of the host to wall off a pathogen by formation of an abcission layer, deposit of callose, or production of occlusions and protection layers goes beyond cell-cell communication and compatibility-incompatibility phenomena. The response may begin because of a signal generated through cell-cell contact, but the communication is obviously at the level of tissue, organ, and even the whole plant. Such long-distance communication involves hormones. Beckman (*in* Horsfall and Cowling, 1980) describes indole acetic acid as traveling "through the tissues for some distance, signaling the intensity of stress, and possibly the direction of stress." The result is that previously quiescent tissues, where cells were in a state of low and mainly catabolic metabolism, are now largely reactivated into a state of reductive anabolism including formation of new cells. This complex series of events is similar to what happens following injury (wound response) or following the breaking of dormancy. The net result is containment of the pathogen and a return of the tissue to the more quiescent state.

CROSS PROTECTION AND INDUCED RESISTANCE

Hamilton (*in* Horsfall and Cowling, 1980) classified the limiting effects of one virus (inducer) on another virus (challenger) into four groups: 1) the challenger produces fewer lesions; 2) the challenger produces smaller lesions; 3) multiplication of the challenger is restricted; 4) symptoms of the challenger are delayed or totally prevented. These four types of limitations also describe the possibilities in cross protection involving bacterium-bacterium, fungus-fungus, nematode-nematode, and mixed combinations of pathogens. To consolidate Hamilton's categories, the pathogen is less active or inactive in tissues following their modification by a prior colonist, as evidenced by slower, less or no growth (or multiplication), and less or no damage to the tissues compared with what happens in tissue not so modified.

The mechanisms of cross protection and induced resistance undoubtedly are different for the different categories of pathogens, vis-a-vis viruses, viroids, bacteria, fungi, and nematodes. Indeed, it should not be surprising if the biophysical and biochemical events leading to limitation of the challenger were unique to each example. For recent and very useful reviews on the state of knowledge on this subject, the reader is referred to several chapters in Horsfall and Cowling (1980), including those of Hamilton for viruses, Goodman for bacteria, Suzuki for fungi, and McIntyre for nematodes. In this section we take a more ecophysiological approach as a framework for better understanding of the subject, and perhaps as a means to point the way for future work and eventual use in biological control.

Direct Antagonism of the Pathogen by a Nonpathogen

Earlier in this chapter, and in Chapter 5, pathogens were portrayed as having the same basic requirements while in host tissue that any microorganism has while in its ecological niche—the physical-chemical environment must not be limiting to growth, there must be a source of nutrients and energy, and the organism must be able to osmoregulate and maintain turgor. The physical-chemical environment includes the availability of space, the suitability of temperature and pH, and the presence or absence of chemical inhibitors. The means to osmoregulate and maintain cell turgor obviously does not apply to viruses but is important to the host and to all microorganisms with semipermeable membranes. The nutritional needs of the unspecialized and often necrotrophic type of parasite are mainly quantitative; these organisms can live on a wide variety of substrates and tend to be limited more by the total supply of available carbon and energy. In contrast, the nutritional needs of the more specialized biotrophs (e.g., rust and mildew fungi) are more qualitative; total carbon and energy are important but special growth factors and probably even metabolic intermediaries may be needed as well. For viruses, a supply of nucleic and amino acids from the host is required for synthesis of DNA or RNA and the protein capsid. The question is: How much of cross protection is direct antagonism, i.e., competition and/or direct modification of the physical-chemical environment of the plant tissue by the first colonist, and how much is indirect, i.e., induction of an active host defense system?

The phenomenon whereby a would-be aggressive colonist is less active or even totally unsuccessful in a niche already occupied or modified by another colonist is the rule in biology. With plant pathogens, the effect is manifest whether the substrate is dead or alive. Thus, *Pythium mamillatum* was a highly successful colonist of wood blocks, provided the blocks were not already colonized by another fungus (Barton, *in* Parkinson and Waid, 1960). Pioneer colonists of the wood (or wheat straw) "protect" these substrates against challengers. Since the reciprocal occurs, i.e., the challenger becomes the protector if first in, we can describe the effect as equivalent to "cross protection." The substrate in this case, being dead, plays no active role. *Antagonism can occur wherever adversaries meet — inside as well as outside the plant.* The antagonism could even be quite subtle, involving no more than a deficiency of one or more key nutrients, perhaps a mild staling effect generated by the previous invader, or interference with ability of the challenger to maintain cell turgor.

Antagonism between two saprophytes in or on a dead substrate is likely to occur if both are similarly adapted to the physical-chemical environment of the substrate and if both use the same nutrients (Clark, *in* Baker and Snyder, 1965). Thus, two fungi that use cellulose are more likely to interact antagonistically in a block of wood or wheat straw than are two fungi of which one uses cellulose and the other lignin. Similarly, in living tissues, two strains of the same pathogen or two organisms pathogenic in the same tissue are most likely to interact antagonistically, because they occupy the same niche both spatially and in time and use the same nutrients. Related strains of viruses are likely to use the same sites or routes of protein synthesis in the cell and also to require a similar quantity and quality of nucleic and amino acids. Such strains are also the most likely to produce a reciprocally antagonistic (cross protection) effect.

Inoculation of a plant susceptible to crown gall with avirulent or even heat-killed cells of *A. radiobacter* pv. *tumefaciens* protects the plant against infection by virulent *A. radiobacter* pv. *tumefaciens* (Cooksey and Moore,

1982a). Competition for attachment sites in host cells – the explanation offered (Lippincott et al, 1977) – is a form of direct antagonism. Cooksey and Moore (1982a) describe it as physical blockage. No active defense system of host origin has been demonstrated in the case of this kind of cross protection.

Cross protection between rust fungi was first studied by Yarwood (1956), who demonstrated that bean rust (*Uromyces phaseoli*) applied to sunflower leaves in advance of *Puccinia helianthi* protected the sunflower against *P. helianthi*. The protection was localized and dosage-dependent. Significantly, when leaves were "cured" of the first ("inducer") rust fungus by mild heat treatment or by a pressure treatment (which caused only slight injury to the host), the leaf was again fully susceptible to heavy infection by the challenger. This observation is evidence against an activated host defense mechanism, which presumably should have exhibited at least some carryover effect in the site occupied by the first fungus. Yarwood (1956) favored the idea of inhibition of the challenger strain by a volatile inhibitor, possibly trimethylene, produced by the first colonist.

Littlefield (1969) demonstrated a localized protective effect of the avirulent race 1 of *Melampsora lini* against the otherwise virulent race 339 on flax. The host carried the *L* gene for vertical resistance to race 1. Littlefield could find no evidence of a volatile inhibitor to explain the cross protection. Rather, the number of penetrations by race 339 was less because the penetration sites were physically blocked by the avirulent race 1. However, successful penetration by the virulent race 339 also produced smaller lesions with fewer uredia, which Littlefield (1969) attributed to a different mechanism, possibly an inhibitory substance or resistance response of the host triggered by the avirulent race 1. Littlefield's results lead to the unavoidable conclusion that the resistance "induced" in flax to race 339 by prior inoculation with race 1 is different from the resistance induced by race 1 to itself. The latter could be a typical incompatible interaction (case 3 in the examples given in the earlier section on membrane integrity) and the former could be the normal result of a disturbed host tissue being physiologically and even nutritionally less receptive or suitable to the otherwise virulent strain.

Antagonism by remote ("systemic") effects is also possible. Estores and Chen (1972) observed that when *Pratylenchus penetrans* was present on half of a tomato split root system and *Meloidogyne incognita* on the other half, the final population of *P. penetrans* was unusually low. By comparison, *P. penetrans* multiplied normally when the other half of the root system was removed completely or when *M. incognita* was absent. The results may be interpreted as evidence of a systemic resistance response, induced in roots exposed to *M. incognita* and then translocated to roots exposed to *P. penetrans* (McIntyre, *in* Horsfall and Cowling, 1980). Another explanation is that roots infected with *M. incognita* represent a stronger sink for photosynthates than do roots colonized by *P. penetrans*. Complete removal of half of the roots would eliminate this sink altogether and thus would not constitute an equivalent constraint on *P. penetrans*. The conclusion from this line of reasoning is that with *M. incognita* on some roots and *P. penetrans* on others of the same plant, *P. penetrans* is less able to obtain food and hence multiplies more slowly.

Hamilton, (*in* Horsfall and Cowling, 1980) proposes five possible mechanisms of cross protection between viruses. Of the five, four fit the category of direct antagonism between the strains involved. Only the fifth raises the possibility of an antiviral substance being produced by the host in response to the inducer but effective against the challenger. The proposed mechanisms are:

1. Challenger RNA becomes encapsidated in protein of the first (inducer) virus. This effectively prevents the challenger from replicating (an equivalent of predation between viruses).

2. A specific RNA-dependent RNA polymerase of the inducer, i.e., the enzyme used by the inducer to assemble its RNA during transcription, sequesters but does not transcribe the challenger RNA. Again, the challenger virus could enter the symplast but would not multiply (an equivalent of a trap plant that attracts but does not permit multiplication of the pathogen).

3. The inducer RNA has genes or coding regions that code for production of an inhibitor (antibioticlike substance) effective against challenger RNA replication (an equivalent of antibiosis).

4. Owing to the drain on the plant because of the action of the first virus, less energy and raw material is available in the cell to support normal replication of a second virus (an equivalent of competition).

5. An antiviral substance is produced by the host in response to the first inoculant but is more effective against the second virus (an equivalent of induced resistance).

Induced Resistance

Evidence for an active defense mechanism of host origin stimulated by prior inoculation with an avirulent or nonpathogenic strain comes from several studies involving *Pseudomonas* spp. in tobacco. A 50:50 mixture of virulent and avirulent cells of *P. solanacearum* introduced into stems of tobacco resulted in only slightly less disease than obtained with the same quantity of virulent cells mixed with water only. In contrast, a 5:95 and a 0.5:99.5 mixture of virulent:avirulent cells produced mild and no wilt, respectively, whereas these same small amounts of virulent inoculum mixed with water produced severe wilt (Averre and Kelman, 1964). In sequential inoculation, protection could be detected as soon as two hours after introduction of avirulent cells but was best after 24 hours (Sequeira and Hill, 1974). Protection also occurred following introduction of heat-killed cells of *P. solanacearum* and was light-dependent, suggesting a role of the host. In a similar study, Lovrekovich and Farkas (1965) demonstrated a localized but highly effective protection against *P. tabaci* if heat-killed cells of a pathogen were introduced 24 hours in advance of live virulent cells. All of these workers concluded that something develops within the tobacco tissue, perhaps within the intercellular spaces (Sequeira and Hill, 1974), that subsequently inhibits the pathogen. In work by Lovrekovich and Farkas (1965), autoclaved cells did not trigger the effect, suggesting that even heat-killed (100°C for 5–15 minutes) cells must be reasonably intact to promote the resistance response or that the factor in the bacteria that initiates the host response is itself sensitive to autoclaving.

Some well-known examples of induced resistance occur with vascular pathogens. These pathogens commonly produce less or even no wilt if their suscept is inoculated first with an avirulent race or a different wilt pathogen (Baker and Cook, 1974). Phillips et al (1967) observed in tomato protected by *Cephalosporium* spp. against *Fusarium oxysporum* f. sp. *lycopersici* that tylose obstructions developed and water flow through the stems was temporarily impeded following inoculation with *Cephalosporium*. *Fusarium oxysporum* f. sp. *lycopersici* was then less successful as a colonist of these stems. They suggested that the obstructions formed by the host in response to *Cephalosporium* also limited *F. oxysporum* f. sp. *lycopersici*. Langston (1969) reached this same conclusion from results using mixtures of *F. oxysporum* f. sp. *pisi* and *lycopersici* to inoculate tomato; living cells of *pisi* induced resistance but dead cells of this pathogen in the mixture gave no protection. Moreover, both the work of K. Ogawa and H. Komada (personal communication and Chapter 8) for sweet potato protected against *F. oxysporum* f. sp. *batatas* by prior inoculation of the plants with a nonpathogenic *F. oxysporum* and that by Gessler and Kuć (1982) for cucumber protected against *F. oxysporum* f. sp. *cucumerinum* by prior inoculation of the plants with *F. oxysporum* f. sp. *melonis* or *conglutinans* show that the induced resistance is of a systemic type and is not limited to the immediate area inoculated with the nonpathogen.

Cultivars differ in the speed with which they can produce callose, an abscission layer, occlusions, zones of phenol deposition, lignins, or other means of sealing off or restricting a vascular infection. The effect of prior inoculation of a plant with a nonpathogenic colonist of the vascular tissue may be to trigger a general state of "greater readiness" or shorter response time, thereby unleashing a resistance potential that resides in these susceptible plants (Kuć, 1980). The cells stimulated into this more defensive state presumably would be the parenchyma cells in contact with the xylem, with their response being systemiclike because hormones released at one site travel through the tissues to other sites (Beckman, *in* Horsfall and Cowling, 1980).

Meyer and Maraite (1971) suggested that competition for nutrients and space was the reason for less fungus growth in melon inoculated with a mixture of two strains of *Fusarium oxysporum* f. sp. *melonis,* one highly virulent and the other of low virulence. Their suggestion was made on the basis that the occlusion reaction requires three to four days whereas the protection was evident much sooner than this. However, three to four days is the reaction time in the susceptible host; in the resistant host, wall swelling has already begun by day one and more generally by day two after infection (Beckman, *in* Horsfall and Cowling, 1980). The work of Meyer and Maraite (1971) does not rule out, but rather supports the possibility that prior presence of an avirulent strain (i.e., a strain to which the melon cultivar was resistant) sets the stage for a more rapid occlusion reaction to an otherwise virulent strain.

Perhaps the example of induced resistance best documented is the remarkable protection of leaves of various cucurbits against *Colletotrichum lagenarium* by prior inoculation of the cotyledons or first true leaf of the plant with *C. lagenarium* (Caruso and Kuć, 1977a; Kuć and Richmond, 1977; Kuć et al, 1975) and by other leaf pathogens of cucurbits (Jenns and Kuć, 1979). The young leaves (those developing above the inoculated leaf) remain resistant for up to four weeks after inoculation of the lower one or two leaves. Moreover, resistance develops within 96 hours after inoculation, after which the inoculated (infected) leaf can be removed with no loss of resistance in the younger leaves. A "booster" inoculation of the older infected leaves can extend the protection beyond four weeks, even into the period of fruiting. The effect can also be induced by inoculation of the lower leaves with tobacco necrosis virus (TNV) or *Pseudonomas syringae* pv. *lachrymans.* In all cases, resistance in the upper leaves coincides with symptom appearance in the infected lower leaves. A systemic-type resistance was also induced in cucumber to *Fusarium oxysporum* f. sp. *cucumerinum* by prior (more than three days prior) inoculation of the leaves with *C. lagenarium* and TNV (Gessler and Kuć, 1982). The induced systemic-type resistance in cucurbits to *C. lagenarium,* as with induced resistance to vascular wilt fungi following prior inoculation with a related but avirulent strain, may involve stimulation of a greater state of readiness or acceleration of a host response that otherwise occurs too slowly to limit the pathogen adequately.

Enhanced Host-Plant Resistance with Mycorrhizae

Improved phosphorus nutrition of the plant can explain much of the influence of vesicular-arbuscular mycorrhizae (VAM) fungi on plant diseases (Chapter

4). This has been demonstrated for phytophthora root rot of citrus (Davis and Menge, 1980), verticillium wilt of cotton (Davis et al, 1979), and take-all of wheat (Graham and Menge, 1982). *Phytophthora citrophthora* and *Gaeumannomyces graminis* var. *tritici* were suppressed as much by phosphorus alone as by VAM-root associations, and verticillium wilt of cotton was favored as much by phosphorus alone as by VAM-root associations. A well-nourished plant will be more resistant to attack by weak parasites favored by a stressed host, but probably is more susceptible to attack by stronger parasites favored by a vigorous host. In addition, onset of verticillium wilt will be favored by maximal rates of water movement up the stem of the host, such as can be expected for plants with a larger root system and larger leaf (and hence transpirational) area, typical of plants with VAM associations. However, with a limiting soil water supply, the opposite effect could occur. The greater leaf area resulting from the VAM association could also lead to more rapid plant water deficit (and hence earlier closure of stomata on plants with VAM associations) if the soil water supply were inadequate to meet evapotranspiration. Perhaps this is why Dehne and Schönbeck (1975) observed less rather than more fusarium wilt of tomato on plants preinfected by *Glomus mosseae*. Clayton (1923) demonstrated 60 years ago that tomato plants with roots in slightly dry soil become "resistant" to fusarium wilt, probably because of reduced rates of transpiration (Cook and Papendick, 1972). Tomato plants with VAM (and with larger leaf areas) could become deficient for water more rapidly than those without VAM. Explaining the complex effects of one microorganism-plant interaction on a second-order microorganism-plant interaction requires attention to the whole plant.

7

THE SOIL ECOSYSTEM AND BIOLOGICAL CONTROL

The enormous diversity of microorganisms found in
nature, which have apparently adapted to specific physical,
chemical and biological variables, can only
be explained by assuming an equally
enormous diversity of ecological niches
—J. S. KUENEN and W. HARDER, 1980

Natural biological control resulting from the action
of antagonistic microorganisms is widespread in
nature and probably responsible to a significant
degree for the health of agricultural crops
—G. C. PAPAVIZAS and D. LUMSDEN, 1980

Antagonistic potential resides in all organisms, and the soil, being the richest source of both numbers and kinds of interacting microorganisms, offers the most opportunities for expression of useful antagonism. Biological control achieved with such practices as crop rotation, burial of crop residue, organic manurial treatments, and flooding results from resident soil microorganisms serving as antagonists, having been provided some advantage such as sufficient time or the proper environment. More gradual biological control sometimes can be achieved by nudging the microbial population in soil, e.g., with treatments such as fertilization to elevate or lower pH. It is also clear, after several decades of discouraging results, that some soilborne plant pathogens can now be or soon will be controlled biologically by introduction of antagonists into soil directly or with the planting material (Chapters 4 and 8), provided the antagonist is properly suited for this role or that some help is provided through manipulations of the soil or rhizosphere environment. An important breakthrough was the recognition of suppressive soils as a source of clues on how the soil ecosystem might be modified to achieve more biological control and as a source of more potent antagonists adapted for introduction into soil or the rhizosphere where needed. The more we learn about why some soils are suppressive, the more we understand which cultural practices can be used to enhance suppressiveness and which must be avoided so as not to upset the natural biological control.

The soil may be the most complex environment occupied by microorganisms, or it may be no more complex than the plant tissues occupied by microorganisms, only less understood. Both soil and plant tissues as environments present a matrix of adsorptive and absorptive surfaces and substances. Both environments present a range of fluctuating redox potentials, pH, water potentials, and temperatures. Both present a dynamic gaseous environment consisting of carbon dioxide that is generally higher than ambient, oxygen that is lower than ambient, and biologically active gases such as ethylene. Both contain a food supply available only to the adapted. This chapter examines soil as a dynamic ecosystem of rapidly changing ecological niches and microbial biomass fluctuating in response to environmental changes and energy supply, and discusses the role of these factors in biological control.

SOIL WATER AND AERATION AS FACTORS IN BIOLOGICAL CONTROL

All physical factors important to the activity of soil microorganisms are in a constant state of flux to one degree or another, but none are more dynamic over time and space than soil water and the associated degree of soil aeration. The gradients in water content and water potential interacting with different concentrations of gases and other materials in solution, sometimes over very short distances, provide a great variety of niches for microorganisms. This, as much as any combination of physical factors, makes possible the diversity of microorganisms in soil.

Of the two—soil water and soil aeration—water is the primary factor, and selectively affects the activity of soil microorganisms in at least four ways: 1) gas exchange (soil aeration); 2) diffusion of solutes, e.g., nutrients toward and wastes away from microorganisms; 3) motility of organisms; and 4) free energy (water potential), which determines availability of water for growth and metabolism. Before discussing these four kinds of effects, a brief treatment of some basic concepts seems necessary, particularly to show relationships between the many terms and units of measure for expressing the water status of soil. Other reviews on the principles involved may be consulted for more details (Cook and Papendick, 1972; Griffin, 1972; Papendick and Campbell, *in* Bruehl, 1975; and Parr et al, 1981).

Some Basic Concepts

Water is held in soil as films and lenses in the pores and around the soil particles (soil matrix) by adhesion and cohesion. Soil water also has various dissolved ions and other solutes. The total free energy (potential energy) of the water, i.e., the *soil water potential*, is the sum of the

Table 7.1. Relationship of water potential to other methods used to quantify water in soils, plants, or media[a]

Water potential[b] (−bars)	Relative humidity (%)	Suction (cm)	pF	KCl (molal)	Diameter of largest water-filled pores (μm)
0.0	100.	0		0	∞
0.001	99.9999	1	0	0.00002	2,908.
0.01	99.9990	10	1	0.00022	291.
0.1	99.9900	102	2	0.00225	29.1
1.0	99.9930	1,020	3	0.02248	2.91
10.0	99.2600	10,200	4	0.22479	0.291
20.0	98.5300	20,400		0.44958	0.145
50.0	96.37	51,000		1.12395	0.058
100.0	92.86	102,000	5	2.24791	0.029
200.0	86.24	204,000		4.49582	0.014

[a] Expanded from Papendick and Campbell (*in* Parr et al, 1981).
[b] 1 bar = 10,000 pascals. Note: Field capacity is about −0.3 bars.

matric component (free energy determined by adhesion and cohesion within the soil matrix) plus an osmotic component (free energy determined by solutes). The gravitational component (height difference from the standard) and pressure component are negligible and can be ignored (Papendick and Campbell, *in* Parr et al, 1981). Soil water potential (Ψ) is therefore expressed as:

$$\Psi_{soil} = \Psi_{osmotic} + \Psi_{matric}$$

Soil water is sometimes expressed as a suction. Suction is only the matric component of soil water potential, with the units having a positive rather than a negative sign. Suction and water content can be related using a hanging funnel apparatus, where the soil is supported as a layer on a porous (sintered glass) plate in the funnel with the funnel exit connected to a flexible tubing of suitable length. The tubing and connecting funnel space beneath the porous plate are filled with air-free water, with the tubing bent in a "U" shape so that the open end is level with the soil. All pores in the soil become filled with water when the height of water in the connecting tube is level with the soil surface. Such soil is saturated and has a matric potential (suction) of zero. A suction develops in the soil when the level of water in the connecting tube is lowered below the height of the soil. For purposes of conversion, a difference of 1,020 cm between the height of the soil and the height of the water level in the connecting tube is 1 bar matric potential (Table 7.1). The term pF used in some literature is the negative log of this height difference, i.e., 10, 100, and 1,000 cm are pF 1, 2, and 3, respectively, (Table 7.1).

The size of the water-filled pores in soil is related to the matric potential (suction) and can be approximated from the capillary rise equation:

$$\text{Pore diameter} = 2.94\Psi_{matric}$$

The approximate size of water-filled pores at different matric potentials is given in Table 7.1.

As the soil drains, a greater proportion of the total pore space becomes air-filled, but the vapor phase in the drained pore space can be assumed to be in equilibrium with the liquid phase. Water potential is related to relative humidity by the equation:

$$\Psi = \frac{RT}{V} \ln RH$$

where R is the ideal gas constant, T is absolute temperature, V is the volume of a mole of water, and ln RH is the natural log of the relative humidity. For purposes of quick calculation, at 20°C the water potential is approximately 14.8 bars lower for each 1% drop in the relative humidity equilibriated below 100% (Table 7.1).

Water Potential Requirements for Growth, and the Relative Competitive Advantage of Microorganisms in Soil

Growth. — The cell of a microorganism in soil will have a water potential in equilibrium with the soil water potential. If the soil water potential is −0.03 bars (field capacity), the cell has a total water potential of −0.03 bars. If the soil is at −15 bars (near that in which most plants wilt), the total water potential of the cell equilibrates to −15 bars — lower still as the soil dries even more. A basic biophysical requirement of the cell for growth is to maintain turgor (Chapters 5 and 6), accomplished mainly by maintenance of a differential between the solute potential of the cell and the total water potential of the environment. The organism will cease growth when this differential (and hence turgor) reaches zero and shrinkage of the cell wall is no longer sufficient to help maintain turgor (Harris, *in* Parr et al, 1981).

The differences among microorganisms in ability to grow at low water potential relates in part to their different methods for osmoregulation. Harris (*in* Parr et al, 1981) divides microorganisms into four classes. Gram-negative bacteria are in classes I and II. The solution potential of the cytoplasm of these microorganisms is about −7 bars, mainly because of basal intermediary metabolites. Microorganisms in class I have no ability to produce or accumulate compatible solutes from internal cellular materials. Bacteria in this class have adequate turgor for growth only if the soil is wetter than −7 bars; growth at water potentials drier than −7 bars is dependent on the influx of solutes from the external environment into the cell to maintain turgor, e.g., amino acids or sugars obtained from organic material or ions from the soil solution. Obviously bacteria in this group are very sensitive to matric potential stress and are likely to cease growth as the water potential drops below −7 bars. *Spirillum spp.* are examples. Class II microorganisms have the added ability to produce and accumulate compatible solutes (e.g., amino acids and polyols) from cellular materials and can thereby osmoregulate below −7 bars in response to water stress. However, such osmoregulation has an energy cost, and growth may be decreased accordingly. Microorganisms in this category include *Pseudomonas aeruginosa, Escherichia coli, Klebsiella* spp., and certain yeasts. Class III microorganisms accumulate compatible solutes such as potassium, glutamate, arabitol, aerythritol, and mannitol constitutively, but cannot induce further production of compatible solutes when exposed to stress. Cells of these microorganisms, typically Gram-positive bacteria, have a normal cell solution potential of about −25 bars. Such cells in equilibrium in soil at saturation (0 bars) would have +25 bars turgor, made possible by their unusually strong cell wall. On the other hand, these organisms are limited in growth to −25 bars matric potential, except to the extent that influx of solutes from the soil solution occurs. *Arthrobacter crystallopoietes* and *Bacillus subtilus* are in class III. Class IV microorganisms are similar to those of class III but have the additional ability to produce and to accumulate solutes. Most streptomycetes, some yeasts, soil fungi, and halophilic bacteria are in this class; they have ability to grow at water potentials well below those of microorganisms in the other three classes.

Relative competitive advantage. — An important factor is that although different groups and species of microorganisms have different lower limits of water potential for growth extinction in soil, the maximum growth rate for most microorganisms in soil usually occurs at or above −1.0 bar matric potential. *Fusarium roseum* 'Culmorum' ceases growth at about −80 bars matric potential and *Gaeumannomyces graminis* var. *tritici* at about −40 bars, but both, given adequate nutrients and freedom from competition, grew maximally at −0.03 bar, the highest matric potential tested (Cook et al, 1972). At the higher matric potentials, microorganisms in soil are able to maintain the greatest turgor with the least expenditure of energy.

As the soil matric potential decreases, all organisms grow more slowly, but the growth rates of some are limited more than others. This affects the *relative competitive advantage* (Baker and Cook, 1974) of the organism, i.e., the advantage of one organism over another as determined by their respective growth rates at one matric potential compared with their respective growth rates at another potential. For example, *F. roseum* 'Culmorum', *G. graminis* var. *tritici*, and a Gram-negative bacterium (typical of Harris' class II, which ceases growth at about −20 bars) all will grow maximally at a matric potential of about −1 bar. The competitive advantage of each relative to the other theoretically is 100% at this matric potential. The outcome in competition among microorganisms at matric potentials of −1 bar or above is therefore determined by factors other than water potential, e.g., temperature, pH, aeration, or availability of nutrients (see subsequent section). At −5 bars, growth rates (and hence the competitive advantage of each organism relative to the others) decrease to about 75% of maximum for the bacterium, and 85–90% of maximum for *G. graminis* var. *tritici,* but remain 95% of maximum for the *Fusarium.* At −10 bars, the respective values become 50, 75, and near 90%, and at −20 bars, 0, 50, and 75%.

Fusarium has the potential to be active at −1 or −5 bars matric potential, given some other selective advantage such as low pH or a favorable temperature, but will be relatively much more active in the drier soils where its competitive advantage is greatest relative to that of most other microorganisms. For example, although −90 bars is near the cardinal minimum for this *Fusarium,* its activity as measured by colonization of wheat straw was maximal between −80 and −90 bars (Cook, 1970), i.e., colonization was greatest where potential growth rate of the fungus was the least. Apparently, the ecological niche most available to this *Fusarium* is between −80 and −90 bars. The relative competitive advantage of *F. roseum* 'Culmorum' is greatest at the lower water potentials because most competitors have no advantage at these water potentials.

Any attempt at biological control of this fungus in soil must either keep the soil water potentials in the tillage layer (where this fungus exists) wetter than −80 to −90 bars, preferably wetter than −15 bars, or provide a mechanism whereby other microorganisms will close the niche otherwise available to this fungus at the lower soil water potentials.

Soil Water Content and Diffusion of Solutes

Soil bacteria tend to be less active or to cease growth at matric potentials several bars above (wetter than) the −7 bars or the −20 to −25 bars expected (Harris, *in* Parr et al, 1981) from their abilities to osmoregulate and maintain turgor (Griffin, 1972). This discrepancy is less apparent or nonexistent with filamentous fungi. The difference is thought to result from limitations on diffusion of nutrients toward the bacterial cells and of wastes away from the cell as the soil water content (and hence thickness of films and length of continous water-filled pores) decreases. Filamentous fungi and probably actinomycetes avoid this problem by their ability to transport nutrients and water through their hyphae. The result is an even greater competitive advantage for fungi (and actinomycetes) over unicellular bacteria than would be predicted by their relative growth rates as the soil dries. *The successful use of bacteria as antagonists, whether introduced or managed as residents, depends on the maintenence of high matric potentials, probably −1 bar or wetter, for the majority if not the entire period during which the biocontrol is needed.*

Dormant spores of fungi are subject to the same limiting effects of dry soil on solute diffusion as those thought to affect bacteria. The advantage of filamentous hyphae is not available to fungi while still in the spore stage.

Thus, chlamydospores of *Fusarium solani* f. sp. *pisi* germinated in soil contiguous to pea seeds at those water contents that provided the greatest supply of exudate (Cook and Flentje, 1967). Similarly, with oospores of *Pythium aphanidermatum*, nutrient availability and not water potential was the limiting factor to germination at water potentials drier than −1 bar (Stanghellini and Burr, 1973). Chlamydospores of *Phytophthora cinnamomi* germinated at −0.25 bars matric potential in soil near roots of avocado in a clay soil but did not germinate at matric potentials below −0.1 bar in a sandy soil. However, when nutrients such as glucose and asparagine were added to the soil solution, chlamydospore germination occurred in both soils at −0.25 bars (Sterne et al, 1977a, 1977b). These examples reveal the great sensitivity and remarkable dependence of the resting spores of plant-pathogenic fungi on a nutrient supply for germination in soil, and they reinforce the premise that any reduction in one or more essential factors (e.g., carbon or nitrogen) in this nutrient supply, whether accomplished by increasing the total active microbial biomass of soil or by introducing an aggressive and competitive root colonist, could result in less infection.

Water-Filled Pores and the Motility of Soil Microorganisms

The distance of movement of soil microorganisms having a motile phase is critically dependent on water-filled pores. As the matric potential decreases, and progressively smaller pores become drained (Table 7.1), the water-filled pathway for movement of motile spores disappears. Different soil microorganisms require a different minimum size pore, depending on their physical size and their method of movement. The zoospores of *Phytophthora* spp. moved 15–25 mm at −0.01 bar but showed no movement at −0.05 bar in a coarse-textured soil mix (Duniway, 1976). The effective diameter of water-filled pores in soil is 291 μm at −0.01 bar and 58 μm at −0.05 bar. Actually, the zoospores of *Phytophthora* move in a helical pathway and, based on amplitude of the movement, have been estimated to require a pore diameter of 200 μm or larger (Griffin, *in* Parr et al, 1981). Movement of motile bacteria is generally limited to soils wetter than −0.5 bar, probably nearer to −0.1 to −0.2 bar or even higher matric potentials (Griffin and Quail, 1968; Wong and Griffin, 1976). Although some migration of antagonistic bacteria by swimming in water-filled pores can be expected, growth along the extending root must also occur if bacteria are to provide the protection needed on roots in soils drier than −0.5 bar.

Soil texture is also important because the relative proportion of sand, silt, and clay determines the frequencies of the different sizes of pores (Papendick and Campbell, *in* Parr et al, 1981). For example, essentially all pores in a clay are 20 μm in diameter or smaller, depending on the degree of aggregation. Microorganisms dependent for motility on pores larger than 20 μm in diameter will be immobile in clay, even when the soil is saturated, because pores of adequate size are rare in soil of such fine texture. In a sand, on the other hand, nearly 60% of the total pore space is larger than 20 μm in diameter and 20% is larger than 200 μm in diameter. Other factors also affect the opportunity for motility, including degree of crumb structure, channels left by earthworms and roots, and the tendency of zoospores to encyst prematurely when they collide with soil particles (Benjamin and Newhook, 1982).

The effects of soil matric potential, pore size, and pore-size distribution are especially evident in the occurrence of perforations of conidia of *Helminthosporium sativum* caused by giant vampyrellid amoebae (Cook and Homma, 1979). No perforations occurred 1 cm deep in either of two saturated soils, but extensive perforations occurred on spores submerged in a thin layer of water in a dish. The difference was attributed to the oxygen supply, as shown for *Phytophthora* (Duniway, 1975). Oxygen would have been limiting to the amoebae 1 cm deep in saturated soil but not in the thin water layer in the dish. Perforations did not occur at −0.01 bar in either soil, unless they were mixed 1:1 with coarse sand. The addition of sand increased the proportion of pore space that drained at −0.01 bar matric potential, whereupon rapid entry of oxygen became possible. In the two soils without added sand, perforations occurred only when they were drained to −0.1 bar matric potential, which corresponded to the water potential for air entry in these soils. Perforations were again prevented in both soils and the soil-sand mixture if the matric potential was drier than −0.2 bar, and perforations were prevented in the dish of water if KCl was added to produce an osmotic potential of −0.4 bar or lower. The −0.4 bar osmotic potential is probably the lower limit for these amoebae, which, at this osmotic potential have only negligible turgor with no cell wall. The limit of −0.2 bar matric potential in the various soils may be the lower limit for osmoregulation by these animals under matric control, or it may be where effective pore diameters become limiting to their movement through the soil to reach the spores.

Several species of amoebae belonging to the Vampyrellidae have been shown or suspected to cause perforations in the pigmented spores and hyphae of fungi in soil (Old and Patrick, *in* Schippers and Gams, 1979). These include *Arachnula impatiens* (Homma et al, 1981; Old, 1977), *Theratromyxa weberi*, and *Vampyrella vorax* (Anderson and Patrick, 1980). Other mycophagous amoebae include *Thecamoeba granifera* s. sp. *minor*, which Alabouvette et al (1981) observed to increase 10-fold in numbers in response to incorporation of biomass of *Fusarium oxysporum* and *F. solani* into soil kept adequately wet. Unquestionably, amoebae play a role in the death rate of pathogen propagules in soil but, like mycophagous nematodes, their potential is limited to very wet soils that are also adequately drained to permit oxygen entry. The lower limit of −0.2 bar for the giant vampyrellid amoebae (Cook and Homma, 1979) is wetter than the −0.30 to −0.35 bar generally accepted as field capacity. Thus, these animals will be inactive when the soil is saturated and also when the soil is at field capacity or drier. This is a serious limitation to their usefulness as antagonists.

Soil Aeration and Biological Control

A well-structured soil consists of crumbs or aggregates of soil particles having small pores inside the aggregates and large pores or channels among the aggregates. The large pores drain freely and have an oxygen status very near that of air above the soil surface. The small pores within the aggregates retain most of the water used by plants. Air within the centers of the aggregates may be near 21% oxygen if the aggregates are adequately drained and if consumption by microorganisms does not exceed the rate of supply. However, since oxygen diffuses only 1/10,000th as fast through water as through air, the path of diffusion in a water-saturated aggregate is effectively 10,000 times longer than in a dry aggregate. Moreover, microorganisms line the path of diffusion and take what they need in accordance with their food supply, temperature, and other conditions that determine their rates of respiration. Where large quantities of organic matter have been added, temperature is ideal for maximal rates of metabolism, and if the aggregates are wet, the consumption rates are likely to exceed the supply, and anaerobic centers will develop in the aggregates (Griffin, 1972). Such aggregates may be as small as a few millimeters in radius. Thus, on balance, there is competition among microorganisms for oxygen in soil.

The concept that soil oxygen may vary between 21 and 0% over a very short distance is a fundamental advance in our understanding of the ecology of soil microorganisms. Measurements of the partial pressure of oxygen in the soil atmosphere are meaningless because they represent little more than averages of extremes and give no idea of the volume of oxygen-free soil. More important, a single aggregate of soil must be visualized not only as a gradient of decreasing oxygen concentrations from the outside inward, but also as a gradient of decreasing redox potentials from the outside inward (Greenwood, in Whittington, 1969; Hattori, 1973). The redox potential drops from approximately 800 mV near the edges of the aggregates, where the oxygen concentration is near 21%, to 200 mV at some distance inside the aggregate, where oxygen is 0%, and may go below −200 mV, possibly even to −420 mV in the center of the aggregate, as nitrate and other electron acceptors are reduced. Aerobic processes such as nitrification occur on the edges of the aggregates. Facultative anaerobes are active at 200 mV (0% oxygen), using nitrate instead of oxygen as an electron acceptor. Their activity occurs mainly along the gradient from 200 down to −200 mV. With exhaustion of nitrate as an electron acceptor, the redox potential poised by the presence of nitrate drops below −200 mV. Strict anaerobes are active along the gradient below −200 mV, using organic acids as electron acceptors. Various products of anaerobic metabolism may thus accumulate, including methane, ethylene, alcohols, aldehydes, and possibly even hydrogen.

Propagules of plant-pathogenic fungi situated inside wet aggregates of soil will probably tend to equilibrate with the redox potential of their environment. The stresses encountered by propagules in this situation would be no different from the stresses encountered in a flooded soil. Equilibration of a propagule with a low redox potential, regardless of what caused it to occur, should predispose the propagule to attack by anaerobes. The important distinctions affecting stress would be the proportion of the total soil volume affected, how long it was affected, and the susceptibility of the propagules to destruction by anaerobes under anaerobic conditions, especially in relation to the effects of flooding, tarping, and organic amendments on the propagules.

Anaerobic bacteria occur abundantly in virtually all agricultural soils (10^3–10^5 colony-forming units per gram, sometimes higher), which is usually taken as proof that the anaerobic conditions necessary for their growth occur to some extent in all agricultural soils. The potential for anaerobes to induce biological control is relatively unexplored and deserves study. Survival for most soilborne propagules of plant pathogens is poorest in flooded or in warm wet soil (Green, 1980; Keim and Webster, 1974; Shokes et al, 1977; Sitton and Cook, 1981), especially if organic materials are also added (Dhingra and Sinclair, 1974).

Predisposition of pathogen propagules to damage from anaerobes by exposing the propagules to low redox potentials also may be part of the reason for their accelerated death rate in soil tarped with transparent polyethylene sheeting. The tarps elevate the soil temperature, increase soil respiration, and serve as a barrier to oxygen diffusion into the soil and carbon dioxide diffusion out of the soil. The presence of fresh organic materials and adequate soil moisture under the tarp would enhance the effect. It would be interesting to know the concentrations of oxygen, carbon dioxide, bicarbonates, and carbonates in tarped soil. It would also be interesting to know the redox potentials at various sites in soil beneath tarp and especially to know the proportion of the total soil volume at redox potentials suitable for growth of anaerobes and production of materials toxic to pathogens. The potential for toxicity of such materials to pathogen propagules would be augmented by the increased soil temperature.

The incorporation of large quantities of organic material into moist warm soil, and the resultant development of a large volume of anaerobic sites, can also lead to problems for plant root growth because of the greater production of acetic and butyric acid, reduced sulfur compounds, and other compounds injurious to plant roots (Russell, 1977). As the readily available energy sources are used, respiration declines and the volume of anaerobic sites shrinks. With this situation, and especially if the soil is repeatedly tilled, organic acids and other phytotoxic compounds eventually become oxidized and hence eliminated. Therefore, fresh organic materials added to soil for biological control should be applied in advance of planting to avoid phytotoxicity. Use of a material sufficiently composted (with readily available compounds already oxidized) also helps avoid the problem of phytotoxins.

Of all the biologically active organic materials produced in anaerobic soil, probably none have wider implications than ethylene for the occurrence or lack of occurrence of biological control (Smith, 1976b). Ethylene is produced in most if not all soils (Smith and Restall, 1971), with the greatest amounts occurring in soils that are flooded or recently amended with fresh organic matter (Burford, 1976; Smith and Cook, 1974; Smith and Restall, 1971; Stevens and Cornforth, 1974). The relationship between the absence of oxygen and the production of ethylene in soil was first demonstrated by Smith and Russell (1969). Ethylene production is inhibited by nitrates but not by ammonium nitrogen (Smith, 1976a), apparently because nitrate nitrogen acts as an electron acceptor in the absence of oxygen and poises the redox potential (Bell, 1969) at a level (+200 mV) too high for ethylene production. Manganous ions also limit ethylene production, whereas ferrous ions are stimulatory, presumably because of their respective effects on the soil redox potential (Smith, 1976a). Ethylene production is also favored by high soil temperatures. Smith and Restall (1971) recorded negligible accumulation of the gas in soil below 15°C but progressively greater amounts as the temperature was increased up to 35°C, the highest temperature studied. Sutherland and Cook (1980) confirmed the effects of low temperatures but showed that even more ethylene production occurs in soil at 50 than at 30°C. Ethylene production is limited at −1 bar and essentially prevented at −5 bars, probably because these water potentials are limiting to the producer organisms (Cook and Smith, 1977).

Seeds of the parasitic plant *Striga* (witchweed) are triggered to germinate by ethylene (Eplee, 1975). In some soils of North Carolina where the weed is a problem, ethylene is injected for witchweed control. In contrast to witchweed seed, ethylene is fungistatic to germination of sclerotia of *Sclerotium rolfsii* in soil (Smith, 1973). Ethylene in the soil has also been shown to suppress take-all of wheat caused by *Gaeumannomyces graminis* var. *tritici* (Cook, *in* Asher and Shipton, 1981). On the other hand, the suggestion that ethylene in soil is responsible for widespread soil fungistasis (Smith, 1973) has not held true (Pavlica et al, 1978; Schippers et al, 1978), nor has the hypothesis that ethylene serves as a regulator of the activity of aerobic microorganisms (Smith and Cook, 1974) been supported by the results of others (Smith, 1978). In general, ethylene has proved to have relatively no effect on germination and growth of fungus and bacterial cells, but more work is needed on this question. One effect of ethylene is to accelerate tissue senesence, which could limit biotrophic parasites but favor necrotrophic and weak parasites.

The actual sources of ethylene in soil are unknown, but soil microorganisms clearly are involved. Soils no longer produce ethylene after sterilization at 121°C (Smith and Cook, 1974; Smith and Restall, 1971). Lynch (1972) and Lynch and Harper (1974a) suggested that fungi produce the ethylene in soil and that *Mucor hiemalis* is one of the main sources. However, fungi cannot be the main source of ethylene in soil. Smith and Cook (1974) showed that treatment of the soil with aerated steam at 80°C for 30 minutes did not destroy the organisms responsible for ethylene production. This finding, and the fact that ethylene production is highly sensitive to water potential in the range of −1 to −5 bars, prompted them to suggest that the ethylene is produced by bacteria, probably the abundant spore-forming anaerobes active in anaerobic microsites. However, while aerobic bacteria have been shown to produce ethylene in pure culture (Primrose, 1976), thus far no anaerobic bacteria have been shown to produce ethylene (Sutherland and Cook, 1980). Nevertheless, bacteria must be involved because the antibiotic novobiocin (effective against bacteria) prevented ethylene production from soil treated at 80°C for 30 minutes and greatly retarded production from raw soil, whereas cycloheximide (inhibitory to fungi) had no effect (Sutherland and Cook, 1980). *Whatever the microbiological source(s) of ethylene in soil, the organism(s) involved must be sensitive to 121°C but resistant to 80°C moist heat for 30 minutes, sensitive to −1 to −5 bars matric water potential, sensitive to novobiocin but not to cycloheximide, and favored by the absence of both oxygen and nitrate so that the redox potential can drop below 200 mV. These criteria are satisfied only by anaerobic spore-forming bacteria.* Clearly, this ecological group deserves much more study for its role in the soil ecosystem and its potential in biological control.

THE MICROBIAL BIOMASS OF SOIL

Amounts and Methods of Measurement

The soil microbial biomass is the total mass of microorganisms in soil. A soil physically suitable for microbial activity (in terms of water potential, aeration, temperature, and other factors), but which has not been disturbed or amended with fresh organic matter for some time, will exhibit a certain rate of basal respiration. This basal respiration is the sum of respiration by all cells metabolically active in the soil at that time, in various states of dormancy or activity and at various ages. A treatment that "stresses" (Jenkinson, 1971) the soil biomass, e.g., air-drying/rewetting, fumigation, or heat treatment, produces a transient increase (flush) in the respiratory activity, which then settles back to the basal level within a few days. Jenkinson (1966) demonstrated that the flush results from the survivors growing on the killed biomass. The total microbial biomass of a soil can be estimated by the magnitude of the flush following destruction of the biomass by chloroform fumigation (Jenkinson and Powelson, 1976). Anderson and Domsch (1978a) added ^{14}C-labeled fungi and bacteria to soils and then exposed the soils for 24 hours to chloroform followed by inoculation with untreated soil; they found that the average percentage of biomass carbon mineralized to carbon dioxide was 43.7% for bacteria and 33.3% for fungi. They used a ratio of 1:3 for the relative proportion of bacterial and fungal biomass in most natural soils (Anderson and Domsch, 1973) and from this, they calculated that 41.1% of the carbon in cells killed in soil is evolved as carbon dioxide within 10 days after chloroform fumigation. This amount is reasonably constant for most soils. Therefore, they divided a K-factor (0.411) into cumulative carbon dioxide (expressed as milligrams of carbon per 100 g of soil) evolved during the 10-day period, less that evolved from nonfumigated soil under the same conditions, and suggested that the result gives an estimate of biomass carbon per 100 g of soil.

Jenkinson and Powelson (1976) calculated that the top 23 cm of soil in the Broadbalk continuous-wheat field at Rothamsted contained 530 kg of biomass carbon per hectare where unmanured and not fertilized, 590 kg of biomass carbon per hectare where fertilized with inorganic nitrogen, and 1,160 kg/ha where farmyard manure had been applied repeatedly during past years. A calcareous woodland soil nearby had an estimated 1,960 kg of biomass carbon per hectare, and unmanured permanent pasture had 2,200 kg/ha. Consider by comparison that 1,000 propagules of *Fusarium roseum* 'Culmorum' per gram in the top 15 cm of soil represents only about 15–20 g of biomass carbon per hectare, approximately 10^{-5} of the total microbial biomass of a manured soil or unmanured pasture as estimated by Jenkinson and Powelson (1976). The continued success of an individual fungus such as Culmorum within such a large microbial biomass can be explained only by the existence of ecological niches where it can effectively compete. Obviously the most important niche to this fungus is host roots. Perhaps it is possible to use the total active microbial biomass of soil to effectively reduce the success of this pathogen or others more generally.

Anderson and Domsch (1978b) showed that the greater the microbial biomass in soil, based on the method of Jenkinson and Powelson (1976), the higher the maximum initial rate of respiration (usually reached in one to four hours) in response to an excess (more than needed to induce the maximum respiratory rate) of glucose or casamino acids. A correlation of $r = 0.96$ was found between milligrams of biomass carbon per 100 g of soil and the maximal initial rate of respiration in response to the added glucose, expressed as milliliters of carbon dioxide per hour per 100 g of soil. Thus, a five- to tenfold difference in total microbial biomass content among 50 soils was reflected in five- to tenfold differences in maximal initial rates of respiration following the addition of an excess of glucose. This finding could explain why the suppressiveness of soils to the activity of plant-pathogenic fungi is often directly related to the level of total microbiological activity. **The larger the active microbial biomass of soil, the greater the sink for carbon, nitrogen, and energy in the soil and the less the likelihood that a finite supply of any one essential nutrient factor from a host will be adequate to satisfy the needs of the pathogen.** (This concept is developed further in the section on suppressive soils, this chapter.)

Several other methods have been tested to determine the total microbial biomass of soil, including the adenosine-triphosphate content (Ross et al, 1980) and heat production measured by microcalorimetry (Sparling, 1981). There is considerable disagreement as to which method of biomass measurement is best. Moreover, correlations among the various methods are sometimes poor (Sparling, 1981). Obviously each method will have limitations. More importantly, the estimates provide some idea of the enormous microbial biomass content of the soil and also show that total microbial biomass is different for different soils, is dynamic with the season and soil condition, and presents the opportunity both for successful gain by a given pathogen and for its suppression.

Clarnhom and Rosswall (1980) used direct counts to estimate biomass of bacteria in a peat layer and in the humus and mineral soil layers of a pine forest podsol in northern Sweden. The peat or soil, blended thoroughly into 1.0% water agar, was transferred to slides, where it was dried and then stained with acridine orange to reveal the live cells. Six size-classes of bacterial cells were distinguished, with density taken as 1.0 and dry weight as 20% of fresh weight. Total bacterial biomass tended to increase with rising soil temperature and to decrease during dry periods. Significant increases in bacterial biomass also occurred within a few days after rainfall, even when water potential had not been limiting before the rain; these increases were thought to result from the leaching of nutrients from the plant canopy into the soil (Chapter 6). Calculations indicated that biomass increased in some cases without an increase in number of cells, especially after a rainfall, when cell size increased without corresponding cell division. Small coccoid cells tended to dominate during dry periods, but many turned into rods during favorable periods. However, only 15–30% of the bacteria were active even during favorable periods. Both the total number of microorganisms and the total biomass declined rapidly soon after periods of growth, which was thought to result from increased rates of grazing by protozoa and nematodes. A crude estimate indicated that about 210 g (dry weight basis) of bacterial biomass was produced per square meter annually in the forest soil to a 10-cm depth.

Measurement of total respiratory response of a soil gives no indication of the kinds of microorganisms that produce the carbon dioxide and that are therefore the main or potential competitors with soilborne pathogens in response to a fresh supply of carbon and energy. Anderson and Domsch (1973) added selective inhibitors together with a large amount of glucose. The inhibitors were streptomycin (inhibitory to protein synthesis by organisms having 70 S ribosomes, i.e., prokaryotes) and cycloheximide (inhibitory to protein synthesis by organisms having 80 S ribosomes, i.e., eukaryotes, mainly fungi). The results indicated that for every 100 ml of carbon dioxide emitted from soil during the first six hours after adding glucose, about 30 ml comes from bacteria and 70 ml from fungi. Soils varied in the proportion of carbon dioxide attributable to bacteria versus fungi, but the greater response attributable to fungi was consistent, whether in arable grassland or forest soil. By inference, soil fungi may be two to three times more important than bacteria as competitors of plant pathogens in soil during the first few hours following release of a fresh supply of nutrients from a host root growing through the soil. This does not imply that fungi have greater potential as introduced antagonists for colonization of the rhizosphere; on the contrary, bacteria seem to be more effective than fungi in this role.

Influence of Heat Treatments and Fumigation

As demonstrated by Jenkinson (1966), dead microorganisms become nutrients for the survivors. Selective heat treatment using aerated steam (Baker, *in* Toussoun et al, 1970) is designed to leave survivors (mostly heat-resistant bacteria nonpathogenic to plants) that use the nutrients. The flush of growth by these bacteria restores and maintains a state in which one or more nutrients is limiting, much as encountered by pathogens in the soil before treatment. Antibiotic production by some of the survivors, e.g., *Bacillus subtilis*, may add to general suppressiveness of pasteurized soil to plant pathogens (Olsen and Baker, 1968).

Soil fumigation can also be selective in the kinds of organisms killed. However, whereas the selectivity of heat treatment results from differential heat tolerances by the microorganisms, the selectivity of soil fumigation results from a combination of differential tolerances and from escape by microorganisms. Fumigants exist as gradients in the soil away from the point of injection, thereby providing sensitive cells a means to escape harm in areas where the dosage is not lethal. Microorganisms also escape exposure

Table 7.2. Influence of fumigating (methyl bromide) soil, air-dry versus moist, on effectiveness of kill of specific groups of soil microorganisms and on subsequent nitrogen status[a]

	Soil air-dry		Soil moistened to ~−0.5 bar	
	Not fumigated	Fumigated	Not fumigated	Fumigated
	(colony forming units per gram)			
Fungi[b]				
Total sporing	4×10^4	$\sim 10^2$	6×10^4	$\sim 10^2$
Bacteria[c]				
Aerobic spores	1.6×10^6	1.3×10^3	2.1×10^6	4.1×10^5
Anaerobic spores	1.6×10^5	$<10^3$	1.4×10^5	5.6×10^3
Gram-negative	1.4×10^5	$<10^3$	0.7×10^5	5.6×10^4
Pseudomonads	0.6×10^5	$<10^2$	0.6×10^5	3.5×10^2
Nitrogen status[d]				
NO_3	21.3 ± 0.6	21.3 ± 0.6	39.3 ± 0.6	21.0 ± 1.0
NH_4	1 ± 0	3.3 ± 1.0	nil	25.0 ± 2.6

[a] 1974 data of A. D. Rovira and R. J. Cook, unpublished.
[b] Estimated by plate counts at 10^{-3}, 10^{-4}, 10^{-5}, and 10^{-6} dilution on Czapek-Dox agar medium plus 0.5% yeast extract (Warcup, 1976).
[c] Estimated by plate counts at 10^{-2}, 10^{-4}, and 10^{-5} dilution for pseudomonads and 10^{-3}, 10^{-4}, 10^{-5}, and 10^{-6} dilution for other groups on media of Ridge (1976) one day after treatment.
[d] Determined by extraction of soil with 2 N KCl at seven days after fumigation.

to methyl bromide when they exist in the centers of wet aggregates, because the gas tends to follow the path of least resistance, diffusing through the drained macropores of soil and leaving some micropores untreated. Failure of methyl bromide fumigation to kill microorganisms is less likely if the soil is fumigated when well-drained and in a state of "good tilth" (Table 7.2).

The recolonization of a fumigated soil following a selective kill is accomplished mainly by those survivors having the shortest response time and the fastest growth rate. Of the fungi, species of *Trichoderma* and *Mucor* are usually the first colonists of fumigated soil. Of the bacteria, *Pseudomonas* spp. are among the first, owing to their short response time and their fast growth rate. A small but adequate amount of inoculum of these microorganisms inevitably is left scattered throughout the soil after fumigation. Unfortunately, root pathogens such as *Pythium* spp. are also rapid colonists of fumigated soil, particularly if a susceptible crop is grown immediately, thereby providing abundant susceptible roots for pathogens of this genus. The evidence now leaves little doubt that much of the plant growth-response to soil fumigation results from the healthier roots made possible by the elimination of numerous root pathogens such as *Pythium* spp. (Anderson, 1966; Bruehl, 1951; Cook and Haglund, 1982; Wilhelm, 1965), which damage absorptive tissues of the roots and thereby impair ability of the roots to obtain nutrient and manufacture growth factors for the tops (Cook, *in* Bezdicek and Power, 1983). The need to refumigate before each planting can be explained by the rapid recolonization of the soil by pathogens, especially fast-growing root colonists such as *Pythium* spp. An adequate delay between soil fumigation and the sowing of the crop might provide a greater advantage to the nonpathogens as colonists of the soil. Establishment through seed treatment of an aggressive, competitive, and preferably antibiotic-producing but nonpathogenic microorganism on the roots to preempt pathogens in the root tip region may be another means to prolong the plant growth-response in fumigated soil (Chapter 6).

SOIL FUNGISTASIS

The concept of soil fungistasis is that for the spores of fungi, dormancy cannot be explained solely on the basis of lack of substrate. Rather, fungus spores are thought to exist in soil in a state of exogenously enforced dormancy, which is counteracted or nullified by an exogenous energy supply such as provided by root exudates. Sclerotia of certain phytopathogenic fungi, e.g., *Macrophomina phaseolina*, illustrate the phenomenon especially well; these propagules are storehouses of food for the fungus, yet they require an external source of nutrients from plant roots or other sources to overcome fungistasis (Hsu and Lockwood, 1973) and germinate. The chlamydospores of *Fusarium solani* f. sp. *phaseoli* require an external source of nitrogen in addition to carbon and energy for germination in soil (Cook and Schroth, 1965), but in axenic culture they germinate with water alone (Griffin, *in* Nelson et al, 1981). Clearly, the dormancy of the sclerotia of *M. phaseolina*, the chlamydospores of *F. solani* f. sp. *phaseoli*, and the propagules of many other fungi in soil is not a straight-forward case of nutrient dependency within the propagules, in spite of the fact that the treatments effective in causing their germination in soil are straight-forward nutrient treatments.

A great deal of research has centered on trying to explain the general phenomenon of soil fungistatis, but it would be just as challenging and perhaps more revealing to explain the basis of dormancy of the propagules of even one fungus in soil. There is probably no one universal explanation for the dormancy of fungus propagules in soil. The mechanisms may be different according to the fungus species and the environmental conditions.

Soil fungistasis has survival value for the pathogen, ensuring that growth will commence only when a host root or other adequate supply of substrate is available. The fungus has little to gain and everything to lose by commencing growth in soil with its own endogenous reserves as its sole supply of energy. Possibly selection in soil has favored energy storage by plant-pathogenic propagules as a means to prolong survival between hosts as well as to support new growth.

Griffin (1970a, 1970b) showed that the carbon requirement for germination of macroconidia and chlamydospores of *Fusarium solani* under axenic conditons is greater when the spores are at a high density than when at a low density. Propagules in high density are well known to limit their own germination by self-inhibitors (Allen, *in* Heitefuss and Williams, 1976; Chapter 5). Perhaps fungus spores in close proximity to (or in physical contact with) the cells or colonies of other microorganisms are subject to inhibition in the same way they inhibit themselves when in dense populations in pure culture.

Cook (*in* Horsfall and Cowling, 1977) suggested that fungistasis in some cases is a form of self-inhibition operative within the propagule in response to stimuli or signals of many kinds from the external environment. The propagules of many soilborne plant pathogens are known to exist in a state of constitutive dormancy in soil and to break dormancy after a certain age (oospores of *Pythium ultimum*), in response to a certain temperature (sclerotia of *Claviceps purpurea*), or perhaps to wetting and drying (sclerotia of *Sclerotium rolfsii*). Perhaps other propagules become "endogenously dormant" in response to some stimuli provided by the complex soil environment and break dormancy in response to other stimuli or signals. Such a mechanism seems likely in nature because the survival value of averting untimely germination is too significant for the fungus to depend on exogenous inhibitors. Such a mechanism could also explain why fungistasis is widespread in soils; the "static" factor is carried to soil with the spore. Ko and Lockwood (1967) have proposed a nutrient sink hypothesis to explain fungistasis, whereby metabolism of spore exudates by microorganisms around the spores creates a nutrient gradient across the membrane and spore wall and causes nutrient-independent spores to become nutrient-dependent. The nutrient-sink hypothesis and a self-inhibitor concept are not incompatible theories, because the nutrient gradient, while it might be sufficiently steep and enduring to drain the propagule of its endogenous reserves, might also activate a mechanism of self-inhibition.

A means to regulate soil fungistasis would be highly significant to biological control. One approach would be to remove the source of fungistasis. Perhaps this is one of the effects of polyethylene tarping of soil. L. J. Ashworth (personal communication) observed that microsclerotia of *Verticillium dahliae* become soft after a while in soil beneath a tarp, suggesting that germination or deterioration had occurred. On the other hand, if fungistasis is a manifestation of a self-inhibitor mechanism activated by various signals from the biotic and abiotic environment, then removal of the source of fungistasis may not be possible. An alternative would be treatments that override fungistasis. The most practical treatment identified thus far for this purpose is the application of organic manures that release nutrients and volatiles stimulatory to germination (Papavizas and Lumsden, 1980; Chapter 4). There is a hazard to this approach if the fungus forms new or secondary propagules and increases its inoculum density in response to the added substrate. Some clones of *Fusarium solani*, in response to such an energy supply, may form new propagules for as long as 10–20 hours after the external nutrient supply has been removed (Myers and Cook, 1972). The challenge is to find a treatment to support germination but not formation of replacement propagules; the result is germination followed by lysis and a reduction in the inoculum density.

Microsclerotia of *Verticillium dahliae* and sclerotia of *Sclerotium rolfsii* germinate in response to organic volatiles released from decomposing alfalfa hay and other residues (Gilbert and Griebel, 1969; Linderman and Gilbert, 1969, 1973, *in* Bruehl, 1975). The volatiles include acetaldehyde, butyraldehyde, valeraldehyde, ethanol, methanol, and probably others (Owens et al, 1969). Methanol vapors from fresh peanut residue in soil have also been shown to stimulate sclerotial germination of *S. rolfsii* (Beute and Rodriguez-Kabana, 1979). Unlike nonvolatile organic materials that stimulate germination, the volatiles apparently serve the propagule as a source of respiratory stimulants that trigger germination and as chemotactic agents that direct mycelial growth to crop residue or host roots.

There are many advantages to the use of volatiles to override soil fungistasis and obtain biological control. Volatiles diffuse through soil rapidly and are also absorbed rapidly by the propagules. Linderman and Gilbert (*in* Bruehl, 1975) showed that only a 15-minute exposure to volatiles from alfalfa hay was as stimulatory to germination of sclerotia of *S. rolfsii* as a five-day exposure. This reflects the speed with which volatiles can penetrate the soil and the cells of the fungus compared with the speed of nutrients supplied in solution. After a two-minute exposure to volatiles, sclerotia of *S. rolfsii* germinated, indicating that these stimulants of germination can enter cells without the help of water. Most important, volatiles are highly ephemeral in soil and, and although they may stimulate germination, the energy available for continuing growth of the pathogen is likely to be insufficient, and lysis may result. In effect, by treatment of soil with only a volatile from plant residue, the germination-stimulating factor of the residue may be separated from the residue itself, thereby inducing suicidal germination by the pathogen. Artificial onion oil has been used experimentally with considerable success in this way, as a stimulant of suicidal germination of sclerotia of the onion white rot fungus, *Sclerotium cepivorum* (Merriman et al, 1981).

When the soil and the sclerotia of *S. rolfsii* were simultaneously exposed to volatiles from alfalfa, sclerotia of *S. rolfsii* germinated, but when the soil only was exposed and then the soil and the sclerotia were both exposed seven days later (i.e., the soil was exposed twice but the sclerotia only once), germination was less than that for sclerotia present on the soil at the time of the first exposure. Such results are similar to those when nutrients are added to soil in successive doses. Those propagules present when the nutrients are first added (and when the microbial biomass has been quiescent for some time) germinate readily, but propagules brought into contact with a soil several days after the soil received nutrients (after microbial growth has been stimulated and some biomass turnover has occurred) are less likely to germinate, even if additional nutrients are provided. The microorganisms stimulated by the first amendment respond even more quickly to a second amendment, leaving less nutrient for the pathogen. Gilbert and Linderman (1971) also demonstrated a "mycosphere" effect (increased number of microorganisms) around sclerotia of *S. rolfsii* pressed into soil. Nearly eight times more total bacteria were found in soil taken from near sclerotia—bacteria thought to have grown on nutrients that leaked from the sclerotia. The mycosphere effect was enhanced if the soil had been exposed previously to volatiles from alfalfa hay.

CROP RESIDUE DECOMPOSITION IN RELATION
TO BIOLOGICAL CONTROL

The speed with which crop residue decays will determine how long pathogens can live within that residue. Two recent studies with labeled residue are discussed here because they give some idea of the dynamics of residue decomposition and biomass turnover following incorporation of fresh residue into soil.

About 50% of ^{14}C medic clover incorporated into different soils in South Australia was lost as $^{14}CO_2$ within the first four weeks. Thereafter, decomposition decreased significantly so that about 30% of the added ^{14}C remained after eight weeks and 15–20% still remained after four years (Ladd et al, 1981). In another study, Jenkinson (1977) estimated on the basis of decay rates for labeled ryegrass under field conditions at Rothamsted that 70% of the ryegrass had a half-life in soil of about three months and the other 30% had a half-life of eight years. The first phase of decomposition involves breakdown of materials such as cellulose, proteins, and possibly even lignins— materials that can be decomposed by a relatively large segment of the soil microbiota. This first and most rapid phase has the greatest effect on pathogens in the residue because the cellulose and proteins are also their main substrate. The second and slower phase of decomposition involves humic portions of the organic material and also complexes of nutrients adsorbed onto clays or in humic acid—substances available to only a small and highly specialized segment of the total microbiota, or not available to any microorganism in soil.

The half-life of three months that Jenkinson (1977) estimated for 70% of the carbon in ryegrass residue is for chopped residue (passed through a sieve having openings 0.5 mm in diameter) thoroughly mixed with soil. The figure of 70% carbon given off as carbon dioxide by eight weeks after addition of medic clover residue to soil was for unchopped fresh residue thoroughly mixed with soil (Ladd et al, 1981). Both studies were done in the field over a period of years. These figures might, therefore, be taken as the maximum rates of decomposition possible for crop residues thoroughly mixed with the soil under field conditions in temperate farming areas. Residue incorporated as large fragments or as clumps concentrated in pockets in soil will decompose more slowly, a situation advantageous to pathogens living in the residue. Crop residues are likely to decompose the most slowly when left entirely on the soil surface (no-till management) or incorporated only partially (minimum-till). This may explain why some pathogens that use residue as a shelter, food base, and springboard for infection are favored by minimum and no-till practices (Cook et al, in Oschwald, 1978; Chapter 10).

The significance of temperature to organic matter decomposition under field conditions is apparent from the results of Jenkinson and Ayanaba (1977), who compared ryegrass decomposition under field conditions in Nigeria and in Rothamsted. The pattern of decomposition was the same at both locations but was four times faster in soil in Nigeria. The average annual temperature at Rothamsted is 8.9°C, compared with 26.1°C in Nigeria. The velocity coefficient for decomposition at 10°C above a given temperature, i.e., $(T + 10)$°C, was two to three times that occurring at T°C. This agrees with the ratio of velocity coefficient for soil respiration, which has been calculated at 2.9 for temperatures in the range of 10–20°C, i.e., 2.9 times greater at $(T + 10)$°C than at T°C. On this basis, a pathogen within crop residue of a given size could be expected to die out with a velocity coefficient nearly three times larger in soil in a tropical area such as Nigeria than in soil in a temperate area such as southern England. Similarly, the rate of supply of organic material needed to maintain the microbial biomass at a given level and condition would probably be two to three times greater in the tropical soil than in an area such as southern England.

The relative rates of decomposition of organic matter in soil are linearly related to the logarithm of soil matric water potential, according to Sommers et al (in Parr et al, 1981), who summarized the available literature through 1981. Decomposition was assessed in terms of carbon dioxide evolution and also nitrogen mineralization; similar trends were indicated by both methods. Decomposition rates were maximal at matric potentials in the range of −0.2 to −0.5 bars. This is not surprising since, as pointed out in the opening section of this chapter, soil microorganisms under control of matric potential grow maximally at potentials near or wetter than −1 bar, regardless of their cardinal minimum water potential, as long as their oxygen needs are met. The range of −0.2 to −0.5 bars matric water potential meets these two requirements. Although the rate of decomposition decreases progressively with decreasing soil water potential, some decomposition occurs even at −200 bars. This probably involves such fungi as *Penicillium* and *Aspergillus* spp., which can grow at −200 bars and probably lower. Limiting water potential is another reason why crop residues tend to decompose more slowly when left on the soil surface, and hence why residue-inhabiting pathogens that are resistant to drying are favored in survival by leaving the residue on the soil surface (Cook et al, *in* Oschwald, 1978). Complete incorporation and thorough fragmentation of the residue accelerates the biological displacement and destruction of these pathogens, much as a campfire bursts into renewed but temporary flame each time the coals are stirred. Unfortunately this kind of tillage favors soil erosion and is therefore not sustainable (Chapter 1).

Composted bark of pine or hardwoods (oak, maple, or poplar) is sometimes used as the organic component of soil mixes for container-grown ornamentals (Hoitink, 1980). The composting is aerobicaly accomplished in three stages: 1) decomposition of easily degradable materials, which takes one or two days; 2) thermophilic phase, in which cellulose is degraded, heating the compost to 40–80°C for up to four months (as with solar tarping of field soil [Chapter 4], such temperature-time combinations are lethal to plant pathogens); 3) stabilization phase, during which decomposition decreases, temperatures decline, and antagonistic mesophilic microorganisms recolonize the material. During this procedure, the compost is piled in large windrows, turned frequently, and aerated with fans through the perforated floors. Usually 1 kg of ammonium nitrogen and 0.3 kg of P_2O_5 are added per cubic meter of bark, but urea or poultry manure may be used. Peat (1:4 [v/v]) may be added. The compost is mixed before use with inert silica sand, perlite, pumice, expanded shale, or Styrofoam when in phase 3. The rooting medium produced by this procedure is well-aerated, nonphytotoxic, and heat-treated. In addition, this composted medium is thought to contain high populations of antagonistic microorganisms and some unidentified water-soluble materials inhibitory (but not lethal) to fungus and bacterial pathogens, making it suppressive to root-disease organisms. It has not been shown how rapidly these inhibitory compounds will be decomposed in this biologically active medium.

PATHOGEN-SUPPRESSIVE SOILS

Pathogen-suppressive soils were defined (Baker and Cook, 1974) as soils in which the pathogen does not establish or persist, establishes but causes little or no damage, or establishes and causes disease for a while but thereafter the disease is less important, although the pathogen may persist in the soil.

Most examples of suppressive soils became apparent initially because the incidence or severity of disease was lower than expected for the prevailing environment and despite ample opportunity for the pathogen to have been introduced. The soil may be suppressive to the pathogen directly, e.g., propagules may die more rapidly or germinate less, or saprophytic growth in the rhizosphere may be suppressed. The result is a shorter life for the pathogen, smaller population, lower incidence or severity of disease for a given population, or some combination of these. Suppressiveness conceivably also operates through the host, e.g., by cross protection or induced resistance caused by the associated microbiota. The terms disease-suppressive and pathogen-suppressive are sometimes used interchangeably, unless the mechanism has been shown to affect growth or survival of the pathogen, in which case "pathogen-suppressive" is preferable.

Suppressive soils are known for the following plant pathogens: *Gaeumannomyces graminis* var. *tritici*, where take-all decline has occurred (Shipton, *in* Bruehl, 1975); pathotypes of *Fusarium oxysporum* that cause vascular wilts (Toussoun, *in* Bruehl, 1975); *F. solani* f. sp. *phaseoli* (Burke, 1965; Furuya and Ui, 1981); the cereal cyst nematode (Kerry, *in* Papavizas, 1981); root-knot nematode (Mankau and Prasad, 1977; Stirling and Mankau, 1978); *Phytophthora cinnamoni* (Broadbent and Baker, *in* Bruehl, 1975), *Pythium* spp. (Hancock, 1977; Kao and Ko, 1983; Bouhot, *in* Schippers and Gams, 1979), *Rhizoctonia solani* (Ko and Ho, 1983; Baker, *in* Schneider, 1982), *Streptomyces scabies*, (Menzies, 1959), and *Sclerotium cepivorum* (Rahe and Utkhede, 1980). There is also evidence that soils may become suppressive to *Phymatotrichum omnivorum* in the southwestern United States if recropped to cotton (Baker and Cook, 1974; King, 1923), and a special fungistatic factor is thought to suppress sclerotia of *Sclerotium oryzae* in certain rice field soils in California (Keim and Webster, 1974).

General and Specific Suppression

Most soils are suppressive in some degree to soilborne pathogens. Gerlagh (1968) referred to the suppressiveness characteristic of all soils to *G. graminis* var. *tritici* as general suppression. General suppression may be the equivalent of the general resistance of living plant tissues to pathogens (Chapter 3); both are quantitative and both relate in some undetermined way to total life in the system. Soil also may be suppressive to plant pathogens in a specific sense (specific suppression), analogous to more specific forms of resistance in plants to pathogens. These two broad categories are defined in more detail below since they may encompass several of the examples of suppressive soil studied thus far. Since highly effective suppressive soils of both types occur, it is clear that they can be equally effective in practical agriculture.

General suppression of a pathogen is directly related to the total amount of microbiological activity at a time critical to the pathogen. A particularly critical time is during propagule germination and prepenetration growth in the host rhizosphere. The kinds of active soil microorganisms during this period are probably less important than the total active microbial biomass, which competes with the pathogen for carbon and energy in some cases and for nitrogen in other cases (Benson and Baker, 1970), and possibly causes inhibition through more direct forms of antagonism. In a sense, general suppression of a pathogen in soil is the equivalent of a high degree of soil fungistasis. No one microorganism or specific group of microorganisms is responsible by itself for general suppression.

Specific suppression operates against a background of general suppression but is more qualitative, owing to more specific effects of individual or select groups of microorganisms antagonistic to the pathogen during some stage in its life cycle. Examples of specific suppression, besides that responsible for take-all decline (Cook and Rovira, 1976), could include the soil from Bogatá, Colombia, suppressive to *Rhizoctonia solani* because of its high population of *Trichoderma hamatum* (Chet and Baker, 1981) and soils from fields in England, where high populations of *Nematophthora gynophila* and other hyperparasites control the cereal cyst nematode. Undoubtedly most pathogen-suppressive soils involve a combination of general and specific suppression. Our objective in biological control should be to find ways to elevate or conserve the level of general suppression, specific suppression, or both.

Soils Suppressive to *Gaeumannomyces graminis* var. *tritici*

The available evidence indicates that a form of specific suppression is responsible for take-all decline. This suppression results from a qualitative change in the soil microbial population following monoculture of the host and probably also the occurrence of severe disease on at least one or two wheat crops early in the sequence (Cook and Rovira, 1976). In contrast, general suppression probably is a function of the nutrient and energy supply available for prepenetration growth by *G. graminis* var. *tritici*, as determined by total active microbiota of the soil acting as competitors with *G. graminis* var. *tritici* for root exudates and for its food base.

Specific suppression. — Several theories have been proposed to explain take-all decline (Hornby, *in* Schippers and Gams, 1979). One of the first mechamisms proposed, hypovirulence, cannot explain take-all decline for reasons discussed in Chapter 5. Antagonism from *Phialophora graminicola* and other avirulent fungi with potential for cross protection of wheat roots against *G. graminis* var. *tritici* likewise cannot explain take-all decline (Deacon, 1976a). Giant vampyrellid amoebae cause perforations of pigmented hyphae of *G. graminis* var. *tritici* (Homma et al, 1979), but their action is too slow and their water requirements too restrictive (Cook and

Homma, 1980) to account for the suppression of *G. graminis* var. *tritici* on roots of wheat. Antagonism from root-colonizing fluorescent *Pseudomonas* species seems to best explain take-all decline for reasons given below.

First, roots of wheat growing in two suppressive soils (each from fields cropped about 20 consecutive years to wheat) had about 10 times more total fluorescent pseudomonads per unit-length than did roots of wheat growing in either of two conducive soils (from a virgin and a multiple-cropped site, respectively). Each of the four soils had been diluted 1:10 with the same fumigated soil (to reduce chemical and physical differences among the soils) and then amended with live inoculum of the pathogen as colonized oat grains (5 g of oat inoculum per kilogram of soil mixture) before the wheat was planted. Besides being present in greater numbers, a greater percentage of the fluorescent pseudomonads from roots from suppressive than from conducive soils were inhibitory to the pathogen in vitro (Figure 7.1). No difference in total number of pseudomonads or of inhibitory types occurred on roots from the same four soils amended with dead oat inoculum of the pathogen. This confirms the report (Gerlagh, 1968) that live virulent inoculum of the pathogen is necessary for expression of the suppression.

Second, several strains and combinations of strains inhibitory to *G. graminis* var. *tritici* in vitro suppressed take-all in both greenhouse pots and field plots (Figure 4.6) when introduced as a seed treatment at sowing (Weller and Cook, 1983). An effective strain marked with resistance to rifampicin and nalidixic acid was shown to colonize wheat roots in natural soils in the field when applied as a seed treatment. At one month after sowing in the field, the introduced strain made up 50% of the total population of aerobic bacteria (measured by a dilution plate-count) on the roots and more than 90% of the total population of pseudomonads. The strain was still detectable on roots nine months after planting. Significantly, the highest population occurred on roots of plants in plots where the take-all pathogen also had been introduced (Weller, 1983), confirming that the pathogen must be present to maintain suppressiveness (Figure 7.2).

Third, the factor suppressive to the pathogen after take-all decline is sensitive to aerated steam treatment of the soil at 60°C for 30 minutes (Gerlagh, 1968; Shipton et al, 1973; Vojinović, 1972). Pseudomonads would be destroyed by this treatment. In addition, the factor in soils suppressive to take-all is highly sensitive to compaction and reduced gas exchange (Cook, *in* Asher and Shipton, 1981), which fits the aerobic nature of fluorescent pseudomonads (Chapter 9).

Fourth, the factor suppressive to *G. graminis* var. *tritici* after take-all decline may not limit the number of infection sites on wheat roots, but rather, may limit the progress of disease after infection (Cook, 1981). This observation supports the hypothesis that microorganisms on the root or in

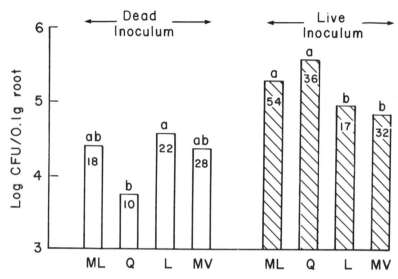

Figure 7.1. The number of antagonistic fluorescent pseudomonads from roots of wheat grown in soils from the indicated fields and amended with live or dead inoculum of *Gaeumannomyces graminis* var. *tritici*. The ML and Q fields had undergone take-all decline and soils from these fields were suppressive to take-all. L and MV were virgin and multiple-cropped sites, respectively, and the soils were conducive to take-all. Each test soil was diluted 1:10 with the same fumigated (L) soil and then amended with the inoculum. Total fluorescent pseudomonads on the roots were estimated by plate counts, and then randomly selected colonies were tested for inhibitory activity on King's medium B. The test with live inoculum was initiated one day later than the test with dead inoculum, and is therefore treated for statistical purposes as a separate experiment. Bars within each test with a different letter were significantly different at $P = 0.05$, according to Duncan's multiple range test. The values inside each bar are the percentage of total fluorescent pseudomonads that were inhibitory to *G. graminis* var. *tritici* in vitro. (Unpublished data from D. M. Weller.)

the "leaky lesions" inhibit the parasitic activity of *G. graminis* var. *tritici* in soils following take-all decline (Cook and Rovira, 1976). Fluorescent pseudomonads are among the most successful of all colonists of roots and lesions.

General suppression. — General suppression probably involves the competitive and other antagonistic effects of the total active soil microbiota during prepenetration growth of the pathogen in soil and the rhizosphere. **The greater and/or the more responsive the microbial biomass to any new source of available nutrients, the more intense the competition likely to be encountered by root pathogens.** Anderson and Domsch (1978b) showed that a soil with a relatively small microbial biomass required relatively less glucose to stimulate maximal respiration than was needed for a soil with a large microbial biomass. This raises the possibility that a given amount of carbon and energy

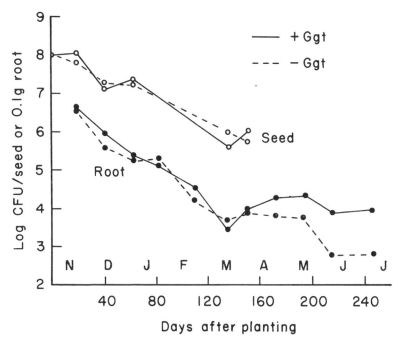

Figure 7.2. The population of a marked strain of a fluorescent pseudomonad, on the seed where introduced, and on the roots in natural soil in a plot at Pullman with and without introduced inoculum of *Gaeumannomyces graminis* var. *tritici*. The strain was inhibitory to *G. graminis* var. *tritici* on agar and suppressive to take-all in the plots. (Reprinted with permission from D. M. Weller.)

as root exudates released into soil having a relatively small microbial biomass is more likely to temporarily exceed the requirements of the local microbial population. A temporary surplus would occur at the root surface and possibly for some distance outward from the root surface, depending on conductivity. The same amount of root exudate released into a soil having a large and/or highly responsive microbial biomass will not persist as long, and no surplus is likely to occur around the root or even on the root surface.

Brown and Hornby (1971) concluded on the basis of their observations that *G. graminis* var. *tritici* requires a "feeding phase" before infecting the seminal roots. At least part of the nutrient for prepenetration growth was provided by the residue fragments. The larger the host fragment containing the pathogen, the greater the amount of infection. Root fragments as a food base were generally insufficient for direct penetration of the seminal root axes. Prepenetration growth of the fungus was also supported by root exudates since growth occurred toward the roots when seedlings were present but no growth occurred beyond the residue in the absence of roots. Growth toward the roots also occurred some distance from the roots. Any reduction in the supply of one or more essential nutrients, caused by competing microorganisms either in the rhizosphere or as coinhabitants with the pathogen in crop residue, is likely to be reflected as less prepenetration growth and hence less infection. This kind of suppression is different from that responsible for take-all decline as discussed below.

Wilkinson et al (1982a) showed that the number of successful infections of roots (recognizable as root lesions) in a Shano silt loam (SSL) from a field that had been cropped 22 consecutive years to wheat (and that had undergone take-all decline) was about the same as in a known conducive Ritzville silt loam (RSL) from an uncropped (virgin) site. The number of infections varied with the size of the residue fragments (infected roots, wheat crowns, or axenically colonized oat grains) containing mycelium of the pathogen. The minimum threshold size of infested crown or oat-grain particles required for infection was 0.35–0.5 mm. Larger particles (0.75–1.0 mm) were required to produce infections in the two soils if root tissues were the substrate, and even the largest infested root particles tested (1.0–2.0 mm) initiated no infections of roots in the conducive RSL. This confirms the observation of Brown and Hornby (1971) that root residue is less effective than crowns as a food base for this fungus. A model developed (Wilkinson et al, 1982a) from the data indicated that the average effective distance for infection (EDI) for all particles, i.e., the average distance between the inoculum source and the root surface (Gilligan, 1980) was 5.9 and 5.3 mm in nonsterile SSL and RSL soil, respectively. However, when the soils were heat-treated (60°C for 30 minutes) or fumigated (methyl bromide), infections occurred with smaller particles, the number of infections per unit mass of infested particles was greater, and the EDI increased to more than 11 mm in both soils. The suppressiveness (specific suppression) known to occur in the SSL (Weller and Cook, 1982) because of take-all decline could not, therefore, be demonstrated on the basis of frequency of infection or EDI. Rather, infection frequency was best explained on the basis of amount of nutrient available to the fungus in the soil and the host residue fragments for prepenetration growth, which now appears to be the basis for general suppression.

The natural SSL and RSL soils used by Wilkinson et al (1982a) were stored air-dry before use and were not wetted until the inoculum particles had been added and the wheat seeds planted. The air-drying and rewetting results in a brief but significant flush of available nutrients, especially in the SSL, which contained more organic matter. A greater flush of nutrients can explain why infectivity of the infested particles was slightly greater in SSL than in RSL. Fumigation or heat treatment would produce an even greater release of nutrients, which can explain the markedly greater infectivity of the inoculum particles in such soil, including the occurrence of infection from greater distances away from the root surface. Axenically colonized oat grains, and to a lesser extent infested crown tissue, would provide a better nutrient source than infested root debris because their higher C:N ratio would tend to attract fewer colonists and hence less competition than would roots.

The classic work of Henry (1932), demonstrating that the optimum soil temperature for take-all is about 25°C in sterile soil but only 12–16°C in nonsterile soil, can be explained by greater general suppression in the warmer than in the cooler soils (Cook and Rovira, 1976). The higher respiratory activity of the warmer soil results in a greater demand for various nutrients, especially for carbon and energy for growth by the associated microbiota, leaving less for prepenetration growth by *G. graminis* var. *tritici*. The same may hold for a soil recently amended with organic materials; such a soil will have a larger total biomass and also a younger biomass. A young or "rejuvenated" microbiota might be a greater sink for nutrients such as are made available by root exudates, thereby resulting in less nutrient to support the pathogen. Virtually any treatment to enhance the total microbial activity of a soil will enhance the levels of general suppression of soil to this pathogen.

The specific and general types of suppression described in this section are distinct from the hypovirulence phenomenon reported for this fungus (Chapter 5) and also from the cross protection demonstrated to occur in certain rotations in England. An effective biological control of take-all should combine all of these if possible.

Fusarium-Suppressive Soils

The existence of *Fusarium*-suppressive soils was first recognized by G. F. Atkinson in 1892 and first studied 50 years ago by Reinking and Manns (1933) in Central America and by Walker and Snyder (1933) in Wisconsin. These early studies pertained to fusarium wilt pathogens, notably *F. oxysporum* f. sp. *cubense* and *F. oxysporum* f. sp. *pisi*. Details of these and other early studies on soils suppressive to the fusarium wilt pathogens are available from Baker and Cook (1974) and Toussoun (*in* Bruehl, 1975). Recent research has been conducted with a suppressive soil from the Chateaurenard region of the Rhône Valley in France, where fusarium wilt-susceptible vegetables have been grown since ancient times with no problem from wilt; with suppressive soils from some areas of the Salinas and San Joaquin valleys of California, where fusarium wilts have not been a problem despite repeated cropping of the fields with wilt-susceptible plants; and certain black humic soils of volcanic origin in Japan that are highly suppressive to radish yellows caused by *F. oxysporum* f. sp. *raphani* (Komada, 1975). The discussion below is limited to this recent work.

Fusarium Wilt Diseases

Biocontrol of the fusarium-wilt pathogens illustrates the use of suppressive soils to inhibit propagule germination and prepenetration growth by the pathogen. It is accomplished through: 1) naturally suppressive soils; 2) transfer of antagonistic microbiota by mixing nontreated suppressive soil with conducive soil; 3) stimulating antagonistic microbiota to greater effectiveness by soil treatments; 4) relatively stable host-plant resistance; 5) cross protection.

Pathogens: At least 76 formae speciales of *Fusarium oxysporum*, several of which have cultivar-specific races (Deuteromycotina, Hyphomycetes, Tuberculariaceae).

Hosts: The species *oxysporum* includes many pathotypes that are more or less specific for a plant genus, species, or even cultivar. Collectively, the host range of the species is very large. The species also include many common soil saprophytes that may colonize plant roots and plant debris but do not cause wilt. The nonpathogenic and pathogenic strains of *F. oxysporum* are morphologically indistinguishable. The host range of some individual formae speciales is now wider than originally thought, but this specificity has proved so remarkably stable that resistance is the principal control today.

Diseases: Various gradations of chlorosis, stunting, vascular discoloration, wilting, and stem streaking are produced on different hosts, usually culminating in premature death. Chlorosis and wilting generally advance upward from the stem base, but symptoms may appear first in the youngest leaves (chrysanthemum). The pathogen may advance from the xylem into the cortex, causing cortical decay or even damping-off of seedlings (watermelon, aster). Plants grown at low soil temperatures may be infected but exhibit few or no symptoms. Plants infected on one side of the root system may show one-sided stem streaking and wilting.

Life Cycle: Once introduced, vascular-wilt fusaria survive in soil for long periods even without a suscept, probably living on roots of nonhosts, as infections of symptomless hosts, or as dormant chlamydospores. Infection is by germ tubes or mycelium that penetrate through the cortex in the region of root elongation or through wounds. Mycelium grows into xylem vessels and spreads upward by growth and by microconidia in the transpiration stream. Plugging of vessels and resultant increased resistance to water movement cause wilting and plant death, at which time the pathogen colonizes the stem more thoroughly. The pathogen produces macroconidia on the dead stem surface; in threshing, these spores may adhere to seed, and rains may wash them into the soil, where they convert to chlamydospores. In stock and flax, the fungus grows into the seed and is internally borne. The pathogen also spreads by infected or infested plants, in soil, on implements, and in water. Inoculum density in soil, soil temperature, and host resistance are interchangeable concomitant factors, an increase in one offsetting decreases of the others in terms of disease incidence.

Resistance is controlled by: 1) a single, qualitative, dominant gene (vertical or type A resistance) that gives stable resistance against specific races of the pathogen; 2) multiple quantitative, additive genes (horizontal or type B resistance) that give little protection at high soil temperature but provide resistance to several races.

Environment: Vascular-wilt pathogens generally grow maximally at 28°C and are inhibited above 33°C, except that at very low substrate osmotic potentials (e.g., -50 to -120 bars) growth can occur at 35°C. Fusarium wilts usually are not important below 17°C. Disease is favored by good soil moisture, probably because of greater transpiration and hence upward distribution of the pathogen. The pathogen can grow at substrate osmotic water potentials down to -100 to -120 bars; this ability probably is advantageous during saprophytic colonization of dead host stems. Acid soils are more favorable than neutral to alkaline soils, and wilt can be controlled by liming the soil to achieve pH 7.0 or above. Sandy soils are more favorable than clay soils, perhaps due to the stimulatory effect of clays (at least of montmorillonitic clays) on development of bacterial antagonists.

Biological Control: Many suppressive soils are known over the world. This suppressiveness is destroyed by moist heat treatment (60°C for 30 minutes), chemical treatment, or gamma radiation and is inhibited by novobiocin treatment, showing that the agent is biological, possibly bacterial. The suppressive microflora can be transferred; as little as 1% (w/w) suppressive soil mixed with conducive soil makes it suppressive in turn. A soil suppressive to one pathotype of *F. oxysporum* is suppressive to other pathotypes as well but not to *F. roseum, F. solani,* or nonpathogenic *F. oxysporum.* Suppressiveness results at least partly from increased competition for ephemeral carbon and energy sources by the larger total active microbial biomass. A strain of *Pseudomonas putida* isolated from the California suppressive soil induced suppressiveness when added to conducive soil; this bacterium may produce siderophores that complex iron, making it unavailable to pathogens, and so suppress disease.

Suppressiveness of soils from the Chateaurenard region in France was thought to involve *F. solani* and saprophytic *F. oxysporum*, perhaps acting as competitors of the pathogen where nutrient supplies were already limiting. Central American soils suppressive to *F. oxysporum* f. sp. *cubense* are high in montmorillonitic clay favorable to growth of bacteria.

The pathotypes of *F. oxysporum* will usually provide temporary cross protection against one another under experimental conditions; possibly some cross protection also occurs between nonpathogenic and pathogenic strains under field conditions. In Japan, cross protection provides effective control of fusarium wilt of sweet potato in the field.

References: Toussoun (*in* Bruehl, 1975), Mace et al (1981), Louvet et al (*in* Nelson et al, 1981), Nelson et al (1981), Scher and Baker (1980, 1982), Alabouvette et al (*in* Schippers and Gams, 1979), Walker (1969).

Fusarium oxysporum behaves differently in suppressive than in conducive soil in several ways. Chlamydospore germination and germ-tube growth are markedly less, commonly only one fourth to one tenth that in conducive soil, depending on the amount of nutrient present (Alabouvette et al, 1980a; Hwang et al, 1982; Komada, 1975; Smith, 1977; Smith and Snyder, 1972). The inoculum density required to produce a given incidence of wilt is considerably greater, sometimes several magnitudes greater (Alabouvette et al, *in* Schippers and Gams, 1979; Smith and Snyder, 1971). The propagule density of the pathogen declines more rapidly in the suppressive than in the conducive soil (Komada, 1975). Finally, soils suppressive to one pathotype of *F. oxysporum* are usually suppressive to other pathotypes of the same species (Alabouvette et al, 1980a; Komada, 1975; Smith and Snyder, 1972). On the other hand, *Fusarium*-suppressive soil from the Chateaurenard region was not suppressive to *Verticillium dahliae, Pythium* spp., *Phytophthora* spp., *Rhizoctonia solani, Sclerotinia* spp., or *Pyrenochaeta lycopersici* (Alabouvette et al, 1980a). Suppressiveness is eliminated by aerated steam treatment at 55°C for 30 minutes (Rouxel et al, 1977; Scher and Baker, 1980), and by 250–500 krad treatment with gamma radiation (Alabouvette et al, 1977), showing the importance of antagonistic microorganisms. As with soils suppressive to *G. graminis* var. *tritici*, the evidence indicates that suppression of *F. oxysporum* relates in part to the total microbial activity of soil but also to antagonism from a specific group of soil microorganisms. Perhaps suppressiveness of soils to pathotypes of *F. oxysporum* is also of either a general or specific type.

That a select group of soil microorganisms is involved is suggested by the fact that in Colorado, the addition of only 1% *Fusarium*-suppressive Metz fine sandy loam from the Salinas Valley to a bed of steamed soil in a commercial glasshouse was sufficient to suppress fusarium wilt of carnation caused by *F. oxysporum* f. sp. *dianthi* (Baker, 1980; Scher and Baker, 1980). The incidence of wilt in a second crop of carnations 17 months after addition of the suppressive soil was 66% in the steamed soil, 50% in steamed soil amended with conducive soil, and 17% in steamed soil amended with suppressive soil. McCain et al (1980) similarly showed that a small amount of suppressive soil from the Salinas Valley mixed with beds of steamed soil has potential to control fusarium wilt of carnation in California. In France, 10% of either of two *Fusarium*-suppressive soils (Siagne and Chateaurenard, respectively) was suppressive to fusarium wilt of cyclamen caused by *F. oxysporum* f. sp. *cyclaminis* when added to steamed soil and to fusarium wilts of tomato and carnation when added to commercial peat mixes. A suppressive soil (also obtained from the Salinas Valley) mixed at only 2% with a conducive Ritzville silt loam from Washington gave detectable suppression of chlamydospore germination of *F. oxysporum* f. sp. *pisi*, and 20% in the mixture was as suppressive as 100% suppressive soil (Hwang et al, 1982).

Two separate lines of investigation indicate that fluorescent *Pseudomonas* spp., possibly siderophore-producing pseudomonads, account at least partly for the suppressiveness of the California soils to *F. oxysporum*. Scher and Baker (1980) showed that each of two isolates of *Pseudomonas* spp. (isolated from mycelial mats of *F. oxysporum* f. sp. *lini* buried in the suppressive soil) was suppressive to fusarium wilt of flax when added to conducive soil at 10^5 cells per gram. Populations of these same bacteria in soil were highly sensitive to aerated steam treatment at 54°C for 30 minutes but not to 49°C for 30 minutes, which corresponds to heat sensitivity of the suppressive factor present in the Metz fine sandy loam. In California, Kloepper et al (1980a) showed for fusarium wilt of flax that addition of either a suspension of cells of strain B–10 of the *Pseudomonas fluoroscens-putida* group or 50 μM of an extracellular siderophore produced by this strain caused a conducive soil to become suppressive to *F. oxysporum* f. sp. *lini*. In contrast, the addition of Fe^{III} but not Fe^{III}EDTA caused the naturally suppressive soil to become conducive. Siderophores are produced mainly or only where iron is limiting, and they provide an important mechanism by which microorganisms obtain iron in competition with other microorganisms. Production of siderophores by root-colonizing fluorescent *Pseudomonas* spp. is thought to account for the reduction in population of total fungi and Gram-positive bacteria on the rhizoplane of plants grown from seed inoculated with the bacteria (Kloepper et al, 1980a). Both fluorescent (ferric-pseudobactin and pseudobactin) and nonfluorescent (ferric-pseudobactin A and pseudobactin A) siderophores have been identified and their structures determined (Figure 9.9) (Teintze and Leong, 1981; Teintze et al, 1981).

The controlling effect of lime on fusarium wilts is nullified by addition of trace nutrients (Chapter 6). Significantly, the suppressive soils from the Salinas and San Joaquin valleys used by Smith (1977) both had higher pH values than the conducive soils used for comparison. Scher and Baker (1980) found that fusarium wilt of flax was significantly less at pH 8.0 than at pH 6.0 in the suppressive Metz fine sandy loam, but was severe at both pH 8.0 and 6.0 in the conducive soil. Possibly iron chelation by siderophore-producing bacteria is involved in the control of fusarium wilts by liming. Any attempt to introduce these bacteria for biological control of fusarium wilt must consider soil pH and the possibility of liming the soil for maximum benefit.

In addition to evidence for a specific form of suppressiveness to *F. oxysporum*, other evidence points to suppressiveness caused by the total active microbial biomass in the soil, serving as a nutrient sink. **The higher the level of microbial activity, the more difficult it will be for any given portion of the microbial population to obtain the nitrogen and/or carbon and energy necessary for their growth.** The result would be less chlamydospore germination and shorter germ tubes in suppressive than in conducive soil, as commonly observed. It is significant in this regard that increasing the amount of nutrient partially nullifies suppressiveness as measured by chlamydospore germination and germ tube growth (Hwang et al, 1982; Alabouvette et al, *in* Schippers and Gams, 1979; Smith, 1977). Germination of chlamydospores of *F. oxysporum* f. sp. *pisi* in suppressive soil from the Salinas Valley was only 10–15% in 24 hours, compared with 80–90% in conducive soil amended with the same concentration of glucose. Doubling the glucose concentration gave 30% germination, and a four- and an eightfold increase in the amount of glucose each gave 50–60% germination in the suppressive soil. One or more factors other than carbon apparently limited germination of *F. oxysporum* f. sp. *pisi* to 50–60% in the suppressive soil from Salinas, but up to that level of germination, carbon apparently was a limiting factor. Neither nitrogen supplied as asparagine nor iron supplied as ferric chloride was effective in increasing the percentage of germination above 50–60%. Increasing the amount of glucose nullified suppressiveness of the Chateaurenard soil in France to chlamydospore germination of *F. oxysporum* f. sp. *melonis*. Carbon and energy must therefore be more limiting in the suppressive than in the conducive soils. Moreover, large amounts of glucose were required to increase germination in the suppressive soils; Anderson and Domsch (1978b) similarly showed that relatively large amounts of glucose were required to stimulate maximal initial rates of respiration in soils with a large microbial biomass.

C. Alabouvette (personal communication) compared carbon dioxide production from a suppressive soil with that from conducive soil, both amended with glucose. The respiratory rise was significantly faster and greater in suppressive soil, and more glucose was required to achieve maximal respiration. Komada (1975) had observed earlier that more carbon dioxide was released from the rhizosphere of radish growing in suppressive soil than from that in conducive soil, owing to greater bacterial activity in the rhizosphere in suppressive soil. He considered that the lower amounts of chlamydospore germination and greater amounts of germ tube lysis were caused by "nutrient deprivation," as a result of this greater bacterial activity. S. F. Hwang (unpublished) has confirmed that the amount of carbon dioxide released from soil in response to added glucose is greater from suppressive soil than from conducive soil, especially during the first 24 hours, and that more glucose was required in the suppressive soil to achieve the maximal respiratory response. These results are consistent with the hypothesis that the microbial biomass in suppressive soil represents a greater sink for nutrients (probably carbon and energy compounds), leaving less to support of chlamydospore germination and germ tube growth by *F. oxysporum*.

Competition between two or more organisms will be greatest if they use the same kinds of substrate and have the same environmental requirements. In the case of *F. oxysporum*, pathogens and nonpathogens of the same or related species are likely to be competitors in soil and the rhizosphere. Where food is more available, as is suspected in the conducive soil, pathogens as well as nonpathogens are more likely to succeed, and infection occurs. Where food is less available, as is suspected in the suppressive soil, only the most competitive will succeed, and the nonpathogens are likely to be more competitive than the pathogens. The greater the population of non-pathogens in soil that is already nutrient-limiting to this species, the more suppressive the soil will be to pathogenic *F. oxysporum*. In the work of Alabouvette et al (1979), nonpathogenic members of *F. oxysporum* and *F. solani* added to steamed soil were suppressive to the pathogen only if the soil was originally (before steaming) suppressive, but not if originally conducive. This result shows that potential competitors have a greater effect in some soils (perhaps in soils potentially more nutrient-limiting) than in others.

Work by Stotzky and Martin (1963) and Stotzky and Rem (1966) revealed that a higher content of montmorillonitic clay in suppressive soils from Central America favored bacterial activity. The addition of montmorillonite to liquid cultures of bacteria enhanced the growth rate of bacteria, whereas kaolinite, present in many conducive soils, did not. It is now well documented (Marshall, 1976; *in* Ellwood et al, 1980) that certain clays may be stimulatory to growth of bacteria, apparently because of the accumulation of nutrients at the clay interfaces, and also because the clay provides more exchange sites for hydrogen ions, thus facilitating a more favorable pH for growth of bacteria. The intensity of competition encountered by propagules of *F. oxysporum* attempting to gain a portion of the finite supply of one or more essential nutrients from a root will relate not only to total active microbial biomass but also to responsiveness of the biomass, as affected by the kind of clay and probably other factors. Perhaps certain clays increase the general suppressiveness of soil to *Fusarium* by enhancing the competitive effects of bacteria. Liming the soil to raise the pH may have a similar effect, by shifting the relative competitive advantage in favor of greater competition from bacteria.

Phytophthora-Suppressive Soils

One of the major plant disease epidemics in the world today is the root rot of *Eucalyptus marginata* in Western Australia and of *E. radiata* in Gippsland, Victoria, caused by *Phytophthora cinnamomi*. The place of origin of this pathogen is somewhat uncertain, but "the evidence appears good for a possible center extending from the New Guinea-Celebes-Malaysian region into northeastern Australia and possibly in other parts of eastern Asia" (Zentmyer, 1980). Controversy (Broadbent and Baker, 1974a; Newhook and Podger, 1972; Shepherd, 1975) seems to center on the time of introduction into Australia from southeast Asia—whether it was so long ago that the pathogen can now be regarded as indigenous to northeastern Australia, or whether it was introduced there "probably no earlier than the late eighteenth century" (Newhook and Podger, 1972) The pathogen spread into Western Australia and Victoria much later, probably with nursery stock from eastern Australia.

Two different approaches to biological control have been used widely and effectively in eastern Australia against *P. cinnamomi* on avocado and pineapple. A third distinctly different method for control of this pathogen on eucalyptus is being developed for forest lands in Western Australia.

Avocado. — A 30-year-old avocado grove was found in Queensland in 1969 in which *P. cinnamomi* was present yet in which only a trace of root rot occurred on the trees despite average annual rainfall of 152 cm. This healthy grove surrounded by many severely diseased neighboring groves had been subjected to a system of continuous legume-maize cover crops, plus application of 0.73 t/ha (two tons per acre) of poultry manure twice a year and dolomitic limestone whenever the soil pH dropped below 6.0. In general, the suppressive soils were red clays of basaltic origin, with abundant organic matter, calcium, and nitrogen (largely in the ammonium form) tied up in the organic cycle typical of tropical rain forests.

This "Ashburner system," developed by Guy Ashburner, is now standard for Queensland and New South Wales avocado growers (Baker, 1978). Old diseased groves may be pulled and new container-grown, pathogen-free trees planted with marked success. Trees of the Ashburner grove were injured by *P. cinnamomi* in the extraordinarily wet year of 1974 (381 cm of rain, 175 cm of it in three days). These trees were then drastically pruned, and heavy applications of straw were made in addition to the usual regimen. The new growth reached 2 m in the first year and formed sizeable trees in the second year.

Ashburner had devised his system empirically in an attempt to maintain rain-forest conditions in the grove, including the high levels of organic matter and calcium, since avocado was said to be a rainforest tree in Central America. Pegg (1977a) found in extensive surveys that severe root rot from *P. cinnamomi* on a range of crops in Queensland was linked with low calcium levels in the soil.

Treatment of the suppressive soil with aerated steam (60°C for 30 minutes) did not destroy the suppressiveness when soil was reinoculated with *P. cinnamomi*, but 100°C for 30 minutes did, indicating that spore-forming bacteria or certain actinomycetes are the biological agents involved.

The suppressive effect appears to result from multiple antagonists, because in 300 attempts no single isolate from suppressive soil produced a suppressive effect. It has also not been possible thus far to transfer the inhibitor effect from a suppressive to a conducive soil (Broadbent and Baker, 1974a) as has been done for soils suppressive to *Gaeumannomyces graminis* var. *tritici* and *Fusarium oxysporum*. The suppressive soils have an exceptionally abundant, active, and heterogeneous microbial population. Such suppressiveness may be of the type referred to earlier as general suppression.

Suppressiveness may be temporarily lost when soil is flooded and waterlogged, probably because the biological balance among antagonists, the pathogen, and components of the background microbiota is altered. Depending on the duration of submersion, a month or more may be required to recover suppressiveness. The addition of excessive amounts of *P. cinnamomi* inoculum also may temporarily swamp suppressiveness and result in severe root rot (Broadbent and Baker, 1974a).

Roots of windfall volunteer avocado seedlings growing in shaded conditions beneath completely healthy avocado trees in suppressive soil are severely rotted, perhaps because of carbohydrate deficiency, but seedlings growing in relatively nonshaded spaces between trees have little or no root decay. Roots of Piccabeen palm (*Archantophoenix cunninghamiana*) are also slightly infected when growing beneath the avocado trees (Pegg, 1977a).

The Queensland rain forest soils have an extraordinary suppressiveness to *P. cinnamomi*. The pathogen cannot be isolated from soils 1 m inside virgin rain forest directly downhill from avocado groves devastated by phytophthora root rot, despite washing and transport by heavy rains for many years. If added directly to rain forest soil, the pathogen dies.

Although *P. cinnamomi* has not been recovered from undisturbed Queensland rain forest soils (Pegg, 1977a), it may be present in amounts too low to be detected by present methods. It is likely to be present in small pockets in favorable sites. Several such areas have been observed in which the lowest spot has no plants susceptible to *P. cinnamomi*. Seedlings of susceptible plants establish in the margins of these spots during the drier years but are killed in wet years. The size of the area thus varies directly with rainfall, and the pathogen survives in the fluctuating margins of the spot. In years of high rainfall, such topographical depressions may overflow and spread the pathogen more widely, but a succession of dry years again restricts the pathogen to the original small area. Such areas are difficult to detect in rain forests because of the fluctuating equilibrium maintained (Baker, 1978). It would be expected that these small wet areas would be attractive to feral pigs, which would wallow there and probably spread the pathogen further (Brown, 1976). These areas sometimes are associated with "patch death" of eucalypts.

Zoospores are the principal infective structures of *P. cinnamomi*, although mycelia can also infect (Malajczuk and Theodorou, 1979), apparently through wounds. Stimulatory material such as produced by *Pseudomonas* spp. is required for the pathogen to produce abundant zoospores. These bacteria, which apparently occur in most soils, with or without *P. cinnamomi*, develop abundantly on the surface of its mycelium, where effective concentrations of the compounds are produced. In constrast, in soil steamed at 40–50°C for 10 minutes, or in soil extract passed through Millipore filters to eliminate live bacteria, the concentration is too low for stimulation. The material from *Pseudomonas* spp. compensates for low light intensity in soil and for an excess of available nutrients, promoting zoospore production underground (Schoulties et al, 1980). Apparently, agents in suppressive soil either inhibit growth of the stimulatory bacteria or decompose the compound produced, but such agents either are not present or occur only in populations too low to be effective in conducive soils. In soil managed by the Ashburner method, *P. cinnamomi* produces very few zoosporangia, and these are attractive to antagonistic bacteria, which attach to and cause lysis and the discharge of undifferentiated contents (Broadbent and Baker, 1974b). In soil of the Ware grove, a few kilometers distant and also suppressive to *P. cinnamomi*, abundant zoosporangia are formed, but lysis is unusually prevalent (Figure 7.3). Soil of the Ashburner grove is also suppressive to *P. citrophthora*, and mycelium of *Pythium ultimum* lyses more readily than in other soils.

Suppressiveness in these soils may operate through several mechanisms, but mainly through a combination of high organic matter, calcium, and ammonium nitrogen, that supports a very large and active lytic microbial population. The soil pH of 6.0–7.0 is favorable to bacteria. The high organic matter and calcium improve soil structure, and hence drainage, to the disadvantage of *P. cinnamomi*. High calcium may also enhance resistance of the avocado to the pathogen. Nitrogen in the ammonium form is inhibitory to growth and sporangium formation by *P. cinnamomi*. A subtle but important contributing factor is that healthy plants remove more water from soil and thereby decrease waterlogging.

Suppressiveness of the soil to *P. cinnamomi* originates in the organic fraction (P. Broadbent and K. F. Baker, unpublished). New fallen avocado leaves are very favorable for growth of *P. cinnamomi*. Fresh avocado leaves can even be used as bait for isolating the pathogen from soil (Pegg, 1977b), as can fresh leaves of eucalyptus, pine, deodar, and pineapple. However, the attractiveness of the leaves diminishes as they become colonized by other microorganisms, and suppressiveness is progressively greater at increasingly greater depths in the decomposed litter layer. The most suppressive material is at the interface between the litter layer and the mineral soil; suppressiveness decreases both upward and downward from that zone. This gradation of suppressiveness in the litter layer probably reflects the succession of microorganisms in the decomposition process, those in the final stages being most suppressive. The most decomposed organic faction is therefore the layer that should be studied most intensively for clues to the nature of suppressiveness to *P. cinnamomi* in rain forest soils. Decomposed, commercial, synthetic mushroom compost is similar to the humus layer resulting from the Ashburner method and should be investigated as a medium with potential for increasing inoculum of the total microflora suppressive to *P. cinnamomi*.

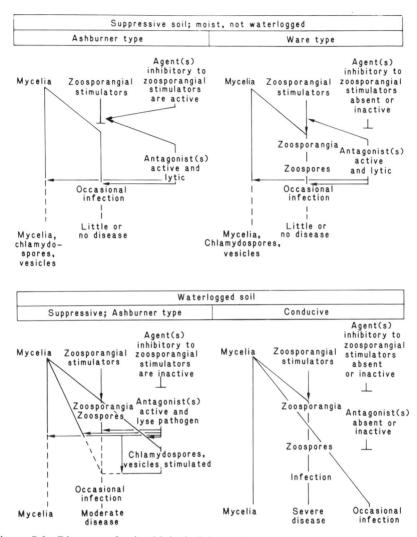

Figure 7.3. Diagram of microbiological interactions of *Phytophthora cinnamomi* and antagonists in suppressive and conducive soils in avocado groves, Queensland, Australia, with different soil moisture levels.

Pineapple.—Root and heart rot of pineapple caused by *P. cinnamomi* cannot be controlled by the Ashburner method because the plants become severely chlorotic if lime is added to the soil. Pegg (1977b) acidified the Queensland pineapple soil before planting, by mixing elemental sulfur (1,200 kg/ha) to a 15-cm depth. It was necessary to add *Thiobacillus thioxidans* with the sulfur to some soils. The soil pH was lowered from 4.0–4.5 to about 3.7, which decreased zoosporangium formation, possibly by reducing

Damping-Off of Seedlings

Biocontrol of the pathogens responsible for damping-off of seedlings illustrates the use of introduced antagonists to suppress prepenetration growth and infection by the pathogen, accomplished by: 1) inoculation of seed or nontreated soil with a specific antagonist; 2) treatment of a soil with moist heat to simplify its microbiota, followed by inoculation of the soil with an antagonist that acts one-on-one and has a higher inoculum density than does the pathogen.

Pathogens: Mainly *Pythium ultimum* and other *Pythium* spp. (Mastigomycotina, Oomycetes) and *Rhizoctonia solani* (*Thanatephorus cucumeris*) (Basidiomycotina, Corticiaceae).

Hosts: Seedlings of all plants are susceptible, some more than others.

Diseases: Preemergence damping-off. Decay of seeds and seedlings before emergence may be due to many soil- or seedborne pathogens but mainly to *Pythium* spp. and *Rhizoctonia solani.* **Postemergence damping-off.** Stems rot near the soil surface, causing plants to fall over; usually caused by *R. solani.* Some seedlings may be so tough or the environment so unfavorable for the pathogen that stems are girdled but plants remain alive; this sore-shin or wire stem effect of *R. solani* is common on Cruciferae. *Pythium* usually rots rootlets from the tips, stunting the plant; its mycelium is fragile and does not hold soil particles. *Rhizoctonia* usually attacks seedlings near the soil surface and has a coarse, tough, brown mycelium that holds soil particles (they dangle when seeds or seedlings and adhering soil are shaken). **Top rot of seedlings or cuttings.** *Rhizoctonia* and sometimes *Phytophthora* spp., under moist conditions, may spread plant to plant through the tops, mycelia forming strands resembling cobwebs. **Cutting and stem rot.** Rot may begin from cut ends, root bases, or wounds; caused by *Rhizoctonia, Pythium, Phytophthora,* or bacteria. **Root rot of mature plants.** *Rhizoctonia* and *Pythium* may carry over from seedling phase and continue activity in the field, resulting in unthrifty plants.

Life Cycles: *Rhizoctonia solani.* The pathogen spreads mainly by mycelium in soil, plant propagules, and equipment. Black hard sclerotia 0.1–10 mm across may adhere to plant or equipment surfaces; these resist drying and remain vaible for up to five years. When mycelia contact a host, they form adherent pigmented infection cushions or hyaline appressoria and infection pegs. Pigmented mycelia resist fungus and bacterial but not amoebal antagonists. Hyaline appressoria and infection pegs may be the site of most hyperparasitism. The pathogen grows saprophytically in soil and spreads through it. The numerous strains of *R. solani* vary in anastomosis group, host range, virulence, age and part of seedling attacked, temperature favorable to attack, and ability to survive in different soils. Basidiospores may form on mycelia at the soil surface and on stem bases. These are forcibly discharged and are airborne, but their importance in damping-off is uncertain. Basidiospores of tropical aerial strains commonly infect leaves. Conidia are not formed. Seed transmission effectively selects for and perpetuates the strain virulent for the given host from generation to generation.

   Pythium ultimum. Its hyaline, fragile mycelium is susceptible to drying and to antagonists. Sporangia form at hyphal tips. The most common type (*P.*

ultimum var. *ultimum*) never forms zoospores, but the sporangia may produce infective germ tubes in one or two hours. They may also remain dormant for more than seven months in moist soil or survive 12 years in dry soil. Oogonia and antheridia are formed with one smooth, thick-walled oospore per oogonium. No sclerotia or conidia are formed. The pathogen survives in soil as dormant sporangia and oospores that germinate very rapidly with suitable soil water potential and host stimulus. Infection may occur within 24 hours, and resistant structures form quickly; the fungus thus escapes antagonists. Spores and mycelia are disseminated with soil or plant propagules. Host specialization of strains is not known. A less prevalent type (*P. ultimum* var. *sporangiferum*) commonly produces zoospores; it is common on wheat roots in Washington. *Pythium debaryanum, P. aphanidermatum,* and *P. arrhenomanes* also produce zoospores but otherwise have similar life cycles. All are common soil inhabitants.

Environments: *Rhizoctonia* damping-off is favored by saline soil, seedling carbohydrate deficiency, low seed vitality, deep planting, and moist soil. The optimum temperature for growth is 25–30°C, with minimum of 8°C and maximum of 31–35°C, depending on the strain. The optimum osmotic water potential for growth is about −5 bars and the minimum is −35 to −40 bars. The incidence of damping-off by *R. solani* and *P. ultimum* is determined by the relative effect of soil temperature on the growth rate of the pathogen versus that of the host; thus a low-temperature crop develops less damping-off at low than at high temperatures. *Pythium ultimum* grows best at osmotic potentials of −1 to −5 bars and grows poorly or not at all at osmotic potentials below −30 to −35 bars.

Biological Control: When soil is chemically or thermally treated, the simplified and decreased soil microbiota makes biocontrol by introduced antagonists (e.g., *Trichoderma* spp., *Penicillium* spp., *Myrothecium verrucaria, Bacillus subtilis*) easier and more effective, but unfortunately these methods are little used. Inoculation of steamed glasshouse soil with *B. subtilis* or *Streptomyces* spp. prevents damping-off of pepper seedlings by *R. solani* and *P. ultimum*. Inoculation of pea and radish seed or infestation of soil with *T. hamatum* protected against *Pythium* spp. and *R. solani* as well as did fungicides. Field tests with corn seed inoculated with *B. subtilis* (in moist soil) or *Chaetomium globosum* (in dry soil) gave good control of seedling blight caused by *Fusarium roseum* 'Graminearum' and *Pythium* spp. Inoculation of pea seed with *Penicillium oxalicum* protected it against decay by *Pythium* spp., *R. solani,* and other pathogens; the *Penicillium* mycelia covered the seed in 48 hours, providing physical protection but not reducing the pathogen population.

References: Baker (1947, 1957), Broadbent et al (1971), Chet and Baker (1980), Ferguson (1958), Harman et al (1980), Leach (1947), Olsen (1964), Kommedahl and Windels (*in* Papavizas, 1981), Parmeter (1970), Walker (1969), Waterhouse (1968).

activity of the stimulator bacteria. The soil cation concentration is increased temporarily by the treatment, and nitrification is inhibited, increasing the ammonium fraction; the increase in cation concentration and in ammonium both inhibit zoosporangium formation. "The population of *Trichoderma* spp. explodes in sulphured soils" (Pegg, 1977a) and probably also is involved in the control, which is now widely used by the Queensland pineapple industry (Figure 1.2).

Eucalyptus marginata (jarrah). — Eucalyptus dieback occurs in Western Australia and Victoria in infertile, poorly drained, microbiologically depauperate soils low in organic matter. Since the pathogen was introduced into these areas in historic times, the soils there may not be as suppressive as those in Queensland and northern New South Wales. However, in Western Australia, where jarrah sustains severe root rot from *P. cinnamomi* in the common pathogen-conducive lateritic soils, damage is less in the moderately suppressive loam soils. When suppressive soil is sterilized, suppression is lost, but it can be restored by the addition of a small quantity of nontreated suppressive soil. *Eucalyptus calophylla* (marri) is more resistant but less valuable as a timber tree. The mycelium of *P. cinnamomi* lyses, and zoosporangium formation is decreased when the fungus is exposed to extracts from the rhizosphere of either tree species grown in suppressive soil, but these effects only occur with extracts from the rhizosphere of *E. calophylla* when the trees are grown in conducive soil (Malajczuk, *in* Erwin et al, 1983; *in* Schippers and Gams, 1979; Malajczuk and McComb, 1979; Malajczuk et al, 1977).

The highly susceptible *Banksia* spp. in the understory provides a root mat through which *P. cinnamomi* grows to reach the jarrah roots. A method is under study that replaces *Banksia* spp. by resistant *Acacia* spp. High-intensity prescription burning kills *Banksia* and its seeds but increases germination of *Acacia* seed, thereby shifting the understory to one unfavorable to the pathogen. The usual low-intensity burning does not kill *Banksia* and does not stimulate *Acacia* germination (Shea, 1979; Shea and Malajczuk, 1977). Because high-intensity burning is injurious to jarrah, it may be necessary to use low-intensity burning and then to sow heat-treated *Acacia* seed. An *Acacia* understory favors litter accumulation. When jarrah is grown in pots with *Acacia*, the population of *P. cinnamomi* is depressed, and extracts from soil in which *Acacia* is growing inhibit zoosporangium formation (Shea and Malajczuk, 1977).

Pythium-Suppressive Soils

Pythium spp. are notorious plant pathogens because of their ability to rapidly exploit fresh plant residue and to quickly colonize soil freed of other soil microorganisms by fumigation or steaming. The severity of diseases caused by *Pythium* spp., as well as those caused by *Rhizoctonia solani*, relates less to the initial inoculum density of the pathogen than to the amount of saprophytic growth of the pathogen before infection (Bouhot, *in* Schippers and Gams, 1979). Accordingly, soils or soil conditions suppressive to saprophytic multiplication by *Pythium* spp. should also be suppressive to these fungi as pathogens.

Bouhot (1975) developed a method to quantify the influence of different soils and their associated microbiota on ability of *Pythium* spp. to attack a host when allowed first to increase saprophytically. Soil containing a natural population of *Pythium* is mixed with steamed soil at various proportions between 0.1 and 30% or is used undiluted. Oatmeal is added as a substrate at 20 g/liter of the mixture. The oatmeal-amended mixture is then placed as a layer around the hypocotyls of six-day-old cucumber seedlings, 60 ml of mixture per 10–12 seedlings growing in a 10-cm pot. *L'unité de potential infectieux* (UPI), or the unit of inoculum potential, is the minimum weight of natural soil in the mixture necessary to cause 10% damping-off of the seedlings, determined after 144 hours at 15°C (Bouhot, 1975). UPI_{50} is the minimum weight of soil necessary to produce 50% plant death (Bouhot and Joannes, 1979; Bouhot, 1980). The more suppressive the soil, the less the saprophytic growth of *Pythium* spp. into the pockets and aggregates of steamed soil in response to the oatmeal in the mixture. The method is therefore weighted so that disease severity is related to saprophytic increase of the pathogen in the soil.

Bouhot and Joannes (1979) found for 600 soils collected from various locations and cropping systems in France that the regression between concentration of natural soil in the mixture and percentage damping-off of cucumber seedlings was nonlinear at concentrations above 10% . With some soils, infectivity increased disproportionately with concentrations above 10% soil in the mixture, but with others, infectivity was less at 30 or 100% than at 10% (Bouhot, 1980). The differences were attributed to competition from the soil microflora during the period allowed for saprophytic activity of *Pythium* spp. (Bouhot and Joannes, 1979). With 10% or less natural soil, the system was dominated by the steamed soil, which contained no antagonists; total disease was therefore proportional to total inoculum of *Pythium* introduced with the nonsteamed fraction. In contrast, with more than 10% nonsteamed soil in the mixture, the system was dominated more by the properties of the nonsteamed soil, which suppressed saprophytic multiplication of the *Pythium* spp. to different degrees depending on its physicochemical properties and microbial makeup.

Without the oatmeal, virtually no damping-off occurred even with 10% or more soil in the mixture. Bouhot and Joannes (1982) concluded that the amount of pathogen inoculum present 72–96 hours after preparation of the mixtures determines the proportion of dead seedlings. Their method, therefore, introduced the variable of time into the concept of inoculum potential. Equally significant is the fact that, given the time, substrate, and space for colonization (steamed soil), the *Pythium* spp. killed 100% of the seedlings with some soils but only half or fewer of the seedlings with other soils (Bouhot, 1980).

Fifteen forest nursery soils compared for suppressiveness to *Pythium* spp., based on the cucumber seedling bioassay, were of three types: 1) conducive, i.e., *Pythium* was able to colonize and the number of seedlings killed increased with time; 2) "tolerant," i.e., *Pythium* was neither favored nor suppressed; and 3) suppressive, i.e., the longer the period between preparation of the mixture and testing, the less the incidence of damping-off of cucumber seedlings.

Suppressiveness of soils to the saprophytic growth of *Pythium* spp., as measured by the cucumber seedling bioassay of Bouhot (1975), probably results from competition from the associated microbiota (Bouhot and Joannes, 1979) more than from any other form of antagonism. This kind of suppressiveness would therefore be of the general type, related to activity of all microorganisms capable of growth from the natural into the steamed soil in response to the oatmeal, and hence potential competitors of *Pythium*. The relative competitive advantage of the *Pythium* spp. would be determined by the physical and chemical environment of the natural soil (Bouhot, 1980). A soil relatively more unfavorable physically to growth of the *Pythium* spp. than to growth of potential competitors would be progressively more suppressive as the amount of that soil in the mix increased. A soil highly favorable to growth of *Pythium* spp. would show no such suppressive effect and might even enhance the relative advantage to *Pythium* spp. as the concentration increased.

Although competition for the substrates in the steamed soil is probably nonspecific, some microorganisms, because of their growth rate and ability to use the same nutrients used by *Pythium* spp., will be more important than others as competitors of *Pythium*. Bouhot (1980) induced suppressiveness in soil by amending the soil with a fermented mixture of various organic materials four days before sowing; the evidence suggested that fungi rather than bacteria were responsible for the suppression, especially the Mucorales. Of four isolates of Mucorales tested for ability to induce suppressiveness, *Actinomucor elegans* was the most effective (Bouhot, 1980). However, a complex of microorganisms — fungi, bacteria, and actinomycetes — were considered more effective than a single species in occupying the free space of the steamed soil and competing for the oatmeal.

In the San Joaquin Valley of California, populations of *Pythium ultimum* are consistently lower in some soils than in others (Hancock, 1977; *in* Schippers and Gams, 1979). When a fresh organic substrate (dried cotton leaves) was added to the different soils, the population of *P. ultimum* increased significantly in those recognized as conducive based on field records but increased little or not at all in the suspected suppressive soils. The suppressive soils were mainly of fine texture with considerable clay content (Hancock, *in* Schippers and Gams, 1979). Clay soils with good aggregation were more conducive than those poorly aggregated. Suppressive soils also had a higher content of sodium chloride (Martin and Hancock, 1981). Repeated applications of green manure, highly favorable to saprophytic growth by *P. ultimum*, tended to convert suppressive soils into conducive soils, possibly because of effects on soil aggregation and also because of the substrate supply for *P. ultimum*. Suppressive soil heated at 50°C for 24 hours became conducive to saprophytic multiplication of *P. ultimum* in response to an added organic substrate. The pattern is therefore not unlike that observed by Bouhot and associates for soils in France; suppressiveness is related both to the soil microbiota and to the physical environment of the soil. Work of F. N. Martin (unpublished) suggests that aggressive saprophytes having a selective advantage over *P. ultimum* in the poorly aggregated clay soils and the saline soils are involved. The *Pythium oligandrum/P. acanthicum* group is commonly isolated from the suppressive soil and could be partially responsible for the antagonism.

A slightly different mechanism of suppression is indicated for *Pythium splendens* in a soil from a tropical grassland in the South Kahala region on the island of Hawaii (Kao and Ko, 1983). This soil was the most suppressive of several tested (Ko and Ho, 1982), based on suppression of germination of sporangia in response to a standardized amount of cucumber root extract added to the soil and of damping-off of cucumber caused by a known density of sporangia of *P. splendens* added to the soil. Suppressiveness of the soil from South Kahala was eliminated either by autoclaving or gamma-radiation. However, the physical-chemical properties of the soil also are important; a nonsterile suspension of either suppressive or conducive soil restored suppressiveness when added to a sterilized South Kahala soil, but not when added to sterilized conducive soil. This observation is similar to that of Alabouvette et al (*in* Schippers and Gams, 1979) for *Fusarium*-suppressive soils in France. Suppressiveness is microbial in origin, but the soil as a matrix or environment apparently determines the potential for the soil population to suppress the pathogen.

Koa and Ko (1983) found that addition to suppressive soil of either 500 ppm rose bengal, 10,000 ppm streptomycin (both inhibitory to bacteria), or 5,000 ppm benomyl (inhibitory to fungi) gave partial nullification of suppressiveness. The suppression could not be shown to involve any specific group or species of microorganisms, but rather it involved the effects of many soil microorganisms. Suppressiveness to sporangial germination could be overcome by increasing the concentration of cucumber root extract above that used in the standard test, suggesting that the failure to germinate in the suppressive

South Kahala soil results, in part, from lack of nutrients. All of these observations suggest that suppression results from more intense competition for nutrients, possibly carbon and energy, by the combined effects of all soil organisms active in the South Kahala soil—a type of general suppression. Suppression is lost after a few months in storage; many microorganisms die during storage and others, because of aging, possibly are less responsive during the first few hours after nutrients and water have been added. Either a reduction in the population of potentially active microorganisms or retardation of their ability to respond to fresh nutrients could provide the advantage needed by sporangia of *P. splendens* to germinate in the soil. In essence, the microbiota of the suppressive soil is like a sponge to any fresh supply of readily available nutrients, and unless enough nutrients are supplied, or the competing microbiota is somehow inhibited or partially eliminated, the sporangia will not germinate and therefore the pathogen cannot infect the host root.

Biological Control of *Rhizoctonia solani* in Suppressive Soil

Like *Pythium* spp., *Rhizoctonia solani* is an aggressive saprophyte that can increase its biomass and energy for attack of a crop when supplied with fresh plant residue. However, *R. solani* also may be more subject than most soilborne plant-pathogenic fungi to attack by several mycoparasites in soil. The classic work of Weindling between 1934 and 1941 showed the potential of *Trichoderma viride* (and/or *Gliocladium virens*) as a mycoparasite of the hyphae of *R. solani*. Recent work reveals that these and related mycoparasites have a significant antagonistic effect on *R. solani* and may be responsible for some *Rhizoctonia*-suppressive soils.

A soil sown to radish every week was suppressive to *R. solani* by the fourth sowing (Henis et al, 1978) and was even more suppressive by the fifth and subsequent sowings (Figure 7.4). The population of *T. harzianum*, antagonistic to *R. solani*, increased with the successive sowings of radish (Liu and Baker, 1980), possibly in response to increases in the amount of *R. solani* in the soil resulting from its parasitism of the radish seedlings. The addition of spores of *T. harzianum* to a conducive soil at the same density found in the suppressive soil caused the conducive soil to become suppressive. The hyphae of *T. harzianum* were observed to coil around those of *R. solani* and to cause lysis, although direct penetration was not observed. In another but related study, a soil from Bogotá, Colombia, highly suppressive to *R. solani* (Figure 7.5) was shown to have a high population of *T. hamatum* (Chet and Baker, 1981). This fungus was highly antagonistic as a necrotrophic mycoparasite, causing exolysis of the hyphae. Still another example has been observed in Holland, where tubers from a field cropped

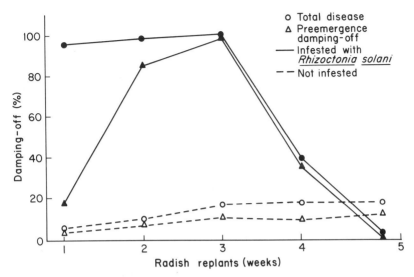

Figure 7.4. The incidence of damping-off of radish caused by *Rhizoctonia solani* in soil that was planted to radish every week for five successive weeks. (Reprinted with permission from Henis et al, 1979b.)

four consecutive years to potatoes showed significantly less damage (Figure 7.6) from *R. solani* than tubers from fields where potatoes were grown only once every three years and cereals and sugar beets the other two years (Jager and Velvis, 1980; Jager et al, 1979). *Verticillium biguttatum* (Gams and Van Zaayen, 1982) (earlier identified as *Gliocladium roseum*), mycoparasitic on *R. solani*, was abundant where potatoes had been grown intensively (Figure 7.7).

A potential limitation on the dependability of suppressive soils to provide biological control of *R. solani* is the evidence that strains of the fungus may differ significantly in their sensitivity to the antagonists and that with sufficient selection pressure, insensitive strains of the pathogen may eventually appear (Baker et al, 1967; Olsen et al, 1967). A strain that causes a bare-patch condition in wheat fields in the Eyre Peninsula, Australia, was grown in massive culture and inoculated into field soil at the time of seeding in Adelaide. The pathogen grew abundantly, rotted roots, and stunted the plants for a time, temporarily swamping the antagonists. However, the pathogen disappeared from the soil in four months and the plants recovered. A crucifer strain of *R. solani* is, however, commonly present in both soils. Comparison of California and South Australia isolates indicated that ability to survive was not correlated with ability to grow, to colonize organic matter in the soil, nor ability to form sclerotia.

The host strains of *R. solani* appear to be sufficiently distinct that antagonists in the soil suppressive to the wheat strains are not suppressive to the crucifer

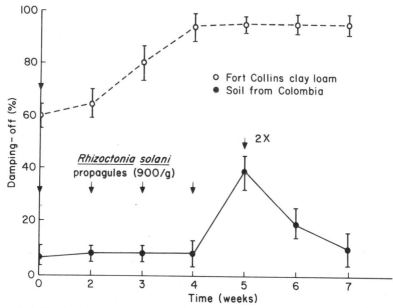

Figure 7.5. The incidence of damping-off of radish caused by *Rhizoctonia solani* in a Fort Collins clay loam, and in a soil from Colombia, where each soil was planted at weekly intervals. The Fort Collins clay loam was infested with *R. solani* one time only, at the time of the first planting, but the soil from Colombia was infested with 900 propagules of *R. solani* per gram of soil with each of the first four plantings (arrows), and 1800 propagules per gram (2x) with the fifth planting. (Reprinted with permission from Chet and Baker, 1980.)

strain, a difference evident even among monobasidial isolates. The situation with *R. solani* is therefore very different from that with *F. oxysporum*, where suppressiveness to one pathotype is also effective against others (Baker, *in* Mace et al, 1981).

Several conditions apparently must be met for induction and maintenance of effective suppressiveness to *R. solani* in soil. There must be an adequate population of the pathogen, which then favors activity of the antagonists, and soil conditions must not be limiting to the antagonists. The former condition is satisfied by repeated sowing of a host susceptible to *R. solani*, causing an increase in amount of hyphae of *R. solani*. However, if the soil pH is near neutrality or above, *Trichoderma* populations do not increase and the soil does not become suppressive, or it may remain only moderately suppressive, a condition most likely to select for insensitive strains of the pathogen. Soil moisture and temperature may also provide a selective advantage to the pathogen over the antagonists, or an opportunity for selection and multiplication of strains not affected by the antagonists. Finally *R. solani* is less susceptible to the suppression if present in the soil as large sclerotia rather than as the

Figure 7.6. Potato tubers with sclerotia of *Rhizoctonia solani* from plots in long-term suppressive soil in The Netherlands, after being held seven days at 20°C under moist conditions. The antagonist, *Verticillium biguttatum*, kills many sclerotia. (Reprinted with permission from Jager and Velvis, 1980.)

more vulnerable small sclerotia (Henis and Ben-Yephet, 1970). These variables point up the potential for use of suppression against a pathogen such as *R. solani*, but they also indicate the delicate balance and even the risks involved.

As revealed in the examples above, and in much of the research on biological control of the past 50 years, **successful biological control in soil depends on the presence of the proper antagonists, on as many different mechanisms of suppression as can be practically combined, and on a soil environment relatively more favorable to the antagonist(s) than to the pathogen.**

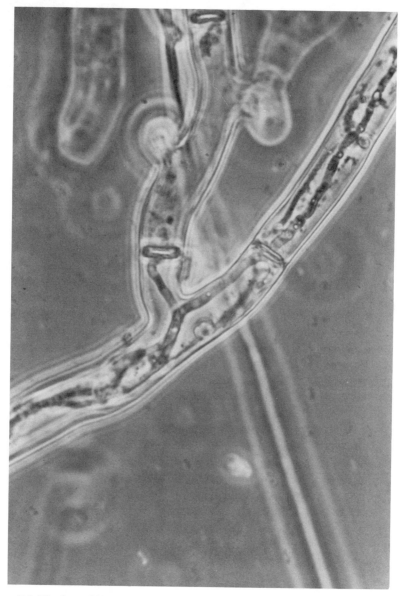

Figure 7.7. Hyphae of *Verticillium biguttatum* in a killed hypha of *Rhizoctonia solani*. The antagonist was originally reported incorrectly as *Gliocladium roseum*. (Reprinted with permission from Jager et al, 1979.)

8

INTRODUCTION OF ANTAGONISTS
FOR BIOLOGICAL CONTROL

*There is now abundant observational evidence that
biocontrol of diseases occurs naturally, but if natural events
are to be manipulated for man's benefit, microorganisms
must be put where they are needed. . . .*
—A. T. K. CORKE and J. RISHBETH, 1981

Plant pathologists traditionally have emphasized the conservation or selective enhancement of resident rather than introduced antagonists for biological control. This was especially apparent in the 1963 symposium *Ecology of Soil-borne Plant Pathogens — Prelude to Biological Control* (Baker and Snyder, 1965). S. D. Garrett there referred to the work of Weindling and Fawcett (1936) on protection of citrus seedlings against *Rhizoctonia solani* by inoculation of acidified soil with *Trichoderma viride*, and to Rishbeth's biological control of *Heterobasidion* (*Fomes*) *annosum* by application of *Peniophora gigantea* to cut stumps of pine. Z. A. Patrick and T. A. Toussoun referred to the work of Ferguson (*in* Baker, 1957) on introduction of saprophytic microorganisms into flats of steamed soil to control *Rhizoctonia solani*. This was virtually the only attention given to introduced antagonists in the symposium. By comparison, there were countless examples by 1963 of biological control by resident antagonists managed through cultural practices and selective treatments.

At the time of publication of our earlier book (Baker and Cook, 1974), Rishbeth's method of biological control was in commercial use and methods to protect pruning wounds by application of antagonists had been reported. Biological control of crown gall had been reported only two years earlier (Kerr, 1972). In addition, the prospects for controlling seedling pathogens with biological seed treatments had been indicated by the success of Chang and Kommedahl (1968) in applying *Chaetomium globsum* or *Bacillus* sp. to seeds of corn to control seedling blight, and of workers in Australia in applying *B. subtilis* A13 to seeds of bedding plants (Broadbent et al, 1971), cereal grains, and carrots (Merriman et al, 1974).

The dearth of examples in 1963, 1974, and even today of biological control with introduced antagonists cannot be attributed to lack of interest by earlier workers. Some of the first attempts at biological control of plant pathogens were with antagonists introduced into soil, namely, the work of Hartley (1921) to control damping-off of conifer seedlings, Millard and Taylor (1927) to control common scab of potato, and Sanford and Broadfoot (1931) to control take-all of wheat (Chapter 2). The antagonists were effective either because the soil had been steamed (Hartley, 1921; Sanford and Broadfoot, 1931) or because large quantities of organic material had been added (Millard and Taylor, 1927) to support the introduced antagonist. The concept that the soil environment must be modified to permit establishment of an alien microorganism was recognized by Weindling and Fawcett (1936), who showed that without acidification of the soil, introduced *T. viride* was ineffective as an agent of biological control.

The lack of attention to biological control by introduced antagonists at the time of the 1963 Berkeley Symposium was largely because, without some drastic shock (such as steam treatment), alien microorganisms could not be established in soil or on plants. The predominant school of thought held that the soil microflora was more or less the same everywhere, varied only quantitatively, and could best be modified by some shock effect. A second but minor school of thought held that the microflora differed qualitatively as well as quantitatively and that therefore antagonists must be added to some soils. Both schools of thought are partly correct. The management of resident antagonists by cultural practices continues to be the area of greatest success, but it also is now clear that antagonists effective against specific pathogens do not reside in all soils and may need to be introduced to achieve biological control.

The time has come for major advances in technology for introducing antagonists and for improving them by selection or genetic manipulation. In addition, the prospects for biological protection of plant surfaces and biological control after infection have expanded our options to include treatment of plant parts, an approach that requires different technology from treatment of soil. If the large number of recent publications on the use of introduced antagonists is any indication, then clearly the outlook for this approach is bright. On the other hand, the problems identified in the earlier work still exist and examples of commercial use of introduced antagonists still number only four or five. This chapter draws mainly on work of the past 10 years to discuss not only some approaches and problems but also the scope of the opportunities available.

ANTAGONISTS AS SOIL TREATMENTS

The direct application of antagonists to soil, either as a broadcast-incorporation or as an application in the seed furrow at the time of sowing, has the greatest potential for use in commercial glasshouse operations, casing soil in mushroom growing, gardens, or other plant production methods involving a limited amount of soil or land. However, we repeat the point made in Chapter 1 that neither introduction of antagonists nor any other approach to disease control should be used in nursery soils merely to suppress a pathogen if, when the stock is transferred to an orchard or garden, the pathogen is still present and can then cause disease. *The essential practice in nurseries is to eliminate pathogens and then use antagonists to prevent their reestablishment.*

Application of antagonists in large fields may be economically feasible in the more distant future, perhaps for high-value crops such as field-grown tomatoes or strawberries, or perhaps if an effective, more efficient, in-furrow method of application can be developed. Direct application to field soil on a large scale also may be feasible in the future for introduction of a highly efficient antagonist that is absent from the soil but that is likely to succeed because of a high population of its host, e.g., *Bacillus penetrans* for control of root-knot nematodes (Mankau, 1980; Stirling and Wachtel, 1980).

Antagonists may be applied to soil to 1) destroy pathogen inoculum, 2) prevent recolonization of treated soil by a pathogen, or 3) protect germinating seeds and roots from infection. If the objective is biological destruction of inoculum, then the most effective antagonists probably will be hyperparasites of the pathogen. If the objective is to fill a biological vacuum left by soil fumigation or steaming, then the most promising antagonists are aggressive saprophytes adapted to the physical environment of the soil. A mixture of aggressive saprophytes may prove superior to a single strain for this purpose.

Antagonists intended as spermosphere or rhizosphere colonists can be applied directly to the soil, but probably will be most effective if introduced on the seed or seedpieces (discussed in the next section). Strains of fluorescent pseudomonads suppressive to take-all were more effective when introduced as a seed treatment than when applied directly to soil at the time of planting (Weller and Cook, 1983; Wilkinson et al, 1982). This reflects the greater competitive advantage to the antagonist when it is introduced on the seed rather than into the soil, just as a seedling pathogen is likely to cause more damage when seedborne than when distributed in the soil. The opportunities for native soil organisms to preempt an introduced antagonist intended as a root colonist are too great if the antagonist is applied at a slight distance from the host plant. On the other hand, Campbell and Faull (*in* Schippers and Gams, 1979) obtained some biological control of take-all in field plots with *Bacillus mycoides* as a liquid cell suspension applied to the soil before planting. In addition, the incorporation of 0.5% (w/w) suppressive soil uniformly into the top 12–15 cm of a fumigated soil at Puyallup, Washington, resulted in take-all decline one year earlier than with five other soils introduced (Baker and Cook, 1974). Apparently, broadcast-incorporation of the right microorganisms has potential to provide biological control of take-all, eventually if not immediately, given a selective advantage such as an abundant supply of wheat roots infected with *Gaeumannomyces graminis* var. *tritici* (a factor favorable to multiplication of the suppressive bacteria). *Talaromyces flavus* added to soil in glasshouse tests resulted in less wilt of eggplant caused by *Verticillium dahliae* and decreased wilt in two field tests by 76 and 67% (Marois et al, 1982). *No approach to biological control with introduced antagonists should be ruled out at this stage of our technology.*

Introduction of Hyperparasites

Control of fungus pathogens. — Most attempts at the introduction of hyperparasites for biological control of plant pathogens have been directed at destroying sclerotial and mycelial inoculum. *Trichoderma* spp. (particularly *T. harzianum*) have been tested most widely, including: tests for control of *Sclerotium rolfsii* on tomatoes in Georgia (Wells et al, 1972); *Sclerotium rolfsii* on peanuts in Alabama; *Sclerotinia minor* on lettuce in France (Davet et al, 1981); *S. rolfsii* and *Rhizoctonia solani* on beans, strawberries, and other crops in Israel (Elad et al, 1980a, 1981a, 1981b; Henis et al, 1979a; Chet et al, *in* Schippers and Gams, 1979); *Sclerotium cepivorum* on onions in Egypt. *Coniothyrium minitans* has been tested in England for control of *Sclerotinia trifoliorum* (Turner and Tribe, 1975), in Australia for *S. cepivorum* (Trutmann et al, 1980), and in Canada for *Sclerotinia minor* on sunflower (Huang, 1980). The efficacy of these hyperparasites is established, but in most cases the approach is not feasible economically, other methods (e.g., a cultural practice or fungicide) are more familiar to the grower or easier to apply, or technology is not available for supplying the antagonist.

Biological control of sclerotia with introduced antagonists theoretically is feasible for at least three reasons. First, of the several hyperparasites that have been identified, most are suited to mass rearing, long-term storage if necessary, and application as a dust or granular product either broadcast or applied in the seed furrow. Second, sclerotia represent a large target compared with chlamydospores or oospores and should therefore be more vulnerable to attack by hyperparasites. Third, in general, the hyperparasites need to destroy sclerotia only in the top 5 cm of soil (or possibly only those sclerotia within rows of the host) in any given crop year. Except for a few pathogens such as *Phymatotrichum omnivorum* (Lyda and Burnett, *in* Bruehl, 1975) and *Sclerotium cepivorum* (Crowe and Hall, 1980), sclerotia deeper than 5 cm rarely are important inoculum.

The limiting factors to use of hyperparasites for destruction of inoculum are largely logistical, namely the amount of antagonist preparation usually needed and the lack of an inexpensive method for mass production and application of the antagonist. Various substrates have been used to produce inoculum for plot work, e.g., wheat bran (Chet et al, *in* Schippers and Gams, 1979), chopped straw moistened by an acid mineral solution (Davet et al, 1981), ground ryegrass seed (Wells et al, 1972), colonized peat-sand mixtures or bark (Gindrat et al, 1977), barley grains (Abd-El-Moity and Shatla, 1981), and milled rice (Turner and Tribe, 1975). All of these materials are readily available and high in stored carbon and energy, yet they are sufficiently fibrous to provide some structure to the final product. Backman and Rodriguez-Kabana (1975) developed a method for mass production and delivery of *Trichoderma* using diatomaceous earth impregnated with molasses as the food base. Commercial products of *Trichoderma* reared by this and other methods are now available (Papavizas and Lewis, *in* Papavizas, 1981; Ricard, 1981). Broadcast rates for these materials (dry-weight basis) are commonly 250–500 kg/ha, sometimes three times these rates. In-furrow treatments obviously would be more economical.

The addition of a readily available substrate with the antagonist can be detrimental if the nutrients then stimulate growth of the pathogen. Kelley (1976) used clay granules impregnated with molasses as a carrier for *T. harzianum* to control damping-off of pine caused by *Phytophthora cinnamomi*, but disease was favored, especially if the soil was also water-saturated. The molasses was stimulatory to *P. cinnamomi*. In attempts to obtain biological control of *Phomopsis sclerotiodes* on greenhouse cucumbers, application of a preparation of conifer bark pellets containing peat and conidia of a *Trichoderma* at the time of transplanting resulted in about half of the seedlings being killed by *Pythium* (Moody and Gindrat, 1977). No damping-off occurred when pellets were added 15–30 days before the seedlings were transplanted. *Phomopsis sclerotiodes* is an important pathogen of glasshouse cucumbers in Europe and has been considered by several workers as a candidate for biological control by introduction of antagonists into the beds of steamed or manured soils or bales of cereal straw (Ebben and Spencer, 1978; Gindrat et al, 1977; Sundheim, 1977), but more work is needed to find a suitable antagonist and delivery system.

A particularly promising hyperparasite of sclerotia is the recently identified *Sporidesmium sclerotivorum*, which invades and destroys sclerotia of *Sclerotinia* spp., *Sclerotium cepivorum, Botrytis* spp., and *Claviceps purpurea* in field soil (Ayers and Adams, *in* Papavizas, 1981). In Maryland, 100 and 1,000 macroconidia of *S. sclerotivorum* per gram of soil introduced in the plow layer in May in field plots heavily infested with *Sclerotinia sclerotiorum* reduced the number of sclerotia by 75 and 94%, respectively, by November. Introduction of one or 10 macroconidia of the mycoparasite per gram of soil had no effect. Where biological control had occurred, the incidence of drop on romaine lettuce the following spring and summer was significantly less.

As an introduced hyperparasite, *S. sclerotivorum* presents advantages but also a problem. The advantages are that the macroconidia are suited to survival in soil, may be applied directly to soil without an accompanying substrate, and will spread from infected to healthy sclerotia up to 1 cm away (Adams and Ayers, 1980). The disadvantage is that the hyperparasite, being specialized for growth on sclerotia, sporulates poorly on conventional substrates. Mass production of macroconidia for testing has been accomplished thus far only by growing the hyperparasite on dead sclerotia of *Sclerotinia minor* in moist sand (Ayers and Adams, *in* Papavizas, 1981). On the other hand, this hyperparasite will persist in soil as long as suitable host sclerotia are available and is thought to already provide considerable natural biological control of sclerotia in field soils (Ayers and Adams, 1979). A major breakthrough will come when an economic method is found for mass producing propagules of this potent antagonist.

In any biological control involving parasitism or pathogenesis of a target species, the biocontrol agent may show a degree of specificity and some agents will be superior to others. *Sporidesmium sclerotivorum* is ineffective against sclerotia of *Macrophomina phaseolina, Sclerotium rolfsii,* and *Rhizoctonia solani* (Ayers and Adams, *in* Papavizas, 1981). Possibly *S. sclerotivorum* prefers sclerotia of Ascomycetes to those of Basidiomycetes. An isolate of *Trichoderma harzianum* with the ability to attack and degrade sclerotia and mycelium of *R. solani* had no effect on sclerotia of *Sclerotium rolfsii,* whereas another isolate of the same species was effective against both pathogens (Chet et al, *in* Schippers and Gams, 1979). Wells and Bell (1979) found highly variable amounts of antagonism among 76 isolates of *T. harzianum* tested in vitro against several soilborne pathogens. An isolate of *T. hamatum* exceptionally effective against sclerotia and mycelium of *R. solani* was found by Chet and Baker (1981) in soil from Bogotá, Colombia, that was highly suppressive to *R. solani.* Collectively, these observations indicate that more powerful strains of hyperparasites of fungus sclerotia can be found or developed by looking for potent antagonists where the pathogen is expected but does not occur, by standard selection procedures, or possibly in the future through recombinant DNA technology.

Soil fungicides may be used in combination with antagonists introduced into soil. A combination of *T. harzianum* and pentachloronitrobenzene (PCNB) at 1–2 mg/kg gave better control than either component alone against damping-off of bean, tomato, and eggplant seedlings caused by *R. solani* (Hadar et al. 1979), and of stem rot of carnation in the glasshouse (Elad et al, 1981b). Papavizas and Lewis, (*in* Papavizas, 1981), Papavizas et al (1982), and Abd-El-Moity et al (1982) have shown further that new biotypes of *T. harzianum* can be selected for tolerance or resistance to fungicides, such

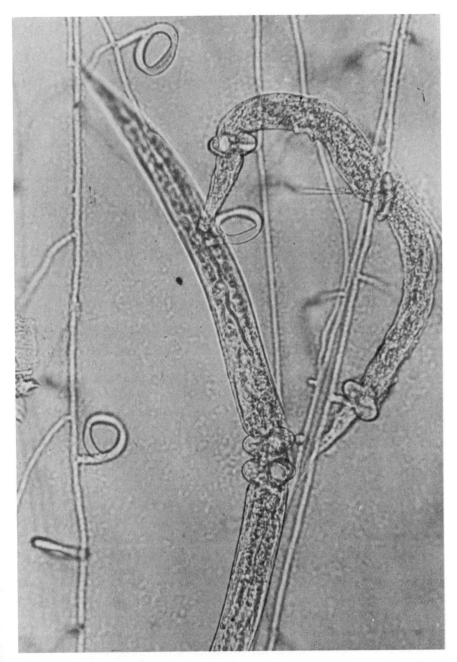

Figure 8.1. *Arthrobotrys dactyloides* trapping larvae of *Meloidogyne* sp. (Reprinted with permission from Mankau *in* Trolldenier, 1971.)

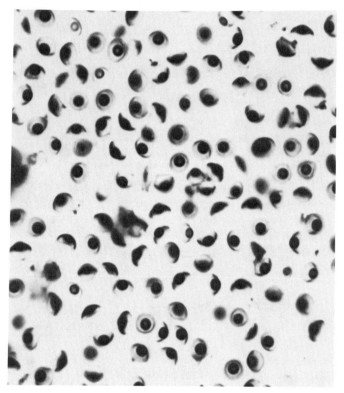

Figure 8.2. Mature parasite spores of *Bacillus penetrans* expelled from an adult female of *Meloidogyne javanica*. The dense central spherules are the endospores within sporangia. (Reprinted with permission from Mankau, 1975.)

as chlorothalonil, iprodione, and benomyl. Some strains with greater resistance to fungicides were less persistent in soil than the parents, but others survived longer than the parents. Particularly significant was the observation that fungicide-resistant biotypes were commonly superior to the parents as biocontrol agents. R. Baker and associates (unpublished) have similarly observed for *T. hamatum* that strains selected for resistance to a fungicide are also commonly more effective as antagonists of *R. solani*. Perhaps resistance to a fungicide also confers resistance to certain inhibitory agents in soil. This research area is only beginning to indicate its potential for practical biological or integrated control and deserves a greatly accelerated research effort.

Control of nematodes. — The role of nematode-trapping fungi (Figure 8.1), egg parasites, and predacious nematodes in the natural regulation of populations of plant-parasitic nematodes has long been recognized. Nematode trappers require a nutrient source to produce the requisite mycelium and traps, and any use of these agents for biological control usually requires that an organic food base also be added (Cooke, 1968). Moreover, fungi antagonistic to nematodes are subject to the same problems of fungistasis and difficulty of establishment as any other alien microorganism introduced into an established soil ecosystem. There is also a problem of density-dependency (Chapter 3), i.e., because the antagonist population depends on the host population, it may become active too late to provide effective biological control. Mankau (1980) pointed out that nematode-trapping fungi usually appear after a period of intense microbial activity in which nematodes have actively browsed on the microbial biomass. The need for a high population of the target nematode to maintain a high population of the antagonist is not a problem with *Dactylella oviparasitica*; this nematode parasite can multiply as a saprophyte in organic matter and maintain its population in the absence of a nematode host.

Bacillus penetrans is one of the more promising candidates as an introduced antagonist for biological control of nematodes. This prokaryote (Figure 8.2), which was described as a protozoan until detailed studies confirmed its identity as a bacterium, "is the most specific obligate parasite of nematodes yet discovered" (Mankau, 1980). Spores attach to the cuticle upon contact with the nematode in soil (Figure 8.3). The cuticle is penetrated by the spore germ tube. A microcolony forms and then fragments into daughter colonies. The body cavity of the host nematode eventually becomes filled with endospores of the bacterium. A single parasitized female root-knot nematode was observed to contain about 2.1×10^6 spores (Mankau, 1980). *Bacillus penetrans* is widely distributed in agricultural soils throughout the world and undoubtedly contributes significantly to natural nematode control, especially nematodes such as *Meloidogyne* spp. to which the parasite is particularly well adapted and synchronized.

Mass production of *B. penetrans* is currently limited to pot cultures of root-knot nematodes maintained on a host. Stirling and Wachtel (1980) reported a method whereby tomato roots containing large numbers of *Meloidogyne* females parasitized by *B. penetrans* are air-dried and then ground into a fine powdery material that is easily stored and handled. Incorporation of this product into soil at 100 mg/kg of soil resulted in 99% of *Meloidogyne* larvae becoming infected by *B. penetrans* within 24 hours. Technology for protection of individual transplants against root-knot nematodes is thus available, but technology for large-scale field application is still to be developed.

An advantage of *B. penetrans* is its ability to infect the sources of nematode larvae, e.g., root-knot nematode females as well as the larvae themselves. *Dactylella oviparasitica* offers a similar advantage. As pointed out in Chapter 4, the use of antagonists to destroy cysts, eggs, and females is more efficient than the use of antagonists that act mainly on larvae after their emergence.

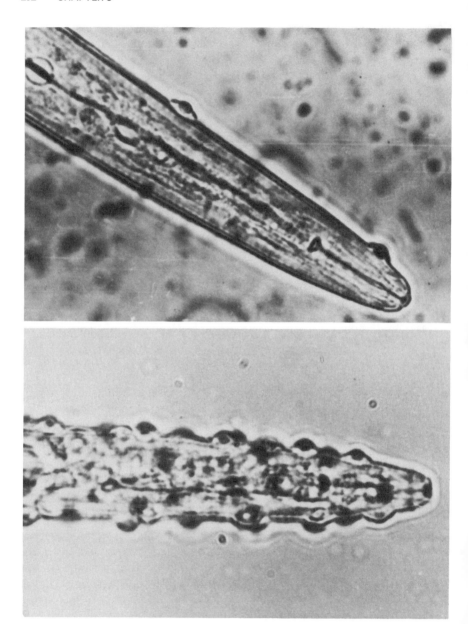

Figure 8.3. Spores of the parasite *Bacillus penetrans* attached to the cuticle of larvae of *Meloidogyne javanica*. Top, anterior of larva infected with two spores such as occurs when larvae move through soil naturally infested with *B. penetrans*. Bottom, anterior of larva heavily infected such as occurs when larvae move through soil artificially infested with the parasite. (Reprinted with permission from Mankau, 1975).

At least two examples of biological control of mushroom pathogens have been achieved by soil incorporation of a commercially available antagonist preparation. A product with the trade name Royale 300, containing a strain of the nematode trapper *Arthrobotrys robusta*, is used to control *Ditylenchus* populations in mushroom beds in France (Cayrol et al, 1978). The antagonist and spawn of *Agaricus bisporus* are added simultaneously to the mushroom compost. Another product containing spores of *Trichoderma* and sold under the trade name BINAB T SEPPIC is applied to casing soil (not to the compost where *A. bisporus* grows) in commercial mushroom houses in France to control *Verticillium malthousei*. The *Trichoderma* is applied as a spray at 1 liter/m² with 1 × 10⁸ *Trichoderma* spores per liter. Good control is achieved.

Introduction of Saprophytes for Colonization of Treated Soil

Soil treatments with live steam, aerated steam, and fumigants are common commercial practices to eliminate plant pathogens in glasshouse crops and in nurseries. Fumigation is also common on a field scale for production of high-value ornamental, vegetable, or fruit crops. The problem of recolonization of the resultant biological vacuum by plant pathogens is well known (Chapter 7). The establishment of aggressive competitive saprophytes following soil treatment but before planting the crop is one way to solve this familiar problem (Baker and Cook, 1974). As pointed out above, antagonists should not be used in place of sanitation and treatments to eliminate pathogens of glasshouse and nursery plants but rather should be used to supplement and extend the effectiveness of these traditional practices.

Fusarium crown rot of tomato caused by *Fusarium oxysporum* f. sp. *radicis-lycopersici* exemplifies a disease important because the pathogen is able to rapidly colonize steamed (100°C) or fumigated soil used for growing tomatoes (Jarvis and Shoemaker, 1978; Rowe et al, 1977). The pathogen is an aggressive saprophyte and rapidly increases its biomass for attack of tomato by growing on nutrients released by the steam or fumigation treatment. Nonpathogenic fungi also colonize the soil from airborne spores and provide limited biological buffering against *F. oxysporum* f. sp. *radicis-lycopersici* (Jarvis and Shoemaker, 1978; Rowe and Farley, 1978), but in Florida a plastic tarp applied during fumigation and maintained throughout the growing season is a barrier to recolonization by airborne saprophytes (Marois and Mitchell, 1981a).

Marois and Mitchell (1981a) isolated fungi from recolonized fumigated soil one week after fumigation and screened them for their ability to increase rapidly when introduced into fumigated soil, to occupy the root zone of tomato, and to protect tomato against infection by the pathogen. Of

Root-Knot Nematode

Biocontrol of this nematode occurs through: 1) destruction of the nematode larvae by bacteria that attach to the larvae as they move through soil; 2) destruction of eggs and larvae by parasitic and trapping fungi; 3) reduction of nematode populations by use of a trap crop in infested orchards to salvage trees; 4) possible increase in the proportion of males, which do not form galls, by alteration of sex ratio.

Pathogens: *Meloidogyne incognita, M. incognita* var. *acrita, M. javanica, M. hapla,* and *M. arenaria* are the most important; others are *M. exigua, M. inornata, M. brevicauda,* and *M. arenaria thamesii* (Nemata, Heteroderidac). Species identification, based on striae in the anus-vulva area of females, is a specialist job.

Hosts: Members of the genus have a very extensive host range, but individual species are more restricted. Cultivar resistance or tolerance occurs in some crops. Some plants (e.g., *Crotalaria spectabilis*) fail to develop giant cells in response to feeding by root-knot nematode; the females then produce few or no eggs and may die. *Crotalaria spectabilis* is therefore a valuable trap crop.

Diseases: Distinctive galls are produced on roots and tubers; spheroid to spindly galls are formed on large and small roots. Tops of affected plants are pale in color, unthrifty, and stunted, and leaves wilt on hot days. Tubers become warty in appearance. Egg masses may be extruded to the root surface. Splitting a gall reveals the pearly white females.

Life Cycle: Eggs are ellipsoidal, extruded in gelatinous masses from the flask-shaped female nematode contained within the host, 200–500 or more eggs per female. Slender wormlike larvae emerge on hatching and molt several times; both females and males penetrate roots behind the rootcap. Females establish in the root with their heads in an intercellular space near the endodermis. Galls may contain numerous females. They feed in rotation on surrounding cells, which become much enlarged; the stylet is thrust through the cell wall, saliva is injected, and the contents are partially sucked out. The stylet is about 10 μm long. Females become obese and immobile. A root tip penetrated by several larvae ceases to grow. Sex is indistinguishable until the third molt. Males are not necessary in the life cycle and do not induce gall formation. Larvae may overwinter in roots, and eggs may overwinter in soil, where they may be 60–70 cm deep; in the north, most larvae are killed by freezing in winter. Larvae move slowly through soil, about 30–60 cm per month, soil moisture and temperature

five isolates that met all three criteria, three were *Trichoderma harzianum,* one was *Penicillium fungiculosum,* and one was *Aspergillus ochraceus.* A composite of all five isolates added to soil increased the inoculum density required for 50% infection to 6,500 chlamydospores per gram, compared with only 300/g required for 50% infection in fumigated soil not

permitting. Wider dissemination occurs on tools, in water, and in infected plant stock.

Environment: Larvae mature to egg-laying females in about 80 days at 14° C or 16 days at 27° C; maturation is severely retarded above 27° C. Roots are infected at 12–35° C; little disease develops below 16° C. Larvae and eggs are killed by a constant temperature of 40–50° C. Larvae require water films and water-filled pores of adequate diameter (60 μm or larger) for movement through soil. A susceptible plant may be grown successfully in heavily infested soil by not allowing pores of the critical size or larger to remain water-filled (e.g., by planting on raised beds between deep irrigation ditches and using frequent light water applications). Sandy soils are much more favorable for root-knot nematodes than clay because a larger proportion of the pores are of a size that permits nematode movement to the roots. The disease is most important in areas of warm soils but is important even in northern climates.

Biological Control: Biocontrol by nematode-trapping fungi, long studied in detail, has not proved effective in practical agriculture. *Bacillus penetrans*, an obligate bacterial parasite, provides considerable natural control and has proved effective in field tests. A method for limited mass production of this bacterium has been devised. A fungus, *Dactylella oviparasitica*, parasitic on eggs of nematodes, has provided good field control of root-knot nematodes on peach in California.

The addition of copious quantities of organic matter has reduced damage from root-knot nematode in some cases, perhaps because populations of trappers and hyperparasites increase or because the nematodes are attracted to the organic matter rather than to the roots.

Female root-knot nematodes, but not the males, initiate galls and cause the disease, and sex reversal is possible up to the third molt. Conditions of crowding or starvation induce an increase in percentage of males, perhaps by some hormonal disturbance. This is a fertile but unexplored area of investigation for control of this important pathogen.

The use of the trap crop *Crotalaria spectabilis* in orchards to reduce nematode populations and rejuvenate infected trees is a valuable biocontrol procedure too little used.

References: Barrons (1939), Christie (1936), Linford (1937), Linford et al (1938), Mankau and Prasad (1977), McBeth and Taylor (1944), Stirling and Wachtel (1980), Thorne (1961), Triantaphyllou (1973), Tyler (1933).

amended with the fungi and 900/g in the nonfumigated soil. Lesions caused by the *Fusarium* were also significantly smaller when the competitor fungi were added. The greater the total number of propagules of competing fungi, whether introduced or as natural colonists, the less the amount of saprophytic growth by the pathogen and the lower the incidence of infection (Marois and Mitchell, 1981b). In the field, biological control of the pathogen was achieved by pouring 25 ml of a suspension of conidia of each of the five isolates (each at 10^4 conidia per milliliter) over the roots of each tomato seedling at the time of transplanting (Marois et al, 1981). At 5,000 chlamydospores of the pathogen per gram of soil, disease incidence was 37% without the antagonists and 7% where antagonists had been introduced.

Elad et al (1981a) showed that introduction of *T. harzianum* into strawberry nursery beds after soil fumigation gave significant protection of the plants from black root rot caused by *Rhizoctonia solani*, not only in the nursery where applied, but also on the plants after they were transferred to commercial fruiting fields. In another study (Elad et al, 1982b), the introduction of *T. harzianum* into the soil in field plots following fumigation with methyl bromide resulted in an 88% reduction in reinfestation of the soil by *Sclerotium rolfsii* and *R. solani*, and gave control of these pathogens on peanuts and tomatoes that was superior to fumigation by itself. Mass application of biological preparations in commercial fields may not be practical on more than a limited scale, but introduction of the antagonist into fumigated nursery soils where transplants are produced, to establish it in the rhizosphere before transplanting, would seem both feasible and practical as a means to retard or possibly prevent recolonization of commercial fumigated fields by pathogens.

ANTAGONISTS APPLIED WITH THE PLANTING MATERIAL

A more economical and often more effective method of biological control is to introduce the antagonist with the planting material. This is an obvious approach where the objective is biological protection of the plant, i.e., protection of germinating seeds, roots, or emerging shoots rather than biological destruction of the pathogen inoculum at large (Chapter 4). This approach may be the only practical way to introduce antagonists for biological control of pathogens on seeds, seedpieces, transplants, or other planting material. This approach—the use of phytosanitizing microorganisms—is unique to biological control of plant pathogens and may ultimately prove to be the most successful means of biological control with introduced antagonists.

Planting material may be treated with any of three kinds of antagonists—fungi, bacteria, or viruses. Antagonistic fungi or bacteria, themselves nonpathogenic or only weakly pathogenic, may be coated on seeds, seedpieces, or roots of transplants. A weak strain of a plant virus, intended for use as a cross protectant, may be inoculated into the seedling sometime before or at the time of transplanting. In general, *protection by introduction of any of these biocontrol agents may be expected only if it is properly es-*

tablished in the infection court in advance of attempted infection by the pathogen.

Biological Seed Treatments

Seed treatments have traditionally involved coating the seed with fungicides to protect them from decay during germination, to eliminate seedborne fungus pathogens, or both. Seed protection is essential for many crops, particularly peas, corn, soybeans, beans, and cotton, which have large seeds highly vulnerable to decay by *Fusarium* spp., *Pythium* spp., and *Rhizoctonia solani*. Chemical seed treatments are usually easy to apply and relatively inexpensive, and they are not dangerous compared with many other pesticides used in agriculture. Nevertheless, a biological seed treatment would have certain advantages over chemicals. Seed treated in excess of that needed for sowing could be fed to poultry or livestock with less risk than if it were chemically treated. In countries where the chemicals are not available, or cannot be purchased, a microorganism could be produced locally and used in place of the chemical. The local production of microorganisms or their products for various uses is currently practiced in the People's Republic of China at the county level and on some communes (Kelman and Cook, 1978).

Fungus antagonists. — The fungus antagonists most effective as biological seed treatments have been species of *Chaetomium, Penicillium*, and *Trichoderma* (Chapters 4 and 9). Treatment of corn seed with spores of *C. globosum* was as effective as thiram or captan against seedling blight caused by *Fusarium roseum* 'Graminearum' (Chang and Kommedahl, 1968), and *P. oxalicum* applied as a coating of conidia on pea seeds protected the seeds as well as did captan during germination and seedling emergence in the field (Kommedahl and Windels, 1978; Kommedahl et al, 1981; Windels and Kommedahl, 1982b). In one trial with peas, the number of emerging plants was 177% and the yield 181% of the control where seeds were treated with *P. oxalicum*, compared with 189 and 113%, respectively, where seeds were treated with captan. Spores of the fungi were applied directly to seeds of corn, peas, or soybeans as a dry coating with no adhesive (Kommedahl et al, 1981). Seed treatment with *P. oxalicum* has also been effective in Washington against damping-off of garbanzo beans caused by *Pythium* spp. (Figure 8.4) (W. Kaiser, unpublished). In tests with *Chaetomium*-coated corn kernels, $2-10 \times 10^6$ ascospores per seed were optimal, with less protection against *F. roseum* 'Graminearum' at spore loads above or below this amount (Kommedahl et al, 1981). Six-week-old spores were more effective than three-week-old spores of *T. harzianum* on soybean seeds, but spore age was not a factor in the protection of pea seeds by *P. oxalicum*.

Figure 8.4. Garbanzo beans grown from seed that had been treated with spores of *Penicillium oxalicum* (right), or that received no treatment (left). *Pythium ultimum* was the main cause of damping-off of the unprotected plants. The row of protected plants in the photo at the right is the same row visible on the upper side of the photo on the left. (Photos courtesy of W. Kaiser.)

The strain of *T. hamatum* isolated by Chet and Baker (1981) from the *Rhizoctonia*-suppressive soil from Bogotá, Colombia (Chapter 7) proved highly effective as a protectant of radish against *Rhizoctonia solani* and also against *Pythium* damping-off of peas when applied as a coating of spores (Harman et al, 1981). However, in soils from New York low in available iron, siderophore-producing bacteria were found to inhibit the *Trichoderma* and to interfere with its effectiveness as a biological seed treatment (Harman et al, 1983). Chitin or cell walls of *R. solani* as an additive to the carrier containing conidia of *T. hamatum* improved the control of damping-off of radish and peas caused by *R. solani* and *Pythium* spp., whereas peat had no effect (Harman et al, 1981).

There has been little evidence that the *Chaetomium, Penicillium,* and *Trichoderma* spp. shown to protect germinating seeds of peas, corn, and soybeans do so by active colonization of the rhizosphere (Harman et al, 1980; Kommedahl et al, 1981; Windels and Kommedahl, 1982a). Rather, the fungus antagonists studied thus far have provided protection only against seed decay and seedling blight, much as do prophylactic chemical seed treatments. Any of these fungi would be well suited to mass production as spores for commercial coating of seeds (Chapter 9).

Magie (1980) has demonstrated the potential for commercial use of a corm treatment for gladiolus using nonpathogenic fusaria to protect against *Fusarium oxysporum* f. sp. *gladioli*. Corm rot "is the most damaging disease of gladiolus in the warm growing areas of the world" (Magie, 1980). The disease produces root rot, stunted plants, and blind flowers. Pathogen-free corms developed from tissue culture are especially vulnerable to colonization by pathogens, but nonpathogens may also colonize these corms and subsequently protect them. Significantly, the work of Magie (1980) demonstrated that certain isolates of *F. moniliforme* 'Subglutinans' (a nonpathogen) give biological control even if applied after the pathogen is established. The corms were dipped for up to one minute in a suspension of $1-2 \times 10^6$ conidia of the nonpathogen per milliliter or for 10–15 minutes in a benomyl-captan, benomyl-thiram, or benomyl-dichloran mixture, stored at 4–6°C in an open dry area for four to six months, and then planted. One isolate (M-865) gave protection equal to that of benomyl, based on the numbers of flower spikes and sound corms at harvest, and it was superior to benomyl in a trial in which the corms were either inoculated with the pathogen 48 hours before application of benomyl or inoculated with the antagonist. A combination treatment with a benomyl-resistant antagonist was suggested as the best possibility for stable control.

Microorganisms with ability as hyperparasites have also been used with success as seed treatments. *Pythium oligandrum*, a highly effective mycoparasite of *Pythium* spp. (Deacon, 1976b), applied as a suspension of oospores (4×10^6/ml) to seeds of sugar beet, was nearly as effective as thiram in protecting beets against damping-off in natural soil (Veselý, *in* Schippers and Gams, 1979). A *Corticium* sp. (later identified as *Laetisaria arvalis*) isolated originally from soil and parasitic to *Rhizoctonia solani* in culture, when applied as a seed treatment to sugar beets in a field study, resulted in 100% greater stand than nontreated seeds (Odvody et al, 1980). This antagonist has also shown potential for protection of table beets against pythium damping-off in New York (Hoch and Abawi, 1979a). It has a shelf life of at least three years when stored as ground air-dried mycelia and sclerotia under nonsterile conditions (Odvody et al, 1980). The sclerotium parasite, *Coniothyrium minitans*, grown on milled rice and applied as a pycnidial dust on onion seed, gave significant biological control of *Sclerotium cepivorum*. The number of healthy plants at 12 weeks averaged 45 and 141 of a possible 200 for the check and the treated seed, respectively (Ahmed and Tribe, 1977).

As pointed out in Chapter 4, an ideal antagonist for use as a seed treatment will combine the attributes of aggressiveness as a competitor and spermosphere colonist with ability as a hyperparasite, producer of antibiotic, or both. This may explain part of the effectiveness of *T. hamatum* isolated from the *Rhizoctonia*-suppressive soil from Colombia; this antagonist is an aggressive colonist of the seed coats of both radish and pea, as indicated by population increases of up to 100-fold in the spermosphere of treated seeds, but it also produces enzymes that destroy the hyphal walls of both *R. solani* and *Pythium* spp. (Chet and Baker, 1980; Harman et al, 1980, 1981). **The greater the number and diversity of methods used by an antagonist to inhibit a pathogen, the more successful it will be in biological control.** Although not yet investigated, the possibility exists that selection for a large number of specific antagonistic traits in a single organism might reduce the fitness of that organism to compete in environments where the specific traits provide no particular advantage. This possibility would be analogous to a race of a pathogen having reduced fitness associated with unnecessary genes for virulence.

Bacterial antagonists. — Most of the effective bacterial antagonists used as seed or seedpiece treatments have been *Bacillus subtilis, Streptomyces* spp., or strains of the *Pseudomonas fluorescens-putida* group. The success with *B. subtilis* A13 and *S. griseus* on cereals and carrots in Australia, with *B. subtilis* on seeds of corn to protect against *Fusarium* seedling blight, and with strain 5406 of *Streptomyces* sp. for cotton in the People's Republic of China (Chapter 4) gives strong evidence for the potential of these bacteria as biocontrol agents. When applied as seed treatments, four strains of *B. subtilis*, which were isolated originally from sclerotia of *Sclerotium cepivorum*, and which produced diffusible antibiotics inhibitory to growth of *S. cepivorum* on potato-dextrose agar, provided significant season-long protection of the partially resistant onion cultivar Festival and one also protected the susceptible cultivar Autumn Spice against natural infection from *S. cepivorum* (Utkhede and Rahe, 1980).

Root-colonizing pseudomonads are common rhizosphere inhabitants in nature and probably contribute to the natural protection of roots against soilborne plant pathogens. Progress in the past few years leaves little doubt that significant use will be found for these aggressive antagonists introduced on seeds or other planting material. Their potential to colonize and protect roots against various pathogens when applied as a seed or seedpiece treatment has been shown for potato (P. Geel and B. Schippers, unpublished; Kloepper and Schroth, 1979; Wadi, 1982), onion (Kawamoto and Lorbeer, 1976), sugar beets (Suslow et al, 1980), and wheat (Weller and Cook, 1983). When introduced on seed, they grow with the advancing root, displacing or preempting the normal rhizoplane microflora and making up a significant portion of the total bacterial population on the rhizoplane, especially at the root tip. Some strains effective against *Gaeumannomyces graminis* var. *tritici* match the characters of known nonpathogenic pseudomonads, but others produce a hypersensitive reaction in tobacco, typical of plant-pathogenic pseudomonads. *Pseudomonas cepacia*, a weak pathogen of onion and inhibitory to *Fusarium oxysporum* f. sp. *cepae* in vitro, provided excellent control of the *Fusarium* in naturally infested soil when introduced on the onion seeds (Kawamoto and Lorbeer, 1976). The bacterium introduced on the seed was recovered from the roots and probably is well adapted to onion roots. Having certain characters of pathogens may account for the success of some of the pseudomonads as root colonists. The possibility that they provide biological control through cross protection or even induced resistance cannot be ruled out. Obviously, it will be necessary to verify lack of virulence to the host before proceeding with any antagonist as an agent of biological control.

Bacteria are usually applied with a carrier or adhesive to adhere cells to the seed. Some effective materials are peat (used for inoculation of legume seeds with rhizobia), methyl cellulose, xanthan gum, talc, or gum arabic. A carrier or adhesive is not essential (cells may be stuck on simply by dipping the seed or seedpieces in a cell suspension) but has the advantage of protecting the bacteria between the time of treatment and planting. Merriman et al (1974) showed that application of *B. subtilis* A13 to carrot seeds in a bentonite-sand pellet produced about one-third more carrots (83.8 versus 63.4 tons per hectare) than did bacteria applied as a water suspension.

Little is known about the minimum or optimum densities for different fungus or bacterial antagonists on seeds, the need for nutrients, shelf life, and other questions basic to commercial use. With fluorescent pseudomonads applied to wheat seeds, root colonization was the same at all densities of 10^7 or greater (W. Howie, unpublished). Wadi (1982) obtained significantly better control of *Verticillium dahliae* on potato with each of five different strains of antagonistic bacteria applied at 10^{10} cells per potato seedpiece than with only 10^3 cells of the strain per seedpiece.

In providing nutrients for the antagonist, care must be taken to ensure that the pathogen is not stimulated. Unwashed cells of bacteria applied to seeds resulted in greater incidence of seed decay caused by *Pythium* spp. (Gindrat, *in* Schippers and Gams, 1979). The normal resident microbiota on seeds may also influence the efficacy of some seed-applied antagonists, and removing this microbiota before applying the desired microorganisms may be necessary. Thus, the seed source and kind of pretreatment (if any) to eliminate the natural microflora must also be investigated.

Research completed during the past 15–20 years on disease control by application of saprophytic or weakly parasitic strains of bacteria and fungi to planting material is opening the way for an important and growing biotechnology that may, in the near future, lead to marked changes in seed treatment formulations and planting procedures. Perhaps this discussion will prompt even more effort to make practical use of biological control agents introduced with the planting material. Moreover, *at this early stage of finding antagonists with suitable spermosphere or rhizosphere competence, it is unwise to rule out any nonpathogenic bacterial or fungus group in soil as a source of strains with potential as biological seed treatments.*

Inoculation of Cuttings and Transplants

Antagonists may be established on cuttings or transplants of ornamentals, vegetable seedlings, young fruit trees, or other plants that are reared in a two-step procedure, i.e., where the plant is started in a nursery and then moved to the field. A common feature of the antagonists included in this section is that all are intended to provide biological protection of infection courts, to give cross protection, or to induce host-plant resistance to the pathogen. The antagonist may be incorporated into the initial rooting medium or bedding or nursery soil, not for the purpose of protecting the nursery stock while in the nursery, but rather to be carried with the stock to the orchard, garden, or field to help keep the plant healthy (Olsen and Baker, 1968). The cutting or transplant may be dipped in a suspension of the antagonists at the time of transplanting, or a more direct method of inoculation of the plant may be used, as when the plant is inoculated with a mild strain of a virus for cross protection. A particularly promising but unexplored opportunity for this approach is with test-tube plantlets started from protoplast or tissue culture; antagonists could be readily established on these propagative units before they are moved to the field.

Biological control of crown gall by strain K84 involves dipping the cuttings, transplants, or sometimes seeds of the susceptible plants into a cell suspension or commercial product of the biocontrol bacterium and then planting or sowing immediately (Kerr, 1980). The recommended concentration is usually 10^6–10^8 bacterial cells per milliliter but, as for other biological materials used for seed treatments, little is known of the best rates or methods to produce or package the bacteria. Strain K84 is used successfully on a commercial scale in Australia, New Zealand, South Africa, Canada, Europe, the United States, and probably elsewhere (Moore, *in* Schippers and Gams, 1979). Commercial formulations use a peat preparation, but cells harvested from fresh agar cultures in water or carboxymethyl cellulose (CMC) are also effective. Strain K84 was shown to survive up to 45 months as a cell suspension in 0.25% CMC, with the viable count dropping only from 9.6×10^8 to 1.5×10^8 colony-forming units per milliliter during this period (Moore, *in* Schippers and Gams, 1979). A CMC suspension mixes easily with water for commercial use at 1:10 or 1:100 dilutions. This biological control has been used successfully as a bare root application for roses and a wide variety of stone and pome fruit and nut trees and on cuttings of rose. Strain K84 has been particularly successful in Australia, where insensitive biotypes of the pathogen have not been found; unfortunately, it is ineffective against certain biotypes of the pathogen present in the United States (on grape) and elsewhere.

Inoculation with mycorrhizal fungi. — The inoculation of plants with mycorrhizal fungi is another method of achieving biological control with introduced microorganisms, either to protect roots against pathogens or to enhance the general resistance of plants to pathogens. Such inoculation is done most commonly during production of the seedling plants or at the time of transplanting. Direct-sown crops such as soybeans may also derive some benefit of pathogen control from the vesicular-arbuscular (VA) endomycorrhizae (Chou and Schmitthenner, 1974), and perhaps the technology for application of VA mycorrhizal fungi as a seed treatment may someday become practical. A serious problem has been how to produce inoculum of these symbionts, which must be grown on roots of a host plant. Either the soil from around the roots or the roots themselves must be used as the source of mycorrhizal inoculum. Commercial soil preparations of this type of inoculum of *Glomus deserticola* are now available in California.

Most work indicates that mycorrhizal fungi must be established on the plant well in advance of exposure of the root to the pathogen. Protection of the woody ornamental *Chamaecyparis lawsoniana* 'Elwoodii' against root rot caused by *Phytophthora cinnamomi* required that the rooted cuttings be inoculated several months before being transplanted into the rooting medium infested with *P. cinnamomi* (Bärtschi et al, 1981). The work revealed further that a mixture of different VA mycorrhizal fungi gave better protection than *G. mosseae* alone, possibly because the mixture increased the chances for establishment of the proper mycorrhiza. Obviously, the potential exists for significant disease control in some cases, but a great deal more information is needed to make the best use of mycorrhizae as biological control agents.

Most agricultural soils contain indigenous species of mycorrhizal fungi, but inoculation may still be beneficial, as shown by Plenchette et al (1981) for apple seedlings. Inoculation of the seedlings with fresh endomycorrhizal roots from *Fraxinus americana* before transplanting them into an orchard site resulted in significantly faster growth than seedlings fertilized with phosphorus at 100 kg/ha. Vesicles of endomycorrhizal fungi were common in roots of inoculated seedlings but uncommon in noninoculated seedlings, in spite of a natural endomycorrhizal flora in the soil. Obviously, the proper amount or kind of VA mycorrhizal inoculum cannot be assumed to occur naturally in the soil.

Good progress has been made on methods to introduce ectomycorrhizal fungi for pine seedlings grown in fumigated nursery soil before being transplanted to plantation or forest sites. Application of either vegetative mycelium in a vermiculite-peat nutrient or basidiospores of *Pisolithus tinctorius* resulted

Figure 8.5. Sweet potatoes grown from cuttings that had been dipped in a spore suspension of a nonpathogenic strain of *Fusarium oxysporum* (top), or that received no treatment. Treatment with the nonpathogen protected the plants against *F. oxysporum* f. sp. *batatas*. (Photos courtesy of K. Ogawa and H. Komada.)

in satisfactory ectomycorrhizae in a test involving 4,000 seedlings of *Pinus taeda* in fumigated nursery soil (Marx and Bryan, 1975). Basidiospores are also effective inoculum for introduction of other ectomycorrhizal fungi and can be preserved by lyophilization (Sinclair et al, 1975). Soil from a plot where mycorrhizae have occurred is also effective when mixed with fumigated nursery soil (Marx and Bryan, 1975).

Agents for cross protection. — Carnation cuttings can be protected biologically against fusarium stem rot caused by *F. roseum* 'Avenaceum' by inoculating them at the time of cutting with any of many different nonpathogens (Chapter 4). The antagonist must be present in advance of inoculation with the pathogen, and all evidence indicates that a form of induced resistance is involved (Baker et al, 1978). The method is used on a limited basis in the propagation process for carnation in Colorado.

Another example of cross protection (probably induced resistance) with potential for commercial use has been developed by K. Ogawa and H. Komada in Japan (personal communication) for control of fusarium wilt of sweet potato (caused by *F. oxysporum* f. sp. *batatas*). Sweet potatoes are grown from cuttings from tubers that have been allowed to produce sprouts in hot beds under glass or clear polyethylene enclosures. The cuttings are highly susceptible to infection by *F. oxysporum* f. sp. *batatas* when transplanted to the field but are protected remarkably well if inoculated first with a nonpathogenic strain of *F. oxysporum* (Figure 8.5). This nonpathogen was first noted as a natural inhabitant of stem tissues of sweet potato plants. Apparently, the fungus is established naturally in some plants, perhaps protecting them from wilt. A dip inoculation of the cut ends with bud cells produced in shake culture, provided better protection than a soil drench with the bud cells (Figure 8.6). The longer the stems were immersed in a suspension of bud cells, up to 16 hours, the better the protection. The entire stem was protected and not just the cut surface where the antagonist was applied, thereby indicating a role of induced resistance. Protection lasted in the field for at least two months.

Plant pathology literature is replete with reports of cross protection and induced resistance achieved by inoculation of the plant with a weak pathogen or nonpathogen (Matta, 1971; Kuć, 1982). However, except for the fusarium stem rot of carnation in Colorado, tobacco mosaic virus (TMV) in tomato in Canada, Japan, Europe, New Zealand, and possibly elsewhere, citrus tristeza virus of orange and lime trees in Brazil and Australia (Chapter 4), and perhaps one or two other virus diseases of plants, none of the known examples of cross protection or induced resistance has been used commercially.

There may be several reasons for the lack of commercial development of this form of biological control, which we think should be progressing more rapidly. First, plant pathologists are generally reluctant to intentionally in-

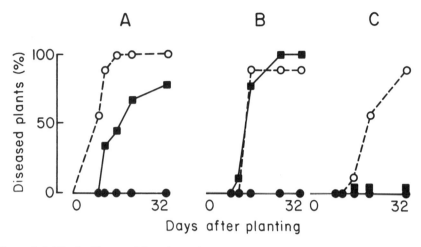

Figure 8.6. The incidence of fusarium wilt of sweet potato caused by *F. oxysporum* f. sp. *batatas* on plants grown from cuttings that received no treatment (○), or the cut ends were dipped in a cell suspension of a nonpathogenic strain of *F. oxysporum* before transplanting (●). The nonpathogen was also tested as a soil drench (■). The soils were (A) autoclaved and then artificially infested with the pathogen, (B) natural soil artificially infested with the pathogen, and (C) naturally infested soil. (Unpublished data of K. Ogawa and H. Komada.)

oculate plants with pathogens. Obviously, caution must be exercised and guidelines are needed (discussed below). Nevertheless, considering the opportunities identified since Chester's (1933) review 50 years ago, perhaps plant pathologists are too conservative. Second, as pointed out earlier (Baker and Cook, 1974), much research on the biological control of plant pathogens has been limited to laboratory studies, and there has been a tendency to await complete understanding before moving toward practical application. This would seem particularly so where the use of cross protection and induced resistance is concerned. However, biological control by this approach need not, should not, and cannot await complete understanding of the host-pathogen interaction. Third, some of the lack of follow-up toward application occurs because the investigator is more interested in basic research and in explaining the results than in seeing the results applied by a user group. A team approach involving some persons oriented toward basic aspects and others oriented toward applied aspects is one solution to this common situation.

At the March 1982 annual meeting of the British Columbia Division of the Canadian Phytopathlogical Society in Vancouver, R. Stace-Smith (unpublished) outlined eight criteria that should be satisfied, if possible, before cross protection is considered for virus disease control: 1) disease incidence is high; 2) the disease is causing severe damage; 3) alternative controls are unsatisfactory; 4) mild strains of the virus are available; 5) the mild strains have proven stability; 6) a simple inoculation procedure is available; 7) inoculation with a mild strain gives a high degree of protection; and 8) a low risk is associated with the technique. These criteria might also apply to inoculation of plants with mildly virulent or avirulent strains of fungi or bacteria to obtain cross protection or induced resistance, namely: the disease is serious and alternative controls are unavailable or unsatisfactory; nonpathogenic or weakly pathogenic strains are available and give a high degree of lasting protection; a method of inoculation is available; and the risk of the method to the crop or other crops in the area or agroecosystem is low.

POSTPLANTING TREATMENT WITH ANTAGONISTS

Biological control achieved with antagonists introduced after the crop is growing is mainly a means to protect foliage and wounds (e.g., pruning wounds) from infections, to retard the production of pathogen inoculum, or to displace a pathogen in a lesion. As pointed out in Chapters 4 and 5, only relatively few species of fungi are specialized for ability as secondary colonists of lesions. These fungi, although nonpathogenic to the plant, are nevertheless specifically adapted for growth in the modified tissues and thereby have potential to starve the pathogen and even to overtake and displace it. Some grow directly on the pathogen as hyperparasites.

Jarvis and Slingsby (1977) obtained significant biological control of powdery mildew on greenhouse cucumbers (caused by *Sphaerotheca fuliginea*) when conidia of *Ampelomyces quisqualis* (*Cicinnobolus cesatii*) were applied in water at 1.2×10^5 conidia per milliliter to the point of runoff every seven to ten days for about a month. *Ampelomyces quisqualis* is well known for its ability to grow through mycelium of powdery mildew fungi and to prevent production of conidia and cleistothecia. Best control was obtained when water was also sprayed on the cucumbers every three to four days. Another fungus, *Hansfordia* sp., colonizes *Cercosporidium* lesions on peanut leaves and prevents formation of secondary inoculum by the pathogen (Taber and Pettit, 1981). Tests are under way in Texas to determine the potential of *Hansfordia* sp. to check epidemic development of cercosporidium leaf spot, when mass-introduced on peanut plants at different times during the disease cycle (R. A. Taber, personal communication). Like other uses of introduced antagonists, more rapid progress can be expected from selecting or breeding better strains from the best possible isolates from nature (e.g., isolates selected from plants protected naturally by the antagonist).

Leaf scars and other natural wounds are an important avenue for infection of apple by the European canker fungus, *Nectria galligena*. Infection occurs when the scar is fresh, generally within 48 hours after the scar is made.

Numerous saprophytic fungi and bacteria cosmopolitan on the apple branches colonize the scars and probably provide some natural exclusion of *N. galligena* (Swinburne, 1973). Of the naturally occurring isolates tested by Swinburne (1973), certain strains of *Bacillus subtilis* inhibited *N. galligena* in vitro through production of antibiotic. When applied as a cell suspension sprayed after leaf fall, these strains gave protection against the pathogen applied 24 hours later. The protection was effective through April, but not into May when the protective layer was shed. Antibiotic-producing strains sprayed onto apple trees in the autumn after leaf fall could be recovered from the leaf scars during the dormant season and into the spring, and they apparently multiplied in the scars (Swinburne et al, 1975). Applied in the fall at 10 and 50% leaf fall, the bacteria reduced the number of autumn infections by *N. galligena* by about one half, equivalent to the protection obtained with phenylmercuric nitrate as a fungicide spray (Swinburne, 1978). The existence of some strains more inhibitory than others, the fact that *B. subtilis* apparently is a natural colonist of leaf scars and hence adapted to this niche, the availability in industry of technology for mass production of *Bacillus* (Kenney and Couch, *in* Papavizas, 1981), and the ability of the bacterium to form spores and hence survive drying and other harsh treatments likely to occur during processing and storage are all advantages to commercial development of this biological control.

Fire blight of pome fruit trees caused by *Erwinia amylovora* is among the most important diseases of pear and apple worldwide. Streptomycin was introduced for control about 30 years ago, but strains of the pathogen resistant to this antibiotic are now common where the antibiotic has been used extensively (Schroth et al, 1979). Strains of bacteria on apple and pear trees with potential to inhibit *E. amylovora* have been known for the past 50 years, but serious attempts at biological control have been made only recently. Particular interest has been focused on nonpathogenic "yellow bacteria," common in blossoms and as a secondary colonist of fire blight lesions. Riggle and Klos (1972) identified this bacterium as *E. herbicola* and obtained partial control of fire blight in the glasshouse and an orchard by inoculating blossoms with it 24 hours before inoculation with *E. amylovora*. They proposed further that *E. herbicola* as a secondary invader of infected branches could "catch up" with *E. amylovora* in the fall, compete with it for nutrients, and thereby reduce the overwintering population of the pathogen. *Erwinia herbicola* grown in a high-sugar medium reduces the pH below that tolerable by *E. amylovora*.

In California, Thomson et al (1976) obtained 68% less fire blight of pear than on nontreated trees in an orchard by applying a mixture of three *Pseudomonas* isolates (all were oxidase-positive and produced a green fluorescent pigment in King's medium B, and two were arginine dihydrolase-positive). They applied the isolates at 3.2×10^7 cells per milliliter for each isolate at 464 liter/ha. The pseudomonads were selected initially because of their ability to colonize and multiply in pear blossoms. The control was equivalent to a chemical treatment with Kocide. In New York, Beer et al (1980) obtained significant biological control of fire blight of apple by spray application of strains of *E. herbicola*. Strains with and without the ability to produce bacteriocins against *E. amylovora* were effective, but as pointed out repeatedly in this book, the ideal antagonist will have several traits, e.g., ability to multiply and compete in the blossoms (Thomson et al, 1976), ability to cause an unfavorable pH for the pathogen (Riggle and Klos, 1972), ability to produce bacteriocin (Beer et al, 1980), perhaps even ability to induce host plant resistance (Goodman, *in* Horsfall and Cowling, 1980).

Earlier (Baker and Cook, 1974), we called attention to the observation of D. W. Dye in New Zealand that the once important fire blight has disappeared as an economic problem. This pattern, when documented and studied for other diseases, has usually revealed a natural biological control, and sometimes a source of antagonists that can be introduced into other plantings. We repeat that workers interested in developing a biological control for fire blight should study the situation in New Zealand, where biological control agents more potent than any tested thus far possibly exist.

Biological control of chestnut blight with hypovirulent strains was developed following an observation in northern Italy that cankers were healing (Chapters 4 and 5). This discovery occurred more than 30 years ago, and only during the past 10 years has a practical application for use of hypovirulent strains been developed. Even today, much remains to be done to perfect the method. Nevertheless, the story of chestnut blight reveals that long study of a natural biological control can open the way for an entirely new biological approach to control of the disease involving introduction of a biological agent.

Dutch elm disease caused by *Ceratocystis ulmi* has defied most attempts at control, but several recent discoveries indicate the existence of biological factors with potential to limit this disease. At least four factors have been identified: 1) colonization of infected trees in northern England and Scotland by *Phomopsis oblongata*, which discourages the beetle vectors of *C. ulmi* and thereby lessens the spread of the pathogen (Weber, 1981); 2) the identification of several saprophytic fungi, notably *Botryosphaeria stevensii* and *B. ribis*, with potential as competitors of *C. ulmi* in wood and ability to suppress inoculum formation (Gibbs and Smith, 1978); 3) the occurrence of both nonaggressive and aggressive strains of the pathogen, and the evidence that nonaggressive strains are infected with a greater number of different dsRNA mycoviruses than are aggressive strains (Pusey and Wilson, 1982); and 4) the evidence of Myers and Strobel (1981) and Strobel and Myers (1981) that a cell suspension or the cellfree filtrate containing an antibiotic from *Pseudomonas syringae*, when injected into one-year-old elm trees, protected the trees against a subsequent challenge inoculation with *C. ulmi*. These clues indicate that the potential for biological control of Dutch elm disease exists and should be pursued on as many fronts as possible.

Most examples of antagonists applied to a growing crop involve biological protection of pruning wounds on shrubs and trees by a competitor or antibiotic-producing nonpathogenic microorganism applied at the time of pruning. The silver leaf disease of plum, peach, and nectarine trees caused by *Chondrostereum purpureum* can be prevented or cured by inoculation of the tree with a strain of *Trichoderma viride* antagonistic to the pathogen (Corke and Rishbeth, *in* Burges, 1981). A method of protecting fresh pruning wounds was developed by Grosclaude et al (1973), who showed that inoculation of wounds with spores of *T. viride* 48 hours in advance of inoculation with *C. purpureum* gave complete protection, whereas inoculation with *T. viride* 48 hours after pruning protected only slightly more than half the pruning wounds. Various modifications of the pruning shears have been introduced to permit automatic inoculation of the wound at the time of pruning (Corke and Rishbeth, *in* Burges, 1981). A biotherapeutic (curative) treatment has also been reported whereby *Trichoderma*-impregnated wood dowels (0.8 × 10 cm) are implanted into the trunk of each infected tree at 10-cm intervals (Corke, 1978). Of 115 infected plum trees so treated in a 10-year-old orchard, 80% of the cultivar Victoria and 92% of the cultivar Czar improved after two years. After three years, 70% of the treated trees had recovered (no silvering), compared with 50% recovery for nontreated trees. Products for the *Trichoderma* treatment against silver leaf of plum are available commercially in Europe.

In Australia, the weak pathogen *Fusarium lateritium*, applied at the time of pruning apricot trees, protects the trees against the gummosis pathogen, *Eutypa armeniacae* (Carter, 1971). Application of benomyl with the benomyl-tolerant *Fusarium* gives even better protection; the benomyl gives immediate protection and the *Fusarium* protects after the benomyl is no longer active. In North America, *T. harzianum* has been tested as a protectant of the summer wounds of red maple (*Acer rubum*) otherwise invaded by wood rotters (Pottle et al, 1977; Smith et al, 1979). Wounds on trees heal and become naturally resistant to colonization within a few days or one to two weeks, but until healed, they are highly vulnerable to infection. The mechanism of protection may be an example of the "possession" principle, where the first organisms, if adequately established, exclude others adapted to the same niche and substrate. Numerous examples of natural protection of wounds and even heartwood (Basham, 1973) of trees have been cited in previous chapters; if studied and tested, most of these probably offer some potential for commercial exploitation.

POSTHARVEST BIOLOGICAL CONTROL

A. Tronsmo and associates in Norway have experimented with *Trichoderma* spp. applied to blossoms of fruits to protect against fruit rot during storage. In one study, spores of *T. viride* and *T. polysporum* in suspension at 10^7/ml sprayed on strawberries at flowering, and every 14 days thereafter, were as effective as sprays of dichlofluanid against subsequent storage rot caused by *Botrytis cinerea* and *Mucor mucedo* (Tronsmo and Dennis, 1977). However, neither the chemical nor the biological treatment gave the level of control desired. *Trichoderma pseudokoningii* was observed to develop naturally in the "dry eye rot" lesions caused by *B. cinerea* on fallen apple fruit and to displace the pathogen in the lesion; when sprayed simultaneously with the pathogen as a mixture into flowers in May, the antagonist gave moderate protection against dry eye rot as measured in September. However, the antagonist was ineffective against natural inoculum of *B. cinerea* (Tronsmo and Raa, 1977). The failure to control natural infections was attributed to the fact that *T. pseudokoningii* did not grow at temperatures below 9°C, whereas the minimum and maximum temperatures during flowering in May were 7.7 and 13.3°C, respectively. Under controlled conditions, the *Trichoderma* controlled dry eye rot at 22°C but not at 4°C. Most isolates of *T. viride* and *T. polysporum* were able to grow at 2°C. Obviously, *the antagonist must be able to grow over at least as wide as or a wider range of physical conditions than the pathogen to be effective under all conditions* (Baker and Cook, 1974).

Although antagonists applied to control pathogens after harvest may seem only remotely practical or feasible, nevertheless the first successful biological control with an introduced antagonist used this approach. Rishbeth's biological control of *Heterobasidion annosum* is achieved by applying the antagonist to a "crop residue," namely the cut surface of the stumps of pine after the trees are harvested. This method, developed through the patience and persistence of J. Rishbeth over a span of 20 years, is now widely used in England and continental Europe (Greig, 1976; Kallio and Hallaksela, 1979). The antagonist is available commercially as sachets of spores that can be easily suspended in water for application to the cut stump, which then remains protected until decayed. The method shows the simplicity and durability possible with biological control. The method and especially the years of research and development reveal further that, although some controls might be discovered and put into operation quickly, most will require a continued long-term effort. Clearly, the end result is worth the effort.

9

ANTAGONISTAE VITAE

For all the pests that out of earth arise
the earth itself the antidote supplies.
— LITHICA, c. 400 B.C.

In our earlier book (Baker and Cook, 1974) emphasis was placed on making maximal use of naturally occurring resident antagonists by selectively modifying cultural practices. We realized, however, that "Sooner or later investigations on the suppressive nature of the soil would lead to determining the organisms important in the suppressive effect, and the mechanisms involved. It is also desirable to isolate specific antagonists from soil for direct use as agents of biological control. . . . "

Remarkable advances have been made in the sophistication of techniques used and in the number of specific antagonists studied in the intervening years (Chapter 8). New examples of successful biological control with resident antagonists continue to appear, but there is little doubt that the future will bring much more work on isolation of significant antagonist(s) and their introduction into a disease complex. This trend is exemplified in this chapter by brief sketches of principal antagonists intensively studied in the past or currently used. The list is not complete; it does not include antagonists that have been casually reported or only superficially studied. For example, many nematode-trapping fungi that have been found only once are not included. When the next *Antagonistae Vitae* is prepared, it certainly will be much longer. However it is hoped that this trend will not reach "fad" proportions to the extent that resident antagonists are ignored or forgotten in the pursuit of individual microorganisms that can be mass-produced, commercialized, and perhaps patented.

The commercial production of inoculum of antagonists for biological control of plant pathogens is a recent outgrowth from the antibiotic industry and from production of *Bacillus thuringiensis* for insect control. Production methods and marketing problems have been discussed by Toyama et al (1970), Underköfler (1976), Kenney and Couch (*in* Papavizas, 1981), and Churchill (*in* Charudattan and Walker, 1982). As in any new field, the deficiency of time-tested information, coupled with bandwagon enthusiasm for the use of introduced antagonists, has its dangers. Opportunity unfortunately is created for unethical, dishonest, opportunistic exploitation of sound work by people who hastily market nostrums said to contain a mixture of potent antagonists. Several of these already have come and gone. The best solution to this situation is to develop effective formulations for sale in the marketplace. Future development of legitimate biological materials for pathogen control undoubtedly will meet this need, much as has already been done with chemicals for pathogen control.

Commercial production has been almost solely concerned with single organisms that can be grown in synthetic media on a large scale. At least two other types of antagonists eventually will require commercial attention.

1. Single antagonists that are obligate parasites (i.e., cannot be grown on culture media) require special methods, but already clues suggest how this can be done. Stirling and Wachtel (1980) have devised a method for mass production and storage of *Bacillus penetrans* on nematodes (Chapter 8 and the treatment of *Bacillus*, this chapter). Moreover, the vesicular-arbuscular endomycorrhizal fungus *Glomus deserticola* is now produced commercially; this fungus produces abundant spores in soil, which are sold as inoculum for citrus in southern California. Parke and Linderman (1980) found that moss (*Funaria hygrometrica*) growing in pot cultures of asparagus contained large numbers of spores of *G. mosseae*, *G. fasciculatum*, or *G. epigeaum* and could be harvested and dried as a source of inoculum.

2. Multiple antagonists can jointly produce suppressiveness in soil not obtainable by single microorganisms. Methods possibly can be devised to increase the volume of suppressive soil from the field in a synthetic soil mixture. If the antagonists survive a treatment with aerated steam (60°C for 30 minutes) that destroys all plant pathogens, such a treated mixture could safely be marketed for inoculation of container soil in which nursery stock is grown; when such stock is planted in the field, the antagonists would be in place for maximum effectiveness. Such a mixture might also be useful for glasshouse crops, as attempted by Alabouvette et al (1980b) for tomatoes and other crops grown in peat in France. Even where one antagonist seems to provide effective biocontrol (e.g., fluorescent *Pseudomonas* spp. for *Gaeumannomyces graminis* var. *tritici* or *Sporidesmium sclerotivorum* for *Sclerotinia sclerotiorum*), future work probably will demonstrate greater effectiveness or stability when multiple microorganisms are used.

As explained in Chapter 1, complexes of adapted microorganisms are more stable and have better biological balance than single microorganisms. In addition, environmental fluctuation may produce a site unfavorable to one antagonist but not to another, which then provides most of the biocontrol. The greater tolerance of *Streptomyces* spp. than of *Pseudomonas* and *Bacillus* spp. to low soil moisture suggests that a combination of these prokaryotes would be more adaptable for field use than any one by itself. Thus, the effective and stable *Phytophthora*-suppressive soil in Queensland has abundant actinomycetes, *Pseudomonas* spp., and *Bacillus* spp., and efforts to find a single microorganism there that will provide effective biological control of *Phytophthora cinnamomi* have been unsuccessful. Indeed, much valuable work is yet to be done in applying biological control to such disease situations, for control probably will not be obtained with single antagonists.

It has been necessary until now to seek antagonists in naturally occurring situations where the disease does not occur despite the presence of the pathogen, a susceptible host, and a favorable environment. We are now on the threshold of a new era in biological control, in which antagonists will be developed by genetic manipulation to fit a specific niche and function. Although the methods presently are being applied to bacteria and yeasts, they may also be applied to actinomycetes, some fungi, and higher plants (Chapter 1). These techniques greatly strengthen the potential of biological control by introduced antagonists.

AGROBACTERIUM CONN

Common soil inhabitant, especially in the rhizosphere.

Morphology and Taxonomy: Eubacteria, Rhizobiaceae; slime-celled. Small, short rods, typically motile by 1–4 peritrichous flagellae; Gram-negative. They do not produce visible gas on ordinary culture media, nor sufficient acid to be detected by litmus. In synthetic media, enough carbon dioxide may be produced to show acid with bromthymol blue or bromcresol purple. Gelatin liquefied very slowly or not at all. Free nitrogen not fixed. Optimum temperatures 25–30°C. For details, see Buchanan and Gibbons (1974).

Agrobacterium radiobacter (Beij. and Van Delden) Conn. Rods 0.4–0.8×1.0–3.0 μm, nonsporeforming, motile by polar flagellae, not acidfast, gelatin not liquefied, starch not hydrolyzed, milk coagulated but not peptomized, slight acid but no gas from glucose, fructose, arabinose, galactose, mannitol, or salicin. Best differentiated from *A. radiobacter* pv. *tumefaciens* by inoculation tests on a susceptible host.

Ecology: Optimum temperature 25–30°C, maximum 37°C, minimum 0°C; thermal death point 51°C. Both *A. radiobacter* pv. *tumefaciens* (cause of crown gall) and *radiobacter* K84 (the antagonist) occur commonly in the rhizosphere of higher plants (Schroth et al, 1971). Other competing rhizosphere microorganisms may inhibit strain K84. Moore (*in* Schippers and Gams, 1979) studied the effectiveness of strain K84 in various environmental situations

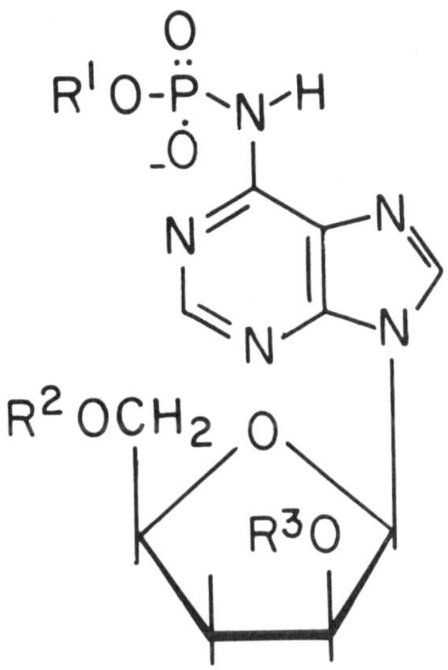

Figure 9.1. Molecular structure of agrocin 84. (Reprinted with permission from Kerr, 1980.)

and indicated the possible existence of soils suppressive to crown gall. There is no evidence that strain K84 can become pathogenic, no decreased plant growth from strain K84 and no effect on *Rhizobium* inoculation of legumes have been reported. In Australia, most agrobacteria around gall-free trees are nonpathogenic, but most around galled trees are pathogenic. When peach seedling roots were wounded, dipped in a suspension of nonpathogenic strains, and planted in soil infested with pv. *tumefaciens*, no galls developed (New and Kerr, 1972). This led to this effective biological control method (Chapter 3).

Role in Biological Control: Strain K84 has been remarkably effective in controlling crown gall over the world, but some insensitive isolates of pv. *tumefaciens* occur in some areas and on some hosts (e.g., on grape). Breakdown of effectiveness of K84 (Panagopoulos et al, *in* Schippers and Gams, 1979) can be minimized by proper application of abundant cells of K84 in the field or by development of a strain of K84 with a deficient plasmid transfer system (Ellis and Kerr, *in* Schippers and Gams, 1979).

Mechanism of Biological Control: Pathogenic strains sensitive to K84 are prevented from transferring DNA from their tumor-inducing (Ti) plasmid (T DNA) to the wounded host by the antagonist, which kills (by production of agrocin 84) or prevents attachment of the pathogen to the host receptor site. Application of strain K84 to roots colonized but not infected by pv.

tumefaciens prevents infection, but latent infections are not inhibited. Strain K84 spreads to noninoculated parts of the plant. Tissue is protected from infection for at least two years. Microorganisms on the plant surface apparently sometimes interfere with the action of strain K84. Roberts et al (1977) characterized the bacteriocin as water-soluble 6-*N*-phosphoramidate of an adenine nucleotide analogue (Figure 9.1). Sensitivity to agrocin 84 is controlled by the Ti plasmid of pv. *tumefaciens*. The host species influences the interaction of K84 and the pathogen. Heat- or chloroform-killed K84 cells do not prevent galling (Kerr, 1980).

Mass Production: Strain K84 is distributed commercially on solid nutrient agar, as aqueous cell suspensions, or in a peat formulation, and applied as cell suspensions (10^6-10^7 units per milliliter) to seeds, bare-rooted plants, graft unions, and cuttings. The half-life of peat preparations is six months at room temperature. Strain K84 is registered by the Environmental Protection Agency (EPA) for commercial use in the United States. Cost of treatment is less than 0.01 dollar per plant (Moore, 1977).

AMOEBAE, MYCOPHAGOUS IN SOIL

Several soil-inhabiting amoebae (Anderson and Patrick, 1978, 1980; Old and Darbyshire, 1978) feed on fungus spores and mycelium, bacteria, algae, flagellates, diatoms, and nematodes, including soil microorganisms pathogenic to plants.

Morphology and Taxonomy: Protozoa, Proteomyxida, Vampyrellidae. *Arachnula impatiens* Cienk., *Biomyxa vagans* Leidy, *Theratromyxa weberi* Zwillenberg, *Vampyrella vorax* Cienk., and *V. lateritia* Leidy have been the most thoroughly studied (Anderson and Patrick, 1978, 1980; Dobell, 1913; Old and Patrick, *in* Schippers and Gams, 1979) of several mycophagous amoebae. The mature trophozoite or amoeboid stage is a large branched, multinucleate, mobile mass with many contractile vacuoles that forms a cytoplasmic network on the substrate. The pseudopodia of this structure contact a spore and the amoeba then flows to it and wholly or partially engulfs it. In the areas of contact, enhanced enzymatic peripheral erosion of the spore wall occurs at the margin of a vacuole and eventually cuts out a disk of wall material. The cell contents spill out or pseudopodia penetrate into the lumen and digest the spore contents. This operation may take from 15 minutes to four to five hours. The amoeba then moves on to attack other spores or to form digestive cysts; undigested cell wall material collects in the central vacuoles of these cysts and is discharged by excystment (Figure 9.2). Reproductive cysts are smaller, have a thick wall, and are more angular in outline; they give rise to small immature amoebae 10–15 μm long

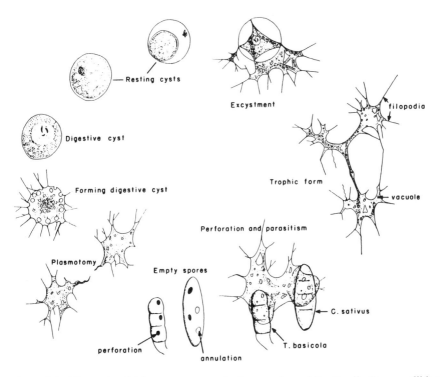

Figure 9.2. Life cycle of *Theratromyxa weberi*, a member of the family Vampyrellidae and one of several species capable of causing perforations in the pigmented cell walls of spores and hyphae of fungi. (Reprinted with permission from Anderson and Patrick, 1980.)

that probably feed on bacteria and eventually fuse to form a mycophagous trophozoite.

Ecology: Water-filled pores and water films around soil particles are necessary for trophozoites to move. Pores must be larger than 1 μm in diameter. The ability of amoebae to perforate spores is greatly impaired in soil at matric potentials wetter than −0.01 bar, apparently because of inadequate oxygen; their activity is maximal at −0.025 to −0.1 bar and again prevented at soil matric potentials less than −0.2 bar. Soil fungi are thus unlikely to be affected when soils are at field capacity (−0.3 bar) or drier. The cysts withstand adverse environmental conditions.

Role in Biological Control: These resident antagonists, common in soil, are especially valuable because they are uniquely able to attack the melanized cell walls of spores. They are important mainly in the destruction of pathogen inoculum during dormant or saprophytic existence of pathogens in soil and are less important or not important as a means to intercept or limit a pathogen during its active growth phase in the rhizosphere. Van der Laan (1954) used *Theratromyxa weberi* in an unsuccessful attempt to control *Globodera rostochiensis* nematodes in pot tests.

Mechanism of Biological Control: Cell walls of spores are penetrated by enzymatic action; since nematode cuticle and cellulosic algal walls also can be penetrated, remarkable nutritional versatility is exhibited.

Mass Production: Soil amoebae are not at present culturable.

AMPELOMYCES CES.: SCHLECHT.

These fungi, specialized hyperparasites of Erysiphaceae, were classed as *Cicinnobolus* until 1959. World wide in distribution.

Morphology and Taxonomy: Deuteromycotina, Coelomycetes, Dematiaceae; slime-spored. Pycnidia superficial, separate, lack setae, thin-walled, pale brown, unilocular; wall 1 cell-layer thick; dehiscence by apical rupture. Conidiogenous cells dolioform to ampeliform, determinate, discrete, hyaline, smooth; channel and collarette minute, formed directly from pycnidial wall cells. Conidia very pale brown, nonseptate, thin-walled, smooth, guttulate, straight or curved, cylindrical to fusiform. For details see Sutton, 1980.

Ampelomyces quisqualis Ces.: Sclecht. Pycnidia 40–105×30–50 μm. Conidiogenous cells 4.5–5.5 μm wide. Conidia 4–6.5×2–2.5 μm.

Ecology: *Ampelomyces quisqualis* is not host specific. It is favored by prolonged moist weather (Philipp and Crüger, 1979). Some strains of *A. quisqualis* are slightly pathogenic to cucumber.

Role in Biological Control: Jarvis and Slingsby (1977) used *A. quisqualis* to control *Sphaerotheca fuliginea* on glasshouse cucumbers; mildew was decreased and yields increased. Sundheim (1978, 1982) sprayed with a combination of triforine and *A. quisqualis* (which is insensitive to triforine) to control powdery mildew of glasshouse cucumbers. Sundheim (1982) and Sundheim and Amundsen (1982) also found that *A. quisqualis* controlled powdery mildew on glasshouse cucumbers.

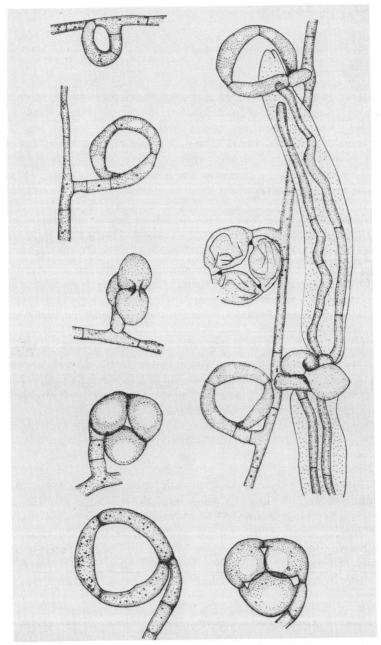

Figure 9.3. Constricting rings of *Arthrobotrys dactyloides*, including from the upper left down, developing ring, unconstricted ring, front and side views of constricted rings, and (right) nematode caught in rings and penetrated by four absorptive hyphae. The two drawings at the bottom are unconstricted and constricted rings of *Dactylella doedycoides*. (Reprinted with permission from Muller, 1958.)

ARTHROBOTRYS CORDA

Commonly occurring predators of nematodes by adhesive networks, loops, or constricting rings.

Morphology and Taxonomy: Deuteromycotina, Hyphomycetes, Moniliaceae; Xerosporae. Conidiophores erect, hyaline, septate, smooth, somewhat swollen at the apex; may grow out at the tip and produce several such swollen sporiferous zones. Conidia borne singly as blown-out ends of conidiophores, obovate to oblong-ellipsoidal, with lower end apiculate, slightly curved, one-septate, smooth, hyaline, distinctly rounded (Rifai and Cooke, 1966). For details see Drechsler (1937).

Arthrobotrys dactyloides Drechsler. Conidiophores rather sparse, hyaline, erect, septate, 200–400 μm long. Conidia borne 4–13 in a loose head, hyaline, 1-septate, elongate, elipsoidal, or digitiform (cigar-shaped), straight or slightly curved, 32–48×7–9.5 μm. Conidia sometimes germinate directly to predacious rings. Constricting rings 3-celled, formed at right angles to mycelium on short 2-celled stalks, 20–32 μm in diameter; these rings triple their volume in 0.1 second when triggered (Figure 9.3). Hyphae penetrate the cuticle of the captured nematode.
Arthrobotrys oligospora Fresenius. Conidiophores 350–450 μm long, with 1–3 successive conidial heads. Conidia plump, obovoid, 22–32×12–20 μm, 1-septate. Chlamydospores yellow, cylindrical to subspherical or ellipsoidal, and about the same size as conidia. Forms simple to complex adhesive networks. Spore germination very sensitive to fungistasis.

Ecology: *Arthrobotrys* spp. are favored by abundant organic matter in soil.

Role in Biological Control: *Arthrobotrys* spp. are residents in soil, causing the nematode population to decrease (Mankau, 1980). Cayrol et al (1978) used *A. robustus* (product called Royale 300) to control *Ditylenchus myceliophagous* on commercial mushrooms. Cayrol and Frankowski (1979) used another *Arthrobotrys* (product called Royale 350) to control *Meloidogyne* spp. on tomatoes in field tests. In lightly infested soil good control was achieved, but in heavily infested soil a preliminary nematicidal treatment was recommended.

Mechanism of Biological Control: Constricting loops or adhesive networks are formed on mycelium, especially in the presence of nematodes; numbers increase greatly from the food base supplied by trapped nematodes. *Arthrobotrys oligospora* conidia are not adhesive and apparently not attractive to nematodes; they are therefore less carried about in soil by nematodes than are those of *Meria coniospora* and *Cephalosporium balanoides* (Jansson, 1982).

ASCOCORYNE GROVES AND WILSON

Former name, *Coryne*, preempted by a conidial fungus. Inhabitant of wood and peaty sphagnum soil in Europe and North America.

Morphology and Taxonomy: Ascomycotina, Discomycetes; slime-spored. Apothecia clustered, gelatinous, violaceous-red to purple, sessile or short-stipitate, of medium size. Asci cylindrical-clavate. Ascospores 1–12 septate, ellipsoid-fusoid, hyaline. Paraphyses abundant, hyaline, filiform, simple or branched. For details, see Groves and Wilson (1967) and Dennis (1956, 1968).

Ascocoryne sarcoides (Jacq.: Fr.) Groves and Wilson. *Imperfect state*: *Pirobasidium sarcoides* von Höhnel. Apothecia superficial, clustered, gelatinous, concave or flat, irregular and lobed, 2–10 mm across, reddish purple, with a low rounded rim, on short ill-defined stalk; flesh of loosely woven mucilaginous hyphae. Asci cylindric-clavate, long-stalked, rounded above, 8-spored, 90–160×8–10 μm. Ascospores uni- or biseriate, elliptical, 1–3 (mostly 1) septate, 10–19×3–5 μm; sometimes germinate in the ascus. Paraphyses filiform, hyaline, unbranched, swollen at the tips. *Pirobasidium* state is commonly associated with *Ascocoryne* and resembles it. Conidia in a violet gelatinous sporodochium; hyphae terminate in short, thick, cylindrical or spherical cells bearing 3–5 sterigmata with cylindric, straight or curved, rodlike conidia 4×1 μm.

Role in Biological Control: Etheridge (1956, 1957) found *A. sarcoides* to be common in decaying wood in living spruce in Canada and also in the heartwood of 7.1% of healthy trees. It is thought to be a pioneer in tree decay but does not itself cause decay. One isolate reduced decay by *Coniophora puteana* and *Polysporus tomentosus* by 50–75% of the rate in the controls. The lack of serious heart rot in 80-year-old black spruce in Ontario, Canada, was reported by Basham (1973) as due to colonization of the trees by *A. sarcoides* antagonistic to the basidiomycetes responsible for heart rot. Ricard (1970) found that *A. sarcoides* could be introduced into fresh logs of Norway spruce and that the lethal action on *Heterobasidion annosum* was due to release of a water-soluble antibiotic. It developed best in the heartwood of spruce.

BACILLUS COHN

Common worldwide in soil.

Morphology and Taxonomy: Eubacteria, Bacillaceae, slime-celled. Spore-forming rods, mesophilic, with maximum growth at 30–45°C. Usually Gram-positive. Uses a wide range of simple organic compounds; growth-factor requirements also simple. Resistant endospores formed in cells may remain dormant for long periods. Selectively isolated by treating aqueous soil suspensions in hot water at 80°C for 10 minutes or by treating soil with aerated steam (60°C for 30 minutes). Motile by peritrichous flagellae. Aerobic or facultatively anaerobic. For details see Smith et al (1946), Gordon et al (1913), Buchanan and Gibbons (1974), and Berkeley and Goodfellow (1981).

Bacillus subtilis (Ehrenberg) Cohn. Rods seldom in chains. Endospore oval, $0.8 \times 1.5 - 1.8 \mu m$, thin-walled, located centrally in the cell but does not distend the walls. Cells relatively small ($0.7 - 0.8 \times 2 - 3 \mu m$); flagella lateral. Does not form poly-β-hydroxybutyrate as reserve material. Strict aerobe that requires no growth factors. Produces extracellular amylases and proteases. Colonies round or irregular.

Bacillus cereus Frankland and Frankland. Most abundant *Bacillus* sp. in soil. Rods $1.0 - 1.2 \times 3.0 - 5.0 \mu m$, Gram-positive, noncapsular, aerobic, motile, and in chains. Endospores $1.0 \times 1.5 \mu m$, central, form in 18–48 hours. Colonies large, usually rough, flat, whitish, form wavy arrangement of chains of cells that may develop into whiplike outgrowths. Requires one or several amino acids, varying with strain; vitamins not required. A few strains fail to grow above 35°C, but some grow at 45–48°C. "Prior to 1940 organisms which have the properties of . . . *B. cereus* were widely regarded as . . . *Bacillus subtilis*." A common transitory variant is called *B. mycoides* or *B. cereus* var. *mycoides* (Flügge) Smith, Gordon, and Clark; cells are formed in chains that determine the characteristic rootlike, long, twisted strands of colonies. For details see Smith et al (1946).

Bacillus penetrans (Thorne) Mankau. Endoparasite of nematodes. Formerly thought to be a sporozoan, *Duboscqia penetrans* Thorne, but now considered to be an atypical *Bacillus;* further study may revise this placement, perhaps to an actinomycete. Sporangia $3.5 - 4.0 \mu m$ in diameter, cup-shaped; consist of central raised area $1.6 \mu m$ in diameter where the endospore is formed. A hole $0.5 \mu m$ in diameter through the nematode cuticle is found beneath the endospore; perhaps formed by enzymes. Vegetative cells dichotomously branched, $1.6 - 1.7 \times 0.6 \mu m$, Gram-positive, septate. Endospore 1.6–1.7 μm in diameter, thick-walled, formed at the apex of vegetative cells.

Role in Biological Control: Infection of nematodes by *B. penetrans* is through a penetration tube formed from the endospore; eventually the nematode body is filled with spores. Spores are released with decomposition of the nematode body and survive at least two years in soil. Larvae are readily infected by moving through infested soil. *Bacillus penetrans* has strong host specificity; *Trichodorus christiei* is infected by one isolate but not by another. *Bacillus penetrans* attacks *Meloidogyne javanica, M. arenaria, M. incognita, M. hapla,* and *Pratylenchus scribneri,* but not 11 other nematodes (Imbriani and Mankau, 1977; Mankau, 1972, 1975; Mankau and Prasad, 1977; Mankau et al, 1976; Mankau, *in* Zuckerman et al, 1981). Widely distributed in soil, it parasitizes mainly plant-parasitic nematodes. Spores firmly adhere to the nematode cuticle as wartlike structures. The sporangia usually attach to the nematode neck region but may occur anywhere on the body. Populations build up slowly in field soil, but Mankau and Prasad (1972) found that applications of nematicides Telone, Temik, Furadan, Nemacur, and Mocap at commercial rates did not affect the bacteria.

Utkhede and Rahe (1980) found that four isolates of *B. subtilis* from British Columbia onion fields gave fairly good season-long protection against *Sclerotium cepivorum* when introduced on the onion seeds; the protection was comparable to the best broadcast chemical treatment with vinclozolin. Braithwaite (1978) found that *B. subtilis* inhibited germination of sclerotia of *S. cepivorum* in soil, perhaps due to antibiosis. Chang and Kommedahl (1968) found that inoculating corn seed with *B. subtilis* controlled seedling blight about as well as captan or thiram when the seeds were planted in soil moderately infested with *Fusarium roseum* 'Graminearum'. Campbell and Faull (*in* Schippers and Gams, 1979) inoculated field plots with *Gaeumannomyces graminis* var. *tritici* and with a suspension of *B. cereus* (*mycoides*) isolated from take-all decline soil; disease severity was reduced and hyphae on roots were lysed. Aldrich and Baker (1970) used *B. subtilis* to protect carnation seedlings against *F. roseum* 'Avenaceum'. Olsen (1964), Olsen and Baker (1968), and Broadbent et al (1971) applied *B. subtilis* to steamed soil and reduced seedling damping-off caused by *Pythium ultimum* and *Rhizoctonia solani*. Broadbent and Baker (1974a, 1974b) found that *B. subtilis* was involved among the resident antagonists to *Phytophthora cinnamomi* in suppressive soils in Queensland, but single isolates were ineffective when introduced into soil. Weinhold et al (1964) found that *B. subtilis* growing on a soybean green-manure crop produced materials that prevented the increase of *Streptomyces scabies*. Swinburne (1973) found that *B. subtilis* sprayed on twigs immediately after leaf fall reduced infection of leaf scars by *Nectria galligena*. Corke and Hunter (1979) also used *B. subtilis* to protect wounds in apple trees against infection by *N. galligena;* 96% fewer *Nectria* conidia were released during 12 months following inoculation with *B. subtilis* than in untreated controls. Broadbent et al (1971, 1977) inoculated *B. subtilis* A13, inter alia, into soil treated with aerated steam, which enhanced growth of a range of plants. Merriman et al (*in* Bruehl, 1975) found that *B. subtilis* A13 applied to oat, carrot, and sweet corn seed increased plant growth in the field. Pelleting inoculum on seed was more effective than dipping in a bacterial suspension. Purkayastha and Bhattacharya (1982) found that *B. megaterium* sprayed on jute plants 24 hours before inoculation with *Colletotrichum corchori* reduced infections from 87 to 10%.

Mechanism of Biological Control: Utkhede and Rahe (1980) found that *B. subtilis* produced antibiotics inhibitory to *S. cepivorum* in culture, but their role in soil is still unknown. The bacterium might also metabolize volatile growth stimulants from onion and thus prevent sclerotial germination. Chang and Kommedahl (1968) thought effectiveness of *B. subtilis* against *F. roseum* 'Graminearum' was due to the short protection period required and the persistence of the bacterium in the rhizosphere. Olsen and Baker (1968) found marked specificity of *B. subtilis* isolates in inhibition of *R. solani* isolates. The control of *N. galligena* in apple is thought to result in part from an antibiotic produced by the effective strains (Swinburne et al, 1975). The effective strain, *B. subtilis* A13, studied by Broadbent et al (1971) exhibited broad-spectrum antibiosis to plant pathogens in vitro, and it was suggested that the inhibition of deleterious root-colonizing microorganisms or their toxins enabled the plant to grow better.

Mass Production: *Bacillus subtilis* and *B. cereus* are very easy to grow in mass culture. *Bacillus penetrans* is not presently culturable, but Stirling and Wachtel (1980) mass-produced it by inoculating tomato plants with large numbers of second-stage *Meloidogyne* larvae infected with *B. penetrans* spores. After seven to eight weeks, the roots were air-dried and finely ground. This material was lightweight and could be stored. When mixed into soil at 100 mg/kg, 99% of *Meloidogyne* larvae were infected within 24 hours.

BDELLOVIBRIO STOLP AND STARR

Bacteria parasitic on other bacteria. Common in soil, with populations up to 10^5/g recorded.

Morphology and Taxonomy: Eubacteria, Spirillaceae; slime-spored. Cells curved or helical, motile by single, thick flagellum.

Bdellovibrio bacteriovorus Stolp and Starr. Minute cells (0.3×0.8 μm), comma-shaped, with single, thick (0.5 μm in diameter), long flagellum. Highly motile; attach with nonflagellated end to surfaces of Gram-negative bacteria. Cells cause lytic plaques in poured plates similar to phage plaques. Different strains are specific to different host bacteria. Encysted resting cells have not been demonstrated in this species but are known for other species.

Role in Biological Control: Scherff (1973) inoculated *Bdellovibrio bacteriovorus* together with *Pseudomonas syringae* pv. *glycinea* at a 9:1 ratio on soybean leaves; both necrotic lesions and systemic toxemia normally caused by the pathogen were suppressed or prevented. Different isolates of *Bdellovibrio* showed a marked difference in effectiveness, apparently related to the number of *Bdellovibrio* cells produced per infected host cell.

Mechanism of Biological Control: Cells of the parasite, following attachment, rotate up to 100 revolutions per second, apparently by swiveling of the posterior with no rotation of the attachment tip. This drilling action penetrates the host bacterial wall. The parasite enters the host bacterial cell through this pore; the contents are then digested, and the parasite leaves the "ghost" behind (Starr and Baigent, 1966). It may take 6–24 hours to complete the cycle, but Scherff et al (1966) found the cycle completed in one hour.

Mass Production: Some strains grow in pure culture on complex media, but most grow only on the bacterial host. The saprophytic culturable forms apparently do not return to the parasitic phase. Scherff (1973) grew *B. bacteriovorus* isolates on a lawn of *P. syringae* pv. *glycinea* for five days, then removed and suspended cells of the parasite in sterile 0.85% saline; this was filtered through Millipore filters (3.0 and 0.45 μm pore size) to remove cells of the host bacterium, thereby providing a preparation of *B. bacteriovorus*.

CANDELABRELLA RIFAI AND COOKE

Nematode-trapping fungi with lateral adhesive networks.

Morphology and Taxonomy: Deuteromycotina, Hyphomycetes, Moniliaceae; Xerosporae. Conidiophores erect, smooth, hyaline, terminated by a small candelabrumlike branching system, on the tips of which single conidia arise as blown-out ends of apices, forming a lax head. Conidia hyaline, obpyriform, ellipsoidal, smooth, 1-septate. For details see Rifai and Cooke, 1966.

Candelabrella javanica Rifai and Cooke. Conidiophores erect, 275–440 μm long, hyaline, smooth. Conidia narrowly obovoid, 1-septate, 35×11.3–13.8 μm, smooth, hyaline.

Candelabrella musiformis (Drechsler) Rifai and Cooke. Conidiophores erect, single, up to 600 μm long, terminating in a candelabralike apex. Conidia single, ellipsoidal, generally curved, smooth, hyaline, 1-septate, 29–45×8–13.5 μm. Chlamydospores thick-walled, tuberculate, globose, terminal on short lateral hyphae.

Role in Biological Control: The fungi are resident antagonists in soil, where they reduce the nematode population. Spores of *C. musiformis* are moderately unaffected by soil fungistasis.

Mechanism of Biological Control: The adhesive network on the lateral mycelial branches traps the nematode, which is then penetrated by hyphae and its contents digested.

CATENARIA SOROKIN

Parasitic on female cyst nematodes; common in Europe and North America.

Morphology and Taxonomy: Mastigomycotina, Chytridiomycetes, Blastocladiales, slime-spored. Thallus composed of nonseptate or sparingly septate hyphae, with few to many rhizoids, swelling at intervals to form zoosporangia or catenulate resting bodies connected by a narrow isthmus. Zoosporangia oval or elliptical; emergence papillae project outward through the nematode cuticle. Zoospores with 1 posterior whiplash flagellum. Resting bodies formed in resting sporangia, oval or oblong; wall smooth, pale brown, 2–3 μm thick; germinate by cracking and forming emergence papillae. For details see Couch, 1945.

Catenaria auxiliaris (Kühn) Tribe. Catenate rhizomycelium develops inside nematode; consists of swollen portions delimited by septa and without isthmuses. Swellings develop into precursor sporangia that are globose, ovoid or oblong, 25–45 μm in diameter; these develop into zoosporangia or resting sporangia. Zoosporangia are not common; they form 6 gelatinous papillae through which zoospores escape. Zoospores globose or ovoid, 3 μm in diameter, with single posterior flagellum 12–15 μm long. Resting sporangia frequent; each develops internally a single plerotic resting spore; wall 1.5–3 μm thick, reticulate with meshlike areole, yellow-brown; germination not observed (Tribe, 1977).

Catenaria anguillulae Sorokin. Zoosporangia hyaline, smooth, spherical. Thallus with narrow isthmuses between swellings. Sporangia uteriform; exit tubes tapering from base to apex; 5–50 zoospores per sporangium, spherical, 1.5–2 µm in diameter.

Role in Biological Control: These are resident antagonists of female *Heterodera schachtii* and *H. avenae* in soil; eggs are never invaded. *Catenaria auxiliaris* is not culturable on agar from either zoospores or resting sporangia (Tribe, 1977). Infection of females is through the anus, vulva, and excretory pore (Birchfield, 1960).

CHAETOMIUM KUNZE: FR.

Common in decaying organic matter and soil.

Morphology and Taxonomy: Ascomycotina, Pyrenomycetes, Sphaeriales; slime-spored. Perithecia superficial, subglobose or elongated, ostiolate; membranous brittle wall with variously modified hairs. Asci thin-walled, stalked, evanescent, club-shaped, linear or cylindrical, 8-spored. Spores nonseptate, olive brown, lemon-shaped. For details, see Chivers (1915) and Ames (1949).

Chaetomium globosum Kunze: Fr. Perithecia rather large, somewhat elongated or subglobose, black, 225–250 µm in diameter, frequently producing short black cirrhi and seated on a thick mass of dark olive to black rhizoids. Lateral hairs numerous, slender, graceful, septate, with minute spines, olive brown, straight or slightly flexed; terminal hairs extremely numerous, interwoven into a compact head, with minute spines, dark olive. Asci irregularly club-shaped, 64×13 µm. Ascospores olive brown, broadly ovate to lemon-shaped or fusiform, 10.5 × 8.5 µm. Very common in organic matter and soil.

Chaetomium cochliodes Palliser. Perithecia globose, 318–360×273–310 µm, with a heavy mass of stout, olive brown to black rhizoids. Lateral hairs numerous, sparsely septate, with extremely fine projections; terminal hairs numerous and finely interwoven, forming a massive shaggy head, coiled, twisted, or undulate. Asci club-shaped, 8-spored, 88×11 µm. Ascospores olive brown, lemon-shaped, 8.9–9.7×6.4–8.4 µm. Produces chaetomin, an antibiotic effective on Gram-positive bacteria (Waksman and Bugie, 1944).

Ecology: *Chaetomium* spp. are especially favored by cellulosic organic matter.

Role in Biological Control: Chang and Kommedahl (1968) inoculated corn seed with *C. globosum* before planting in a field having a moderate inoculum density of *Fusarium roseum* 'Graminearum'. Seedling stand in the control was 34.7%; in corn inoculated with *Chaetomium*, it was 53.7% and in corn treated with captan and thiram it was 55.3 and 63.7% respectively. Tveit and Moore (1954) found that *C. gobosum* and *C. cochliodes* occurred naturally on oat seed from Brazil and were the basis of reported "resistance" of Brazilian varieties to *Helminthosporium victoriae*. Wood and Tveit (1955) controlled *Fusarium nivale* on oats by sowing oat seed heavily contaminated with *F. nivale* in the field along with an oat-straw culture of *C. cochliodes*. Plant stand with *Chaetomium* was 38–40%, compared with 63% for seed treated with organic mercury dust and 25% for the check. Kommedahl and Mew (1975) found that *C. globosum* was an effective protectant of corn seedlings when soil was relatively dry; *Bacillus subtilis* was the more effective antagonist when the soil was moist. Mew and Kommedahl (1972) showed that spores of *C. globosum* applied dry to corn seed reduced the percentage of kernels yielding *Fusarium*, *Penicillium*, and *Mucor* spp. from 60% (nontreated) to 9% (treated), compared with 19% for seed treated with *B. subtilis*. *Chaetomium* sp. reduced ascospore production by *Venturia inaequalis* about 30% when sprayed on detached apple leaves and held overwinter in the field; *Athelia bombacina* prevented ascospore formation when similarly applied (Heye and Andrews, 1982).

Mechanism of Biological Control: Control of fusarium seedling blight of corn by *C. globosum* was thought to result from the antibiotic cochliodinol, although effective strains did not inhibit *F. oxysporum* in culture (Meiler and Taylor, 1971). Ability to colonize the seed coat and use the seed exudates otherwise stimulatory to the pathogen probably also is involved.

Mass Production: *Chaetomium* spp. are readily grown on agar culture and on cellulosic material.

CICINNOBOLUS EHRENBERG

See *Ampelomyces*.

CLADOSPORIUM, LINK: FR.

Common ubiquitous saprophyte on above-ground plant parts, crop residues, and in soil.

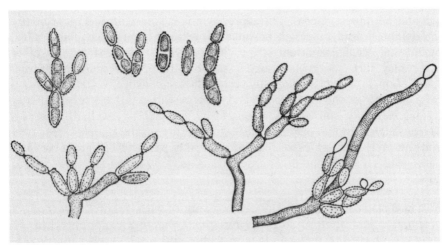

Figure 9.4. *Cladosporium herbarum* conidia and conidiophores. (Reprinted with permission from Ellis, 1971.)

Morphology and Taxonomy: Deuteromycotina, Hyphomycetes, Dematiaceae; Xerosporae. Conidiophores erect, pigmented, irregularly branched at the apex. Branches produce conidia in acropetal succession by apical budding. Conidia in branching chains, hyaline or pigmented, smooth or rough, mostly nonseptate; chains very fragile, breaking up at maturity and frequently shedding branches, which have well-marked dark attachment scars. For details, see Ellis, 1971.

Cladosporium herbarum (Pers.) Link: Fr. *Perfect state, Mycosphaerella tassiana* (de Not.) Johans. Conidiophores straight or flexuous, often nodose, olivaceous brown or brown, smooth, up to 250 μm long. Conidia in fairly long, often branched chains, ellipsoidal or oblong, rounded at ends, olivaceous brown, rather thick-walled, distinctly verruculose with low warts, 0–1 septate (sometimes several septate), 5–23×3–8 μm, with small but protuberant scars at ends (Figure 9.4).

Cladosporium cladosporioides (Fries) de Vries. Conidiophores up to 350 μm long but generally shorter, pale to mid-olivaceous brown, smooth or verrucose. Conidia in long branched chains, mostly nonseptate, ellipsoidal or limoniform, 3–11×2–5 μm, pale olivaceous brown, smooth or verruculose.

Ecology: Airborne conidia (or ascospores of the sexual stage) collect on leaves, flower parts, stems, and other aboveground plant parts. Growth occurs on pollen and other nutrients present naturally on the foliage. Hyphae may also become established endophytically, e.g., in substomatal chambers or intercellularly in the epidermis or deeper tissues. As the tissues age and become senescent, this saprophyte is ideally positioned to colonize the dying or dead tissues in advance of competitors. Wheat plants killed by take-all, cephalosporium stripe, barley yellow dwarf, frost, or other cause may turn sooty black under suitable humidity because of asexual sporulation by *Cladosporium* spp. established epiphytically and endophytically on and in the leaf blades and sheaths earlier in the season. Crop residue enters the soil already colonized by this saprophyte. Perithecia of a *Mycosphaerella* stage form on wheat straw in the spring, release ascospores, and the cycle repeats. *Cladosporium* is weakly parasitic but probably nonpathogenic on wheat and other crops. *Cladosporium herbarum* grows best at 20°C.

Role in Biological Control: As an early colonist of wheat straw, *Cladosporium* preempts colonization of straw by *Fusarium roseum* 'Culmorum' when the straws are buried in the soil. It is probably also a significant natural competitor of leaf pathogens. Corke and Hunter (1979) treated fresh pruning wounds of apple trees with *Cladosporium cladosporioides;* and found 96% fewer conidia of *Nectria galligena* in the next 12 months, equal to control obtained with benomyl applied to wounds. Bhatt and Vaughan (1962, 1963) found that three applications of a heavy suspension of *C. herbarum* spores in 1% glucose on strawberries in the field did not reduce ripe fruit decay caused by *Botrytis cinerea*, but the yield of marketable berries was increased, perhaps because blossom blight and green fruit rot were reduced. Zuck et al (1982) found *Cladosporium* spp. sporulating on perithecia of *Venturia inaequalis* on fallen apple leaves; the perithecia lacked normal contents.

Mechanism of Biological Control: *Cladosporium* spp. are mainly competitors with plant pathogens; established first, they preempt the nutrient supply of pathogens. Bhatt and Vaughan (1963) thought that the yield increase of strawberries in Oregon from application of *C. herbarum* resulted from raising the pH (making it unfavorable for pectolytic enzymes of *B. cinerea*), from producing toxins, or from colonizing dead or senescent floral organs that would otherwise be occupied by *B. cinerea*.

Mass Production: *Cladosporium* spp. can be grown readily on solid media.

CONIOTHYRIUM CORDA

Some species are plant pathogens, but one is an important mycoparasite of sclerotia.

Morphology and Taxonomy: Deuteromycotina, Coelomycetes, Sphaerop-
sidales; slime spored. Pycnidia separate, globose, pale or dark brown, immersed, unilocular, thin-walled; ostiole circular, central, sometimes papillate.
Conidiogenous cells holoblastic, discrete, dolioform to cylindrical, hyaline to
pale brown, smooth, formed from inner walls of pycnidium. Conidia brown,
thick-walled, 0–1 septate, verruculose, cylindrical, spherical, elliptical, or
broadly clavate. For details see Sutton, 1980.

Coniothyrium minitans Campbell. Pycnidia globose, superficial, smooth, with
a black carbonaceous covering, 200–700 μm in diameter; ostiole central. Pycnidiospores exuding from ostiole as black slimy mass; spores brown, ellipsoidal, smooth or minutely roughened, 4–6×3.5–4 μm. Produces pycnidia in
infected sclerotia but is nonpathogenic to green plants.

Ecology: *Coniothyrium* is widely distributed in temperate regions of the world.
The optimum temperature for spore germination, infection of sclerotia, and
destructive parasitism by *C. minitans* is 20°C; the rate of decay can be sub-
stantial at 15–24°C, but growth and infection are greatly retarded below 7°C.
The minimum relative humidity required for growth is 95% (about −75 bars).
Coniothyrium minitans survives in soil 18–24 months but is not considered
to be a soil inhabitant (Tribe, 1957). It survives well aboveground. Slugs,
mites, and Collembola contribute to the spread of *C. minitans* in the field.
Field-grown sclerotia of *Sclerotinia sclerotiorum* contained *C. minitans* and
other fungi (Trutmann et al, 1980).

Role in Biological Control: *Coniothyrium minitans* has been effectively used
for control of *Sclerotinia sclerotiorum* on sunflower (Huang, 1980) and beans
(Trutmann et al, 1980), of *S. trifoliorum* on clover (Tribe, 1957; Turner
and Tribe, 1975), and as a seed treatment against *Sclerotium cepivorum*
on onion (Ahmed and Tribe, 1977). It also attacks sclerotia of *Sclerotinia
minor, Botrytis cinerea, B. fabae, B. narcissicola*, and *Claviceps purpurea* but not those of *Sclerotium draytonii, S. gladioli, S. delphinii, S.
rolfsii*, or *S. tuliparum* (Turner and Tribe, 1976). Up to 65% of sclerotia
of *Sclerotinia trifoliorum* on the surface of soil were infected when pycnidial dust of *C. minitans* was added to field plots (Turner and Tribe, 1973).
Huang and Hoes (1976) showed that 97% of sclerotia of *S. sclerotiorum* were
destroyed when *C. minitans* was added to soil. Sclerotinia wilt of sunflower
was reduced by 42 and 40% in two years when inoculum was added in the
furrow at time of seeding, giving yield increases of 23 and 24%, respectively.
In naturally infested soil, *C. minitans* parasitized sclerotia on roots and inside roots and stems. In sunflower field plots inoclated with *S. sclerotiorum*,
Huang (1980) found 1,514 sclerotia in 60 m of row in the checks but only
36 sclerotia per 60 m in rows where 10 kg of inoculum of *C. minitans*
had been added, 401 sclerotia where *Gliocladium catenulatum* had been
used, and 843 where *Trichoderma viride* had been used. The percentage of

wilted sunflowers was decreased from 43 to 25 by *C. minitans*. Ahmed and Tribe (1977) applied *C. minitans* to the seed furrow or as a seed dressing in glasshouse experiments and obtained 57–61% control of *Sclerotium cepivorum* on onion.

Sclerotinia sclerotiorum* was effectively controlled by *C. minitans* applied on detached bean leaves. Because the sclerotia contained a sizeable population of *C. minitans*, and the growth rate of germinating sclerotia was decreased, the long-term effectiveness may have been better than it appeared. The treatment may not reduce disease in the current crop but would favor early active decay of sclerotia when in soil, thus diminishing the amount of disease in the following year. Possible use of benomyl spray with subsequent inoculation with *C. minitans* was suggested for beans (Trutmann ct al, 1982).

Mechanism of Biological Control: Hyphae of *C. minitans* penetrate and decay sclerotia, reducing the amount of inoculum of the pathogen. Mycelium grows parallel to that of the pathogen, and the short side branches of the hyphae penetrate the cells, which then die. Jones and Watson (1969) and Jones et al (1974) showed that enzymatic preparations of *C. minitans* contained ecto- and exo-β-(1-3)-glucanase and chitinase that lysed mycelia of *S. sclerotiorum*.

Mass Production: *Coniothyrium minitans* is easily grown on potato dextrose agar at 20°C. A mixture of melanized pycnidia, hyphal fragments, and spores are obtained after two-weeks of growth, by scraping the culture surface in distilled water. This suspension, diluted to give 5×10^7 spores per milliliter, was sprayed on the soil surface where sclerotia of *S. sclerotiorum* occurred (Trutmann et al, 1980; Turner and Tribe, 1976). Ahmed and Tribe (1977) produced masses of pycnidia by growing *C. minitans* on autoclaved milled rice in plastic bags held at 20°C for six weeks. This was air-dried for one week, ground to pass a 600- μm sieve, and stored at 20°C until used. Onion seed was pelleted with 10 g of spore dust to 20 g of seed and 50 ml of methyl cellulose sticker; after drying for one week, seed was again coated with another methyl cellulose layer. Huang (1980) grew *C. minitans* for four to six weeks on an autoclaved mixture of barley, rye, and sunflower seed in polypropylene bags; this was applied moist directly to soil or was air-dried for two to four weeks before use.

CORTICIUM FRIES

See *Laetisaria*.

CORYNE TULASNE

See *Ascocoryne*.

CROTALARIA LINNAEUS

A trap crop with capacity to diminish the population of *Meloidogyne* spp. in soil.

Morphology and Taxonomy: Spermatophyta, Angiospermae, Leguminosae. Annual and perennial herbs and shrubs. Leaves alternate, simple or palmate with 3, 5, or 7 leaflets. Flowers yellow, brownish yellow, sometimes blue or purplish, solitary or racemose, papilionaceous; standard prominant and larger than the wings; keel beaked and curved; stamens 10, connate; style strongly incurved or reflexed, somewhat bearded. Fruit a globose or oblong pod, inflated, the many seeds loose at maturity. Plants and seeds poisonous to livestock and poultry. For details see Bailey and Bailey (1976).

Crotalaria spectabilis Roth. Subshrub up to 120 cm tall. Leaves simple, long, and narrow. Flowers yellow, up to 2.5 cm across, in racemes up to 30 cm long, standards streaked purplish; pods smooth, 5 cm long.

Role in Biological Control: Its potential as a trap crop for root-knot nematode was demonstrated by C. W. Mcbeth in 1942. Mcbeth and Taylor (1944) showed that growing *C. spectabilis* as a cover crop in infested peach orchards increased, over a five-year period, the average annual yield per tree from 10.4 kg where a susceptible cover crop was grown, to 60 kg where *Crotalaria* was grown. The tops of the trees increased about twofold in size, and trunk diameter increased from 6.3 to 11 cm. A test with sugar beet in California gave a 200% increase in yield following one summer planting to *C. spectabilis* (Baker and Cook, 1974). It is valuable when planted either preceding a susceptible crop or as an intercrop in an orchard, to decrease the population of nematodes in soil and in roots of susceptible crops. In Brazil, *C. spectabilis* was as effective as either Nemagon or Basamid applied to the soil in decreasing the larval population of *M. incognita* and, upon return to a susceptible crop, the nematode population increased more slowly where *C. spectabilis* had been grown than where either of the two nematicides had been applied (Huang et al, 1981).

Mechanism of Biological Control: Female larvae of *Meloidogyne* spp. penetrate roots of *C. spectabilis*, but because giant cells are not formed there, the immobilized nematodes starve or at least do not lay eggs.

DACTYLARIA SACCARDO

Nematode-trapping fungi with adhesive loops, networks, and knobs.

Morphology and Taxonomy: Deuteromycotina, Hyphomycetes, Moniliaceae; Xerosporae. Conidiophores erect, simple, hyaline, septate, denticulate, sometimes swollen at the apex. Conidia hyaline, 1–4 septate, cylindrical, clavate or filiform, borne singly in acropetal succession at the apex of conidiophores on somewhat prominent teeth.

Dactylaria vermicola Cooke and Satch. Conidiophores erect, unbranched, up to 250 μm long, bearing up to 7 conidia as a capitate group; sometimes grows further to form 1–3 more groups of conidia. Conidia elongate-ellipsoidal to fusiform, 27.5–47.5×13.5–17.5 μm, rounded at the apex, tapering to a narrow base, 1–4 septate (usually 2-septate). Produces adhesive loops and networks. *Dactylaria thaumasia* Drechsler, *D. clavispora* Cooke, *D. scaphoides* Peach, *D. gampsospora* Drechsler, *D. pyriformis* Juniper, and *D. polycephala* Drechsler also have adhesive networks and capitate or whorled groups of conidia. Spores of *Dactylaria thaumasia* are highly sensitive to soil fungistasis.

Dactylaria haptotyla Drechsler. Conidiophores erect, 115–325 μm long, with 1–7 apical branches. Conidia form singly at the apex of each branch, forming a loose capitate group, 18–45×6.5–13.5μm. Conidia hyaline, 33–55×7.4–13.3 μm, fusiform, 1–4(usually 3) septate. Produces adhesive knobs 7–10×6–8.5 μm on stalks 7–27 μm long (Drechsler, 1950).

Role in Biological Control: *Dactylaria* are resident antagonists in soil; they reduce the nematode population.

Mechanism of Biological Control: Adhesive networks and knobs trap nematodes. Hyphae then penetrate the nematodes and digest their contents.

DACTYLELLA GROVE

Parasites and predators of nematodes, using adhesive networks, lobes or knobs, constricting rings, or knobs and constricting rings. They can also be parasitic on eggs.

Morphology and Taxonomy: Deuteromycotina, Hyphomycetes, Moniliaceae; Xerosporae. Conidiophores erect, simple, hyaline, septate, on short hyphal branches, bearing one or several spores acropetally. Conidia ellipsoidal to fusoid or cylindrical, several- to multiseptate, hyaline, tapering toward the ends, with one cell (usually the centermost cell) wider and longer than the others.

Dactylella oviparasitica Stirling and Mankau. Conidiophore a hyaline hyphal branch rarely more than 2 μm long, extremely short for *Dactylella* spp. Conidia thin-walled, hyaline, fusiform, 31–60×2.7–5.0 μm, 4–7 septate, borne singly (rarely with 2) and apically as blown-out ends of conidiophore apex. Resistant structures and sexual spores not known. Parasitizes nematode eggs; hyphae in eggs are thickened and convoluted. No nematode-trapping organs are formed (Stirling and Mankau, 1978). This species is considered to be closely related to the *Dactylella* species that parasitize oospores, rhizopods, and amoebae.

Dactylella doedycoides Drechsler. Conidiophores hyaline, erect, septate; 225–500 μm long, with a knoblike tip bearing a single conidium. Conidia hyaline, top-shaped, 28–39×15–24 μm, usually 2-septate, the middle cell swollen and barrel-shaped. Constricting rings circular, 20–36 μm in diameter, 3-celled, on 2-celled, curved or straight stalks (Figure 9.3). Rings are sometimes also adhesive (Drechsler, 1940).

Dactylella lobata Duddington. Conidiophores erect, septate, 250 μm long, bearing single apical conidia. Conidia hyaline, fusiform, 4-celled, 32–54×8–12 μm, often germinate directly to form an adhesive lobe 9–13×8–9 μm.

Ecology: *Dactylella oviparasitica* requires light for sporulation, and spores germinate readily on various media. The restricted capacity of *Meloidogyne* females to produce eggs on Lovell peach allows the parasite to hold the nematode in check, but on grape and tomato an equal control does not occur (Mankau, *in* Zuckerman et al, 1981).

Role in Biological Control: Resident antagonists in soil, reducing the nematode population, affecting both females and egg masses. No evidence of specificity of *D. oviparasitica*, since it attacks four species of *Meloidogyne* as well as *Heterodera schachtii*, *Trichodorus semipenetrans*, and *Acrobeloides* sp. (Stirling and Mankau, 1979). Because it kills nematodes while they are still congregated in the egg stage, use of *Dactylella oviparasitica* as an introduced parasite may be effective.

Mechanism of Biological Control: Constricting rings have a specialized stimulus-sensitive area that triggers a change in wall structure increasing permeability to water and possibly the amount of osmotically active material in a ring cell (Muller, 1958). The sensitive area of the wall may be a site of intimate interrelation of protoplast and cell wall that produces this generally irreversible inflation. The reaction requires about 0.1 second to inflate the three cells about 300%. Heat will trigger the mechanism as well as physical contact, but light, ultraviolet light, pressure, electric shock, and several chemicals will not. Ring cells are not adhesive (Muller, 1958).

Mass Production: *Dactylella oviparasitica* sporulates heavily on a solid medium containing yeast extract, soluble starch, peptone, egg yolk, filtered V-8 juice, and mineral elements when held in light (Stirling and Mankau, 1978). *Dactylella lobata* grows well on cornmeal agar and potato dextrose agar (Duddington, 1951).

DARLUCA CASTAGNE

See *Sphaerellopsis*.

DATURA LINNAEUS

A decoy crop that causes resting spores of *Spongospora subterranea* to germinate and die.

Morphology and Taxonomy: Spermatophyta, Angiospermae, Solanaceae. Bushy annuals, shrubs, or trees. Leaves alternate, large, simple, entire or coarsely toothed. Flowers large, axillary, solitary, erect or pendulous, white, red, violet, or yellow. Calyx long-tubular, 5-toothed; corolla trumpet-shaped, with spreading limb broadly lobed; 5 stamens inserted near base of corolla; style long, filiform, and 2-lobed. Fruit a large 2-celled, prickly or spiny capsule. For details see Bailey and Bailey (1976).

Datura stramonium Linnaeus. Annual, up to 150 cm high, glabrous, foul-smelling. Leaves coarsely toothed, to 20 cm long. Flowers white, single, 5–12 cm long. Capsule erect, spiny. Plant is strongly narcotic and poisonous to man and animals.

Role in Biological Control: White (1954) found that a dense planting of *D. stramonium* turned under at the time of flowering caused the resting spores of *Spongospora subterranea* (cause of powdery scab of potato) to germinate and die. The potato crop then planted had only 7% of tubers infected and an average disease rating of 1, compared with 37% and an average rating of 4 for plants grown in fallowed soil.

Mechanism of Biological Control: Apparently the residue of *D. stramonium* stimulates the resting spores to germinate, and because the residue is not a host of the pathogen, the germlings then die. *Datura* is thus a decoy plant.

ERWINIA WINSLOW ET AL

Common on leaf surfaces of many plants and in apple buds; a natural microbial epiphyte.

Morphology and Taxonomy: Eubacteria, Enterobacteriaceae; slime-celled. Peritrichous-flagellated motile rods that normally do not require organic sources of nitrogen for growth. Produce acid with or without visible gas from a variety of sugars. May or may not liquefy gelatin or produce nitrites from nitrates. Rods straight, $0.5-1.0 \times 1.0-3.0 \,\mu m$. Gram-negative. Facultative anaerobes. For details see Buchanan and Gibbons (1974).

Erwinia herbicola (Geilinger) Dye pv. *herbicola* (Löhnis) Dye. Gram-negative rods with peritrichous flagella. Facultative anaerobe. Does not produce indole, but does produce phenylalanine deaminase and hydrogen sulfide. Does not produce acid from sorbitol. Produces a yellow, water-insoluble pigment on some media (Dye, 1969).

Erwinia uredovora (Pon et al) Dye. Gram-negative rods, $1.5 \times 0.6 \,\mu m$, motile with peritrichous flagella. Produces a yellow, water-insoluble pigment on some media. Facultative anaerobe. Liquefies gelatin. Produces indole but not phenylalanine deaminase or hydrogen sulfide. Produces acid from sorbitol. May be the same as *E. herbicola* pv. *herbicola*. Parasitic on *Puccinia graminis* f. spp. *tritici, avenae*, and *secalis*, and on *P. rubigovera* f. sp. *tritici*.

Ecology: *Erwinia herbicola* pv. *herbicola* occurs commonly as an epiphyte on apple, cherry, and apricot but less commonly on pear leaves; it is an inhabitant of blossoms infected by *E. amylovora*. Some strains of pv. *herbicola* are ice-nucleation active (INA) and promote frost injury on sensitive plants; other strains are not INA and have the potential to protect plants from INA strains and thereby prevent frost injury (Lindow, 1983). The temperature optimum for growth of *E. uredovora* is 30°C; some growth occurs at 37°C. Growth requires high humidity or free moisture.

Role in Biological Control: Riggle and Klos (1972) inoculated pear blossoms in the field with *E. herbicola*, and 24 hours later with *E. amylovora*. In two tests, fire blight infections were 40 and 67% on the controls and 20 and 58% on the treated blossoms. Beer et al (1980) applied *E. herbicola* to apple blossoms one day before inoculating them with *E. amylovora*, using 10^8 cells per milliliter as a spray. Infection was 36% compared with 76% in controls; with 10^6 cells per milliliter, 57% of the trees were infected.

Mechanism of Biological Control: *Erwinia herbicola* pv. *herbicola* produces large quantities of acidic byproducts on high sugar media (and on nectar in blossoms), perhaps inhibiting *E. amylovora* or competing with it for nutrients in the blossoms. The isolates of *E. herbicola* capable of protecting plants against INA strains probably operate by competition, antibiosis, or both (Moore, *in* Papavizas, 1981). Chatterjee et al (1969) found that *E. herbicola* pv. *herbicola* breaks down the arbutin of pear trees to form hydroquinone, which is toxic to *E. amylovora*.

FUSARIUM LINK: FR.

Ubiquitous in soil and decaying organic matter. Most are saprophytes, but some are important plant pathogens.

Morphology and Taxonomy: Deuteromycotina, Hyphomycetes, Tuberculariaceae; Gloiosporae. Conidiophores solitary, simple or aggregated, or with complex branching to form a sporodochium; ultimate branches terminate in phialides. Phialides taper distally, sometimes with apical collarette. Conidia of two types: large macrocondia one to several septate, fusoid, hyaline, cylindrical or curved, with a well-marked foot cell at the attachment end, and produced in mucus-forming gloeoid masses. Microconidia nonseptate or 1-septate, ovoid to short cylindric, produced in short chains or small heads. Chlamydospores common, single or in clumps or chains, thick-walled. For details, see Booth (1971), Gerlach and Nirenberg (1982), Nelson et al (1981), and Toussoun and Nelson (1976).

Fusarium roseum Lk.: Fr. A large heterogeneous group; contains at least three pathogens considered by Snyder et al (1957) to be the equivalent of horticultural cultivars (Culmorum, Graminearum, and Avenaceum). Most strains are nonpathogenic, and some give biological control. Microconidia generally not formed. Macroconidia formed on mycelia or sporodochia. Conidiophores are lateral hyphal branches that branch several times to form metulae, each with 2–4 apical cylindrical phialides and an apical pore surrounded by a collar, 16–22×4–5 μm. Conidia curved, fusoid, generally with pointed apex and well-marked foot cell, 3–5 septate, mostly 30–55×4.5–5.5 μm. Chlamydospores single globose cells, but more commonly in chains or clumps intercalary in hyphae or in macroconidia, sometimes terminal, 6–11 μm in diameter. Perfect states, when known, are in *Gibberella*.

Fusarium lateritium Nees: Fr. *Perfect state, Gibberella baccata* (Wallr.) Sacc. Conidiophores are simple lateral branches of hyphae terminating in 2–4 metulae, each of which terminates in 1–3 phialides 10–30×2.5–4 μm. Loosely formed sporodochia usually appear as the culture ages; conidiophores in sporodochia are reduced in size. True microconidia generally absent. Macroconidia falcate to straight, narrowly fusoid, 3–7 septate, beaked at the apex, and with a marked pedicellate foot cell. Short conidia are 3–5 septate, with a short beak, 22–48×3–4 μm; long conidia 5–8 septate, 40–75×2.5–5 μm. Chlamydospores generally sparse, may form in macroconidia or intercalary in hyphae; 7–10×7–8 μm. Perithecia 175–265×140–227 μm; asci 4–8 spored, 65–80×8–11 μm; spores smooth, hyaline, 12–18×4.5–7.5 μm, 1–3 septate.

Fusarium oxysporum Schlecht. Microconidia abundant, oval-ellipsoid, cylindrical, straight to curved, 5–12×2.2–3.5 μm. Macroconidia sometimes sparse, borne on sporodochia or elaborately branched conidiophores, thin-walled, 3–5 septate, fusoid-subulate, pointed at both ends, 27–66×3–5 μm (usually 27–46 ×3–4.5 μm, and 3-septate). Chlamydospores smooth or rough-walled, generally abundant, terminal or intercalary, solitary or in pairs or chains.

Ecology: These generally aggressive, highly competitive saprophytes are able to degrade a wide variety of substrates, including cellulose, pectins, lignins, and other complex materials. They are well adapted to survival in soil. Most agricultural soils contain between 10^4 and 10^5 *Fusarium* propagules per gram, mainly as *F. roseum, F. solani*, and *F. oxysporum*. Nonpathogens as well as pathogens are successful colonists of the cortical tissues of roots and belowground stems, where they exist as epiphytes and endophytes until the organ dies. The fungus is then ideally positioned to colonize the tissue more thoroughly in advance of competing soil microorganisms. The cereal pathogens (e.g., Culmorum) require only that the plant tissue approach death (i.e., to be stressed), whereupon they grow aggressively deeper into the plant, taking possession in advance of competitors. Nonpathogens related to pathogens commonly occupy the same niche; thus nonpathogenic members of *F. roseum* occur in mixed populations with pathogens as colonists of grasses, nonpathogenic members of *F. oxysporum* occur with pathogens in plant roots, and nonpathogenic members of *F. solani* occur with pathogenic members of *F. solani* in stems of beans, peas, or other hosts. *Fusarium* spp. are generally favored by dry soil, being most active at soil water potentials between −15 and −100 to −120 bars. Airborne ascospores or water–splash macroconidia are adapted to aboveground habitats on plant parts if moisture is adequate.

Role in Biological Control: Carter and Price (1974) inoculated fresh pruning wounds on apricot trees with a mixture of *Fusarium lateritium* and benomyl to control cankers caused by *Eutypa armeniacae*. Benomyl is toxic to *Eutypa* but not to the *Fusarium*, and so protects the wound from infection by *Eutypa* until the *Fusarium* occupies the site. Mower et al (1975) used an isolate of *F. roseum* 'Sambucinum' to "almost completely control ergot with the use of a rather inexpensive spray at a relatively low concentration." The *Fusarium* spores produced on the sphacelial stage of ergot are spread by insects seeking the honeydew, thus providing secondary spread. Apparently the effective *Fusarium* strain does not attack cereals, has no subacute mammalian toxicity, and breaks down ergotamine to less toxic compounds. Cunfer (1975) obtained biological destruction of ergot sclerotia with the related *F. roseum* 'Heterosporum'. Other cultivars of *F. roseum* have been effective against other pathogens, e.g., *F. roseum* 'Semitectum', which parasitized the oospores of *Sclerospora graminicola* in the "green-ears" of pearl millet (Rao and Pavgi, 1976), and which invaded and killed hyphae, conidiophores, and conidia of *Cercospora* in lesions caused by this pathogen on mulberry, phalsa, and nightshade (Rathaiah and Pavgi, 1973). Dimock and Baker (1951) showed that *F. roseum* under moist New York conditions infected the centers of pustules of *Puccinia antirrhini* at 10°C, but at 21–32°C it advanced into healthy tissue. This *Fusarium* produced no lesions on healthy plants. Alabouvette et al (*in* Schippers and Gams, 1979) found that suppressiveness of a soil in France to *F. oxysporum* f. sp. *melonis* was eliminated when the soil was treated with moist heat (55°C for 30 minutes) but was restored when soil was inoculated with nonpathogenic clones of *F. oxysporum*. *Fusarium solani* was partially effective but *F. roseum* was ineffective in restoring suppressiveness. *Fusarium oxysporum* and *F. solani* both have been implicated as pathogens of the cysts of *Heterodera glycines* in soils in Alabama (Morgan-Jones and Rodriguez-Kabana, 1981) and North Carolina (Gintis et al, 1982).

Mechanism of Biological Control: Mainly an aggressive competitor in substrates, *Fusarium* has the ability to preempt pathogens in tissues or overtake and displace them as secondary colonists in lesions. Some strains of *F. roseum* have ability as mycoparasites. *Fusarium lateritium* is a low-grade pathogen of apricot wounds that prevents their infection by *Eutypa*. *Fusarium roseum* invades tissue of snapdragon that has been physiologically modified by rust invasion, thus depriving *Puccinia* of living cells.

GENICULARIA RIFAI AND COOKE

Nematode-trapping fungi with adhesive network and mycelium.

Morphology and Taxonomy: Deuteromycotina, Hyphomycetes, Moniliaceae; Xerosporae. Conidiophores erect, septate, at first straight but becoming flexuous, geniculate (elongating by repeated subapical renewal of growth), hyaline, smooth. Conidia arise singly as blown-out ends of conidiophores; a new growing point starts next to the conidium, and a new conidium forms, displacing the first to a lateral position. Conidia obpyriform, smooth, 1-septate, the lower cell much smaller than the distal one, hyaline. Adhesive network formed on nematode-infested agar. For details see Rifai and Cooke, 1966.

Genicularia cystosporia (Duddington) Rifai and Cooke. Conidia borne in a widely-spaced panicle, obpyriform, 1-septate, hyaline to pale pink, smooth, 24–35.5×18–25 μm. Forms adhesive, three-dimensional hyphal network in which nematodes are caught.

Genicularia perpasta Cooke. Conidia plump, obpyriform, 21.5–30×12.5–19 μm, smooth, thin-walled, hyaline, 1-septate. Simple, lateral, septate, adhesive hyphal branches formed at right angles, up to 100×4 μm, sometimes curved into a loop, but they rarely anastomose. Main hyphae are also adhesive, showing the beginning of differentiation of hyphal traps.

Genicularia paucispora Cooke. Conidia sparse, usually 1–2 per conidiophore, 25–33.8 × 12.5–16.3 μm. Adhesive loop forms from anastomosis of branch with main hypha.

Role in Biological Control: Resident antagonists in soil, *Genicularia* reduce the nematode population.

Mechanism of Biological Control: An adhesive network or lateral mycelial branches trap nematodes. These are then penetrated by hyphae, and their contents are digested. This food base then stimulates formation of more mycelia and traps.

GLIOCLADIUM CORDA

Morphology and Taxonomy: Deuteromycotina, Hyphomycetes, Moniliaceae; Gloiosporae. Condiophores more or less erect, hyaline, terminally branching in a penicillate fashion. Phialides divergent or appressed, bottle-shaped; phialospores hyaline or pigmented, nonseptate, slimy (forming globose or loose columnar heads). Common in soil. The taxonomy of the genus is discussed by Morquer et al (1963).

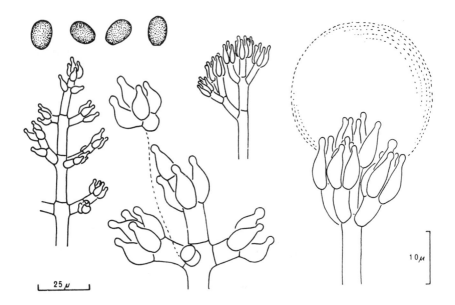

Figure 9.5. *Gliocladium virens*, showing phialospores, branching system, phialides, and large drops of slime containing conidia. (Reprinted with permission from Rifai, 1969.)

Gliocladium virens Miller, Giddens and Foster. *Perfect state, Hypocrea gelatinosa* (Tode: Fr.) Fr. Often confused with *Trichoderma* spp. Side branches of erect conidiophores closely approach the supporting conidiophore and have closely appressed apical flask-shaped phialides bearing a large drop of slimy conidia. Phialospores smooth, green, elliptical, 4–6×3–4 μm. Chlamydospores globose, thin-walled, 6–8 μm in diameter (Figure 9.5). For details see Miller et al, 1957.

Gliocladium roseum Bainier. *Perfect state, Nectria gliocladioides* Smalley and Hans. Conidiophores erect, bearing brush-shaped conidial-bearing apparatus. Conidia borne in slimy droplets. Conidiophores are *Verticillium*-like in young culture.

Gliocladium catenulatum Gilman and Abbott. Conidiophores 1–2 branched, coarse, pitted or rough, 50–125 μm long. Heads of conical chains in long, close columns, enveloped in slime, up to 150 μm long. Phialides 10–20 μm long. Conidia elliptical, smooth, pale green, 4–7.5× 3–4 μm.

Ecology: *Gliocladium* spp. are common soil fungi. *Gliocladium roseum* occurs particularly in neutral to alkaline soil and in marshy areas, but also occurs in the rhizosphere. Optimum temperature for growth of *G. roseum* is 25°C.

Role in Biological Control: *Gliocladium* spp. are resident antagonists in soil, reducing populations of fungi. Barnett and Lilly (1962) showed that *G. roseum* parasitized *Ceratocystis fimbriata, Helminthosporium sativum, Trichothecium roseum,* and *Thamnidium elegans* in agar cultures. Moody and Gindrat (1977) showed that sclerotia of *Phomopsis sclerotioides* (cause of black root rot of cucumber) were rotted by *G. roseum,* and that root rot was less in the first two weeks and was not significant thereafter. Huang (1980), working in sunflower field plots inoculated with *Sclerotinia sclerotiorum,* obtained a decrease from 1,514 sclerotia in 60 m of row in checks, to 401 sclerotia when 10 kg of inoculum of *G. catenulatum* was added and 36 when *Coniothyrium minitans* was added. Tu (1980) showed that *G. virens* inhibited formation of sclerotia of *S. sclerotiorum* and parasitized those already formed. It also parasitized *Rhizoctonia solani* (Tu and Vaartaja, 1981) but not *Phytophthora megasperma* var. *sojae* or *Pythium ultimum.* Howell (1982) found that *G. virens* parasitized *R. solani* but not *Pythium ultimum,* which it inhibited by antibiosis. Damping-off of cotton seedlings was reduced when the antagonist was placed in soil with the seed. *Gliocladium catenulatum* attacked mycelium and spores of *Fusarium* spp. (Huang, 1978), and *G. roseum* attacked sclerotia of *Botrytis allii* (Walker and Maude, 1975).

Mechanism of Biological Control: Hyphae grow around mycelia and spores of susceptible fungi, killing them by enzymatic or toxic action, and sometimes then penetrating cells of the victim. The action does not occur at a distance (Barnett and Lilly, 1962), and may be another example of hyphal interference. *Gliocladium virens* produces potent antibiotics and probably was the agent involved in many instances in mycoparasitism attributed to *Trichoderma* spp. (Webster and Lomas, 1964). *Gliocladium catenulatum* kills cells of *S. sclerotiorum* without direct penetration (Huang, 1978; Huang and Hoes, 1976).

Mass Production: Flasks containing a sterile mixture of peat, soil, leaf compost, sand, and Czapek-Dox broth were inoculated with *G. roseum,* held four to seven days at 25°C, and then used at 500 ml of preparation per 1,500 ml of soil. Conifer bark pellets 0.5–1.0×0.6 cm were made from cultures grown on barley kernels plus conifer bark and fresh barley flour; they were used at 50 g per 1,500 ml of soil.

GLOMUS TULASNE AND TULASNE

Vesicular-arbuscular (VA) mycorrhizal fungi, common over the world, that form extensive loose hyphal networks extending several centimeters outward from roots of a wide range of plants, sometimes interconnecting roots of the same or different plant species (Whitington and Read, 1982).

Morphology and Taxonomy: Zygomycotina, Mucorales, Endogonaceae; dryspored. Chlamydospores borne on undifferentiated nonseptate, persistant, nongametangial hyphae, terminal or intercalary. Spore walls single or double. Mature spores contain oil droplets. Chlamydospores formed in loose sporocarps, in roots, or usually free in soil, generally hypogeous. Zygospores and sporangia not observed. Common in grasslands and cultivated fields. Form VA mycorrhizae with a wide range of plants. Mycorrhizal roots of plants of the same or possibly different species may be interconnected by the hyphae, which facilitates exchange of certain mineral nutrients and sometimes carbon. Arbuscules form in cortical cells of host, but later degenerate. Vesicles also may form in cortex; some may develop into chlamydospores. For details, see Gerdemann (1968), Gerdemann and Trappe (1974), Sanders et al (1975), and Trappe (1982).

Glomus fasciculatum (Thaxt. sensu Gerd.) Gerdemann and Trappe. Chlamydospores borne free in soil, in loose aggregations, in small compact clusters, and in sporocarps up to 8×5 mm, globose or flattened, tuberculate, gray-brown, without peridium. Chlamydospores $35-150 \mu m$ in diameter; globose to ellipsoid, cylindrical, or irregular, smooth or roughened from adhering debris; walls hyaline to light yellow or yellow-brown.

Glomus mosseae (Nicol. and Gerd.) Gerdemann and Trappe. Chlamydospores yellow to brown, globose to ovoid, obovoid, or irregular, $105-310 \times 110-305$ μm, with $1-2$ funnel-shaped bases $20-50 \mu m$ in diameter; with a wall $2-7 \mu m$ thick (consisting of a thin hyaline outer membrane and a thick yellow-brown inner layer). Spores ectocarpic or borne in sporocarps up to 1 mm in diameter, with $1-10$ spores, enclosed in a peridium of hyaline, loosely interwoven, septate hyphae.

Ecology: Roots of plants grown under high light intensity and with moderate deficiency of nitrogen or available phosphorus are more susceptible to infection; partial defoliation decreases infection. Harley (1969) suggested that "any external factor which causes a slow growth-rate of roots, or which reduces the proportion of actively growing tissue on a root system will appear to increase infection," a factor to be considered in their relation to root disease. *Glomus* spp. "may very well be the most common of all soilborne fungi" (Gerdemann and Trappe, 1974). Chlamydospores germinated in agar, but growth did not continue (Mosse, 1962); inoculation of surface-sterilized *G. mosseae* spores on plants aseptically grown was accomplished only when a *Pseudomonas* sp. or ethylenediaminetetraacetic acid (EDTA) was present.

Role in Biological Control: The role of VA mycorrhizae in plant growth was reviewed by Mosse (1973) and Harley and Smith (1983), and methods to study VA mycorrhizae are available in Schenck (1982).

Hussey and Roncadori (1982) and Strobel et al (1982) considered that VA mycorrhizae decreased nematode injury to plants by: 1) improving plant vigor, offsetting nematode damage; 2) altering or reducing root exudates; 3) retarding nematode reproduction in plant tissue. Reproduction of *Meloidogyne incognita* was suppressed by *Gigaspora margarita* only when both occupied the same root. Taber (1982a) found that weed seeds in soil were invaded by *Glomus fasciculatum*, which there formed chlamydospores, another site for reproduction of the mycorrhizal fungus. *Gigaspora* spores were also associated with weed seeds in soil (Taber, 1982b). Bärtschi et al (1981) found that a natural mixed population of VA mycorrhizae applied to *Chamaecyparis lawsoniana* six months before inoculation with *Phytophthora cinnamomi* gave 93.5% control of root rot after 12 weeks, but gave no control when applied at the time of inoculation with *P. cinnamomi*. Inoculation with *G. mosseae* eight months before or at the same time as *P. cinnamomi* gave no control. Mataré and Hattingh (1978) found that *G. fasciculatum* somewhat stimulated growth of avocado seedlings but had no effect on infection of roots by *P. cinnamomi*. The effect of *G. fasciculatum* on *P. parasitica* on citrus roots was to increase host tolerance of the pathogen through greater absorption of phosphorus and other minerals (Davis and Menge, 1980).

Schenck and Kellam (1978) concluded, from a literature review, that VA mycorrhizae increased disease severity in seven cases, decreased it in 22, and had no effect in nine cases. "Interactions between VA mycorrhizal fungi, plant pathogens, and host plants may . . . vary with specific combinations. . . . To minimize . . . losses, considerably more studies . . . are necessary." Dehne (1982) similarly concluded that VA mycorrhizae increased diseases caused by soilborne fungi or nematodes in four cases, decreased them in 17 cases, and had an uncertain effect in 29 cases.

Mechanism of Biological Control: Infection by VA mycorrhizae may provide biological control indirectly, by enhancing uptake of phosphorus and perhaps other nutrients, which thereby increases host plant resistance to pathogens favored by nutrient deficiencies in the host. VA mycorrhizae may also protect roots against pathogens by using surplus carbohydrates from the roots, by secreting antibiotics, by favoring protective rhizosphere microorganisms, or by inducing morphogenic and biochemical changes in host tissue unfavorable

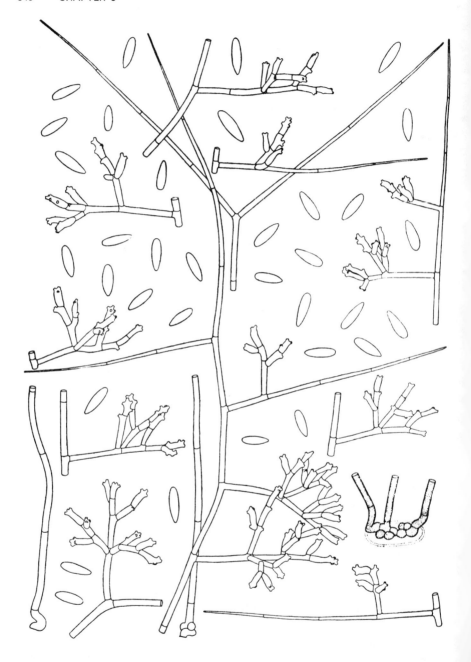

Figure 9.6. *Hansfordia togoensis*, showing conidiophores and conidia. (Reprinted with permission from Hughes, 1951.)

to the pathogen (Schenck and Kellam, 1978). Barea and Ascón-Aguilar (1982) found that *G. mosseae* synthesized auxins and two gibberellinlike and four cytokininlike substances that may affect interaction of host and pathogen.

Mass Production: *Glomus* spp. are obligate or near-obligate parasites and must therefore be grown on living roots. The potential for growth of *Glomus* spp. in culture is doubtful. Chlamydospores washed from soil by wet-sieving may be added to pots of nontreated soil in which asparagus is growing. The asparagus roots are infected and chlamydospores form on mycelium in the soil. Moss (*Funaria hygrometrica*) growing on the soil surface provides a favorable medium for chlamydospore formation in a *G. epigaeum*-asparagus association and may provide a medium for commercial production of inoculum (Parke and Linderman, 1980). A California company sells soils that contain *G. deserticola* chlamydospores that have been tested and are said to be free of *Phytophthora, Pythium, Rhizoctonia,* and *Fusarium.*

HANSFORDIA HUGHES

Saprophytic and mycoparasitic fungi; not recorded from soil.

Morphology and Taxonomy: Deuteromycotina, Hyphomycetes, Dematiaceae; Xerosporae. Conidiophores hyaline or pigmented, erect, straight or bent; lateral branches fertile, single or in pairs; branches terminate in 1–3 cylindrical sporogenous cells. Conidia arise singly from truncate dentricles, are nonseptate, hyaline, spherical or ovate to fusoid, smooth or slightly roughened, borne acropetally (Figure 9.6). For details, see Hughes (1951).

Role in Biological Control: *Hansfordia* sp. occurs as a natural secondary colonist of certain leaf spots, e.g., cercosporidium leaf spots on peanut in Texas. Production of secondary inoculum of the pathogen is then diminished by the antagonist, which penetrates the stromatic cells and hyphae. It did not parasitize lesions caused by *Cercospora* or *Puccinia* when applied to peanut leaves to control leaf spot (caused by *Cercosporidium personatum*) (Taber and Pettit, 1981). *Hansfordia* sp. produces masses of conidia rapidly on suitable media and could be mass-produced readily.

LAETISARIA BURDSALL

The *Corticium* sp. studied in 1977–1980 as a biological control agent is now placed in this genus.

Morphology and Taxonomy: Basidiomycotina, Hymenomycetes, Corticiaceae; slime-spored. Basidiocarps effused, pellicular, smooth, with broad septate hyphae that usually lack clamp connections. Basidia cylindrical to clavate, with 4 sterigmata. Basidiospores ovoid, hyaline, thin-walled. See Burdsall (1979) and Burdsall et al (1980) for details.

Laetisaria arvalis Burdsall. Teleomorph has not been found in nature. Hymenium in culture is up to 200 µm thick, membranous, with yellowish white fertile areas, smooth to farinaceous, formed on hyphal strands and sclerotia. Hyphae septate, with occasional clamp connections; strands up to 50 µm wide consisting of 2–8 parallel or entwined hyphae. Basidia 40–60×8–12 µm, broadly clavate, hyaline, thin-walled with 4 sterigmata up to 6×3.5 µm. Basidiospores 9.5–12.5 × 5.5–7 µm, ellipsoid, hyaline, thin-walled, smooth. Sclerotia 400–600 µm wide and dark reddish brown; rind 20–40 µm thick. Ananomorph is *Isaria fuciformis* Bak.

Ecology: Biocontrol effective at 15–25°C over a wide range of soil water potentials (Hoch and Abawi, 1979). *Laetisaria* is common in soil throughout the temperate world. *Laetisaria arvalis* was not inhibited by benomyl or thiabendazole soil treatment, but *Rhizoctonia solani* was (Papavizas et al, 1982).

Role in Biological Control: Lewis and Papavizas (1980) found that *Corticium* sp. added to soil as mycelium and sclerotia at 706 kg/ha decreased cucumber fruit rot caused by *R. solani* from 75% (checks) to 50% (treated) in field tests. Odvody et al (1980) showed that using *Corticium* sp. as a seed coating reduced damping-off of beans, soybeans, and sugar beets in soil infested with *R. solani*. *Corticium* sp. has not shown evidence of pathogenicity to crop seedlings (Hoch and Abawi, 1979a). Allen et al (1982) found that "*R. solani* and *L. arvalis* formed a stable equilibrium which could survive on an alternate host. Application of the hyperparasite reduced *R. solani* temporarily. . . . temporal alteration of hyperparasite application and weed control may prove to be an effective biological control technology."

Mechanism of Biological Control: *L. arvalis* almost prevented saprophytic activity of *R. solani* (Lewis and Papavizas, 1980). The fungus is parasitic on *R. solani* (Burdsall et al, 1980; Odvody et al, 1980) and *Pythium* spp. (Burdsall et al, 1980).

Mass Production: *L. arvalis* lived three years stored as air-dried mycelium and sclerotia under nonsterile conditions at 24–26°C (Odvody et al, 1980). The fungus may be grown commercially on diatomaceous earth granules ready for field inoculation or produced on pellets of sugar beet pulp.

LEUCOPAXILLUS BOURSIER

Common in subtropical areas. Common ectomycorrhizal fungus on a wide range of trees.

Morphology and Taxonomy: Basidiomycotina, Hymenomycetes, Agaricaceae; dry-spored. Carpophores fleshy; lamellae decurrent or sinuate. Spore print white; spores hyaline, rough to warty or smooth, up to 10 μm long, short, ellipsoid to ovoid. Cystidia none. Stipe central, thick, fleshy; veil none. For details, see Singer (1975).

Leucopaxillus cerealis (Lasch.) Sing. var. *piceina*. Spores rough, warty, hyaline. Fleshy cap with central stipe, lamellae decurrent.

Role in Biological Control: Marx (*in* Bruehl, 1975) found that *L. cerealis* var. *piceina* protected roots of *Pinus echinata* against infection by *Phytophthora cinnamomi* by the physical barrier of the mantle covering the short roots and by secreting an antibiotic, diatretyne nitrile.

Mechanism of Biological Control: The antibiotic diatretyne nitrile produced by *L. cerealis* var. *piceina* inhibits germination of zoospores of *P. cinnamomi* at 50–70 ppb and kills them at 2 ppm (Marx, 1969). The ectomycorrhizal hyphal mantle also provides a partial physical barrier. *Leucopaxillus cerealis* var. *piceina* inhibited 92% of the root pathogens tested in cultures. Marx and Davey (1969) found diatretyne nitrile in the ectomycorrhizal short roots of *Pinus echinata*.

MYROTHECIUM TODE: FR.

Widespread cellulolytic fungi; common in soils, especially soils high in organic content. Some are parasitic on plants.

Morphology and Taxonomy: Deuteromycotina, Coelomycetes, Tuberculariaceae; Gloiosporae. Sporodochia small, discoid, sessile or with short stalks, often confluent, dark-green to black. Spore mass viscous and green, later becoming hard and black. Conidiophores erect in a compact layer, septate, hyaline,

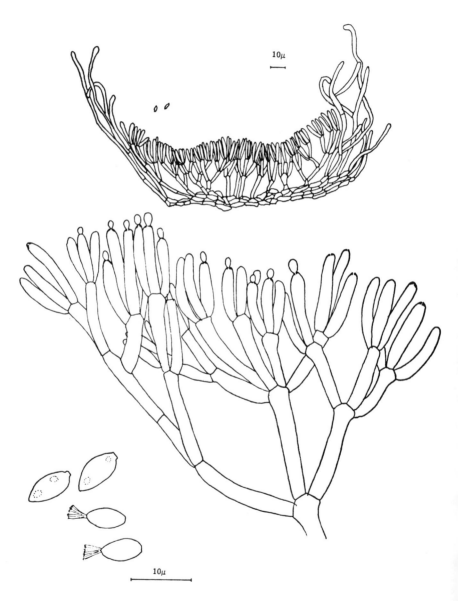

Figure 9.7. *Myrothecium verrucaria*, showing a sporodochium, conidiophores, and conidia with fantailed appendages. (Reprinted with permission from Tulloch, 1972.)

branched, tipped with slender, hyaline, verticillate phialides. Conidia hyaline to olive brown, ovoid, fusoid or cylindrical, nonseptate, in slimy masses. For details, see Preston (1943, 1961) and Tulloch (1972).

Myrothecium verrucaria (Alb. and Schw.) Ditm.: Fr. Sporodochia 50–150 μm thick, rarely well developed, usually a thin layer of hyaline cells 3.5–5 μm in diameter; marginal hyphae curling, usually verrucose, hyaline, septate, cells 10–25×1.5–2.5 μm. Spore mass wet, black but surrounded by white floccose margin. Conidiophores branch repeatedly, the ultimate branches bearing hyaline, septate phialides 10.5–14.5×1.5–2 μm, 3–6 in a whorl. Conidia fusiform, one end pointed and the other protruding, fan-tailed, truncate, 6.5–8×2–3.5 μm (Figure 9.7).

Role in Biological Control: *Myrothecium verrucaria* suppressed damping-off of seedlings caused by *Rhizoctonia solani* (Ferguson, 1958). Its activity was augmented by the addition of cellulose (e.g., planting pepper seed), but when used to suppress *R. solani* the amount of toxin produced was so great that pepper seedlings often were stunted. White et al (1948) studied the cellulolytic activity, and Brian and McGowan (1946) isolated a strong, highly specific fungus antibiotic from *M. verrucaria*. Numerous antibiotics (verrucarin, roridin, myrothecin, dehydroverrucarin, muconomycin) have been isolated from *M. verrucaria* and *M. roridum*. Some diseases of sheep, horses, and pigs have been reported as possibly caused by *M. verrucaria*.

NEMATOPHTHORA KERRY AND CRUMP

Obligately parasitic fungus on females of *Heterodera avenae*.

Morphology and Taxonomy: Mastigomycotina, Oomycetes, Leptolegniellaceae; slime-spored. Mycelium filamentous, mainly intramatrical, much branched; extramatrical hyphae rarely branched and function as sporangia. Zoospores laterally biflagellate, encysting inside or outside zoosporangia. Oospores on lateral mycelial branches, very thick-walled. For details, see Kerry and Crump, 1980.

Nematophthora gynophila Kerry and Crump. Mycelium much branched, eventually occupying much of the substratum, thin-walled; extramatrical hyphae up to 350 μm long, function as zoospore-discharge tubes. Zoospores 7×11 μm, about 75 per zoosporangium. Oospores 35 μm long, spherical, with thick, pitted wall, produced on undifferentiated lateral segments, long-lived. Hyphae penetrate the adult or juvenile nematode through cuticle; each nematode may support production of 12,000 zoospores and 3,000 oospores.

Life cycle takes less than seven days at 13°C. Lives at least two years in the soil.

Ecology: *Nematophthora* is common in soil in Europe, particularly in soil infested with *Heterodera avenae*, the cereal cyst nematode. In England, an equilibrium of cereal cyst nematode eventually establishes at 5–10 eggs per gram of soil in fields cropped intensively to cereals; thereafter the nematode causes little yield loss, "the only known example of natural agents giving effective, long-term control of a cyst-nematode" (Kerry and Crump, 1980). There are 200–400 spores per gram of soil (50–100 million/m² to the plow depth) in wheat fields where cyst nematodes are controlled. Movement of zoospores through soil probably requires pores 60 μm or larger in diameter (Duniway, 1976), which is nearly the same or larger than required for movement of the nematodes. Infection of *H. avenae* by *N. gynophila* is inhibited in soil drained to a water content at which the nematodes remain active. Low rainfall reduces the field population of *N. gynophila* (Kerry et al, 1980). In England, soils with the lowest population of *N. gynophila* are coarse sands and free-draining soils over chalk.

Role in Biological Control: A resident antagonist in soils infested with the cereal cyst nematode, *H. avenae*, the fungus also attacks *H. carotae, H. cruciferae, H. goettingiana, H. schachtii*, and *H. trifolii*, but not *Globodera rostochiensis*. Formaldehyde applied as a soil drench (3,000 liter/ha) increased the rate of multiplication of *H. avenae* because it killed many of the parasitic fungi but was weakly nematicidal (Kerry et al, 1980, 1982a, 1982b; Williams, 1969). In soil with *N. gynophila*, cyst and egg numbers after harvest were reduced by 95 and 97%, respectively, compared with the numbers in soil where parasites (but not nematodes) were almost eliminated by formaldehyde treatment (Kerry et al, 1980).

Mechanism of Biological Control: Zoospores infect female nematodes by invasion through the cuticle. The cuticle is destroyed, and nematodes rapidly lose moisture (Kerry, *in* Papavizas, 1981).

Mass Production: Use is restricted by the inability to culture the fungus apart from the host nematode and the inability to germinate the oospores. "It seems unlikely that natural enemies could be cultured artificially and added to soil in sufficient numbers to give immediate control." (Kerry, *in* Papavizas, 1981).

PENICILLIUM LINK: FR.

Common in soil and on decaying plant materials.

Morphology and Taxonomy: Deuteromycotina, Hyphomycetes, Moniliaceae; Xerosporae. Conidiophores erect, hyaline, septate, with characteristic brush-like branching penicilli. Phialides borne in groups at the apex or on branches, flask-shaped, hyaline. Conidia in long chains, globose to ovoid, hyaline to darkly pigmented, smooth or rough. Sclerotia sometimes formed. For details see Raper and Thom (1949).

Penicillium lilacinum Thom. Conidiophores 100–600 μm long, smooth or finely roughened, hyaline or slightly yellow; penicilli bear tangled masses of conidia in chains up to 50–70 μm long, arising at two or more levels; sterigmata 5.0–6.0 μm long, thin, abruptly tapering. Conidia elliptical, 2.5–3.0×2.0 μm, smooth, light vinaceous in mass. One of the most abundant of soil penicillia.

Penicillium nigricans (Bainier) Thom. Conidia borne on short branches of aerial hyphae, with penicilli terminal, bearing diverging branchlets with cluster of 6–12 sterigmata-bearing chains up to 50–75 μm long. Conidiophores rarely more than 200 μm long, smooth. Conidia 3.0–3.5 μm in diameter, globose, echinulate or spiny, appearing olive brown under high power.

Penicillium frequentens Westling. Conidiophores up to 100–200 μm long, smooth or finely roughened; penicilli almost entirely monoverticillate; sterigmata 10–12 per verticil, 8–12 μm long. Conidia in chains up to 150 μm long. Conidia globose to subglobose, thin-walled, smooth or finely roughened, 3.0–3.5 μm in diameter.

Penicillium oxalicum Currie and Thom. Conidiophores smooth, 100–200 μm long, occasionally mono- but normally biverticillate, with conidial chains in columns 500 μm or more long; sterigmata in terminal clusters of 6–10, 9.0–15.0 μm long. Conidia elliptical, smooth, 4.5–6.5×3.0–4.0 μm.

Penicillium chrysogenum Thom. Conidiophores in a dense stand 150–350 μm or more long, walls smooth, hyaline; penicilli biverticillate, asymmetrical, terminate in verticils of 2–5 metulae 10–12 μm long, bearing sterigmata in verticils of 4–6, 8–10 μm. Conidial chains in 200- μm columns. Conidia elliptical, rarely subglobose, 3–4×2.8–3.5 μm, smooth, yellowish green in mass. Produces penicillin.

Role in Biological Control: Ghaffar (1969) used cultures of *P. nigricans* to control *Sclerotium cepivorum* in glasshouse soil tests, but seed treatment with spores did not protect the seedlings from *S. cepivorum*. Utkhede and Rahe (1980) found *P. nigricans* to be a natural inhabitant of *S. cepivorum* sclerotia, but field tests did not give control as effective as that with *Bacillus subtilis*. Bruehl and Lai (1968) found that *Cephalosporium gramineum* in wheat stubble was unaffected by bacteria in moist soil, but was displaced in the straw in dry (−150 to −210 bars) alkaline soil by *Penicillium* spp. Moore (*in* Papavizas, 1981) found that *Penicillium* spp. decreased crown gall on Mazzard cherry seedlings from 61.9 to 4.1% for *Agrobacterium radiobacter* pv. *tumefaciens* biotype 1 and from 81.4 to 32.8% for biotype 2. Ebben and Spencer (1978) used *P. lilacinum* to control black root rot of cucumber (caused by *Phomopsis sclerotioides*) in pot tests; dry weights of unprotected plants were 26% lower than those of noninoculated plants but only 9% lower when *P. lilacinum* was also added. Disease control was short-term and did not result in increased yields. *Penicillium lilacinum* did not compete with *P. sclerotioides* in the rhizosphere.

Windels (1981) found that pea seed dusted with spores of *P. oxalicum* and planted in field soil germinated, the fungus sporulated, and the hyphae covered the seed by the third day, protecting the seeds and seedlings from infection by a complex of *Aphanomyces, Fusarium, Pythium*, and *Rhizoctonia* spp. There was minimal damage to the cotyledons. Spores of *P. oxalicum* on roots did not germinate and did not protect against postemergence damping-off. Kommedahl and Windels (1978) found that *P. oxalicum* was more effective than other penicillia and that it could be improved by selection; it was the best antagonist tested under field conditions over several years. Windels and Kommedahl (1978) found that when pea seed was coated with *P. oxalicum* and planted in infested soil in a pea nursery, the total rhizosphere population of *Penicillium* was increased but the rhizosphere population of *Fusarium*, actinomycetes, or bacteria was not affected. Control of preemergence damping-off of peas with 6×10^6 spores of *P. oxalicum* per seed was as effective as with captan. In field trials with five pea cultivars, inoculation of seed with *P. oxalicum* improved stands for four of the five cultivars in one of two years and, in one test, gave stands and pod weights equal to those produced from captan-treated seed and significantly better than those from nontreated seed (Windels and Kommedahl, 1982a). Liu and Vaughan (1965) used *P. frequentens* in glasshouse tests to control damping-off of beet seedlings caused by *Pythium ultimum*. In nontreated field soil, the stand with *Penicillium*-treated seed was 46% of the untreated control (autoclaved soil), but this was improved to 66% by adding K_2HPO_4. and urea. Dutta (1981) showed that a root-dip application of *P. chrysogenum* protected tomato seedlings from infection by *Verticillium albo-atrum* in infested soil.

Mechanism of Biological Control: *Penicillium* operates mainly as a competitor of pathogens on the host and as a secondary colonist of lesions or infested crop residue; it is especially successful under dry conditions. Its ability to produce antibiotic may also be important. Dutta (1981) suggested that production of antibiotic substances by *P. chrysogenum* inhibited the normal growth of *V. albo-atrum* in soil and the rhizosphere. Windels (1981) thought that *P. oxalicum* on pea seed provided physical protection against a complex of seed pathogens by the layer of hyphae formed on the coat and by using the seed exudates otherwise used by the pathogens. Moore and Cooksey (1981) found a field in Washington in which crown gall was not important, although apple seedlings had been continuously grown there for many years; the soil contained 10^6 colony-forming units of a *Penicillium* sp. that was highly antibiotic to *Agrobacterium tumefaciens* in vitro.

Mass Production: Ebben and Spencer (1978) found that *P. lilacinum* grown in shake culture and mixed into composts persisted for more than seven weeks, but gave only limited control of *Phomopsis sclerotioides*. Kommedahl and Windels (1978) grew *P. oxalicum* on potato dextrose agar in petri dishes for one to two weeks at 24°C and then harvested the dry spores. The spores of *P. oxalicum* produced on agar media or oat grains were effective for inoculation of pea seeds and were equally effective whether 20 or 125 days old (Windels and Kommedahl, 1978). Treated seed should be stored at 5°C, instead of 24°C, before planting.

PENIOPHORA COOKE

Common in wood and bark in most temperate areas.

Morphology and Taxonomy: Basidiomycotina, Hymenomycetes, Corticiaceae; dry-spored. Fructifications waxy, coriaceous, cartilaginous, membranous, floccose, or filamentous, always resupinate, effused, even. Basidia simple, with 2–4 white ellipsoidal spores. Cystidia may or may not be encrusted and are more or less immersed in the hymenium. Distinguished from *Corticium* by the presence of true pointed cystidia. For details, see Burt (1925).

Peniophora gigantea (Fr.) Karst. Fructifications 3–30 cm in diameter, 100–500 μm thick, broadly effused, hyaline, white, waxy, swollen when wet, and hornlike when dry. Hymenium even, pale pinkish buff, pale olive buff, or pallid mouse-gray; margin white, fibrillose, radiating. Cystidia encrusted, 40–50×8–12 μm, confined to the hymenium. Spores hyaline, even, 4.5–5×2.5–3 μm. Usually seen on coniferous bark as large, white, or pink–buff, cartilaginous fructifications (Figure 4.6).

Ecology: Inoculation of pine stumps with oidia of *P. gigantea* immediately after the tree is felled gives the fungus priority as a pioneer colonist, and it then protects the stump against colonization by *Heterobasidion annosum*. It grows into the stump but does not infect roots of neighboring trees grafted to those of the stump, as *H. annosum* will do if allowed to colonize the stump. The stump decays rapidly, and sporophores of *P. gigantea* form within a year.

Role in Biological Control: Rishbeth (1963, 1979, *in* Bruehl, 1975) in a series of admirable studies on inoculation of freshly cut stumps by a low-grade pathogen, *Peniophora gigantea*, established a highly successful biological-control program. In addition to protection of pine, *P. gigantea* also proved effective on Norway spruce; specific strains of *P. gigantea* in 5% ammonium sulfamate have been used on Sitka spruce and Douglas-fir. It is ineffective on European larch.

Mechanism of Biological Control: The biological control of *H. annosum* by *P. gigantea* probably results from various combinations of hyphal interference (Ikediugwu et al, 1970), a short-range antibioticlike effect (Rishbeth *in* Bruehl, 1975), and exclusion of *H. annosum* from the stump surface physically and through competition for nutrients.

Mass Production: *Peniophora gigantea* is grown on malt agar and the oidia are washed off with a sucrose solution. Talc and Cellophos B600 are added to the suspension and the mixture is poured into molds and desiccated to form tablets (sachets). Each tablet, dispersed in 100 ml of water, gives at least 1×10^6 viable oidia per milliliter and will inoculate about 100 stumps (Rishbeth, 1963). Artman (1972) added the oidia to lubricating oil placed on the chain saw to inoculate the fresh cut. In England, the oidial suspension is now commercially available and applied to stumps at a cost of 0.6–1.2 p per stump. It had been applied to about 62,000 ha by 1973 (Greig, 1976; Rishbeth, 1979).

PHIALOPHORA MEDLAR

Common in grassland and forest soils.

Morphology and Taxonomy: Deuteromycotina, Hyphomycetes, Dematiaceae; Gloiosporae. Conidiophores short or lacking, smooth, frequently with sporogenous cells directly on hyphae. Phialides usually short, cylindrical or inflated, frequently flask-shaped, single or in groups (rarely penicillate), pigmented, with pronounced apical collarette. Phialospores nonseptate, hyaline or pigmented, spherical to ovoid, smooth, produced in a collarette and gathered in slimy balls. *Phialophora*, being a form genus, contains many unrelated species.

Of interest in biological control are those species associated with roots of grasses and cereals. For details on the genus, see Cain (1952) and Ellis (1971, 1976).

Phialophora radicicola Cain. Conidiophores hyaline or dark-walled, with conidia produced in slimy heads at the tips of phialides that arise laterally from conidiophores. Phialides flask-shaped, often irregular and curved, with hyaline walls; older phialides have conspicuous flared collarette, 5–20×1–3 μm. Conidia hyaline, nonseptate, pointed at the attached end, rounded at the apex, 3–7×1–2 μm. It grows as dark runner hyphae on roots, with narrow hyaline hyphae extending into the root cortex. Reported by McKeen (1952) as a weak parasite on roots of corn. According to Walker (*in* Asher and Shipton, 1981), this name has been misapplied on numerous occasions (e.g., for the fungus on roots of cereals and grasses in England that protects roots against *Gaeumannomyces graminis* var. *tritici*). Deacon (1974a) gave the name *P. radicicola* var. *graminicola* to the fungus suppressive to take-all in England; this automatically erected the fungus described by Cain as var. *radicicola*. Walker (1980) proposed the name *P. graminicola* for *P. radicicola* var. *gramicola*.

Phialophora graminicola (Deacon) Walker. Conidiophores hyaline to brown. Phialides single and lateral on hyphae or in loose clusters at the ends of branches; hyaline or occasionally pale brown, 5–20×2–4 μm. Conidia hyaline to slightly tinted yellow, variable in size and shape, rounded at the apex, in groups of 5–11(14×1.5–2.5 μm) (Figure 9.8). May be wider and more oval in shape if grown in liquid media. Hyphopodia produced on wheat coleoptiles; brown, subglobose to oval, 6–15 μm in diameter, with slightly lobed margins. Lignitubers form around the penetration peg that grows from the hyphopodia into underlying coleoptile tissue of wheat seedlings. Cultures are slow-growing (4–6 μm per day), white at first but becoming pale gray to dark greenish gray. *Perfect stage* is *Gaeumannomyces cylindrosporus* Hornby et al. For details, see Walker (*in* Asher and Shipton, 1981).

Phialophora sp. (lobed hyphopodia). Indistinguishable from the asexual stage of *G. graminis* var. *graminis* (Walker, *in* Asher and Shipton, 1981).

Ecology: *Phialophora radicicola* is most effective at soil temperature of 20°C, and is ineffective at low soil temperature.

Role in Biological Control: No role of *P. radicicola* in biological control has been demonstrated, because the fungi that protect wheat against take-all and turf against ophiobolus patch are now identified as *P. graminicola* or *Phialophora* sp. with lobed hyphopodia. Scott (1970) isolated *P. radicicola* (later named *P. radicicola* var. *graminicola* and now *P. graminicola*) from

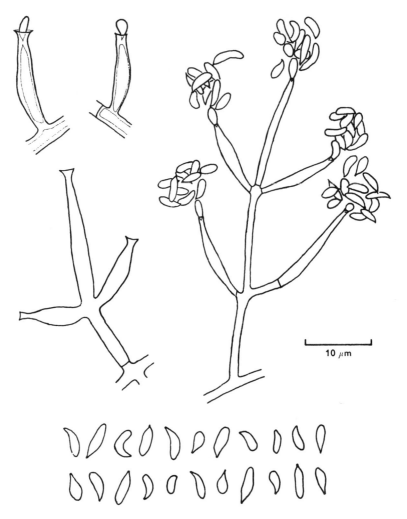

10 μm

Figure 9.8. *Phialophora graminicola*, showing conidiophores, phialides, and conidia. (Reprinted with permission from Scott, 1970.)

roots of grasses and cereals in England. He showed that the fungus infects but is avirulent on cereal roots and raised the possibility that the avirulent fungus may explain why take-all is seldom serious in England in wheat following grass in spite of the presence of susceptible grasses. Balis (1970) showed that infections by avirulent *P. radicicola* (*graminicola*) protected wheat against *G. graminis* var. *tritici*. Deacon (1973b) found that the fungus was widespread in grass rotations in England and provided significant natural biological control of *G. graminis* var. *tritici* on wheat grown in the first or second year, and also that *P. graminicola* protected turf (*Agrostis* sp.) against ophiobolus patch. Wong and Siviour (1979) showed that both a *Phialophora* sp. with lobed hyphopodia from England and *P. graminicola* from Australia controlled ophiobolus patch of turf in pot experiments when both the antagonist and the pathogenic *G. graminis* var. *avenae* were introduced into sterilized soil. Wong and Southwell (1980) showed that *Phialophora* sp. with lobed hyphopodia and other related avirulent fungi suppressed take-all in wheat field plots in Australia. The antagonist was grown on oat grains and this source of inoculum was sown one to one (v/v) with the wheat seed into soil infested with *G. graminis* var. *tritici*. This method was effective in moderately infested soil but not in soil heavily infested with the take-all pathogen; thus it is perhaps best used following a break crop. Because the take-all pathogen is better adapted to cooler soils, strains of *Phialophora* spp. that colonize wheat roots at soil temperatures of 7–15°C should be selected.

Mechanism of Biological Control: Protection occurs in the infection court and even within the root cortex of wheat. Deacon (1973a, 1973b, 1974a, 1974b, 1976a, 1976b) found that *P. graminicola* may invade the root cortex but not the endodermis of living wheat. Speakman and Lewis (1978) demonstrated that *G. graminis* var. *tritici* does not grow longitudinally in the stele if the surrounding cortex is colonized by *P. graminicola*, suggesting that infection by the avirulent fungus somehow modifies the root, making it more resistant or otherwise unacceptable to *G. graminis* var. *tritici*. Deacon (1976b) suggested that induced resistance may be involved. Competition for nutrients during prepenetration growth on the root also may be involved.

Mass Production: Sterilized moist oat grains in glass jars are inoculated with the fungus and incubated at 25°C for four weeks. The fully colonized grains are air-dried and stored at 1.4°C.

PISOLITHUS (PERS.) COKER AND COUCH

Common puffballs in forested areas, forming ectomycorrhizae with a wide range of trees over the world.

Morphology and Taxonomy: Basidiomycotina, Gasteromycetes, Lycoperdales; dry-spored. Puffballs with moderately thick hard peridium and dark gleba. Distinct peridioles or irregular bodies make up the gleba.

Pisolithus tinctorius Pers. Coker and Couch. Fruiting bodies oval to club-shaped; may be more than 18 cm high, with a well-developed rooting base. In young specimens the gleba is gelatinous and has an inky fluid; later develops oval to globose pockets that become dry and crumbly. When peridium breaks, a powdery mass of spores is released. Capillitial threads few or lacking. Basidia bear 2–6, almost sessile, globose spores that are brown, irregularly warted, about 7 μm in diameter. Dehiscence by attrition of apex of sporophore. For details, see Coker and Couch, 1928.

Ecology: *Pisolithus tinctorius* grows well on sites of low fertility, high soil temperatures, and low soil pH (e.g., coal spoil banks of pH 3.4) (Marx, 1977; Trappe, 1977). It is common in well-drained, gravelly, or sandy soil.

Role in Biological Control: Ross and Marx (1972) showed that *P. tinctorius* ectomycorrhizae on roots of *Pinus clausa* protected the roots from *Phytophthora cinnamomi*. Only 40% of nonmycorrhizal seedlings survived after two months, whereas 70% of mycorrhizal seedlings survived, although *Pisolithus tinctorius* infected only 25% of the feeder roots. Inoculum density of *P. cinnamomi* was reduced in soil around seedlings of *Pinus echinata*, of which 70–89% were infected by *P. tinctorius* or *Cenococcum graniforme* (Ross and Marx, 1972). Marx and Davey (1969) found that lateral roots of pine seedlings that had *P. tinctorius* ectomycorrhizae on other roots were completely susceptible to *P. cinnamomi*, suggesting nontransfer of chemical inhibitors (Marx, 1972).

Mechanism of Biological Control: Marx (1969) found that *P. tinctorius* was not antibiotic to any of 48 root pathogens tested in cultures. *Pisolithus tinctorius* was found (Szaniszlo et al, 1981) to produce hydroxyamate siderophore iron chelators that greatly increased the uptake of iron from the soil, benefiting the plant.

Mass Production: Spores are copiously produced in nature and are easily collected for inoculum. However, the thick-walled waxy spores are extremely difficult to germinate. Mycelium may be used for inoculation in nurseries (Marx and Bryan, 1975; Trappe, 1977). Smith (1982) found that "By optimizing growth rate in a medium suitable for deep tank fermentation, and with proper formulation, it is feasible to consider producing this ectomycorrhizal fungus for commercial use in forestry." Yu and Trione (1982) devised a method for determining viability of *P. tinctorius* spores.

PSEUDOMONAS MIGULA

Common inhabitants of soil, plant debris, and especially the rhizosphere. Producers of potent antibiotics.

Morphology and Taxonomy: Eubacteria, Pseudomonadaceae; slime-celled. Cells single, straight or curved rods, generally 0.5–1.0×1.5–4 μm; motile by polar monotrichous or multitrichous flagellae. No sheaths or prothecae produced; no resting stages known. Gram-negative. Metabolism respiratory, never fermentative; strict aerobes. Catalase positive. Frequently develop fluorescent, diffusible pigments of green, blue, violet, lilac, rose, or yellow, particularly in iron-deficient media; many species develop no pigment. For details, see Buchanan and Gibbons (1974).

Pseudomonas fluorescens Migula. Cells have multitrichous flagellae. Gelatin liquified; pyocyanine absent; oxidase positive; arginine dihydrolase positive. Seven biotypes recognized by Palleroni and Doudoroff (1972). Produces soluble fluorescent pigments. No growth-factor requirements; uses glucose and aromatic compounds but not starch. Lysed by ethylenediaminetetraacetic acid (EDTA). Maximum growth at 35–37°C.

Pseudomonas cepacia Ballard et al (*P. multivorans*). No fluorescent pigments. Flagellae multitrichous. Highly omnivorous; uses more than 100 organic compounds as carbon sources, including cellobiose, D-arabinose, and D-fucose.

Pseudomonas putida (Trev.) Migula. Some cells more than 4 μm long. Gelatin not liquified; produces no pyocyanin; oxidase positive; arginine dehydrolase positive. Maximum growth at 35–37°C.

Ecology: *Pseudomonas* spp. are favored by moist soil high in organic matter and are especially prevalent in the rhizosphere and on the rhizoplane. They are capable of growth over a wide temperature range, including at near-freezing temperatures and at 35–37°C, depending on the strain. Some strains of *P. syringae* are ice-nucleation active and promote frost injury on sensitive plants; other strains are not ice-nucleation active and inhibit strains that are, protecting plants against frost injury (Lindow, 1983). Spurr and Sasser (1982) found *P. cepacia* to be common on tobacco roots in South Carolina.

Role in Biological Control: *Pseudomonas fluorescens* strain C12 and *Pseudomonas* sp. strain A2 were used by Nair and Fahy (1976) for commercial control of bacterial brown blotch of mushroom (caused by a toxic metabolite produced by *P. tolaasii*), giving 8.0–16.6% increased yields of blotch-free caps.

Olivier et al (1978) and Olivier and Guillaumes (1981) used nonpathogenic strains of *P. tolaasii* and *P. fluorescens* sprayed on the casing material at the time it was applied to the mushroom beds; yield was not increased, but the incidence of blotched caps was decreased by 40–60%. *Pseudomonas fluorescens* and *P. putida* have been implicated in natural suppression of *Gaeumannomyces graminis* var. *tritici* in the wheat rhizosphere (Sivasithamparam et al, 1979; Smiley, 1978a) and in areas where take-all decline has occurred (Cook and Rovira, 1976). Strains of *P. fluorescens* inhibitory to *G. graminis* var. *tritici* in vitro were suppressive to take-all in field trials when introduced on the wheat seed at sowing. *Pseudomonas putida* obtained from a California *Fusarium*-suppressive soil induced suppressiveness to *F. oxysporum* f. sp. *lini* in a conducive Colorado soil (Scher and Baker, 1980, 1982). This fluorescent *Pseudomonas* sp. produced a peptide siderophore (pseudobactin) under iron-deficient conditions that complexed the iron in the soil, making it less available to the *Fusarium* than to the host. Application of Fe^{III}ethylenediaminedi-o-hydroxyphenylacetic acid (Fe^{III}EDDHA) to conducive soil produced the same effect as *P. putida*.

Kloepper et al (1980a, 1980b) and Kloepper and Schroth (1981a, 1981b, 1981c) showed that strains of the *P. fluorescens-putida* group applied to potato seedpieces rapidly colonized and spread along the roots, displaced native rhizosphere microorganisms, and increased plant growth. The increased growth occurred in natural soil but not under gnotobiotic conditions, thereby indicating that the group has a role in biocontrol of harmful rhizosphere microorganisms (Suslow and Schroth, 1982a, 1982b). Noninhibitory mutants from growth-promoting rhizobacteria also gave no increased growth. Loper et al (1982) showed that some deleterious rhizosphere *Pseudomonas* spp. that caused root deformity and decreased elongation in sugar beets produced indoleacetic acid (IAA). Suslow (1982) found that some soilborne phages prevented plant-growth promotion by phage-sensitive rhizobacteria, introducing another factor to be considered in selecting beneficial rhizobacteria. Schroth and Hancock (1982) suggested that long-term soil suppressiveness is due to "an assortment of microorganisms" but that temporary suppression may result from "some of the key microorganisms . . . used under conditions which favor their activities." Pseudobactin-producing *Pseudomonas* spp. decreased the rhizosphere fungi by 23–64% and Gram-positive bacteria by 25–93%. A strain effective in vivo against take-all was shown to colonize wheat roots when introduced on seed (Weller and Cook, 1983). Austin et al (1977) showed that *P. fluorescens* on leaves of *Lolium* reduced spore germination and germ tube growth and caused lysis of hyphae of *Drechslera dictyoides*.

Mechanism of Biological Control: Nair and Fahy (1972) showed that the antagonistic strains of *P. fluorescens* or *P. multivorans* reduced brown blotch of mushroom caps from 100% down to 7.6–11.1%, perhaps by competition with *P. tolaasii* for nutrients. No antibiosis against *P. tolaasii* (or *P. putida*, a stimulator of cap formation) was shown in culture. The action of the fluorescent pseudomonads in plant growth enhancement is thought to result from their production in the rhizosphere of a siderophore, pseudobactin (Figure 9.9), that deprives pathogens of iron, thereby permitting the plant to grow better. Misaghi et al (1982) showed that *P. fluorescens* was inhibitory in culture to *Rhizoctonia solani, Sclerotinia sclerotiorum, Phymatotrichum omnivorum, Phytophthora megasperma,* and *Pythium aphanidermatum* and that this effect could be overcome by adding iron to the medium, which prevented fluorescent pigment formation. Howell and Stipanovic (1979, 1980) showed that isolates of *P. fluorescens* produced two chlorinated phenyl pyrrole antibiotics. One (pyroluteorin) decreased *Pythium* damping-off of cotton seedlings but was ineffective against *Rhizoctonia solani,* and the other (pyrrolnitrin) was effective against *R. solani* but not *P. ultimum.* Pyroluteorin is released from the cells during growth, pyrrolnitrin only after lysis of the cells. Pyrrolnitrin was also effective against *Thielaviopsis basicola, Verticillium dahliae,* and *Alternaria* sp. It lasted for up to 30 days in moist soil without loss of activity. The frost-inhibiting action of some isolates of *P. syringae* appear to operate by antibiosis or competition with the ice-nucleation active bacteria.

Mass Production: Nair and Fahy (1976) placed neutralized gamma-irradiated peat at 50–60% moisture content (wet weight) in polyethylene bags, inoculated them with *P. fluorescens* C12 and *Pseudomonas* sp. A2, and then incubated them for four to five weeks at 25°C. Commercial life of the peat inoculum was extended to four months by storing the cultures initially at 1–5°C for 12 weeks, then at 25°C for four weeks. The inoculum was mixed with the neutralized horticultural-grade peat used in casing mushroom beds. Kloepper and Schroth (1981a) devised a dry powder formulation for inoculating seed with *Pseudomonas* sp.; xanthan gum was autoclaved and added to the bacterial cell suspension (10^9 cells per milliliter). After the mixture had set for 20 minutes, dry talc was added and this mixture was then dried as a thin layer at 12°C for three to four days. A mixture containing 20% xanthan gum maintained populations of bacteria for two months at 40°C and was easily applied to seedpieces. Weller and Cook (1983) prepared suspensions of bacterial cells in 1.5% methylcellulose (10^8–10^9 cells per milliliter), which was applied directly to wheat seeds (10^7–10^8 cells per seed) and the treated seeds then dried. Populations remained near 10^7 per seed for up to five weeks at 5°C. Essentially all work to date has used solid media (e.g., solid agar media in petri plates) for mass production of the *Pseudomonas* spp. For commercialization, it will be necessary to grow the bacteria in fermenters, but tests must be conducted first to ensure that cultures produced in liquid media are as effective as those from solid media.

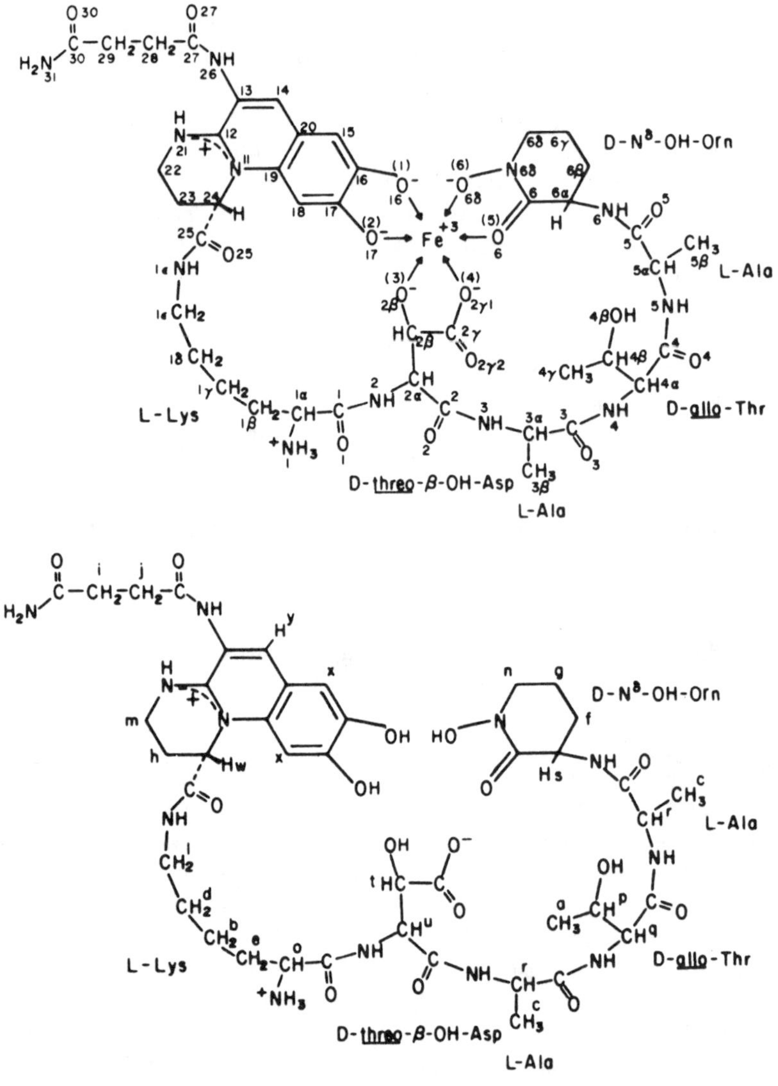

Figure 9.9. Molecular structure of (top) ferric pseudobactin, and (bottom) pseudobactin, siderophores produced by strains of fluorescent *Pseudomonas* spp. (Reprinted with permission from Teintz and Leong, 1981 and Teintz et al, 1981.)

PYTHIUM PRINGSHEIM

Morphology and Taxonomy: Mastigomycotina, Oomycetes, Pythiaceae; slime-spored. Sporangia at the ends of hyphae or intercalary, threadlike, spherical, or lemon-shaped; contents usually empty into a vesicle, where they differentiate into zoospores before release. Zoospores kidney-shaped, biflagellate, hyaline; movement uniform. Oospores single in the oogonium, spherical, thick-walled, smooth or spiny, yellow or gray; germinate by zoospores or germ tube. For details, see Waterhouse (1968), Middleton (1943), and Van der Plaats-Niterink (1981).

Pythium oligandrum Drechsler. Sporangia formed promptly and abundantly, intercalary or terminal, subspherical, 25–45 μm in diameter; evacuation tube 35 μm or more long. Zoospores usually number 20–50 in a vesicle, longitudinally grooved, reniform, biflagellate, moderately active, 9–10 μm. Oospores hyaline or yellow, subspherical, 13–30 μm in diameter, wall 0.9–2.2 μm thick with 15–125 spiny pointed protuberances 3–7 μm long (Figure 9.10).

Ecology: *Pythium* spp. are pioneer colonists of fresh (green) crop residue, responding mainly to simple sugars and other readily available nutrients that diffuse from the fresh substrate. Oospores germinate, the hyphae colonize, and new oospores form very quickly under proper temperatures if soil is suitably moist.

Role in Biological Control: Vesely̆ (1978, *in* Schippers and Gams, 1979) treated sugar beet seed with a suspension of mycelia or oospores of *P. oligandrum* before sowing; damping-off was 31% without and 9% with the treatment in nonsterile soil. *Pythium oligandrum* is parasitic on other *Pythium* spp. as well as on other fungi. Deacon (1976a) found that *Botryotrichum piluliferum, Phialophora radicicola, Fusarium nivale,* and *Psalliota* spp. were highly susceptible; *F. roseum* f. sp. *cerealis, Gaeumannomyces graminis* var. *tritici, Phialophora graminicola, Arthrobotrys superba,* and *Pythium ultimum* were moderately susceptible; *G. graminis* var. *graminis* was resistant; most Basidiomycetes and *A. musiformis* were antagonistic to *P. oligandrum.* Hyphae are susceptible when young, become resistant at maturity, and then become susceptible again. This mycoparasite may grow inside the hyphae of its host fungus or in close proximity but exterior to host hyphae, obtaining nutrients made available from the substrate by exoenzymes from the host. Oospores germinate readily in the presence of a host fungus (Vesely̆, 1978). In the San Joaquin Valley of California, *P. oligandrum* was among the principal competitors of *P. ultimum* in clay soils naturally suppressive to saprophytic growth and multiplication by *P. ultimum* (Martin and Hancock, 1981). The concentrations of soluble ions, particularly sodium and chloride, were up to five times higher in the soil suppressive to *P. ultimum* than in the conducive soil, which apparently favored *P. oligandrum* and others antagonistic to *P. ultimum. Pythium oligandrum* is more tolerant of chloride ions than is *P. ultimum* (Martin and Hancock, 1982).

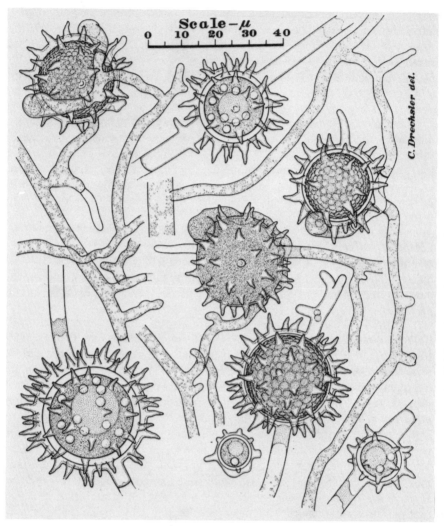

Figure 9.10. *Pythium oligandrum*, showing developing oogonia and antheridia, well-developed oospores, and small intercalary and terminal oogonia. (Reprinted with permission from Drechsler, 1946.)

Mechanism of Biological Control: Control is mainly by mycoparasitism in the case of *P. oligandrum* (Deacon, 1976b). Parasitism by *P. oligandrum* possibly is some type of hyphal interference as described by Ikediugwu et al (1970) for coprophilous basidiomycetes, since the hyphae of the parasite commonly grow contiguous to but not inside the host. Deacon (1976b) noted that hyphal coiling around the host mycelium occurs mainly if the hyphae

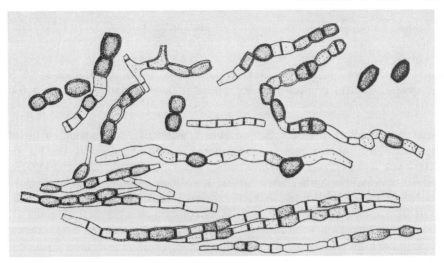

Figure 9.11. *Scytalidium lignicola*, showing hyphae with darker swollen intercalary or terminal arthroconidia. (Reprinted with permission from Ellis, 1971.)

are relatively resistant to the mycoparasite but not if the hyphae are highly susceptible.

SCYTALIDIUM PESANTE

Morphology and Taxonomy: Deuteromycotina, Hyphomycetes, Dematiaceae; Xerosporae. Hyphae smooth, hyaline to mid-brown, with occasional darker swollen cells and thick, very dark brown septa, often parallel in bundles. Stroma none; setae and hyphopodia absent. Conidiophores branched or not branched, straight or flexuous, hyaline or brown, smooth. Conidiogenous cells fragmenting and forming arthroconidia, intercalary or determinate, cylindrical or ellipsoidal. Conidia catenate, separating, simple, smooth, nonseptate, thin-walled, cylindrical or oblong; or ellipsoidal, thick-walled and mid- to dark-brown (Figure 9.11). For details, see Ellis (1971).

Scytalidium uredinicola Kuhl., Carm., and Miller. Hyphae in infected aecia of rust fungus give a bright greenish glaucous color to the aecia as they break through the bark; color fades thereafter. Septate hyphae break up into cylindrical arthroconidia, 1.5–2×2–4 µm, later swelling to form 3–4 µm ovoidal spores with thick, dilute olive walls (Kuhlman et al, 1976).

Ecology: In South Carolina, the time of maximum sporulation by *S. uredinicola* and of decline in sporulation of *C. quercuum* f. sp. *fusiforme* was April 7–14. *Scytalidium* sp. grew well in agar at pH 4–7, but scytalidin is more active in acid (pH 5 or less) than in alkaline conditions. Scytalidin applied to seeds of spruce, lettuce, or radish caused no delay in germination, even at 1,000 ppm.

Role in Biological Control: *Scytalidium uredinicola* is a mycoparasite of perennial rust cankers caused by *Endocronartium harknessii* in the Pacific Northwest and by *Cronartium quercuum* f. sp. *fusiforme* on slash and loblolly pine in the southeastern United States. It reduces the amount of aeciospore inoculum available to infect oak, the alternate host (Kuhlman et al, 1976). It is also a colonist of wood and, if established as a pioneer in poles of Douglas-fir, it prevents colonization of the poles by decay fungi. It produces a thermostable antibiotic, scytalidin (Stillwell et al, 1973). *Scytalidium uredinicola* attacks galls of *C. quercuum* f. sp. *fusiforme* on pine, suppressing aeciospore production by the rust fungus and also suppressing germination of aeciospores. In one study, 28% of the galls were parasitized, and parasitized galls produced 72% less aeciospore inoculum per unit of surface area (Kuhlman, 1981a). Ricard and Laird (1968) inoculated standing conifers with *Scytalidium* sp. by firing metal-tipped, inoculated birch dowels into the trunks with a rifle. Ricard et al (1969) drilled holes in Douglas-fir poles and pushed softwood dowels impregnated with *Scytalidium* sp. into the holes to protect the pole against decay caused by *Poria carbonica*, or inoculated dowels were shot into trees with a nail gun. Not all isolates of *Scytalidium* were effective in control of wood decay.

Mechanism of Biological Control: *Scytalidium* sp. shallowly permeated the wood of Douglas-fir power poles in Oregon and prevented decay by *Poria carbonica* by formation of scytalidin (Ricard and Bollen, 1968; Stillwell et al, 1973; Strunz et al, 1972). The antibiotic permeated the wood and, being highly stable, remained inhibitory to *Poria*. At 25°C, scytalidin is soluble in water at less than 400 ppm. Of 61 fungi tested, 52 were sensitive to scytalidin at 50 μg per disk; among those very sensitive in culture were *Peniophora gigantea, Phytophthora infestans, P. lateralis,* and *Polyporus schweinitzii* (Stillwell et al, 1973). Tsuneda et al (1980) showed that *S. uredinicola* destroyed hyphae of *Endocronartium harknessii* in wood 300 μm below the sori and disintegrated the spores and basal-cell region of active sori. On agar, spores were disintegrated without penetration, perhaps by toxins, lytic enzyme, or both.

Mass Production: *Scytalidium* spp. grow readily, but slowly, on agar media.

SPHAERELLOPSIS COOKE

Formerly placed in *Darluca* Cast. Ubiquitous rust mycoparasites.

Morphology and Taxonomy: Deuteromycotina, Coelomycetes, Dematiaceae; slime-spored. Pycnidia eustromatic, immersed but becoming erumpent, uni- or multilocular, each locule with separate simple ostiole, walls of brown thick-walled cells. Conidiophores hyaline to pale brown, septate, smooth, formed from inner cells of locular walls. Phialides indeterminate, cylindrical to dolioform, hyaline to pale brown, smooth. Conidia hyaline but later very pale brown, irregularly verrucose, 1-septate, straight, ellipsoidal, apex with gelatinous cap. For details, see Sutton (1980).

Sphaerellopsis filum (Biv.-Bern.: Fr.) Sutton. *Perfect state, Eudarluca caricis* (Fr.) Eriksson. Pycnidia up to 200 μm in diameter. Conidiophores 7.5–20×5–7 μm. Conidia 17–20×5.5–6.5 μm. Occurs on a wide range of rust fungi (Keener, 1934) with some evidence of specialization.

Role in Biological Control: Swendsrud and Calpouzos (1972) found that infection of sori of *Puccinia recondita* on wheat by *S. filum* varied from 26 to 70%, depending on the length of the postinoculation mist period. When *S. filum* spores were applied three days before inoculation with the rust fungus, no control was obtained. Swendsrud and Calpouzos (1970) found that an unidentified substance from urediospores of *P. recondita* increased the viability and germination of *S. filum*. Six other fungi did not enhance germination of this mycoparasite. Kuhlman et al (1978) found that 93% of telial sori of *Cronartium strobilinum* on oak in Florida were infected by *S. filum;* only 0.8% of the sori produced telia in this pure stand of oak, but 32% were infected and 26% produced sori in a sparse stand. Rust was controlled in the pure oak stand but not in the sparse stand. On *C. quercuum* f. sp. *fusiforme* on oak, *S. filum* produced some infection with only four hours of moist incubation, but 16–24 hours gave maximum infection. Pycnidia of *S. filum* formed four days after inoculation. The potential for biocontrol of rust on oak was considered greater for *C. strobilinum* than for *C. quercuum* f. sp. *fusiforme*.

SPORIDESMIUM LINK: FR.

Morphology and Taxonomy: Deuteromycotina, Hyphomycetes, Dematiaceae; Xerosporae. Conidiophores single or fasciculate, brown or dark brown, often with successive terminal proliferations. Conidia solitary as blown-out ends at the apices of conidiophores; simple, straight, curved, or sigmoid, cylindri-

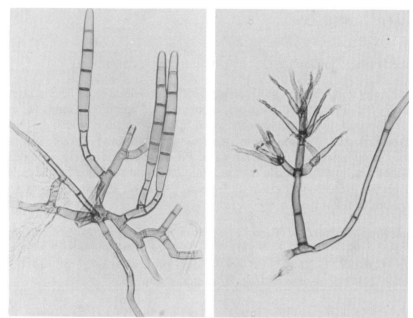

Figure 9.12. *Sporidesmium sclerotivorum*. Conidiophore (left) with mature conidia and an empty conidiogenous cell left when first conidium was released. *Selenosporella* state (right) of *Sporidesmium sclerotivorum*, showing multiple whorls of conidiogenous cells; lower whorl branched. (Reprinted with permission from Uecker et al, 1978.)

cal, fusiform, obclavate; straw-colored to dark brown; smooth or verrucose; transversely septate or pseudoseptate (Ellis, 1971).

Sporidesmium sclerotivorum Uecker, Ayers, and Adams. Colonies form loose or thick mats of conidia and mycelia on sclerotia of *Sclerotinia sclerotiorum;* aerial mycelium pale to mid-brown, immersed mycelium contorted and dark brown to black. Conidiophores simple or branched, erect, superficial, smooth, pale to dark brown; bearing terminal branches of 2–3 cells 14–25×4–5 μm, apical cells conidiogenous. Produces two asexual spore forms: 1) single conidia that develop as blown-out tips of short conidiogenous cells, light to mid-brown, smooth, tapered gently toward base and apex, 5–7 septate, 60–92×6–8 μm. 2) a *Selenosporella* state that develops on the same or separate hyphae. Conidiophores of the *Selenosporella* state are simple or branched, 25–100×4–9 μm; conidiogenous cells 2–5 in whorls, 12–35×4 μm; conidia hyaline, banana-shaped to fusiform or cylindric, 2–8×0.9–1.0 μm (Figure 9.12). Chlamydoɔpores formed in aging macroconidia; microsclerotia sometimes formed (Uecker et al, 1978).

Ecology: *Sporidesmium sclerotivorum* is active as a mycoparasite on non-melanized cells of sclerotia of *Sclerotinia sclerotiorum, S. trifoliorum, S.*

minor, Sclerotium cepivorum, and *Botrytis cinerea,* but not *Macrophomina phaseolina, Sclerotium rolfsii,* or *Rhizoctonia solani* (Ayers and Adams, *in* Papavizas, 1981). It is active as a mycoparasite of sclerotia in soil at 15–25°C, over a pH range of 5.5–7.5, and at soil water potentials of −8 bars or wetter. Sclerotia are destroyed maximally at 25°C, soil pH 6.3, and −3 bars soil water potential. Macroconidia germinate in three days in soil near sclerotia but do not germinate in the absence of sclerotia. The mycoparasite is widely distributed in the United States and probably elsewhere.

Role in Biological Control: This resident mycoparasite in soil may be responsible for widespread natural destruction of sclerotia of susceptible fungi. Adams and Ayers (1981, 1982) showed that *S. sclerotivorum* added to field soil at 100 and 1,000 conidia per gram killed 75–96% of sclerotia of *Sclerotinia minor* and gave a 40–83% control of lettuce drop in the next two years. The unprotected checks were 24–66% diseased. *Sporidesmium sclerotivorum* grown on sclerotia in a sand culture and added to soil infested with sclerotia of *Sclerotinia minor* (1% w/w) infected 100% of the sclerotia in five to six weeks.

Mechanism of Biological Control: Antagonist invades the sclerotium directly, penetrating rind cells intracellularly and then spreading through the sclerotium, causing it to disintegrate. Threadlike branching hyphae spread 1 cm or more from sclerotium to sclerotium in soil.

Mass Production: *Sporidesmium* cannot be cultured readily on synthetic media unless nutrients from the host sclerotia are added (Ayers and Adams, *in* Papavizas, 1981). It can be grown readily on sclerotia of *Sclerotinia minor* on quartz sand at 24–26°C; condia can be harvested in 6–12 weeks for application to soil.

STREPTOMYCES WAKSMAN AND HENRICI

Worldwide; common soil bacteria; some plant and animal pathogens but mostly saprophytes; producers of potent antibiotics.

Morphology and Taxonomy: Eubacteria, Actinomycetales; dry-spored. Prokaryotic microorganisms with filamentous cells, usually producing a characteristic mycelium 0.7–0.8 μm in diameter. Spores formed by fragmentation of straight or spiral-shaped aerial hyphae, becoming long chains of spores (usually >20) at the ends of branches. *Streptomyces* grows readily but slowly on synthetic media; surface of colonies velvety and covered with aerial mycelium. Substrate mycelium nonseptate, not fragmenting; forming hard, densely textured, raised colonies. Spores oblong, oval or spherical; smooth, hairy or spiny; nonseptate (Waksman, 1967). Cell walls contain LL-2,6-diaminopimelic acid, glycine, glutamic acid, alanine, but little or no pentose. Majority of antibiotic-producing actinomycetes are in *Streptomyces*. Aerobic, mesophilic, nonacidfast, Gram-positive. For details see Buchanan and Gibbons (1974).

Streptomyces griseus (Krainsky) Waksman and Henrici. Sporophores straight, flexuous, or fascicled, never in spirals, often in tufts. Aerial mycelium water green; substrate mycelium hyaline to yellow to olive buff. Produces streptomycin and cycloheximide; some strains produce no antibiotics. Melanin negative. Colonies gray in color.

Streptomyces praecox (Mill. and Burr.) Waksman. Aerial mycelium well developed, gray with greenish tinge; open spirals. Spores spherical or ellipsoidal, 0.8 μm in diameter. Very strong odor. Grows well at 37.5°C. Saprophytic species; shown in the classical studies of Millard and Taylor (1927) to effectively compete with *S. scabies* when green manure was added to soil.

Streptomyces lavendulae (Waks. and Curt.) Waksman and Henrici. Aerial mycelium cottony, red-gray to lavender-cinnamon. Spores smooth, in spirals at the ends of long straight sporophores, sometimes open, sometimes compact. Melanoid pigment produced. Produces streptothricin.

Ecology: *Streptomyces* favored by alkaline and neutral soils; pH 7.0–8.0 is optimum; 4.8–5.0 is minimum. Optimum temperature is 25–37°C. Organic matter is favorable for actinomycetes. *Streptomyces* are active in dry soils, being able to grow at soil water potentials down to −60 bars, possibly drier. Spores and mycelia become airborne. Common in the rhizosphere. Ebben and Spencer (1978) found that *S. griseus* populations could not be maintained in peat composts.

Role in Biological Control: *Streptomyces* are potentially very effective antagonists of plant-pathogenic fungi, especially in environments too dry for other bacteria. Yin and associates (1957, 1965) in the People's Republic of China selected *Streptomyces* sp. 5406 from among several thousand soil and rhizosphere microorganisms tested in vitro. When the strain was mass produced and introduced with cotton at planting, marked improvement in stands and plant vigor were observed. This strain has been used on an estimated 6 million hectares of cotton over the past 30 years (Cook, *in* Papivazas, 1981). Merriman et al (1974) applied a strain of *S. griseus* to wheat and carrot seed in a water suspension or pelleted with bentonite-sand in a bacterization experiment, but the benefits were not significant, and yields of the crops were less than in comparable tests with *Bacillus subtilis* A13. On the other hand, tests with *S. griseus* on oats increased dry weight of tops by 25%. Ebben and Spencer (1978) found that *S. griseus* would grow well in mushroom compost, where it provided some control of *Phomopsis sclerotioides* on cucumber, but other actinomycetes normally present in the decomposing hay bales used to grow cucumbers were equally or more effective. *Streptomyces griseus* might be used to fortify effects of the native actinomycetes. Tschudi and Kern (*in* Schippers and Gams, 1979) found that *S. lavendulae* lysed living mycelium of *Gaeumannomyces graminis* vars. *tritici* and *avenae*, as well as that of several *Fusarium* spp.; activity was inversely related to melanization of the host. The activity was thought to be specific and enzymatic in nature.

Because *Streptomyces* spp. are better able to tolerate dry soil than are *Pseudomonas* and *Bacillus* spp. and are potent antibiotic producers, they should be more actively studied for use in biological control and bacterization, either alone or in combination with other prokaryotic microorganisms.

Mass Production: Techniques have been developed for large-scale production of pure cultures of microorganisms, and actinomycetes have been largely involved in this program. Some 3,034 metric tons of commercial antibiotics were produced in the United States in 1963 (Waksman, 1967). Should the need arise for mass production of inoculum of *Streptomyces* spp., there should be no developmental problem. Strain 5406 of Yin and associates is grown on a solid substrate consisting of cotton seed cake, soybean meal, and other materials (S. Y. Yin, personal communication).

TAGETES LINNAEUS

Morphology and Taxonomy: Spermatophyta, Angiospermae, Compositae. Annual, branching, erect or diffuse, glabrous, strongly scented plants with oil glands on leaves and involucres. Leaves opposite, pinnately dissected. Flower heads radiate, solitary or clustered; involucral bracts in one series, united nearly throughout into a long tube. Achenes with pappus of 3–6 unequal scales or bristles. Much interspecific hybridization in commercial ornamental varieties. For details see Bailey and Bailey (1976).

Tagetes erecta Linnaeus. Stout, erect, branching annuals about 60 cm high. Leaves pinnately dissected with large glands near the margins. Heads yellow or orange, 5–10 cm across, the many rays long-clawed, the stout peduncle swollen just below the head; involucre campanulate; pappus with 1–2 long-awned scales and 2–3 shorter blunt ones. African marigold.

Tagetes patula Linnaeus. Bushy annual, 39–45 cm high. Leaves pinnately divided, teeth of segments with large gland at the base. Heads 4 cm across, rays numerous, yellow with red markings; involucre thick, oblong, with acute teeth, glandular-dotted; pappus of 1–2 long-awned scales and 2–3 shorter blunt ones. French marigold.

Role in Biological Control: *Tagetes* spp. produce terthienyls that inhibit nematodes and reduce their population. They are inhibitory to *Pratylenchus, Haplolaimus,* and *Tylenchorhyncus* spp. but not to *Globodera rostochiensis.* Terthienyls are released from growing roots, even without their decay, but benefits require three to four months to become evident. *Meloidogyne* larvae that penetrate roots of *Tagetes* do not develop beyond the second larval stage. There is some evidence that a-terthienyl is inhibitory to some plant-pathogenic fungi (Baker and Cook, 1974).

TRICHODERMA PERSOON: FR.

Members of this group have been a source of much confusion since the erection of the genus by Persoon. Of the original four species, only one (*Trichoderma viride*) is now in the genus. Harz first clearly delimited the genus in 1871. Oudemans and Koning in 1902 first reported it in soil, where it is now known to be ubiquitous. In 1926, Abbott considered that there were four species in the genus, and this was accepted by Gilman in 1957. Bisby (1939) confirmed the connection between the pyrenomycete *Hypocrea rufa* and *T. viride,* and considered that the conidial state of *H. gelatinosa* was a *Trichoderma.* He found no reliable distinguishing characters among *Trichoderma* isolates and therefore considered it to be a monotypic genus. Most workers from 1939 to 1970 followed Bisby, applying *T. viride* to any green-spored *Trichoderma.*

Webster and Lomas (1964) showed that the *T. viride* reported by Weindling and Emerson (1936) to produce gliotoxin and by Brian (1944) and Brian and McGowan (1945) to produce gliotoxin and viridin actually was *Gliocladium virens (perfect state, H. gelatinosa). Trichoderma* spp. are more common in soil than *Gliocladium* spp. but are not considered to produce antibiotics; rather, they are effective mycoparasites.

Rifai (1969) revised the genus *Trichoderma* to include nine species aggregates, and this has now generally been accepted. The production of useful new biotypes of *T. harzianum* by Papavizas and Lewis (1981), Papavizas et al (1982), and Abd-El-Moity et al (1982) by ultraviolet radiation may further complicate classification problems of these mycoparasites.

Morphology and Taxonomy: Deuteromycotina, Hyphomycetes, Moniliaceae; Gloiosporae. Conidiophores erect or straggling, highly ramified, more or less conical, weakly or strongly verticillate, bearing divergent flask- or ninepin-shaped phialides singly or in clusters, from which subglobose to ellipsoidal, slimy, nonseptate phialospores are borne, often gathering in balls at the openings of the phialides. Chlamydospores commonly formed, intercalary or rarely terminal, globose to ellipsoidal, hyaline, smooth-walled. Colonies in culture usually grow rapidly, floccose, tufted, white to green. For details, see Rifai (1969).

Trichoderma viride Pers.: Fr. *Perfect state, Hypocrea rufa* (Pers.: Fr.) Fr. Type species of the genus. Widespread in soil. Many references to *T. lignorum* are probably to *T. viride*, and references to *T. viride* actually may be to other species. Conidiophores terminate in phialides. Phialospores are rough-walled under oil-immersion lens, green, subglobose to short obovoid, 2.8–5.0×2.8–4.5 μm (Figure 9.13). Colonies fast-growing on malt agar (cover 9-cm petri dish in four days at 20°C); have a coconut odor when old.

Trichoderma hamatum (Bon.) Bainier. Widespread in soil. Conidiophores terminate in sterile elongations. Phialospores smooth, green, often more than 4–5×2.5–3 μm, obovoidal to subcylindrical (Figure 9.14). Colonies fast-growing on malt agar (cover 9-cm petri dish in five days at 20°C) and bear compact areas or tufts of conidiophores.

Trichoderma harzianum Rifai. Common in soil. Conidiophores terminate in phialides. Phialospores smooth, green, subglobose to short obovoid, 2.4–3.2×2.2–2.8 μm (Figure 9.15). Colonies fast-growing.

Trichoderma polysporum (Link: Fr.) Rifai. Common in soil. Conidiophores terminate in flexuous sterile elongations. Phialospores smooth, oblong ellipsoidal, hyaline, 2.4–3.8×1.8–2.2 μm (Figure 9.16). Colonies slow-growing and have white conidial areas scattered over the surface.

Trichoderma koningii Oud. Common in soil. Conidiophores terminate in phialides. Phialospores smooth, green, elliptic-subcylindrical, 3–4.8×1.9–2.8 μm (Figure 9.17). Colonies fast growing.

Ecology: Species of this genus occur in soil worldwide and are reported to be very efficient mycoparasites on a wide range of plant pathogens, including *Armillaria mellea, Pythium* spp., *Phytophthora* spp., *Rhizoctonia solani, Chondrostereum purpureum, Sclerotium rolfsii,* and *Heterobasidion annosum. Trichoderma viride* is an active antagonist in moist soil but is inhibited under very wet conditions when in soil at pH 5.4 or above (Anderson, 1962–1964). Chet and Baker (1980) found that *T. harzianum* was most active as an antagonist in soil at pH 6.5 or lower. The efficiency of *T. harzianum* in control of *S. rolfsii* on bean decreased as the soil temperature rose above 22°C

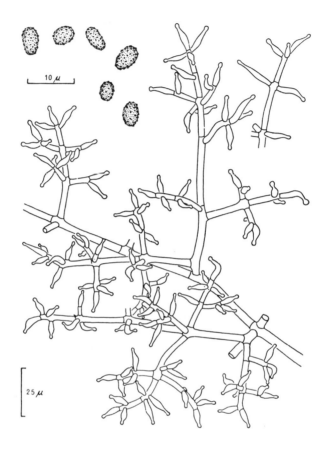

Figure 9.13. *Trichoderma viride*, showing conidiophores and rough-walled phialospores. (Reprinted with permission from Rifai, 1969.)

(Elad et al, 1980a). Liu and Baker (1980) found that the suppressiveness that developed to *R. solani* in soils during monoculture of radish, and which was associated with the activity of *T. harzianum*, developed more rapidly in acid (pH 5) than in alkaline (pH 7–9) soils and lasted longer in moist soil (−1.35 bars) than in dry soil (−87 bars matric potential). Liu and Baker (1980) concluded that "*Trichoderma* spp. commonly inhabit soils having high moisture content," although strains probably grow down to −80 bars (Baker and Cook, 1974).

Because of the extremely rapid growth and copious production of spores, *Trichoderma* spp. rapidly colonize substrata in soil after chemical or thermal treatment of the soil. Steamed and fumigated soil quickly becomes colonized

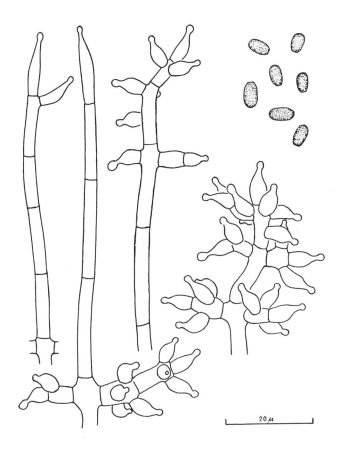

Figure 9.14. *Trichoderma hamatum*, showing conidiophores with reduced or modified sterile hyphal elongation, phialides, and phialospores. (Reprinted with permission from Rifai, 1969.)

by the fungus. Soil treated with carbon disulfide to control *Armillaria mellea* develops a high *Trichoderma* population; *Trichoderma* is only moderately resistant to carbon disulfide (Saksena, 1960) but its growth is so rapid that it quickly outgrows other fungi. Davet et al (1981) found that growth of *T. harzianum* in soil was depressed by benomyl and stimulated by thiram. The fungus is tolerant to metalaxyl and to ammonia; *T. hamatum* is sensitive to ammonia (Schippers et al, 1982).

Papavizas (1982) found that *T. harzianum* survived in soil without a food base for at least 130 days but did not survive well or increase in the rhizosphere of bean or pea seedlings. It may be necessary to ob-

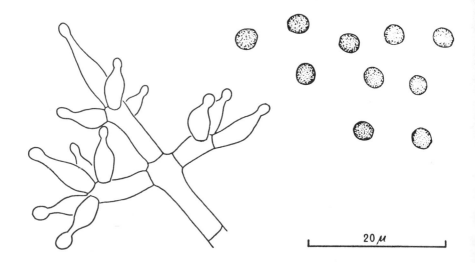

Figure 9.15. *Trichoderma harzianum*, showing phialides and phialospores. (Reprinted with permission from Rifai, 1969.)

tain isolates from the rhizosphere to find strains adapted to the rhizosphere.

Role in Biological Control: Perhaps the earliest use of *Trichoderma* in biological control was inadvertent — the control of *A. mellea* in California by soil treatment with carbon disulfide (Horne, 1914), later found (Bliss, 1951) to be due to enhanced mycoparasitism of the pathogen by *T. viride*. Perhaps because phytopathologists emphasize fungi in disease etiology, they naturally turned first to fungi for biocontrol and found one of the most prevalent soil fungi involved in many cases. *Trichoderma* spp. therefore have been studied by many workers for biological control.

Weindling (1932) and Weindling and Fawcett (1936) acidified citrus seedbeds to enhance the effect of *T. viride* in controlling damping-off of seedlings caused by *Rhizoctonia solani*. Marshall (1982) found in glasshouse tests that damping-off of snap beans by *R. solani* in soil of pH 3.5 was 65% less when seed was inoculated with *T. harzianum*. Elad et al (1982a) showed that coating cotton seed with *T. hamatum* or *T. harzianum* decreased damping-off in field trials. Henis et al (1978) and Chet and Baker (1980) added spores of *T. harzianum* to soil to increase its suppressiveness to *R. solani*. A soil naturally suppressive to *R. solani* near Bogotá, Colombia, was found by Chet and Baker (1980) to have *T. hamatum;* this fungus transferred to a conducive Colorado soil made the conducive soil suppressive to *R. solani*. Harman et al (1980, 1981) treated pea and radish seeds with spores

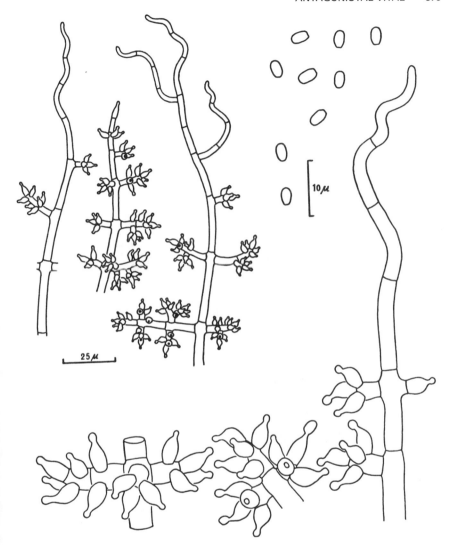

Figure 9.16. *Trichoderma polysporum*, showing conidiophores with and without sterile hyphal elongations, phialides, and phialospores. (Reprinted with permission from Rifai, 1969.)

of *T. hamatum*, protecting the seedlings from infection by *Pythium* spp. and *R. solani*. However, the treatment for peas was ineffective in soils from New York because of interference from siderophore-producing fluorescent pseudomonads.

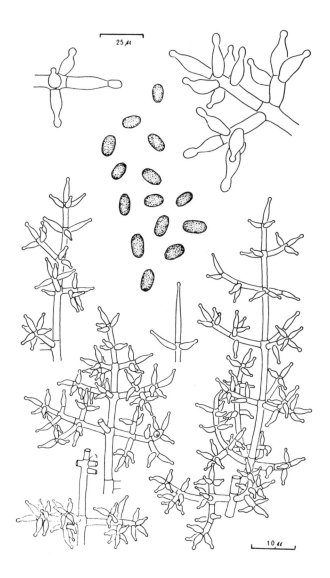

Figure 9.17. *Trichoderma koningii*, showing conidiophores, phialides, and phialo-spores. (Reprinted with permission from Rifai, 1969.)

Wells et al (1972) applied *T. harzianum* along rows of tomato seedlings; the fungus rotted the sclerotia and partially protected the tomatoes from *Sclerotium rolfsii*.

Grosclaude et al (1973) inoculated plum trees in the field with a spore suspension of *T. viride* at the time of pruning, using special inoculating shears.

When the pruning wounds were inoculated two days later with *Chondroster-eum purpureum*, the trees were protected from silver-leaf disease. Ricard (1970) found that *T. polysporum* and *Ascocoryne sarcoides*, which occurred naturally in spruce trees, prevented decay of the trees by *Heterobasidion annosum*. Pottle et al (1977) and Smith et al (1979) inoculated wounds of red maple with *T. harzianum* to protect against infections by wood-rotting Hymenomycetes.

Huang (1980) obtained a decrease from 1,514 sclerotia to 843 per 60 m of row when *T. viride* was added to the seed furrow in sunflower field plots infested with *Sclerotinia sclerotiorum*, but *Coniothyrium minitans* reduced the number of sclerotia to 36 per 60 m of row. Locke et al (1982) applied *T. viride* to soil to control *Fusarium oxysporum* f. sp. *chrysan-themi*.

Elad et al (1980a) used *T. harzianum* in field plots naturally infested with *Sclerotium rolfsii* and *Rhizoctonia solani*, reducing disease caused by the two pathogens on beans from 25 to 10% and from 13 to 7.5%, respectively. Punja et al (1982) failed to obtain control of *S. rolfsii* in turf in California with *Trichoderma* spp., although the antagonists apparently established in the soil.

Mechanism of Biological Control: *Trichoderma* spp. control mainly by being mycoparasitic and aggressive competitors with pathogens. Growth of mycelia of *Trichoderma* spp. along and coiled around hyphae of host fungi has been observed by many workers (e.g., Chet et al, 1981; Liu and Baker, 1980; Weindling, 1932). Penetration of host mycelia may or may not occur, but susceptible hyphae become vacuolated, collapse, and finally disintegrate. The mycoparasite then grows on the hyphal contents.

Much of the early literature attributed the biological control by *T. viride* to production of the antibiotics gliotoxin and viridin, but the fungus is now con-sidered to have often been incorrectly identified and probably was *Gliocladium virens* (Webster and Lomas, 1964). However, some *Trichoderma* isolates do produce antibiotics, especially at low pH (Dennis and Webster, 1971a, 1971b).

Chet and Baker (1980) showed that *T. harzianum* and *T. hamatum*, act-ing as mycoparasites of *Rhizoctonia solani* and *Sclerotium rolfsii*, produced β-(1-3)-glucanase and chitinase that caused exolysis of the host hyphae; an-tibiosis was not observed. The optimal pH levels for these enzymes are 4.5 and 5.3, respectively. *Trichoderma hamatum* also produces cellulase, which perhaps explains its ability to parasitize *Pythium* spp. (Chet and Baker, 1981). *Trichoderma harzianum* is not cellulolytic. Jones and Watson (1969) found that β-(1-3)-glucanase produced by *T. viride* solubilized mycelia of *Sclerotinia sclerotiorum*.

Mass Production: *Trichoderma harzianum* and *T. polysporum* have been produced as pellets (BINAB T SEPPIC) (Ricard, 1981), in wood dowels (Corke and Rishbeth, *in* Burges, 1981), and as other formulations (Papavizas and Lewis, *in* Papavizas, 1981). The solid substrate method (Toyama et al, 1970) is used in preference to the submerged liquid fermentation method (Templeton et al, 1980) for reasons outlined by Underköfler (1976). Material sufficient to protect 20,000 trees against silver-leaf disease had been sold in Europe by 1981. *Trichoderma viride* was grown by Huang (1980) on autoclaved barley, rye, and sunflower seeds at room temperature for four to six weeks; it was applied to soil when moist or was air-dried for two to four weeks. Backman and Rodriguez-Kabana (1975) used molasses-enriched clay granules as a food base for *T. harzianum*, which they applied at 140 kg/ha; this material has low bulk, no residue, and can be applied with standard agricultural granule machinery. Kelley (1976) used the same mixture and species to inoculate pine seedbeds in an unsuccessful attempt to control *Phytophthora cinnamomi*. Elad et al (1980b) grew *T. harzianum* on an autoclaved 3:1:4 mixture of wheat bran:sawdust:tapwater for inoculation of field plots.

TRICHOTHECIUM LINK: FR.

Worldwide saprophytes found on decaying vegetable matter and in soil.

Morphology and Taxonomy: Deuteromycotina, Hyphomycetes, Moniliaceae; Xerosporae. Conidiophores single or in loose groups, erect, straight or flexuous, simple or branched, hyaline, septate, with tips bearing conidia basipetally in chains. Conidia arise as blown-out ends of conidiophores, forming a fragile chainlike cluster; 1-septate, hyaline (tinted yellow or pink in mass), smooth-walled, oblong-ellipsod, rounded distally and with an obconical basal cell, curved and with a well-marked truncate attachment point (Rifai and Cooke, 1966).

Trichothecium roseum (Pers.) Lk.: Gray. Conidia distinctly pink in mass, 16–20×9–11.8 μm, with characteristic "ear." Conidiophore simple, erect, 100–250 μm, 2–3 septate, rough with minute granules, forming spores in a two-ranked chain of 20–30 spores (Figure 9.18), each borne horizontally (Ingold, 1956).

Ecology: *Trichothecium roseum* is ubiquitous on apple trees and often infects scab (*Venturia inaequalis*) lesions on fruit.

Role in Biological Control: Corke and Hunter (1979) found that inoculating pruning cuts of apple with *T. roseum* prevented infection of 54% of the

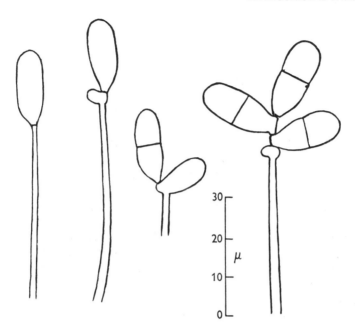

Figure 9.18. *Trichothecium roseum*, showing conidiophores and conidia in two-ranked chains, each borne horizontally. (Reprinted with permission from Ingold, 1956.)

wounds, curtailed lesion development by 25%, and suppressed production of spores of *Nectria galligena* by 70%. Because *Trichothecium* causes pink rot of apple fruit, its use in biocontrol is unlikely.

TUBERCULINA TODE: SACCARDO

Morphology and Taxonomy: Deuteromycotina, Coelomycetes, Tuberculariaceae; Xerosporae. Sporodochia hard, glabrous. Conidiophores simple. Conidia not capitate, acrogenous, without mucus, catenate.

Tuberculina maxima Rostr. Conidiophores large, globose, lilac-colored. Conidia 8.4–13.2×12.1–20.9 μm, lilac-colored, produced singly at the ends of sporogenous hyphae. Spores from sporodochia are windborne.

Ecology: *Tuberculina* invades cankers and galls caused by *Cronartium coleosporioides, C. pyriforme, C. cerebrum,* and *Uredinopsis mirabilis* in the United States (Hubert, 1935). Wicker (1981) reported it on *C. comptoniae, C. commandrae, C. ribicola,* and *Endocronartium harknessii.* This fungus has also been observed to infect galls produced by *C. quercuum* f. sp. *fusiforme* on southern pine (Kuhlman, 1981b). Freezing for 19 months does not destroy spore viability. Spores germinate at 5°C; optimum temperature range is 20–25°C, and thermal death point is 25–30°C. Fungus grows on media at pH 3.0–8.0, with an optimum at pH 5.2.

Role in Biological Control: For many years *Tuberculina maxima* was known to attack the perennial aecial cankers of white pine blister rust caused by *C. ribicola* in the western United States, but it was thought to be unimportant in control. R. T. Bingham and J. Ehrlich in 1941 and J. W. Kimmey in 1965 observed an apparent increase in *Tuberculina*-infected cankers from 1935 to 1965 (Kimmey, 1969). Extensive treatment of cankers with the antibiotics phytoactin and cycloheximide in the late 1950s were thought (Wicker, 1968) to "have been detrimental to blister rust control" because of toxicity to *T. maxima* (Baker and Cook, 1974). *Tuberculina maxima* infects sporulating pycnia and aecia of *Cronartium* and sporulates there about two weeks later. It only invades tissue already penetrated by rust, and it dies when the rust fungus is dead. *Tuberculina* overwinters as mycelium in active rust cankers and as conidia and produces spores whenever cankers are erumpent (Wicker and Woo, 1973). Hungerford (1977) examined cankers in the Pacific Northwest from 1966 through 1972; 19% of the lethal-type branch cankers were inactive in 1966, but 56% were inactive in 1972, an increase of 6% per year. Of inactive lethal cankers in 1966, 78.4% were still inactive in 1972. The antagonist has not been observed on the uredial or telial states of *C. ribicola*. Wicker (1981) concluded that *Tuberculina* "has delayed blister rust damage and prolonged the life of infected trees, but it has not controlled the disease." In the Southeast, galls of *C. quercuum* f. sp. *fusiforme* infected by *T. maxima* one spring produced fewer aecia the following spring (Kuhlman, 1981b).

Mechanism of Biological Control: *Tuberculina maxima* infects through erumpent cankers of *C. ribicola*, grows into the pycnia and aecia, and kills the mycelium. It apparently ramifies through cankered pine tissue but not through healthy bark tissue. The cells of infected tissue are rapidly destroyed, "enzymatically destroying the food source vital to the survival of the obligate parasite" (Wicker, 1981).

Mass Production: *Tuberculina maxima* can be readily grown on agar media.

VERTICILLIUM NEES VON ESENBECK: FR.

In addition to including important plant pathogens *(V. albo-atrum* and *V. dahliae)*, this genus includes important hyperparasites of nematodes, insects, and fungi.

Morphology and Taxonomy: Deuteromycotina, Hyphomycetes, Dematiaceae; Gloiosporae. Conidiophores erect, septate, hyaline or pigmented, simple or branched, usually in primary, secondary, or higher order verticils or phialides. Phialospores hyaline or subhyaline, nonseptate, produced in slimy balls at the apices of phialides, ovoid or short-cylindric, sometimes flattened on one side or even allantoid.

Verticillium chlamydosporium Goddard. Has an aleuriospore state, *Diheterospora chlamydosporia* Goddard (Barron and Onions, 1966). Conidiophores erect, verticillately branched, tapering to a knobbed end that bears a single spore; phialides cylindrical, 15–30 μm long, in whorls of 3–5 (Figure 9.19). Phialospores hyaline, 2–5×1.5–2 μm in diameter, ovoid to short-cylindric, smooth. Aleuriospores very common, 4–9 celled, thick-walled, globular, 15–30×10–20 μm, muriform, slightly lobed, with granular contents, borne terminally on short side branches 15–30 μm long, very persistent. Older mycelium completely converted to aleuriospores, becoming a creamy or ochraceous powder. An important egg parasite of cyst nematodes.

Verticillium lecanii (Zimmermann) Viegas. A heterogenous species aggregate. Phialides solitary or few in whorls of 3–4 on erect conidiophores or arising from slightly differentiated prostrate aerial hyphae, aculeate, 12–40×0.8–3.0 μm. Conidia in heads or parallel bundles, cylindrical or ellipsoidal, 2.3–10×1.0–2.6 μm. Chlamydospores absent. Perfect state unknown. (For more information, see Domsch et al, 1981; CMI Descriptions of Pathogenic Fungi and Bacteria No. 610.)

Verticillium biguttatum Gams. Conidiophores erect, arising from submerged hyphae, 150–280 μm long, 1–4 septate, with several whorls of 3–5 phialides near apex. Phialides 18–30 μm long, tapering, appearing roughened in air but smooth in water. Conidia regularly cylindrical, straight, smooth, 5.5–9.5×1.5–2.2 μm, with two conspicuous guttules (Figure 9.20). Chlamydospores and perfect state absent (Gams and Van Zaayen, 1982).

Verticillium nigrescens Pethybridge. Conidiophores erect, hyaline, bearing 1–2 nodes of 1–3 verticillate phialides. Phialides 20–50×1.5–3.0 μm, subulate. Conidia single, hyaline, ellipsoidal to short-cylindrical, simple or rarely 1-septate, 4.0–8.5×1.5–2.5 μm. Chlamydospores abundant, terminal or intercalary, single or in chains of 2–5, olivaceous brown, globose or pyriform, 5.5–10 μm in diameter. Microsclerotia absent. (For more information, see Domsch et al, 1981; CMI Descriptions of Pathogenic Fungi and Bacteria No. 257).

Ecology: *Verticillium chlamydosporium* is less host-specific for *Heterodera avenae* than is *Nematophthora gynophila;* it also attacks females and eggs

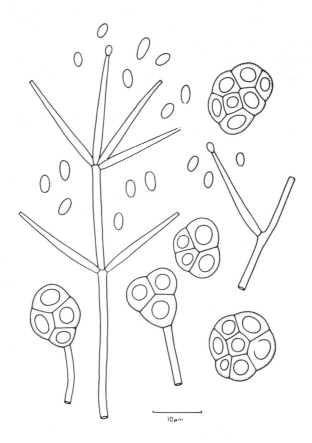

Figure 9.19. *Verticillium chlamydosporium*, showing conidiophores, conidia, and aleuriospores. (Reprinted with permission from Morgan-Jones et al, 1981.)

of *Meloidogyne arenaria* (Morgan-Jones et al, 1981), *Ascaris*, and a snail. Activity of *V. chlamydosporium* is favored by moist soil, but the fungus is more active in drier soils than are *N. gynophila, Catenaria auxiliaris*, or an unidentified Lagenidiaceous fungus (Kerry, *in* Papavizas, 1981). Mendgen (1979) found that *V. lecanii* strongly infects *Puccinia striiformis;* optimum temperature for development was 15°C. Plant tissue was invaded only after the rusted tissue had degenerated. *Verticillium lecanii* is one of the most important and common entomogenous hyphomycetes; apparently there is no difference in pathogenicity among isolates from rusts or aphids. It can decompose cellulose, chitin, pectin, and starch and is strongly proteolytic; it has high tolerance of fluorides in solution. The temperature optimum for *V. biguttatum* is 24–27°C, and the maximum is 30°C.

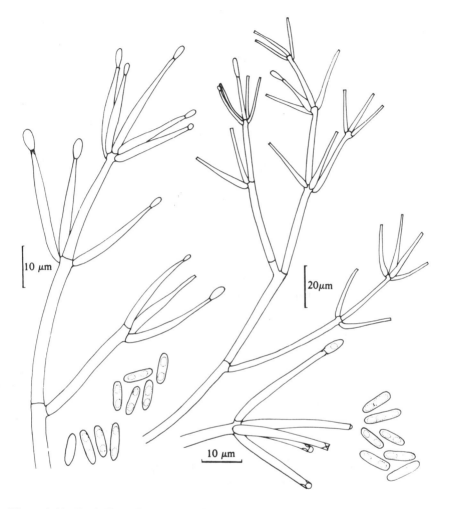

Figure 9.20. *Verticillium biguttatum*, showing conidiophores with whorls of phialides, and conidia. (Reprinted with permission from Gams and Van Zaayen, 1982.)

Role in Biological Control: *Verticillium chlamydosporium* is mainly a resident antagonist. It has been identified as an inhabitant in British soils, including soils not infested with the cereal cyst nematode, *Heterodera avenae*. It attacks mainly eggs of *H. avenae* at all stages of egg development but also attacks females (Kerry et al, 1982a, 1982b). Infected females form small cysts with few eggs.

When carnations were inoculated with a suspension of mixed spores of *Uromyces dianthi* and *V. lecanii*, less rust resulted than when they were inoculated with the rust fungus alone (Spencer and Atkey, 1981); Spencer (1980) found 50% less rust when *V. lecanii* was sprayed on the plants. Rust (*Uromyces appendiculatus*) of bean, when sprayed with spores of *V. lecanii* from rust pustules or from aphids, were "smothered . . . and urediniospores were invaded" (Allen, 1982). Jager and Velvis (1980) and G. Jager (personal communication) found that *V. biguttatum* infected a high percentage of sclerotia of *Rhizoctonia solani* on potato tubers from suppressive soils in The Netherlands. Melouk and Horner (1975) found that the weak pathogen, *V. nigrescens*, inoculated into peppermint and spearmint five to nine days before infection with *V. dahliae*, greatly decreased wilt symptoms and reduced the number of propagules in the plant to 5.0–6.6% of the level of the controls. Skotland (1971) found that *V. nigrescens* from pea was only slightly pathogenic on peppermint, eggplant, cantaloupe, and tomato. Schnathorst and Mathre (1966) used a mild strain of *V. albo-atrum* for cross protection of cotton against virulent strains of the pathogen.

Mechanism of Biological Control: *Verticillium chlamydosporium* infects eggs in egg masses of cyst nematodes, killing them, but is not a specific parasite. Mycelium of *V. lecanii* spreads among the urediospores deep within the rust sori and sometimes invades the spores (Spencer and Atkey, 1981) but is not host specific. In water droplets containing *V. lecanii*, germination of *Uromyces dianthi* urediospores was not affected, but germ tubes of *U. dianthi* were shorter and those of *V. lecanii* were longer than when *U. dianthi* was grown alone. *Verticillium lecanii* was not observed to penetrate hyphae of the rust fungus. When sprayed on carnation plants two days before inoculation with *U. dianthi*, *V. lecanii* produced no effect because of the short life of its spores; but if applied from one day before to seven days after inoculation with the rust fungus, it reduced the number of pustules from an average of 107.2 to 41.6. Hänssler and Hermanns (1981) showed that, within 48 hours, *V. lecanii* grew over the cysts of *Heterodera schachtii*, the sugar beet nematode, and penetrated and destroyed the cysts. The fungus also attacks aphids.

Mass Production: *Verticillium lecanii* is grown in submerged culture, then transferred to husked millet where it forms hyphae; the millet is ground in a ball mill for application.

VIRUSES AND VIRUSLIKE ENTITIES

There are several examples in which viruses are important in biological control of plant pathogens.

1. Phages of bacteria and actinomycetes are common and undoubtedly reduce the number of bacteria in nature, but science has so far been unable to effectively use them in control of plant diseases (Vidaver, 1976).

2. Mycoviruses are common in fungi but generally do little or no harm (Hollings, 1982). Some, however, can diminish virulence of the fungus or result in loss of vigor or death. The dsRNA in *Endothia parasitica* (Chapter 8) is thus controlling chestnut blight in Italy and France and has potential to provide biological control of this disease on an individual tree basis in the United States (Day and Dodds, *in* Lemke, 1979). *Rhizoctonia solani* develops sickness in culture and becomes hypovirulent when infected with one or more dsRNA mycoviruslike agents (Castanho and Butler, 1978a, 1978b; Castanho et al, 1978). The nonaggressive strains of *Ceratocystis ulmi* were found by Pusey and Wilson (1982) to contain a larger number of dsRNA infections than the aggressive strains, although both types of strains were infected.

3. Cross protection of the plant from virus damage has been provided by prior inoculation with a mild or avirulent virus (Chapter 8). Tobacco mosaic virus on tomato in The Netherlands (Rast, 1979) and New Zealand (Mossop and Procter, 1975) and tristeza virus of citrus in Brazil (Costa and Müller, 1980) and Australia (Fraser et al, 1968) are examples of successful field use of the method.

10

AGRICULTURAL PRACTICES
AND BIOLOGICAL CONTROL

*It seems characteristic of man's grappling with nature
that we often learn to manipulate a new aspect of our world
before we learn fully to understand it or even
how best to use these new powers.*
—ALLEN HAMMOND, 1980

The method of tillage, the selection of planting date or plant spacing, the choice of crop sequence, the kind or placement of fertilizer, and the intensity or timing of irrigation provide almost an infinite number of ways to modify the soil and plant environments or the plant itself. The result may be the creation of a niche for one or more antagonists, elimination of the niche of a pathogen, enhancement of host plant resistance to pathogens, or all three possibilities. Until recently, all attempts to enhance biological control of plant pathogens were achieved through cultural practices, plant breeding, or both. Farming practices and plant breeding will continue to provide important means to achieve biological control as aids to the introduction of antagonists as well as to manage resident antagonists and host plant resistance.

Biological control achieved with help from agricultural practices may be discussed in the same sequence in which a grower might use the practices. The first decision for any given field concerns which cropping system to use and, if mixed, which crops to grow and their sequence or rotation. The next decision usually concerns the method of tillage, if any, and whether to use a preplant treatment with the potential to promote biological control. The field ready for sowing, the grower may now vary the planting date, planting rate, row spacing, or method and depth of sowing to improve biological control of aerial as well as of soilborne pathogens. The cultivar or cultivar mixture is also chosen at this stage. The kind, amount, and placement of inorganic fertilizer and whether to apply it before, during, or after planting is also decided. After the crop is sown, if the field is irrigated, the grower must decide on the intensity, timing, and frequency of irrigation or whether to withhold irrigation. This sequence, which follows the cropping season, is used here as our format for discussion of the use of agricultural practices for biological control purposes.

CROPPING SYSTEM

The cropping system is the sequence or combination of crops grown in a single field or wider area. The corn-soybean rotation in the corn belt states of the United States, the double-cropping of wheat and soybeans in the same field each year in the southeastern United States, and the continuous growing of wheat in the Great Plains states are distinct cropping systems. The practice in the Great Plains of growing one wheat cultivar with one source of genetic resistence to *Puccinia graminis* f. sp. *tritici* in one area and other cultivars with different sources of resistance to this pathogen in neighboring areas to interrupt long-distance movement of given races is an example of the use of a cropping system over a wide geographic area to achieve biological control. Similarly in the People's Republic of China, different sources of resistance to *P. striiformis* are used in different regions across the North China Plain, to reduce the spread of specific races of the pathogen from the early-sown wheat in the west to later-sown wheat in the east (Kelman and Cook, 1977). The cropping systems in use around the world are mainly the result of trial and error and of the gradual recognition by farmers of the most dependable and productive sequences and combinations to meet the needs of their crops.

Crop rotation. – The trend with most major food and fiber crops (e.g., cereal grains, soybeans, cotton, and potatoes) is toward the shortest possible rotation necessary to maintain an acceptable level of productivity. More and more frequently, crops are grown with no rotation. In eastern Washington, cephalosporium stripe of winter wheat (caused by *Cephalosporium gramineum*) can be controlled by a three-year rotation, in which land is fallowed or sown to any spring crop (spring wheat, spring barley, peas, or lentils) for two consecutive years, followed by a crop of winter wheat (Bruehl, 1968). Because of the potentially greater economic return from winter wheat, growers repeatedly risk sowing winter wheat after only a one-year break. A significant portion of the winter wheat in eastern Washington was grown in this short, high-risk rotation in 1980 when winter conditions favored the disease. Losses of 50–75% were common in many fields.

Take-all of wheat caused by *Gaeumannomyces graminis* var. *tritici* can be controlled in the irrigated Columbia Basin and Snake River Plains districts of the Pacific Northwest by any of several two-year rotations, e.g., wheat-potato, wheat-corn, wheat-beans, or wheat-sugar beets. The disease is also controlled by alfalfa-wheat, provided the alfalfa stands are not infested with grass hosts of the take-all pathogen, and provided that wheat is not grown more than one year before returning to alfalfa or some other nonhost crop. Nevertheless, growers repeatedly risk a second sowing of wheat, to control verticillium wilt of potato or stem nematode (*Ditylenchus dipsaci*) of alfalfa, or for some other economic reason. In these cases, take-all becomes important. The wheat-soybean double-crop system used in the southeastern United States and in southern Brazil is highly attractive as a way to make greater use of capital but allows only four months between wheat crops, which is inadequate for take-all control.

Some crops, though not seriously damaged by a particular soilborne pathogen, nevertheless serve as better hosts for maintenance of the pathogen population than do other available crops. This can result in more disease when a fully susceptible crop is again grown. Barley is suspected to behave in this way with *G. graminis* var. *tritici*; in some areas, although usually damaged less than wheat, it can maintain the pathogen population at a level potentially very damaging to a subsequent crop of wheat. Soybeans behave this way with *Cylindrocladium crotalariae*, the cause of cylindrocladium black rot of peanuts. Soybeans become infected and support increases in microsclerotium populations of *C. crotalariae*, but growth and yield are not suppressed (Phipps and Beute, 1979). A susceptible peanut cultivar grown after soybeans can be severely damaged. Corn, cotton, and tobacco all are immune to this pathogen and hence are the preferred rotation crops with peanuts to achieve biological control of *C. crotalariae*.

Cultivar rotation. — A system of cultivar rotation for cotton has been developed in California to maximize yields in fields infested with *Verticillium dahliae* (Ashworth and Huisman, 1980). Acala SJ-4 is relatively tolerant to verticillium wilt and outyields the more susceptible SJ-2 in fields with high (9–13 microsclerotia per gram) inoculum densities of *V. dahliae*. However, SJ-2 has a higher yield potential and commonly outyields SJ-4 in fields with relatively low inoculum densities of *V. dahliae*. Fields are monitored for populations of *V. dahliae*, and SJ-4 is planted only where the inoculum density of *V. dahliae* approaches the threshold level. When the population declines to a lower density, the more susceptible but potentially higher yielding SJ-2 is again grown. This practice permits continuous cropping to cotton. It should be possible, in situations such as those described by Ashworth and Huisman (1980), to eventually combine the traits of resistance and high yield potential in the same cultivar. This has been accomplished for fusarium wilt of cotton caused by *F. oxysporum* f. sp. *vasinfectum* (Kappelman, 1980). A great deal of progress has been made in breeding cotton for resistance to verticillium wilt and several sources of resistance are now identified (Wilhelm, *in* Mace et al, 1981); it should only be a matter of time before high yield and resistance to verticillium wilt are available in the same cultivar.

A cropping system used to control soybean cyst nematode (*Heterodera glycines*) in the southeastern United States involves growing a susceptible soybean cultivar in the field after a resistant cultivar has been grown, to reduce selection pressure for new races (Riggs et al, 1980). McCann et al (1982) demonstrated for one sample of soil from a Missouri soybean field that six resistant lines of soybean each selected a population of the cyst nematode capable of parasitizing and multiplying on the roots of that line. Race 3 dominates the population when a fully susceptible cultivar is grown. A typical rotation for nematode control is grain sorghum followed by a soybean cultivar resistant to race 3, and then a soybean cultivar susceptible to race 3.

In some cases, cultivar rotation can be used to supplement crop rotation to control a soilborne pathogen. In Australia, the wheat cultivar Festiguay, planted in fields infested with the cereal cyst nematode (*H. avenae*), not only outyields other wheat cultivars in the presence of the nematode but also supports development of fewer nematode cysts (Rovira and Simon, 1982). About four years is normally required between wheat crops before the natural death of eggs of *H. avenae* lowers the population below the economic threshold. If Festiguay is grown and the field is then planted to pasture or some nonsusceptible crop, the field can be safely replanted to a nematode-susceptible wheat cultivar in two to three years. In England, populations of this nematode eventually decline below the economic threshold in fields where susceptible cereal crops are grown continuously (Kerry, *in* Papavizas, 1981). This decline in the cereal cyst nematode populations has been associated with an increase in the populations of up to four species of soil fungi parasitic on the nematode (Chapters 4 and 7). A method to grow these fungi and introduce them would greatly shorten the time required to achieve effective biological control of this nematode (Kerry, *in* Papavizas, 1981). Such a method might then open the way for biological control of other cyst nematodes on other crops.

Parasites and pathogens of the cysts and eggs of *H. glycines* occur in the Mississippi Valley and elsewhere in the United States where soybeans are grown (Gintis et al, 1982; Morgan-Jones and Rodriguez-Kabana, 1981), but these antagonists apparently are not present in sufficient numbers to provide the degree of biological control of this nematode that has been reported for *H. avenae* in England. On the other hand, plant protectionists working in Jilin Province in northeastern China, where soybeans have been cultivated for centuries, reported to the U.S. Biological Control Study Team in 1979 (personal communication) that *H. glycines* occurs but is rarely of economic importance in Jilin Province. Serious outbreaks of the nematode do occur in China, but mainly where soybeans were introduced more recently, as in Heilongjiang Province in the far northeast and in provinces farther south along the coast. Perhaps a nematode decline phenomenon has occurred with cultivation of soybeans over the centuries in Jilin Province, which, if studied, would reveal effective antagonists for biological control of this nematode in other areas.

Figure 10.1. Wheat growing in adjacent plots under irrigation at Lind, Washington. The plot in the upper photo had been planted to irrigated wheat every year from 1968 to 1977 (7 years), alfalfa from 1975 to 1977 (3 years), and wheat in 1978 (the year of the photo). The plot in the lower photo was in the 11th consecutive year of wheat. The foreground of each plot had been fumigated with methyl bromide to eliminate any suppressiveness to take-all, the back half of each plot was not fumigated, and the take-all pathogen, *Gaeumannomyces graminis* var. *tritici* then introduced as colonized oat grains into the seed furrows of the four rows on the left. The soil in plots cropped 11 consecutive years to wheat was suppressive to take-all, unless fumigated, whereas a break to alfalfa for three years resulted in soil conducive to take-all, whether or not it was fumigated (after Cook, 1981).

Crop monoculture. — The discovery that some soilborne pathogens controlled by crop rotation can also be controlled by long-term monoculture of the host offers a practical alternative to biological control through the cropping practice. Disease (or pathogen) decline with crop monoculture has been documented for take-all, cereal cyst nematode, common scab of potato, and rhizoctonia seedling blight of radish (Chapter 7). For each of these examples, the decline has been shown to begin within 5–10 years after initiation of the monoculture system, and with rhizoctonia seedling blight of radish, decline was evident already with the fourth and fifth sowing of radish in as many weeks (Figure 7.4). Disease decline beginning after a few weeks, a few years, or even 10 years is sufficiently rapid to observe and document. Perhaps other pathogens eventually decline in importance as well, but not until after several decades or a century. Obviously, a decline that requires a long time is too slow to be of much practical use. The value may come from finding even one example of disease decline, which could then be studied as a source of clues on how to initiate or accelerate the process in other fields. Detailed study of even the few examples of disease decline associated with a specific cropping system is important to understanding biological balance and the biological control that eventually may occur if the cropping system is allowed to stabilize.

After disease decline is achieved with monoculture, it becomes important to know which crops not to grow, so that the biological control is not upset. Crops not susceptible to *G. graminis* var. *tritici* will control take-all when grown in rotation with wheat, but they also may counteract the suppressiveness of soil associated with take-all decline. Thus, alfalfa, oats, potatoes, and soybeans, each grown for three years after seven years of wheat, were about as effective as soil fumigation with methyl bromide in causing the soil to become conducive to take-all (Cook, 1981). By comparison, soil in adjacent plots sown to wheat for 11 consecutive years was suppressive to the disease (Figure 10.1). In California, the rotation of potatoes with barley, cotton, or sugar beets seemed only to retard the increase in common scab of potato and therefore also to retard scab decline (Baker and Cook, 1974). In The Netherlands, *Rhizoctonia solani* was more severe when potatoes were grown only one year in three (with cereals and sugar beets in the other two years) than when grown in monoculture (Jager et al, 1979). It seems that for take-all, common scab of potatoes, and possibly certain other diseases, *either adequate rotation must be used to control the disease through diminution of inoculum, or monoculture must be strictly practiced to favor antagonists suppressive to the pathogen.* Alternating between the two kinds of cropping systems can be counterproductive to both methods of biological control.

Mixed cropping and interplanting — Mixed cropping and interplanting, ancient and still widely used cropping systems in Asia, Africa, South America,

Powdery Mildew of Barley

Biocontrol of the pathogen responsible for this disease is accomplished by planting a mixture of host cultivars, each having a different genetic resistance, thereby reducing the potential for population increase by the pathogen.

Pathogen: *Erysiphe graminis* f. sp. *hordei* (Ascomycotina, Pyrenomycetes, Erysiphales).

Hosts: Limited mainly to wild and cultivated *Hordeum* spp.

Disease: Colonies of the pathogen on the blades and sheaths of susceptible leaves have a gray, powdery appearance, owing to profuse production of conidia on the surface. The parasite absorbs photosynthetic products and other nutrients from host cells through lobed haustoria that extend into the cells. The mildew colony is a strong sink for photosynthates, and carbohydrates may be translocated toward the colony; the result is less carbohydrate for filling grains and for translocation to the roots. Plants with severe mildew may have roots shorter than those of healthy plants and may be more vulnerable to drought.

Life Cycle: The pathogen, being an obligate parasite, declines rapidly in population in the absence of a suitable host. Within a barley-growing area, the period between maturity of one crop and establishment of the next may be bridged by parasitism of wild barley present as weeds in the area or of volunteer plants of cultivated barley that commonly grow after harvest. Cleistothecia with ascospores may form in the mildew colony as the host matures and possibly serve as a means of overwintering and a source of inoculum the following spring. The pathogen sporulates rapidly and profusely on a susceptible host

and elsewhere (Harlan, 1976), give significant control of some diseases. Van Rheenen et al (1982) demonstrated in Kenya for beans grown in association with maize that the incidence of halo blight, common bean mosaic, anthracnose, common blight, scab, *Phoma*, mildew, bollworm, and angular leaf spot all were less, "sometimes remarkably less," than the incidence on beans in a uniform sowing. White mold was more severe on beans intercropped with maize, and rust and aphids were erratic. Intercropping and mixed cropping provide methods whereby one plant can be used to significantly modify the microclimate of another plant. The microclimate created may be less favorable to the pathogen, more favorable to the host or potential antagonists of the pathogen, or any combination of these. Nonsusceptible plants may also intercept or filter out inoculum of the pathogen, as is thought to occur with cultivar mixtures (Wolfe and Barrett, 1980). Nonsusceptible plants may also affect the activity of a vector of the pathogen and thereby provide biological control of inoculation. Van Rheenen et al (1982) suggested that the greater disease

with suitable temperature and moisture; a complete cycle of infection, colony development, and sporulation can be completed approximately every 7–10 days during active vegetative growth of the host. Inoculum production is of the "compound interest" type, sensu van der Plank. Infections early in the life of the plant can lead to a severe epidemic by the time of heading or shortly thereafter if the pathogen is not restricted in its rate of infection, sporulation, or both.

Environment: Powdery mildew of barley is favored by the same cool weather ideal for growth of the host. Humid weather is essential for an epidemic, but the pathogen is less dependent on leaf wetness than are many other leaf-infecting fungi.

Biological Control: Powdery mildew can be controlled by resistant cultivars, but each new source of resistance selects for new virulent races. M. Wolfe and associates in England have obtained significant biocontrol by planting a mixture of three or four barley cultivars in the field, rather than a single cultivar. Each component in the mixture is susceptible to a different race or races in the pathogen population. Inoculum produced on the plants of one cultivar may be intercepted by neighboring nonsusceptible cultivars. Interrace competition on any given plant is also possible, and some biocontrol results from induced resistance achieved when the avirulent inoculum attempts to infect the resistant plant in the mixture.

References: Wolfe et al (*in* Jenkyn and Plumb, 1981), Wolfe and Barrett (1980).

control with mixed-cropping and intercropping may account for the reluctance of growers in some areas of Africa to abandon these ancient but highly effective cropping systems.

PREPLANT SOIL TREATMENTS

Tillage

The method and amount of tillage is probably second only to cropping system in importance as an agricultural practice to influence biological control of plant pathogens. Even before the plow, primitive tools of various types were invented for turning the soil and preparing a seedbed. Methods of tillage evolved through the centuries to meet many needs besides seedbed preparation, e.g., the ridge and furrow method in England in areas where soil drainage was a problem, and the rigolen process in Holland to deeply bury inoculum of pathogens. As with the cropping system, the tillage used for a given crop in a particular area today represents choices made largely through trial and error and many small discoveries and advances in mechanization, each leading to further modification in timing, frequency, depth, method, or some other aspect of the tillage system. Most fungus pathogens of the facultative type, as well as plant-parasitic nematodes, plant-pathogenic bacteria, and the antagonists of these pathogens, complete all or part of their life cycles in the soil. Thus, how and when the soil is tilled have obvious relevance to biological control.

Tillage may affect the activity of a soilborne pathogen and its interaction with antagonists and the host through effects on: 1) the physical environment for biological activity; 2) the position of the crop residue occupied by the pathogen, i.e., whether the residue is deeply buried or left on the surface; and 3) the position of the pathogen itself. The latter two kinds of effects are distinct when the pathogen exists as resting propagules independent of crop residue but cannot be separated when the pathogen exists in the crop residue.

Effects on the physical environment — Burke et al (1972) found for irrigated beans in Washington that loosening the soil by tillage to a depth of about 30 cm resulted in less yield loss caused by *Fusarium solani* f. sp. *phaseoli*. The tillage provided at least three benefits: roots were able to penetrate below the layer of soil where most pathogen inoculum occurred; plants with deeper roots were less likely to develop low water potentials and become water-stressed during days of high evaporation demand; and infiltration of irrigation water was more rapid and hence less likely to interfere with oxygen movement through soil to the roots. *Fusarium solani* f. sp. *phaseoli* can cause lesions on the hypocotyls and roots of nonstressed beans but, like other *Fusarium* pathogens of cortical and stem tissues, becomes more aggressive and produces more damage when the host is stressed (Cook, *in* Nelson et al, 1981). Low plant water potentials caused by inadequate uptake of water to satisfy the evaporation demand is one kind of stress. Reduced oxygen supply in the root zone (such as occurs with temporary flooding) is another form of stress on beans (Miller and Burke, 1975). Subsoiling can therefore assist biological control of fusarium root rot of bean by allowing the roots to grow beyond reach of the pathogen (Chapter 5) and by assisting the limited but significant general resistance mechanisms of bean to remain more functional against the *Fusarium* (Chapter 6).

The development over time of a tillage pan or plow sole (better known as a pressure pan) is a major problem for root growth of crop plants in most areas where farming has been intensive. In addition to restrictions on root growth (Russell, 1977), the higher bulk density of the pressure pan affects soil drainage and hence the physical environment of soil microorganisms.

R. R. Allmaras, J. M. Kraft, and J. L. Pikul, Jr. (unpublished) measured the effects of compaction on a Walla Walla silt loam in a field not cropped or tilled since 1913, a field rotated between wheat and peas for 30 years with conventional tillage, and a field rotated between wheat and fallow with conventional tillage for 30 years. The two fields rotated between wheat and peas or wheat and fallow had a dense layer at a depth of about 20 cm. Bulk density was nearly 1.2 g/cm in this layer, compared with 1.1 g/cm above and below this layer (Figure 10.2). The field not tilled since 1913 had a layer only slightly compacted at a depth of 18 cm, corresponding to a pressure pan started before tillage was discontinued nearly 50 years earlier. Soil pH in the two cropped (and tilled) fields was 6.0–7.0 at 30 cm and deeper, but was 5.0–5.5 above the pressure pan, because of long-term use of ammonium fertilizer. The rate of movement of water through the soil as a saturated wetting front was 1.6 cm/day through the pressure pan, compared with 34.4 cm/day beneath the pressure pan. In the field not cropped or tilled since 1913, the figures were 7.2 and 31.7 cm/day 18 and 36 cm deep, respectively. Tensiometers placed above and below the pressure pan showed that drainage was severely retarded, but once through the pan, water penetrated the soil rapidly.

The population of *Pythium ultimum* was nearly all confined to depths above the pressure pan (Figure 10.2). The population of *Fusarium solani* f. sp. *pisi* was greatest above the pan and absent in the pan layer, but some propagules were detected in soil below the pan down to 60 cm, the deepest level sampled. The absence of *F. solani* f. sp. *pisi* in the pan layer was thought to result from the death of propagules caused by anaerobic conditions during the wet winter months. The occurrence of *Fusarium* below the pan layer reveals the potential of this pathogen to establish wherever the roots of peas occur. Significantly, in a survey of fields not planted to peas for 10–20 years, *F. solani* f. sp. *pisi* occurred only below the pressure pan. This finding indicates that abiotic and biotic stresses are sufficient to eliminate this pathogen from the tillage layer, where organic matter is greatest and wetting and drying the most extreme, but not from soil 30–60 cm deep, where conditions for long-term survival apparently are better. These results confirm previous observations that *elimination of a pathogen such as a chlamydospore-forming Fusarium by crop rotation is nearly impossible* and that alternative approaches such as suppression of chlamydospore germination, prevention of increases in the amount of inoculum, and elimination of conditions that stress and predispose the host are more likely to succeed. We need to know much more about the effects of stress factors on root growth, regeneration, and tolerance to root rot.

Ecofallow is a system of reduced tillage developed for the semiarid Central Great Plains of the United States whereby winter wheat, grain sorghum, and fallow are used in a three-year rotation, with the grain sorghum sown directly into the undisturbed wheat stubble (Doupnik and Boosalis, 1980).

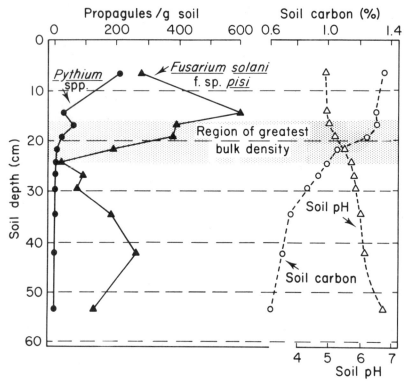

Figure 10.2. The populations of *Pythium* spp. and of *Fusarium solani* f. sp. *pisi* at different depths in plots at Pendleton, Oregon, where wheat and peas had been alternated as crops for many years. A pressure pan (region of greatest bulk density) developed over the years at the depth of tillage, and unsaturated conductivity of water through this layer is greatly impeded relative to soil above and below the pan. Soil pH has dropped over the years, because of the repeated usage of ammonia as a source of nitrogen. The concentration of soil carbon is highest in the tillage layer where most crop residues occur. (Chart courtesy of R. R. Allmaras, J. M. Kraft, and J. L. Pikul, Jr.)

Various terms have been applied to the practice of direct-sowing into crop residue with no seedbed preparation, "direct drill" and "no-till" being the most common. Leaving the wheat residue on the soil surface lowered soil temperatures during early growth of sorghum. It also trapped more snow during the winter months, increased water infiltration, and resulted in less loss of water by not disturbing the topsoil. The net gain of water in the soil profile for the sorghum crop averaged 5 cm per crop. The incidence of stalk rot of sorghum caused by *Fusarium moniliforme* was 11% where ecofallow was used and 39% where sorghum was sown in a prepared seedbed and less total water was available for the crop. Average yields increased from 6.1

to 8.7 hl/ha (43–61 bu/A). Stalk rot is favored by high temperature and plant water stress. Ecofallow therefore creates a soil environment physically more favorable for the sorghum than for the *Fusarium*. Sorghum is most susceptible to the *Fusarium* when predisposed, and ecofallow helps ensure that the predisposition does not occur. *Fusarium moniliforme* exemplifies the pathogens described in Chapters 3, 5, and elsewhere, as weak parasites that exist as endophytes in their host, waiting for or causing the host to become stressed, whereupon they grow more aggressively to take possession of the tissue. The result is severe disease.

Crop-residue effects — Unfortunately, although some plant pathogens can be controlled by reduced tillage systems, many others are favored if the soil is not tilled adequately. Moore (1978) demonstrated in Washington that take-all occurred in higher incidence and plants were more severely infected in no-till (direct-drilled) plots than in adjacent conventionally tilled plots. The top 10 cm of soil from no-till plots contained nearly five times more infested crop residue than the same soil layer from adjacent tilled plots, and most of the infested residue was of a unit size retained by a 1.25-cm screen. This confirms an observation of Cunningham (1967) in Ireland that take-all was less where wheat stubble was plowed 20 cm deep than where it was plowed only 10 cm deep. When reduced or shallow tillage is practiced, the large and intact fragments of pathogen-infested residue, mainly basal stems and adjoining roots of the previous wheat crop, are ideally positioned for the pathogen to grow directly into the lower stem tissue of the succeeding crop (Hornby, 1975). Thorough fragmentation of the residue and mixing of the soil expose more surfaces to colonization by saprophytes. Thorough mixing and deeper burial also limit the fungus to root infections; the fungus established in a root lesion must then grow along the root axis to reach the stem base but in so doing is more vulnerable to antagonists on the roots. Thorough tillage rather than no tillage is thus the best way to achieve biological control of take-all in consecutive crops of wheat in Washington.

In contrast to the observations of Moore (1978), direct-drilling of wheat in England into undisturbed grass killed by herbicide resulted in less take-all than where the grass sward was tilled to prepare a seedbed for wheat (Brooks and Dawson, 1968). In Scotland, take-all in continuous barley was less severe with direct-drilling than with conventional tillage (Lockhart et al, 1975). Conditions between harvest of one crop and sowing of the next are drier in Washington than in England and Scotland, and the pathogen may survive better on the soil surface in a drier area. Yarham (*in* Asher and Shipton, 1981) suggests that under British conditions, antagonism may be intensified for this fungus in the top few centimeters of nondisturbed soil. Just as the environment of each region or country is unique, so must the cultural practices that give best control of a given pathogen be unique. The ideal niche for *G. graminis* var. *tritici* may be in the top 1–2 cm of soil or on the soil surface in Washington but deeper in soil in Great Britain. Tillage practices to eliminate the niche of this pathogen must be adjusted accordingly.

In general, root or leaf-attacking pathogens totally dependent on plant residues for survival and sporulation are favored by reduced tillage systems. Thorough burial of the infested residue shortens the life expectancy of pathogens such as *Helminthosporium maydis* and *Septoria tritici* contained within the residue (Chapter 7) and preempts their opportunity for dispersal above ground (Cook et al, *in* Oschwald, 1978).

Pathogens that can increase their biomass and energy through saprophytic growth on crop residue can be partially controlled by tillage that places the residue physically beyond their reach. Thus, the "nondirting method" was developed by Garren and Duke (1958) to control *Sclerotium rolfsii* on peanuts and other crops in the southeastern United States. After leafy residue was thoroughly buried to preempt saprophytic colonization of the residue by the pathogen, cultivation was carefully practiced to avoid bringing soil and residue into contact with the plant. Sclerotia of *S. rolfsii* do not germinate at soil depths greater than 1–2 cm.

In eastern Washington and adjacent northern Idaho, direct drilling of winter wheat through the surface residue of a previous wheat crop increases damage from *Pythium* spp. (Cook and Haglund, 1982; Cook et al, 1980). Deep burial of fresh residue is thought to limit the saprophytic multiplication of *Pythium* spp. As shown by R. R. Allmaras, J. M. Kraft, and J. L. Pikul, Jr. (unpublished), the *Pythium* propagules in this region are mainly in the top 15–20 cm of soil (Figure 10.2). Burial of the straw beneath this infested layer denies the pathogen access to this food base. Minimum tillage and no-till leave the residue within and above the layer of the greatest *Pythium* population and hence easily accessible to it. Chaff, which is freer than most other plant parts of the usual array of epiphytes and endophytes, is particularly effective in promoting *Pythium* attack of wheat roots (C. Chamswarng, unpublished). A practice whereby all wheat residue, including the chaff, would become leached, weathered, and colonized by saprophytes before coming into contact with the soil might help reduce *Pythium* damage where deep burial of the residue cannot be practiced.

Gudmested et al (1978) noted a different effect of residue incorporation on stem and stolon damage to potatoes caused by *Rhizoctonia solani*. Where potatoes followed wheat in North Dakota, plowing the wheat stubble 26 cm deep increased disease, but mixing the straw into the top 10 cm of soil suppressed disease. The pathogen normally is active mainly in the top 10 cm of soil. The disease suppression obtained with wheat straw resulted from the intensification of microbiological activity and antagonism. Management of the residue of potato plants infected with *R. solani* was a different matter, because such residue serves as an important source of inoculum for infection of the next potato crop. Gudmested et al (1978) recommended disking as the best tillage method where potatoes follow wheat, to maximize the suppressiveness of the soil to *R. solani;* they recommended deep plowing where potatoes follow potatoes, to bury the infested host residue beyond the reach of stolons and stems of the next potato crop.

Most soilborne pathogens have little or no ability as saprophytic colonists of crop residue in soil. *Cylindrocladium crotalariae* is thought to form microsclerotia only in living roots of the infected host and is a poor saprophytic colonist of dead plant material in soil (Phipps and Beute, 1979). *Sclerotium oryzae* may continue to form sclerotia after infected rice stubble is buried in soil but does so only in lesions formed on stems at the water line while the plants are still alive (Bockus et al, 1979). Cutting the rice plants 0–7 cm above the ground (below the site of infection) and removing this portion of the residue controlled the inoculum density of this pathogen as effectively as did burning. This confinement of facultative-type parasites to that portion of the residue colonized through parasitism results, in part from noninfected plant parts becoming colonized quickly by saprophytes that had existed as epiphytes and endophytes on the living plants.

Vascular wilt fungi such as *Verticillium dahliae, Fusarium oxysporum*, and *Cephalosporium gramineum*, already contained within the xylem of their host, need grow only a short distance (radially into the stem) as saprophytes to thoroughly colonize the stems of their host, much of which they accomplish while the dead plant is still standing in the field. Perhaps some of the saprophytic growth from infected xylem into surrounding stem tissue could be prevented by plowing at the earliest possible time after harvest, thereby giving soil saprophytes greater opportunity to colonize the stems in advance of these pathogens.

Organic Amendments

Organic amendments promote biological destruction of inoculum through the germination-lysis mechanism and by the decay of propagules that follows the intensification of microbiological activity (Chapters 4 and 7). Organic amendments can also favor biological protection of the plant. The biological control achieved with organic amendments results, in part, from enhanced competition among the soil microorganisms for nitrogen, carbon, or both, and may be expressed as fewer propagules germinated or less prepenetration growth of the pathogen in the infection court.

The use of an organic manurial treatment for disease control is an old practice largely abandoned in much of the United States. The cost, and especially the logistics, of adequately treating the large acreages of corn, soybeans, wheat, potatoes, and other major crops with organic residues are

Figure 10.3. A field in Hunan Province, People's Republic of China, showing piles of compost ready for spreading and incorporation into the soil as a source of fertilizer.

generally prohibitive on large modern farms. Organic amendments are perhaps more practical for biological control of plant pathogens on smaller, diversified farms, especially for pathogens of vegetable crops and in orchards, as exemplified by the Ashburner method for controlling *Phytophthora cinnamomi* in avocado orchards of Queensland, Australia. On the other hand, practical methods for using organics on a large scale have been developed. Streets (1969) described a method to permit annual cropping of cotton in Arizona without serious damage from *Phymatotrichum omnivorum* by incorporating a winter crop of papago peas as green manure in the spring before planting the cotton. In the People's Republic of China, about 80% of the fertilizer requirements for crop production are met with organic sources (Figure 10.3), consisting mainly of composted crop residue, green manure, night soil, and livestock manure. Compost in excess of 100 metric tons per hectare annually is not uncommon as a preplant treatment for rice, cotton, wheat, and other crops. Kelman and Cook (1977) suggested that the widespread and intensive use of organic materials in China was one reason for the general absence of important root diseases on crops in that country.

While the evolution of conventional farming in the United States since World War II has been away from the use of organic manures, renewed interest in "organic farming" began sometime during the mid-1970s (Lockeretz et al, 1981). The U.S. Department of Agriculture recognized this movement when it established a special task force in 1978 that produced the *USDA Report and Recommendations on Organic Farming* (Papendick et al, 1980). The American Society of Agronomy at their 1981 annual meeting sponsored a popular symposium on organic farming (Bezdicek and Power, 1983), and in May 1982, the House Subcommittee on Department Operations, Research, and Foreign Agriculture of the House Committee on Agriculture initiated congressional hearings concerning the "Organic Farming Act of 1982." The companion Senate version was labeled the "Innovative Farming Act of 1982." Support for these two pieces of legislation was much stronger than it would have been only a few years earlier, and both were only narrowly defeated. Terms such as "low-energy biological farming," "alternative farming," and others have arisen with this movement.

Organic farming involves frequent use of crop rotation. The rotation usually includes a legume hay crop about every third year, which is plowed under as a source of nitrogen for the other crops in the rotation. This practice is also beneficial for pathogen control. In addition, organic farms maintain livestock, and the manure is applied to the land, probably contributing further to disease control. Finally, organic farms use more mechanical cultivation (mainly for weed control) than conventional farms (Lockeretz et al, 1981), as discussed in the preceding section, thorough tillage is generally more advantageous to disease control than reduced tillage.

Nevertheless, in spite of the probable disease control benefits from organic farming, no one has yet experimentally compared the relative importance of plant diseases on organic versus conventional farms. Conventional farms also make use of crop rotations (Lockeretz et al, 1981). Moreover, the residue left by a bumper crop, whether produced by the conventional or organic method, is more likely to provide the quantities of organics needed for disease suppression than are standard applications of proprietary organic amendments. Conceivably, some farms managed organically could have more disease problems, especially from seedborne pathogens, because there may be a greater tendency to sow seed produced on that farm rather than to obtain clean seed from an outside source. *There is no substitute for sanitation and good crop husbandry, whether the method of farming is organic or conventional.*

Flooding

Where it can be practiced, flooding the soil during a warm period, when the oxygen demand of the soil microbiota is greatest, can provide highly effective biological control of soilborne plant pathogens. Flooding is mainly a way to biologically destroy inoculum, which is probably accomplished by anaerobes that decay the metabolically inactive propagules and produce toxic materials inhibitory to the propagules. Unfortunately, root-infecting pathogens intro-duced into the soil after flooding may be more destructive, because of the biological vacuum that results from the indiscriminate destruction of soil organisms by the treatment. For several years, the United Fruit Company used flooding on a large scale in Central America in an effort to eliminate *Fusarium oxysporum* f. sp. *cubense* from the soil to permit replanting of wilt-susceptible bananas (Stover, *in* Holton et al, 1959), but it abandoned the practice because of rapid recolonization of the fields by the pathogen (Chapter 4).

Flooding seems most effective if used in combination with crop rotation. Rotation of cotton with paddy rice — the wet-dry method (Chapter 4) — is widely used to control fusarium wilt of cotton in the People's Republic of China (Cook, *in* Nelson et al, 1981). A one-year rotation with paddy rice was the only rotation in a California study that reduced the population of *Verticillium dahliae* (Figure 10.4) below the level of detection (Butterfield et al, 1978). Moore (1949) demonstrated in Florida that flooding the soil is highly effective in eliminating sclerotia of *Sclerotinia sclerotiorum*. Sclerotia were completely decayed in 23–45 days regardless of the depth of burial, and alter-nating between flooding and drainage at three-day intervals was as effective as continuous flooding. Stoner and Moore (1953) subsequently demonstrated that sclerotia of *S. sclerotiorum* rot in 20 days under the conditions achieved with rice farming in the Florida Everglades and that rice-vegetable rotations provide economic control of pathogens of vegetables. Flooding the soil is an ancient but still important and widely used method of biological control. The possible mechanisms involved in biological control by flooding of the soil are discussed in Chapter 7.

Solar Heating of Soil

The discovery by Katan et al (1976) that pathogen inoculum can be eliminated from soil by solar heating beneath a transparent polyethylene tarp has been described by J. DeVay in California as "the biggest breakthrough in plant-disease control in many years" (Yost and McGill, 1982). The treatment is now used commercially in established pistachio orchards in California (Figure 4.2) to control verticillium wilt, with no serious injury to the trees (Ashworth et al, 1982; 1983). In Israel, solar heating was approximately equal to methyl bromide for control of *V. dahliae, Rhizoctonia solani*, and *Sclerotium rolfsii* on potatoes in field plots (Elad et al, 1980b). Many pathogen propagules are destroyed by the heat alone, but some destruction probably also results from predisposition of the propagules to accelerated decay by soil microorganisms (Chapters 4 and 7). The method is more effective if soils are wet and might also be enhanced if used in combina-tion with organic amendments to accelerate oxygen consumption in the

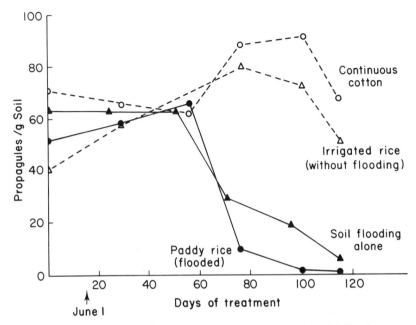

Figure 10.4. Population density of *Verticillium dahliae* in fields and following exposure to different crops and irrigation and flooding treatments at Five Points, California. (Reprinted with permission from Pullman and DeVay, 1981.)

soil. Such a combination of treatments might be more like flooding and would shorten the time required for effective control. Polyethylene tarping is safe, and the only residue is the remnant tarp. Used alone or in combination with other soil treatments, this approach to disease control has great potential in areas of adequate sunshine if the area to be tarped is not too large.

Fertilization and Adjustment of Soil pH

The use of inorganic fertilizers and the adjustment of soil pH with lime or sulfur are ways to achieve biological control mainly through effects on host-plant resistance, the environment around the plant, and antagonists on or near the plant. However, a few cases are known of inorganic fertilizers (or products formed from the inorganic compounds by microorganisms), affecting pathogens independently of the living host: 1) withholding nitrogen from the top layer of soil containing crop residue infested with *Gaeumannomyces graminis* var. *tritici* weakens the possession of the straw by the pathogen and facilitates its displacement by saprophytic colonists in the soil (Garrett,

Take-All of Wheat

Biocontrol of the wheat take-all pathogen illustrates the use of agricultural practices to favor resident antagonists, enhance resistance of the host, or both. The control includes: 1) decreasing the amount of inoculum in soil through tillage that encourages displacement of the pathogen in the crop residue by saprophytes; 2) suppressing growth of the pathogen on the host root and/or within lesions or infested residue fragments by antagonistic bacteria in suppressive soil; 3) limiting the pathogen by prior establishment on the root and in the root cortex of avirulent fungi related to the pathogen; 4) restricting disease development after infection through proper nutrition of the host to maximize its resistance.

Pathogen: *Gaeumannomyces graminis* var. *tritici* (Ascomycotina, Pyrenomycetes, Diaporthales).

Hosts: Wheat, barley, and many grasses, including those in *Bromus, Agropyron,* and *Festuca.*

Disease: The first symptoms usually appear as well-defined dark (black) lesions on seminal and crown roots. Lesions expand along the root axis because of ectotrophic growth of the pathogen as pigmented runner hyphae on the root surface. The pathogen also grows in the stele. Root infections lead to crown and stem infections, also characterized by a dark, often black discoloration where decay is most advanced. Severe infections can kill plants in the seedling stage, but usually the disease develops more slowly and the plants die at various stages after heading. Plants killed after heading are recognized by their "white heads" against a background of green plants. Diseased plants commonly die in patches.

Life Cycle: The pathogen exists in soil as mycelium in fragments of roots and stems colonized earlier through pathogenesis. Two kinds of hyphae are produced: pigmented macrohyphae (runner hyphae) generally resistant to microbial decay (but susceptible to perforation by giant amoebae), which grow from the food base to the root and then superficially (ectotrophically) on the root surface, and hyaline microhyphae, which grow into the cortex and stele. Runner hyphae also permit the fungus to grow from plant to plant if roots contact one another; this

---

1956); 2) applying urea to soils in alder stands in western Oregon accelerates the displacement of *Poria weirii* by *Trichoderma* spp. in the infested alder wood (Nelson, 1975); 3) lowering the soil pH with sulfur prevents sporangium formation by *Phytophthora cinnamomi* and favors *Trichoderma* populations, thereby controlling root rot of pineapple (Pegg, 1977b); 4) compounds formed by sulfate-reducing bacteria inhibit rice nematodes (Jacq and Fortuner, 1979).

Fusarium foot rot of wheat in dryland areas of the Pacific Northwest can be controlled by reducing the nitrogen supply for the crop, which delays the onset of the plant water stress that is necessary for the dis-

explains the occurrence of disease in patches. A thick, highly pigmented layer of mycelium may develop on the surface of stems beneath the leaf sheaths at or slightly above the soil surface; this stromalike layer supports development of perithecia if moisture is adequate.

Environment: Infection and disease progress are favored by cool (5–15° C), moist (near field capacity during most of the early life of the plant), alkaline (pH 7.0–8.0) soil. Development of white heads is most likely during hot, dry weather when water transport to the tops cannot keep pace with transpiration because of crown rot.

Biological Control: Decay or displacement of the pathogen in crop residue by saprophytes begins almost as soon as the host is dead. A single year of rotation of wheat or barley with a nonsusceptible crop (potatoes, beets, corn, alfalfa, beans, and many others) is usually enough time for most of the pathogen inoculum to die because of the action of competitors for the crop residue. The maximum time possible should be allowed between harvest of one crop and sowing of the next, even with cereal monoculture. The life of the pathogen in crop residue can be shortened by frequent, thorough tillage and by keeping available soil nitrogen low. Disease may be suppressed when the crop is provided ammonium rather than nitrate N; ammonium lowers the rhizosphere pH, which increases availability of trace nutrients and favors antagonists. Growth suppression by antagonists (probably pseudomonads) on the roots and carried over in the infested crop residue is also achieved naturally after prolonged monoculture. Some biocontrol of take-all has been achieved in field trials by introducing one or more strains of antagonistic pseudomonads as a seed treatment. Cross protection by avirulent fungi (*Phialophora graminicola* and possibly others) related to *G. graminis* var. *tritici* is achieved in England by growing wheat after a grass rotation. Host resistance can be maximized by providing adequate phosphorus, potassium, magnesium, and trace nutrients.

References: Asher and Shipton (1981), Nilsson (1969).

ease (Cook, 1980). Stalk rot of corn caused by *F. moniliforme, Gibberella zeae,* and *Diplodia maydis* is also less with lower rates of nitrogen; among other effects, a lower rate of nitrogen results in fewer kernels, leaving more carbohydrate for the stalks, which then senesce more slowly and maintain their "metabolically dependent defense systems" longer (Dodd, 1980).

The potential for disease control by use of the proper form of nitrogen was made clear by Huber et al (1968) for take-all of winter wheat; ammonium was suppressive to take-all, but nitrate favored the disease. Ammonium nitrogen has also been shown to suppress take-all in Western Australia (MacNish,

1980) and Germany (Trolldenier, 1981). The effect relates to rhizosphere pH; ammonium lowers the rhizosphere pH, which is suppressive to take-all, and nitrate elevates the rhizosphere pH, which favors take-all (Smiley and Cook, 1973). The altered rhizosphere pH, in turn, may actually affect the susceptibility of wheat to take-all, by reducing the availability of essential trace nutrients (Reis et al, 1983; Sarkar and Wyn Jones, 1982; Chapter 6). Wheat roots fertilized with ammonium also have larger populations of pseudomonads antagonistic to *Gaeumannomyces graminis* var. *tritici* (Smiley, 1980a, 1980b).

Ammonium forms of nitrogen combined with N-serve (to inhibit nitrification) are especially suppressive to take-all in the Willamette Valley of western Oregon, where soils are naturally acid. R. L. Powelson and T. L. Jackson (unpublished) have shown further that ammonium chloride or ammonium chloride plus potassium chloride is even more effective than ammonium sulfate in western Oregon. The chloride lowers the plant water potential (Christensen et al, 1981), which is suppressive to take-all. Ammonium chloride also results in greater activity of fluorescent pseudomonads antagonistic to *G. graminis* var. *tritici* on wheat roots than occurs on roots fertilized with ammonium sulfate (Halsey and Powelson, 1981a, 1981b).

Take-all almost invariably becomes more difficult to control when the soil is limed (Cook, *in* Asher and Shipton, 1981). In some cases, liming the soil is essential, e.g., the acid soils of Brazil must be limed to avoid aluminum toxicity, but caution must be taken to avoid raising the pH more than necessary. Besides causing nutrient imbalances or deficiencies that may increase host plant susceptibility to take-all, lime is also thought to delay the onset of take-all decline. In western Washington, take-all decline occurred in the third consecutive crop in soil at pH 5.0–5.5, but not until the sixth or seventh crop in blocks at the same site in which the pH had been raised initially to 7.0 with lime (Figure 10.5). Take-all decline is also slow to occur in some fields of the Columbia Basin, where the natural soil pH is 7.0–7.5. The long-term use of ammonium fertilizers, which tends to acidify the soil (Figure 10.2), probably is advantageous to take-all suppression.

Smiley (*in* Bruehl, 1975) has pointed out a significant relationship that **soilborne pathogens favored by acid soils are favored by ammonium, suppressed by nitrate, or both, and the converse holds for those favored by neutral or alkaline soils.** For example, besides *G. graminis* var. *tritici* on wheat, *Phymatotrichum omnivorum* on cotton, *Thielaviopsis basicola* on tobacco, *Verticillium dahliae* or *V. albo-atrum* on tomato, and *Streptomyces scabies* on potato all are suppressed by an acid environment — and by ammonium fertilizer. In contrast, *Sclerotium rolfsii* on many crops, *Fusarium solani* f. sp. *phaseoli* on bean, *F. roseum* 'Culmorum' on wheat, and the fusarium wilt pathogens of tomato, cotton, and chrysanthemum are suppressed by an alkaline environment — and by nitrate fertilizers. Similarly, Woltz and Jones

Figure 10.5. Wheat at Puyallup, Washington, in a plot where wheat had been grown the previous two years, and where take-all caused by *Gaeumannomyces graminis* var. *tritici* had been severe both previous years. The lower right corner (within the white line) was typical of five such areas in the plot where lime had been applied three years earlier (soil pH elevated from 5.5 to 7.0 at the time of liming, but was back to 6.0 at the time of this photo). Take-all decline occurred in the nonlimed areas of this plot by the third year (yield was 7,200 kg/ha) but did not occur until the sixth or seventh crop in the five limed subplots.

(*in* Nelson et al, 1981) point out that lime for control of fusarium wilts may be ineffective when large amounts of ammonium are applied. Dobson et al (1983) have added club root of cabbage, caused by *Plasmodiophora brassicae*, to this correlation by showing that the disease suppression obtained by liming the soil is enhanced by nitrate fertilizer, but tends to be counteracted by ammonium.

In Florida, where the soils are sandy and highly conducive to fusarium wilts, uniform liming of the soil to pH 6.5–7.5 gives disease control. Woltz and Jones (*in* Nelson et al, 1981) have shown further that *F. oxysporum* is less pathogenic in limed soils, possibly because certain micronutrients are less available. Wilt suppression with liming was reversed by application to the soil of iron, manganese, and zinc in various combinations. The deficiencies apparently are not so serious as to adversely affect the host. Apparently, requirements of the pathogen for these trace nutrients exceed requirements of the host.

Dobson et al (1983) have shown that the variable control of club root with liming results from inadequate mixing of the lime with the soil. Microsites of acid soil in the limed plots may permit infection and nullify benefits of the treatment (Chapter 6).

Hubbard et al (1982) found that the failure of *Trichoderma hamatum* to provide biological control of damping-off of pea in New York state when applied as a seed treatment was due to starvation of the antagonist for iron, caused by certain strains of pseudomonads thought to produce siderophores in the soil that was already low in iron. When the iron content of the soil was increased, *T. hamatum* was not inhibited and it protected the seeds against damping-off. The siderophore-producing bacteria are being considered for their potential as antagonists of plant pathogens, but we can expect negative as well as positive effects from these microorganisms. Perhaps fertilization practices offer a valuable means to manage the effects of these antagonists.

Providing the proper amount and balance of mineral nutrients is especially important to suppression of damage by weak parasites, whether root or leaf invaders. In this respect, one benefit of the organic fertilizers used by organic farmers may not be the amount of nitrogen or phosphorus provided by the products but rather in the number of different nutrients and especially trace nutrients that tend to be lacking in the pure forms of commercially available inorganic fertilizers, and that may be subclinically deficient in plants grown in conventional agriculture. The possibility exists that subclinical deficiencies, although not apparent as classical deficiency symptoms, nevertheless predispose plants to greater attack by weak parasites. Dodd (1980) points out the greater resistance of corn to the complex of weak parasites responsible for stalk rot when the corn is supplied with adequate potassium; this nutrient is involved in stomatal function, and deficiencies of it are thought to slow the rate of photosynthesis, thereby reducing the ability of the plant to maintain the high sugar content associated with stalk rot resistance. With take-all, virtually any nutrient deficiency that interferes with basal cell metabolism and maintenance predisposes the plant to greater disease (Chapter 6).

METHOD AND DATE OF PLANTING

By planting time, the grower can no longer be concerned with the elimination of primary inoculum (as accomplished by crop rotation and preplant soil treatments). Rather, the grower now becomes concerned with how to manipulate the host so as to avoid infection by the primary inoculum or retard production of secondary inoculum. For many host-pathogen-antagonist-environment combinations, the host is susceptible for only a certain period (e.g., in the seedling stage or during flowering), the pathogen is capable of responding for only a certain period (e.g., following certain temperatures), and the antagonists and the environment favor infection by the pathogen for only a certain period (e.g., during a cool rainy season). An epidemic results when the periods of host susceptibility, pathogen response, and suitable biotic and abiotic environment coincide. Row spacing, plant density, date of planting, and depth of planting mainly determine the ability of the crop to escape or avoid infection by pathogens.

The many ways that seeding date, row spacing, and seeding depth can influence plant infection or disease severity are illustrated by the control possible for pathogens of winter wheat in the Pacific Northwest states, mainly in eastern Washington and adjacent Oregon and Idaho.

1.   Foot rot (caused by *Pseudocercosporella herpotrichoides*) can be suppressed or completely controlled in fall-sown wheat by seeding in late September to early October, rather than in early to mid-September. The result is less crop canopy cover during November to March, when temperature and moisture are ideal for pathogen sporulation and dissemination. The more open plant canopy favors more rapid drying of the soil surface where the inoculum resides; infection is limited accordingly. The ideal temperatures for sporulation by the pathogen are 5–10°C. During periods in this temperature range, rainfall and humidity are sufficiently marginal that the drying made possible by even a slightly smaller canopy can be significant. Late planting, by delaying the primary infections, also reduces the chances for secondary spread by the pathogen. Unfortunately, late sowing on summer fallow also favors soil erosion. For this reason mainly, the fields are sown as early as possible and protected by benomyl fungicide applied in the spring.

2.   Fusarium foot rot (caused by *F. roseum* 'Culmorum') is a problem mainly of winter wheat sown in late summer and early autumn on summer fallow in low-rainfall areas (40–45 cm annual precipitation). The disease is less severe when the seeding date is delayed until mid-September or later, and where rows are spaced at least 40 cm apart. These two practices result in less fall growth and fewer plants. Practiced in addition to the use of less nitrogen, they result in a slower rate of use of the finite water supply in the soil profile and hence delay the onset of water stress – and of fusarium foot rot.

3.   Cephalosporium stripe (caused by *C. gramineum*) is most important on winter wheat sown early in the higher rainfall areas in soils with poor drainage, and where soil freezing and heaving are common (Bruehl, 1957). Soil heaving breaks the roots and opens avenues for *C. gramineum* to enter the plant. The disease is less severe on winter wheat sown in late

September or in October than on wheat sown in early to mid-September. The resultant smaller root system (with the later sowing) is thought to be less prone to breakage and hence infection in the poorly drained soil.

4. Dwarf smut of winter wheat (caused by *Tilletia controversa*) is a problem mainly for winter wheat at higher elevations or in the more northerly portions of the region where the crop may exist beneath a snow on unfrozen ground for three to four months during the winter. Infection exceeded 50% in susceptible cultivars seeded in October, but the same cultivars seeded in late August showed only a trace of smut (J. T. Waldher and R. J. Cook, unpublished). Teliospores of the pathogen responsible for infection exist on or very near the soil surface, and they germinate only when the temperature approaches 5°C. Infection is mainly through new shoot initials. With early seeding, the shoot initials are established as tillers by the time the pathogen is capable of infection, and they are no longer available for infection. Deep seeding (4–6 cm or deeper) also helps prevent infection, possibly because the shoot initials are some distance away from the inoculum during their period of greatest susceptibility.

5. Flag smut (caused by *Urocystis agropyri*) is a problem in south-central Washington and Oregon for winter wheat sown deep (5–10 cm) and in early September. The warm soil and greater depth of soil that must be penetrated by the emerging shoots are ideal for infection by this fungus. Late and shallow sowings can provide excellent control of flag smut.

6. Pythium seedling blight and root rot of wheat (caused by *Pythium ultimum* var. *ultimum* and var. *sporangiferum, P. aristosporum,* and *P. torulosum* [C. Chamswarng, unpublished] and possibly other species) are most important on winter wheat sown so late that the seedlings must establish in cold, wet soil (Cook and Haglund, 1982). The earlier the wheat is sown and the warmer and drier the top layer of soil during emergence of the seedlings, the less is the damage from *Pythium* spp.

7. Snow molds are caused by *Typhula idahoensis, T. incarnata, Sclerotinia borealis*, and *Fusarium nivale*, and snow rot is caused by *Pythium iwayamai* and *P. okanoganense* (Lipps and Bruehl, 1980). This complex of winter pathogens causes less loss on wheat seeded either very early (e.g., mid to late August) or very late (mid to late October). With early sowing, as practiced in northern Washington, plants enter the winter with many tillers and a large leaf area and are consequently more tolerant of leaf mold and rot. The leaves are destroyed, but new shoots emerge in

the spring from surviving crowns. The cultivar Sprague is superior to other commercially available wheats, because it stores large carbohydrate reserves in the crown and hence makes the strongest recovery in the spring in spite of loss of leaves (Kiyomoto and Bruehl, 1977). With late sowing (as practiced in southern Idaho), plants enter the winter in the one- to three-leaf stage and apparently escape infection by these fungi.

8.   Take-all is less severe in the irrigation districts and west of the Cascade mountains on consecutive crops of winter wheat if the crops are sown in late October to early November. Late sowing allows maximum time for breakdown of infested crop residue and minimum time for exposure of the crop to the pathogen. Planting on ridges so that the infection court on the plant (and the surrounding soil) dries more quickly also helps limit take-all (R. L. Powelson and T. L. Jackson, unpublished).

Obviously, no one planting date is ideal for control of all soilborne pathogens of Pacific Northwest wheat. Fortunately, each disease is somewhat limited to distinct areas within the region, and different sowing dates are therefore practiced to the extent possible in the different areas of the region.

One of the most common uses of timing and method of planting is for control of preemergence and postemergence seedling blight. Corn planted in the early spring when soils are cold is likely to sustain serious damage from damping-off fungi, mainly *Pythium* spp, whereas corn seeded later, when soils are warmer, escapes such damage. The damping-off and stem infection caused by *Rhizoctonia solani* on beans and many other warm-temperature crops grown in the Salinas Valley of California are considerably less in crops sown after the soil has warmed to 18.4°C, when germination is faster (Baker, *in* Parmeter, 1970). Planting depth is also critical; the greater the depth of sowing, the longer the time required for emergence and the longer the seedling is vulnerable to damage by *R. solani.*

Sclerotinia sclerotiorum, cause of white mold, infects beans, potatoes, and many other crops by ascospore inoculum released from apothecia at the soil surface. The use of plant spacings and plant genotypes that produce a more open canopy limits white mold (Steadman, 1979) in much the same way late seeding of winter wheat limits pseudocercosporella foot rot. Reduced plant density within the rows or a wide row spacing also reduces sclerotinia wilt of sunflower, but for a different reason (Huang and Hoes, 1980). Infection of sunflower in Manitoba results from myceliogenic germination of sclerotia of *S. sclerotiorum* in soil and from colonization of the plant through roots or through the stem at or near the soil surface. Secondary infections occur by the pathogen growing from plant to plant. "Four to five times as many plants were killed at the 10-cm spacing than at the

30-cm spacing" (Huang and Hoes, 1980). A single primary infection eventually killed as many as eight neighboring plants spaced 10 cm apart in the row. Secondary spread was minimized by spacing the plants 30–40 cm apart.

IRRIGATION PRACTICES

Irrigation is among the oldest of all agricultural practices, having been used as early as 4000 B.C. in Mesopotamia, 2600 B.C. in China, and 1000 B.C. in Peru, and by other ancient civilizations during these same periods. An estimated 61 million acres of land were irrigated in the United States in 1981. Unfortunately, further expansion of irrigation onto additional land or even the continuation of current irrigation practices will be difficult because of limited water supplies, the increased cost of energy for pumping, the salinity problem, and the unavailability of land for new projects (Chapter 1). A more encouraging trend is the development of new irrigation technology for more efficient use of water, including drip or trickle irrigation to supply a small amount of water to the root zone of each plant, the use of electronic timers and computers to obtain more precise irrigation and remote sensing to determine when to irrigate (Clawson and Blad, 1982). Although irrigation engineering research has advanced rapidly in the past 30 years, research on the biological aspects, including the optimal amount of water, the timing of irrigation for the crop, and the use of water management for disease control, has progressed more slowly.

Probably no other single agricultural practice has a greater influence on the plant-microorganism environment than does irrigation. The use of overhead sprinkler irrigation in a previously dryland wheat field in east central Washington is equivalent to relocating that field somewhere in the Midwest of the United States. Diseases of sprinkler-irrigated wheat in Washington are similar to the diseases of wheat in the Midwest, e.g., leaf rust caused by *Puccinia recondita* and scab caused by *Fusarium roseum* 'Graminearum.' In addition, irrigation upsets the biological balance present before the irrigation, particularly in the soil, where the microbial balance is the result of long-time adjustments to arid and semiarid conditions. The "biological buffering" of these once desert but now irrigated soils has proved generally inadequate to prevent or even retard the establishment and increase of soilborne pathogens such as *Verticillium dahliae, Gaeumannomyces graminis* var. *tritici, F. solani* f. spp. *phaseoli* and *pisi*, nematodes, and many others (Baker and Cook, 1974).

Just as the production of irrigated crops favors many pathogens, so should it be possible through irrigation management to limit these pathogens. Irrigation water management, where possible, is among the few agricultural practices, sometimes the only one, available to the grower to influence biological control after the crop is growing. Irrigation affects biological control of plant pathogens through effects on the physical environment of the soil and the plant surface. This environment can be made relatively more favorable for antagonists than for the pathogen. Irrigation also may enhance the resistance of plants to pathogens, by alleviating predisposing stresses during critical periods in the life of the plant.

Modification of the Environment to Favor Antagonists

Common scab of potatoes caused by *Steptomyces scabies* can be controlled by keeping the soil at −0.4 bar or wetter during tuber formation (Lapwood and Adams, *in* Bruehl, 1975). This method is used in England, The Netherlands, and the western United States and probably in many areas where soil water for potatoes can be managed by irrigation. Lenticels are the main avenue for infection by *S. scabies* on the developing tuber, but these become suberized and hence resistant to the pathogen as they mature. Tubers grow at the tips of stolons as a succession of expanding internodes. The oldest internode is nearest the stolon end; this internode is therefore the first region to develop susceptible lenticels and the first region where lenticels mature and become resistant to *S. scabies*. The internodes expand at the rate of one every three to five days, depending on temperature, distance from the stolon end, and other factors. Scab can be controlled by keeping the soil wetter than −0.4 bar until all lenticels are mature, generally for four to six weeks. If the soil is allowed to dry to a water potential below −0.4 bar, even for a few days, the internodal region that has susceptible lenticels during that period will develop scab.

The protection of susceptible lenticels from *S. scabies* in wet soil is thought to result from domination of the infection court by bacteria that exclude the pathogen through competition for space, nutrients, and oxygen (Lapwood and Adams, *in* Bruehl, 1975). Relatively more bacteria than tyrosinase-positive actinomycetes (including *S. scabies*) were found in lenticels in wet soil, and the reverse was found in dry soil (Lewis, 1970). The wet soil is thought to favor swimming by bacteria (at −0.4 bars, effective pore necks in soil are 8–9 μm and larger in diameter), which reach the lenticels before the slower-growing actinomycete hyphae. In addition, oxygen diffusion rates would be restricted and hence more favorable to the bacteria, especially to facultative anaerobes. Metabolites leaking from the unsuberized lenticel would be the main if not the only important source of nutrient, and if they were dominated first by the bacteria, little would be left for prepenetration growth by the pathogen (Labruyére, 1971). However, any limitation to bacterial activity, e.g., drainage of pores required for their motility, would favor the pathogen.

Conceivably, the suppressiveness of soil to common scab that develops after potatoes have been grown for many years is the same factor suppressive to scab in moist soils. Menzies (1959) showed that a transferable biological factor was responsible for suppression of scab on irrigated potatoes in the Yakima Valley of Washington. If bacteria are involved, then obviously the method of soil irrigation is also important. Common scab still occurs in some fields but is no longer a problem on 90% of the 450,000 acres of irrigated potatoes in the Pacific Northwest states (R. E. Thornton, personal communication), all of which are irrigated intensively. Possibly the suppression of common scab by irrigation is more effective if the field also has a long history of potatoes, or conversely, the decline in common scab with successive crops of potatoes depends also on adequate moisture for expression of the antagonism. The interrelationships between scab-suppressive soils and irrigation deserve study.

Many diseases caused by fungi are suppressed when the infection courts are kept moist enough for maximal bacterial activity. The general absence of fusarium foot rot of wheat in wetter soils of eastern Washington in low-lying moist areas is attributed, in part, to greater lysis of germlings of the pathogen *F. roseum* 'Culmorum' by soil bacteria (Cook and Papendick, 1970). Smith (1980) recommends for best control of the fairy ring fungus *Marasmius oreades* that the grass first be killed or the sod removed, the soil then thoroughly mixed to a 15- or 20-cm depth to break up mycelium of the fungus, and the area kept well watered. This treatment, especially keeping the soil wet, not only reduces the amount of inoculum of *M. oreades* but also effectively closes the niche of this fungus in the sod by giving bacteria the competitive advantage. Colbaugh and Endo (1972) similarly concluded for the melting-out disease of Kentucky bluegrass that keeping the leaves wet and hence relatively more favorable to bacteria provided better control of *Helminthosporium vexans* than allowing the leaf surfaces to dry regularly.

The advantage to bacterial colonization of water films on the plant surface and of keeping the soil matric potentials near field capacity or above emphasizes the need to pay careful attention to the irrigation program whenever bacteria are used as antagonists. Burr et al (1978) showed for pseudomonads introduced on potato seedpieces that root colonization did not occur in soils drier than −0.5 bar matric water potential. The best colonization of wheat roots by a fluorescent *Pseudomonas* sp. occurred at −0.3 to −0.5 bars, and some colonization occurred at −0.7 bars, but −1.0 bar was considered too dry (W. Howie, unpublished). Conceivably, even a brief period of limiting water potential in the rhizosphere could permanently prevent colonization of roots by bacteria introduced on the seed, even after the water potential is again within the range required for colonization. A brief dry period could allow the establishment of native fungi and actinomycetes; these microorganisms, established as pioneer colonists on parts of roots formed while the soil was too dry for bacterial colonization, could then persist as bands around the root axis (Figure 10.6), much as common scab develops as a band around potato tubers. Such a band on a root would effectively isolate the introduced bacteria to that portion of the root between the seed and the barrier, preventing them from advancing by multiplication along the root when the soil was again suitably moist. The best method of irrigation to prevent this problem would be the center pivot system, set to make a complete revolution every 12–24 hours, or the drip system, which never allows soil around the roots to become dry.

Pathogens with limited ability to compete at low water potentials can be controlled biologically by allowing the infection courts to dry. Competition from, and preemption of the infection court by nonpathogenic fungi and actinomycetes probably contribute to protection of roots against *Pythium* and *Phytophthora* in drained or slightly dry soils. These pathogens have a strong preference for root tips not already dominated by other microorganisms, and any advantage to competitor fungi, e.g., dry soil relatively more favorable to saprophytic fusaria, penicillia, aspergilli, and others capable of growth at low water potentials, will probably limit *Pythium* and *Phytophthora*. This may be part of the benefit to growing susceptible plants on raised beds, where the soil containing most of the pathogen inoculum is maintained at a matric potential relatively more favorable to colonization of the rhizosphere by saprophytic fungi than by *Pythium* or *Phytophthora* spp. Most water for the plant is obtained by roots deeper in the soil, where inoculum density of the pathogen is lower. In addition, plants tend to become more resistant to these pathogens as they mature. Some damage may continue to occur on juvenile feeder roots at the greater depths, but this is relatively mild compared to damage possible to seedlings that must establish and develop in wet soil.

With irrigated wheat in the Pacific Northwest, significant control of take-all can be achieved by applying the water as fewer applications but in larger quantities with each application. By this means, the top 15–20 cm of the root and stem surfaces and contiguous soil is allowed to dry between irrigations. *Gaeumannomyces graminis* var. *tritici* cannot grow at matric potentials below −40 to −45 bars and its growth rate is reduced by half at about −20 bars. Growth-limiting water potentials for this fungus develop quickly in the topsoil during periods of high evaporation demand, typical of much of the growing season in the irrigation districts of this region. The soils and plant surfaces are densely colonized by nonpathogenic *Fusarium* spp. and other fungi better able to grow in the range of −15 to −120 bars. These fungi are thought to preempt the nutrient supply of *G. graminis* var. *tritici* as the soil dries, thereby limiting the ectotrophic growth of this pathogen at water potentials that otherwise might permit its growth and infection. The best irrigation method for take-all suppression is either the furrow (rill) system or overhead sprinklers that can be left in one place until 10–15 cm of water have been

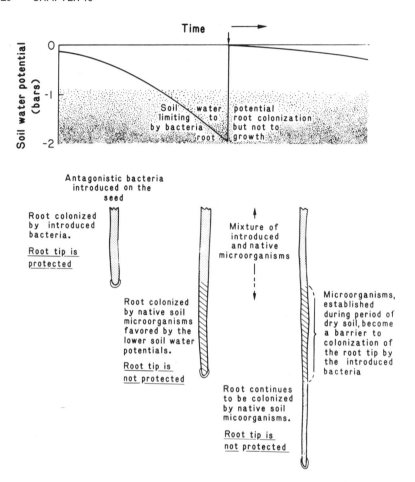

Figure 10.6. Diagrammatic representation of the potential effect of a brief period of low soil water potential on colonization of roots by bacteria applied with the seed. The water potential, being too low for root colonization by the bacteria but not too low for root elongation, may favor establishment of native soil organisms on the root, and then become a barrier to further colonization by the introduced bacteria when the soil is again wet.

applied. Unfortunately, take-all control by allowing the top 15–20 cm of soil to dry is diametrically opposed to conditions best for root colonization by antagonistic bacteria. Such bacteria are thus unlikely to provide much, if any, take-all control where drying of the top soil around the plants is the control method in use. The center pivot system, which is ideal for the pathogen, is also best for root colonization by seed-applied bacteria. The biological control system must be tailored to work with the predominant system of irrigation, or the system of irrigation must be modified to meet the needs of the biological control.

Alleviation of Predisposing Plant Water Stress

Fusarium foot rot of wheat caused by *F. roseum* 'Culmorum' ceases to be a problem in fields managed traditionally as dryland wheat or fallow if the field is brought under irrigation. The many factors that influence disease severity in the absence of irrigation all are of secondary or no importance if the field is irrigated. The major difference is that plant water potentials remain relatively high with irrigation and hence unsuitable for fusarium foot rot. Even a single application of 10 cm of water, if applied during the boot stage or about heading, has been known to stop the disease where essentially all plants were already infected (Cook, 1980). Some growers with large acreages of wheat rotated with one year of fallow attempt to apply a uniform 10–15 cm of water by sprinkler irrigation to each field every two years, rather than to irrigate a few fields intensively each year. Total production for the ranch is as great or greater than when the water is concentrated on a few fields. Fusarium foot rot is less or nonexistent, but soils are still too dry for take-all.

Macrophomina phaseolina, cause of stalk rot of sorghum, similarly may establish in the root and stem tissues of its host but causes damage only if the host becomes stressed for water (Edmonds, 1964). Stress at flowering is more critical than stress at other times in the life of the plant, and water applied at this time greatly limits the disease. Chapter 6 discusses several diseases favored after infection has occurred, if the host then develops low plant water potentials, e.g., *Phytophthora cryptogea* in an otherwise resistant cultivar of safflower, *P. cinnamomi* in rhododendrons, and certain weak pathogens of woody ornamental trees and shrubs. In all cases, a relatively small amount of irrigation water applied at a critical period can be as effective as and certainly more efficient than intensive irrigation during the entire life of the crop. A system selected on the basis of best and most efficient control of a weak pathogen might also be the best and most efficient system for the crop, even when the pathogen is absent. A weak pathogen can serve as an indicator of conditions favorable or unfavorable to the crop. If the plant subjected to a particular irrigation program does not become more susceptible to a weak pathogen, it might be assumed that the irrigation is adequate for the plant.

In contrast to the diseases discussed above, which are favored by water stress on the host, pythium root rot of sugar beets can be controlled by withholding water late in the growing season (M. Stanghellini, personal communication). The predisposing factor in this case is high root temperatures (27°C or above) as the crop is nearing maturity (June–July). The pathogen, *Pythium aphanidermatum*, exists in remarkably high population densities (one to five oospores per 0.1 cm³ of rhizosphere soil), which is enough inoculum to produce 170 lesions per beet root, assuming 100% infection efficiency of inoculum within 1 mm of the root surface (Stanghellini et al, 1983). The pathogen is essentially inactive from planting in the fall (September–November) until the following summer. Even one infection is sufficient to rot a beet root at temperatures above 27°C. Fields are assayed at planting time and assigned a "risk factor" based on the population density of *P. aphanidermatum*. Internal root temperatures of the beet are monitored by measuring soil temperature at the 10-cm depth (root temperature and soil temperature at 10-cm depth are equal). High-risk fields are harvested first, preferably before the root temperature reaches 27°C. Fields that cannot be harvested are also not irrigated, because oospores of the pathogen germinate poorly or not at all at matric potentials below −0.1 bar. Thus, while the beet root may be fully susceptible, infection does not occur at the lower matric potentials, and crop damage is minimized.

SELECTION OF THE BEST COMBINATION OF PRACTICES

The environment of any given field is determined by the combination of agricultural practices applied over a period of years. Thus, a field where conventional tillage is used will present a different environment for the crop, pathogens, and antagonists than will a neighboring field where minimum or no tillage is used. A field consistently plowed in the fall will present a different environment to the crop and associated biota than one consistently plowed in the spring. Even subtle differences in management, e.g., in planting date, choice of cultivar, or method of fertilization, can be important; the ecological effects of such slight differences may not be apparent immediately but continued over many years the effect can be significant. Consider, by analogy, the average summer temperature of a lake being permanently elevated by only 2°C; the effects on the ecology of the lake would be highly significant, even though the direct effects of a 2°C rise on any one species of fish or plant would be relatively minor in the short term.

Diseases, nematodes, insects, and weeds adapted to the niche created by the particular management will gradually increase in importance until counterbalanced by limiting biotic and abiotic factors. Attempts to explain an apparently sudden occurrence of a disease or pest problem in one field and its absence from the neighboring fields only by comparing the current management differences are inadequate. The current year's management is obviously important, but often the combination of practices of the past several years is more important. When the important contributing factors are known, a plan must be devised to best reduce the problem. Most diseases will subside only if the crop management is stabilized or some adjustment in the management is made, but the benefits may not occur the first year. Just as the problem may develop after several years of disease-favorable management, so several additional years of disease-suppressive management may be needed to nudge the pathogen out of the agroecosystem. The alternative practice of applying a pesticide can be effective but usually is less permanent. Moreover, where the pesticide kills nontarget organisms as well as the target pathogen, greater fluctuations in disease can be expected.

A major challenge for research on biological control achieved with agricultural practices is how to identify the best combination of practices to either 1) close the niche of the pathogen by making the environment relatively more unfavorable for the pathogen than for the host, antagonists, or both, or 2) avoid making a niche available to the pathogen in the first place. The following guidelines are offered to help identify the best management for pathogen control. These guidelines will be useful mainly for pathogens that have only limited or no aerial dispersal phases and will be of less value for pathogens that can sweep across several fields or entire regions planted to a susceptible host. Linderman et al (1983) outline strategies for finding and characterizing systems of biological control.

First, verify by observation that the plants in some fields have distinctly less disease than plants in other fields of comparable management. Verify further that the difference in amount of disease cannot be explained on the basis of differences in cultivar, seed source, soil type, rainfall patterns, topography, climate or some other obvious factor.

Second, choose two fields that show the greatest difference in amount of disease for the least difference in management, i.e., the cropping history, method and timing of tillage, fertilization, and irrigation (if any) are as much alike as possible, yet plants in one field have distinctly more disease than plants in the other field. Verify to the extent possible whether the difference in amount of disease relates to a difference in cultural practices used in one field compared with those used in the other.

Third, identify the most likely cultural practice responsible for the difference in disease, e.g., timing of the primary tillage after harvest, the date of seeding, whether one field has consistently received more fertilizer than the other. Identify the second most likely factor, the third, and a fourth if necessary. Consider the possibility that the difference in cultural practices may have occurred in a previous year.

Fourth, observe other fields that either have or do not have severe disease, to confirm that the cultural practice identified as the most influential does, indeed, relate consistently to presence or absence of severe disease. Throughout this process, *as much time and effort should be spent inspecting fields where the disease does not occur or is mild as is spent inspecting those where the disease is severe.*

Fifth, proof of the hypothesis that a particular cultural practice or combination of practices is suppressive to the disease requires experimentation. The experiments should be conducted in the area or in one of the fields initially observed. Preferably only one variable, but no more than two or three, should be tested as replicated treatments in the experiment. Since the effect of the particular cultural practice may be subtle or slow, it may be necessary to carry on the trials in the same field or on the same site for several years to obtain proof of the effect.

Identification of the most likely cultural practice(s) responsible for the healthier (or more diseased) crop will be based on knowledge of the life cycle of the pathogen, but allowance must be made for the unexpected. Nature produces some significant surprises. As an example, leaving the residue of the previous winter wheat crop on the soil surface, as accomplished with a "stubble mulch" tillage system for control of soil erosion, was predicted to favor pseudocercosporella foot rot of the next winter wheat crop more than burying residue with a mold-board plow. The inoculum source, infested stubble, must be on the soil surface to be effective. However, a nine-year experiment established that foot rot occurred in highest incidence and was most severe on wheat in plowed plots (Cook and Waldher, 1977). The wheat made greater fall growth in the plowed plots, possibly because the soil was warmer. Although the amount of inoculum on the soil surface probably was less, even the smaller amount of inoculum was more effective because of the denser crop canopy.

Approximately 10 years was required between the first outbreaks of fusarium foot rot in the Pacific Northwest and the determination of the best management for its control. Acute fusarium foot rot caused by *F. roseum* 'Culmorum' first appeared in eastern Washington in 1962 (G. W. Bruehl and C. S. Holton, personal communication). Severe disease occurred in some fields but not in others. One observation was that the disease occurred in wheat crops of the "best farmers." It was determined that these farmers were among the first to adopt the high-intensity management for the new semi-dwarf wheat cultivars; namely, they seeded slightly earlier in the autumn and they applied 20–25% more nitrogen for the crop than did other growers in their area. Proof that nitrogen favored the disease required an additional three to four years of experiments in the field (Papendick and Cook, 1974). Pathogen-suppressive management was then recommended (Cook, 1980).

An experiment to establish the influence of break crops and especially of alfalfa on persistence of the factor(s) responsible for take-all decline lasted 11 years (Cook, 1981). Experiments that established that take-all is suppressed by tillage lasted only three years but were conducted at three widely separated locations (Moore, 1978).

In many respects, detection of a management system or combination of agricultural practices suppressive to a particular disease is like finding a naturally suppressive soil. The number of combinations of different practices used by the many growers is almost without limit. To experimentally test all of these different combinations is impossible. It is infinitely more practical to find and verify that combination of management practices used by even one grower that results in the least disease and then work out the mechanisms as necessary to extend the suppression principle to other fields. Experiments in the field can be slow and risky, owing to variable weather, but are also extremely productive, since once complete, even one successful trial can point the way to a biological control that works in the field.

11

PERSPECTIVES

Plant pathologists, in their effort to control diseases biologically, are learning to combine useful features of the environment, host, and antagonists with vulnerabilities of the pathogens. It is clear that acceptable control can be achieved through minor adjustments in the soil and plant environment, increase of general as well as more specific forms of host resistance, selective enhancement of antagonists, and increased vulnerability of the pathogen, accomplished by specific management practices, plant breeding, and introduced antagonists. The manipulation of any one of these components by itself may contribute relatively little toward complete disease control, but together they can provide control that is economical, stable, and acceptable to growers.

As pointed out in Chapter 2, agriculturalists became interested at an early date in cultural control of plant disease, and this method of control has been emphasized to this day. When chemical control was developed about 1800, it was added to cultural practices. Breeding for resistant cultivars was added beginning about 1900, and resistant varieties and chemicals became major defenses against plant diseases after 1930. All three methods assumed that the pathogen was present. When biological control with antagonists began to be investigated after 1930, it was only natural to consider that antagonists might also be present as residents. Consequently, most of the early studies (except Hartley [1921], Sanford [1926], and Sanford and Broadfoot [1931]) dealt with resident antagonists.

The "one-on-one" syndrome (Baker and Cook, 1974) then came into play as investigators understandably tried to achieve biological control by adding a single antagonist to treated soil. Such experiments were often successful, at least temporarily, and this result had two effects on biocontrol research. First, it overemphasized the possibility that biological control can be achieved with a single organism selected from among the many resident antagonists present in nontreated soil or on plant surfaces. This strategy still impedes recognition by investigators of the significance of general suppression by the total microbiota active during times critical to the pathogen (Chapter 7). Second, the emphasis on a soil treatment or a shock effect as a prerequisite to the use of introduced antagonists led to the interpretation that microorganisms could not be introduced successfully into nontreated soil. These half-truths led workers away from seeking naturally occurring biological controls such as suppressive soils, studying the mechanisms of their operation, and considering how to transfer populations of antagonists from a suppressive to a conducive soil or from plant to plant.

The successful studies of Rishbeth after 1950, using the introduced *Peniophora gigantea* to control *Heterobasidion annosum* on newly cut pine stumps, revealed the commercial potential for using introduced antagonists in biological control. This accomplishment also helped plant pathologists to recognize the potential use of antagonists on the host itself (established in specific infection courts) as an alternative to their use in soil. Some suppressive soils are now thought to involve antagonists operative at the plant-soil interface (Rovira et al, 1983), raising the possibility that antagonists obtained either singly or as groups from these soils can be introduced into conducive soil with the planting material, a more practical method than mixing them with soil more generally.

Perhaps the single greatest recent advance in biological control has been the realization that effective antagonists or antagonistic populations can be found by seeking, studying, and ultimately exploiting situations where the pathogen should be important but is not (Baker and Cook, 1974). Suppressive soils are only one example of this phenomenon. These antagonists, used as they are or perhaps improved by genetic manipulation, when combined with our best efforts in breeding and habitat management by cultural practices, can provide highly stable and effective biological control.

Chemical control, once considered the ultimate weapon, is being integrated into disease management procedures that also involve altered cultural practices, use of pathogen-free plant propagules, a measure of host resistance, thermal or chemical disinfestation of plant propagules or soil, inoculation of plant propagules or soil with antagonists, and sanitary methods, among others. For many diseases, especially those of field-grown crops, the objective also has shifted from absolute control to economically acceptable control, as it was realized that elimination of the last trace of disease costs more than any benefits can return. It is in this context that biological control of plant pathogens is being generally accepted and used today. We must seek the best control possible within economic constraints.

Floricultural crops, grown for their beauty, must be free of blemishes and require perfect disease control. Growers of these and other high-value crops such as strawberries and horticultural nursery stock can afford to use expensive procedures such as soil fumigation that virtually eliminate pathogens. However, even under such special conditions, biological control may play a significant role by providing complementary support to other measures. Moreover, as biocontrols improve, it may be possible to replace some chemical treatments without sacrificing the degree of control.

Detection of Effective Antagonists

Antagonists that provide biological control of plant pathogens are far more common than is generally appreciated, for when disease is not present its absence is often casually accepted or attributed to other phenomena. Earlier (Baker and Cook, 1974), we advised that "Antagonists should be sought in areas where the disease . . . does not occur, has declined, or cannot develop, despite the presence of a susceptible host, rather than where the disease occurs." This triad of *exclude, extinguish,* or *expunge* has, for investigators in many areas, proved productive as an approach to detecting potential biological control. Linderman et al (1983) and Baker and Cook (1974) have presented detailed suggestions for discovering these examples of effective antagonists. Some examples of probable natural biological control by antagonists are cited below to reveal their diversity and also to make clear that the phenomenon of significant natural pathogen suppression is not limited to suppressive soils.

1. *Phymatotrichum omnivorum* on alfalfa and other crops in Arizona (King, 1923).
2. *Fusarium oxysporum* f. sp. *pisi* on pea in Wisconsin (Walker and Snyder, 1933); f. sp. *cubense* on banana in Central America (Reinking and Manns, 1933); f. sp. *melonis* on melon in France (Louvet et al, 1976).
3. *Gaeumannomyces graminis* var. *tritici* on wheat in Kansas (Fellows and Ficke, 1934), in England (Glynn, 1935), and now in many wheat-growing regions.
4. *Botrytis cinerea* on lettuce (Brown and Montgomery, 1948; Newhook, 1951), tomato (Newhook, 1957), and chrysanthemum leaves (Blakeman and Fraser, 1971) in England.

5. *Helminthosporium victoriae* inhibited by *Chaetomium globosum* and *C. cochlioides* on oat seed from Brazil (Tveit and Moore, 1954; Wood and Tveit, 1955).

6. *Streptomyces scabies* on potato in soils subjected to intensive potato cropping in Washington (Menzies, 1959) and elsewhere (Baker and Cook, 1974).

7. *Rhizoctonia solani* on pepper in California (Olsen, 1964), on wheat in South Australia (Baker et al, 1967), and on carnation in Colombia (Chet and Baker, 1981).

8. *Fusarium solani* f. sp. *phaseoli* on beans in Washington (Burke, 1965) and in Japan (Furuya and Ui, 1981).

9. *Poria carbonica* by *Scytalidium* sp. on Douglas-fir poles in Oregon (Ricard and Bollen, 1968).

10. *Endothia parasitica* (hypovirulent) on chestnut in Italy and France (Grente and Sauret, 1969a, 1969b).

11. *Cronartium ribicola* by *Tuberculina maxima* on white pine in Idaho (Kimmey, 1969).

12. *Heterobasidion annosum* by *Trichoderma polysporum* and *Ascocoryne sarcoides* on spruce logs in Sweden (Ricard, 1970).

13. *Phytophthora cinnamomi* on avocado in Queensland, Australia (Broadbent et al, 1971).

14. *Fomes pini* by *Ascocoryne sarcoides* in the heartwood of black spruce trees 75 years old and older in Ontario, Canada (Basham, 1973).

15. *Eutypa armeniaceae* inhibited by *Fusarium lateritium* in wounds on apricot trees in South Australia (Carter and Price, 1974).

16. *Fusarium oxysporum* on pine seedlings in coniferous forests in California (Toussoun, *in* Bruehl, 1975).

17. *Meloidogyne javanica* on peach by *Dactylella oviparasitica* in California (Stirling et al, 1979).

18. *Sclerotium cepivorum* on onion in British Columbia (Utkhede et al, 1978).

19. *Pythium ultimum* on sugar beet in Czechoslovakia (Veselý, 1978), in forest soils in France (Bouhot and Perrin, 1980), and in soils in the San Joaquin Valley of California (Hancock, *in* Schippers and Gams, 1979).

20. *Globodera rostochiensis* on potatoes in India by an unidentified fungus (Goswami and Rumpenhorst, 1978).

21. Tristeza virus on citrus in Brazil (Costa and Müller, 1980).

22. *Heterodera avenae* on wheat by *Nematophthora gynophila* and *Verticillium chlamydosporium* in England (Kerry and Crump, 1980).

23. *Ceratocystis ulmi* by *Phomopsis oblongata* as a saprophyte and deterrent to beetles in elm logs in northern England and Scotland (Weber, 1981).

Each of these examples (and others still waiting to be discovered) offers an exciting opportunity for both basic and applied research in biological control. The few examples that have been studied in some depth (e.g., hypovirulence in *E. parasitica*, and take-all decline) show the great value of seeking antagonists where the disease has failed.

Unexploited Opportunities for Biological Control

A number of opportunities for biological control of plant pathogens noted in the literature have thus far not been exploited. Sailer (*in* Papavizas, 1981) also enumerated "examples of missed opportunities" for applications of biological control of insects. Some unexploited opportunities for biological control of plant pathogens are given below.

1. The common occurrence of *Chaetomium globosum* and *C. cochlioides* on oat seed from Brazil and their inhibitory effect on *Helminthosporium victoriae* (Tveit and Moore, 1954) suggest that *Chaetomium* spp. are prevalent in oat-producing areas of Brazil and that *H. victoriae* may be unimportant there. The biological control of other seedborne pathogens there should also be investigated.

2. The ease with which the environment can be controlled for glasshouse crops, and the common practice of steam or chemical treatment of the soil used for these crops make this an excellent opportunity for biological control by introduced antagonists (Ferguson, 1958; Olsen, 1964). That the method has not been commercially developed on glasshouse crops has been attributed to the need for near-perfect disease control for these high-value crops, but this presupposes that biocontrol is to be used as a single procedure. Integration of biological control by introduced antagonists with soil treatments, pathogen-free planting stock, and sanitation as practiced by the best growers has potential for stabilizing and improving disease control of glasshouse crops.

3. Common scab of potato (caused by *Streptomyces scabies*) declines in fields cropped continuously to potato, a phenomenon first noted in Washington nearly 25 years ago (Menzies, 1959). Menzies (1959) also demonstrated that a transferable, biological, suppressive factor occurred in soils from fields where potatoes had been grown for many years. The implications of this phenomenon as a source of clues to biological control of common scab are obvious, but thus far the mechanisms of suppression of this disease have not been studied. N. A. Anderson (personal communication) recently showed in Minnesota that scab-suppressive soil from one area of the state, when mixed into plots of scab-conducive soil in another area of the state, gave significant suppression of scab on potatoes subsequently grown in

the plots. This confirms the transferability of a factor in the field and provides further evidence that significant clues to biological control of *S. scabies* are waiting to be uncovered in scab-suppressive soils.

4. Males of the root-knot nematode (*Meloidogyne* spp.) do not cause galls in roots. Sex reversal, common in lower animals, may occur in these nematodes into the third larval stage, and stresses such as nutrient deficiencies greatly increase the percentage of males in the nematode population (Triantaphyllou, 1973). Study and application of this potential method for controlling this important pathogen should be increased.

5. *Crotalaria spectabilis* planted in orchards or in rotation with vegetables as a trap crop for root-knot nematodes (*Meloidogyne* spp.) is a demonstrated effective biological control (Baker and Cook, 1974; Huang et al, 1981) but has been little used. Chemical soil treatments are used, even though none has proved very useful in planted orchards.

6. The anomalous unimportance of fire blight (caused by *Erwinia amylovora*) in New Zealand since the disease nearly annihilated the pome fruit industry there in 1919–1929 (Baker and Cook, 1974) has continued into 1983, even though the virulent pathogen is generally present there. Because of the unimportance of the disease in New Zealand, local pathologists have not undertaken the investigation, affording foreign investigators an excellent opportunity for research on biological control of this important disease.

7. Strains of *Fusarium roseum* isolated from pustules of *Puccinia antirrhini* on snapdragon in humid climates could be screened for ability to colonize rust-infected tissue without producing toxins that kill healthy host tissue (Baker and Cook, 1974). This approach could yield effective isolates for field inoculation and biocontrol of this and possibly other rusts.

8. *Taphrina deformans*, which causes peach leaf curl, persists as yeastlike cells on the bark of trees and should be vulnerable to epiphytic antagonists that could be applied (Baker and Cook, 1974).

9. The continuous cropping of fields to lettuce and other leafy crops in the Santa Maria Valley of California, and the abundance of sclerotia of *Sclerotinia sclerotiorum* in the crop debris would lead one to expect severe losses in leafy vegetable crops there (Baker and Cook, 1974). However, losses are consistently minor despite the favorable environmental conditions and crops. This would appear to be a good place to seek mycoparasites of sclerotia of *S. sclerotiorum*.

10. African and French marigolds (*Tagetes erecta* and *T. patula*) act as inhibitory plants to *Pratylenchus penetrans, Hoplolaimus* sp., and *Tylenchorhynchus dubius* nematodes by producing nematicidal ter-

thienyls that exude from growing roots. Unconfirmed studies indicate that these terthienyls are also toxic to *Fusarium oxysporum* f. sp. *callistephi, Verticillium albo-atrum, Septoria tageticola*, and *Helminthosporium sativum* (Baker and Cook, 1974). The 33 species of *Tagetes* are a group of potentially inhibitory plants that should be studied and used in biological control. Other plants in the Compositae are also inhibitory or lethal to nematodes (Gommers, 1973) and may have potential in biocontrol.

Besides these examples of proven or obviously effective biological controls that have not been exploited, or in some cases even studied beyond the original report, there are some unique types of resident antagonists in soil that hopefully will attract more research attention in the future. Each of these types has serious limitations ecologically or biologically, but each also is undoubtedly already important in natural biocontrol of soilborne plant pathogens. More information on the occurrence, ecology, and ways to overcome the limitations of these microorganisms is needed if we are to find ways to enhance their effectiveness as resident antagonists. The types of antagonists and their known limitations are listed below:

1. *Bdellovibrio bacteriovorus*, limited to Gram-negative bacteria, and highly host-specific.

2. Mycophagous amoebae (mainly in the family Vampyrellidae), active mainly or exclusively in soils with water-filled pores at least 1 μm in diameter and those sufficiently drained to permit entry of oxygen.

3. A large group of predaceous fungi able to trap larval and adult nematodes but poor as competitors in soil and susceptible themselves to antagonists (their effectiveness in predation is not related to the size of the nematode population).

4. Bacteriophages, highly host-specific but thus far not of proven usefulness in controlling bacterial numbers.

5. Mycophagous nematodes such as *Aphelenchus avenae*, restricted in mobility by the necessity of a water film and of water-filled soil pores of adequate size and probably also restricted in soil where oxygen is limiting.

6. Anaerobic bacteria, abundant in soil (Chapter 7) and producers of resistant spores and potent antibiotics and toxins, but inactive except in oxygen-free sites or sites that are oxygen- and nitrate-free and of low redox potential; nevertheless deserving of serious investigation for their possible role in biological control of soilborne plant pathogens.

Unusual Approaches to the Discovery of Biological Control

Many new examples of biological control are discovered when a new concept or way of looking at a problem is developed or when a new technique reveals new approaches. Thus, when soils in England under continuous cropping to cereals were treated experimentally with formalin (3,000 liter/ha), injury from *Heterodera avenae* increased because of the low nematocidal and high fungicidal effect of the formalin (Kerry et al, 1980). This led to the discovery that *Nematophthora gynophila* and *Verticillium chlamydosporium* were effective in long-term control of the cereal cyst nematode in fields cropped intensively to cereals. The nematocides ethylene dibromide (EDB) and dibromochloropropane (DBCP) do not reduce activity of nematode-parasitic fungi, but formalin and the dichloropropane-dichloropropene mixture (DD) do (Mankau, 1968).

Routine culturing of pruning wounds of apricot in South Australia revealed the consistent presence of *Fusarium lateritium* in wounds free of *Eutypa armeniacae*. Treatment of the pruning wounds with benomyl gave short-term (but not long-term) protection against infection by *E. armeniacae* but showed that *F. lateritium* was tolerant of benomyl and that it occurred in the benomyl-treated wounds (Carter and Price, 1974). These observations led to one of the first combination chemical-biological control procedures, whereby an antagonist and chemical are applied as a mixture, thereby providing control more effective than either can provide alone.

Studies on the incidence of pathogenic and nonpathogenic strains of *Agrobacterium radiobacter* in South Australia showed that the ratio of the pathogenic (pv. *tumefaciens*) to nonpathogenic (*radiobacter*) types was high around trees with crown gall and low around healthy trees. Inoculation of tomato with nonpathogenic isolates (to decrease the ratio of pathogenic to nonpathogenic types) reduced crown gall. One of these isolates (K84) has since proved to be effective in tests in many countries as a commercial control of crown gall (Kerr, 1974, 1980).

Chestnut blight (caused by *Endothia parasitica*) rapidly destroyed the native chestnut stands in the United States following introduction of the pathogen in 1904, and no control has yet been developed in the United States. Introduced into Italy in 1938, the pathogen caused similar losses, but by 1950 many cankers were spontaneously healing and the rate of disease spread was decreasing. This observation by Biraghi (1951) prompted Grente to culture the healing cankers to determine whether *E. parasitica* was still present in them. This led to a series of discoveries by a succession of workers in Europe and the United States showing that dsRNA, virus-like transmissible determinants occur in the fungus and are associated with hypovirulence and healing cankers (MacDonald et al, 1979). Hypovirulence is now in use for biological control of chestnut blight in France and is under

Nonparasitic Exopathogens

Biocontrol of these pathogens is the use of microorganisms to preempt, displace, or inhibit: 1) nonparasitic but pathogenic rhizosphere microorganisms that restrict plant growth by impairment of root health, 2) epiphytic bacteria that initiate frost damage of plants at temperatures between -2 and $-5°$C.

Pathogens: Nonparasitic exopathogens include many types of microorganisms that are able to cause disease without having first established a parasitic relationship with the plant. These microorganisms include: 1) ice-nucleation active (INA) bacteria (*Pseudomonas syringae, Erwinia herbicola,* and *P. fluorescens,* the main INA bacteria, commonly make up 0.1–10% of the bacterial flora), 2) nonparasitic pathogenic bacteria that produce mild phytotoxins or hormonelike substances harmful to the plant (*Enterobacter, Klebsiella, Citrobacter, Flavobacterium, Achromobacter, Arthrobacter,* and *Streptomyces* spp., *Erwinia herbicola, Bacillus cereus, Pseudomonas cichorii, P. syringae, P. fluorescens, P. viridiflava,* and probably others), and 3) weakly parasitic but strongly pathogenic fungi that produce potent phytotoxins (e.g., *Penicillium oxalicum, Aspergillus flavus, A. tamarii, A. wentii, Alternaria alternata, Periconia circinata*).

Hosts: Growth of all plants probably is diminished under some conditions by class 2 and 3 pathogens above. Frost-sensitive plants may be injured at temperatures above $-5°$C if INA bacteria are present as epiphytes on the leaves.

Diseases: Decreased plant growth is the most common symptom of the effect of these pathogens on roots, probably because the damaged roots perform fewer functions for the tops. Other effects include decreased seed germination, shorter hypocotyls and radicles, shorter roots and laterals, inhibition of root hair development, browning of roots, and root distortion. Increased infection by root-colonizing fungi may result. Frost sensitivity of most plants is increased by the epiphytic presence of INA bacteria that initiate ice formation on and in plants at temperatures between -2 and $-5°$C.

Life Cycle: These soil microorganisms have strong saprophytic capabilities and are not parasitic. They are prevalent in most field soils, where they grow and attach to elongating roots and shoots. The deleterious rhizosphere biota are possibly favored selectively by root exudates and multiply as roots elongate. Roots usually have relatively few microorganisms in the meristematic region but are more fully colonized in the region of elongation, the root hair zone, and farther back from the tip. The initial effects of the exopathogens on plant roots include greater loss of energy-containing compounds from the roots, possibly because of the diffusion gradient created or by modification of root cell membrane permeability; inhibition of root growth, possibly by production of ethylene or indole acetic acid; and direct injury to the root by production of phytotoxins that enter the plant and affect various physiological processes. These various kinds of root damage may predispose roots to decay by other pathogens. Fungi in class 3 may exist as epiphytes and endophytes of healthy host tissue and then colonize the tissues more thoroughly at or near the time of the host's death. They then sporulate copiously and increase their inoculum density. In California, some bacterial isolates reduced weight of sugar beets by 21–48%. From 8 to 15% of the rhizosphere bacteria are thought to be deleterious to the plant.

Environment: Because of the multiplicity of microorganisms and crops involved and the newness of this subject, predisposing factors have not been studied. It can be assumed, however, that deleterious plant-inhabiting microorganisms occur for most environments where plants grow.

Biological Control: Inoculating with phyto-, rhizo-, or phyllosanitizing micro-organisms (bacterization, plant growth-promoting rhizobacteria, beneficial microorganisms) is a very promising means of decreasing damage from these pathogens and therefore of increasing crop growth and yield without increased energy or land demands or environmental pollution. A culture of selected bacteria is inoculated on seed (or seedpieces) before planting; the bacteria migrate along the roots and may spread from root to root and plant to plant in the field. *Pseudomonas* spp. are well adapted to rhizosphere occupancy but are sensitive to drying. Spore-forming *Bacillus* spp. are more durable than *Pseudomonas* spp. but less specialized for the rhizosphere. Both of these bacterial groups have given excellent results in field tests. Marketable roots of carrots increased 48%, sugar beets yielded 4–8 tons more per hectare, potatoes gave 5–33% greater yield, and the yield of radishes was increased 60–144%. Cultures are best obtained from roots of the given crop in the given area and then selected on agar culture for broad-spectrum antibiosis; candidates selected by this process are then tested in the field. Rhizosanitizing bacteria inhibit and thus preempt or displace deleterious rhizosphere biota; antibiotic-negative mutants produced no increased growth-response of inoculated plants. Rhizosanitizing bacteria sometimes are ineffective in acid soil. *Pseudomonas putida* and *P. fluorescens* produce siderophores (secondary metabolites that sequester iron) that deprive the deleterious biota of this essential element. There is a possibility that the rhizosanitizing biota may also increase growth by hormonal (gibberellin, IAA) secretion, mineralization of phosphate, or nitrogen fixation, but the evidence favors control of root pathogens as the principal factor.

Decreasing the population of INA epiphytic bacteria by use of competitive non-INA strains of *Erwinia herbicola* or *Pseudomonas syringae* effectively protects plants against frost injury at temperatures down to $-5°C$; ability to produce antibiotic is not an essential characteristic of these beneficial strains, suggesting competition as the main method by which the INA strains are controlled. Application of bactericides or ice-nucleation inhibitors is also helpful.

References: Broadbent et al (1971, 1977), Brown (1974), Merriman et al (1974; *in* Bruehl, 1975), Kloepper and Schroth (1979, 1981b, 1981c), Kloepper et al (1980a, 1980b), Lindow (1983), Salt (*in* Schippers and Gams, 1979), Schroth and Hancock (1981, 1982), Suslow and Schroth (1982a, 1982b), Woltz (1978).

intensive study for application to chestnuts in the United States (Chapters 4 and 5).

The introduction of antagonists on seeds to protect roots against pathogens is a new and promising approach to biological control and is likely to attract wide commercial interest as a component of the new biotechnological agriculture. Because of its potential significance, some of the key events in its development thus far are mentioned here.

Broadbent et al (1971) in New South Wales noted in soil inoculation tests with bacteria to control pathogens of bedding plants that plants in series inoculated with the antagonist but not the pathogen frequently were much larger than those in the noninoculated series. The increased growth was similar to that from seed bacterization reported earlier by Soviet workers and studied in England, Australia, and Finland (Brown, 1974). Of special significance was the observation (Broadbent et al, 1971, 1977) that the most growth-promoting strains of bacteria were those selected initially because of their broad-spectrum inhibitory activity against plant pathogens in culture plates. One strain, *Bacillus subtilus* A13, was inhibitory to all nine pathogens used in the screen and produced significant plant-growth responses when used as a seed treatment in field tests (Merriman et al, *in* Bruehl, 1975). This breakthrough was the first indication of a relationship between broad-spectrum inhibition and increased plant growth.

California workers (Burr et al, 1978; Kloepper and Schroth, 1979; Kloepper et al, 1980b; Suslow and Schroth, 1982a; Chapters 4 and 8) screened their "plant growth-promoting rhizobacteria" for ability to inhibit a broad spectrum of plant-pathogenic bacteria and fungi in culture plates. In addition, the bacteria were isolated initially from roots, which helped ensure that they would be adapted for growth in the rhizosphere. These bacteria were then shown to protect roots against growth suppression caused by root pathogens, including exopathogens. At least part of the protection may be due to siderophores that sequester iron from the rhizosphere microorganisms that inhibit plant growth. The increased growth of bedding plants in the work of Broadbent et al (1971, 1977) is now thought to involve such root protection, because the possibility cannot be ruled out that one or more undetected root pathogens (including the rhizosphere-inhabiting exopathogens) occurred in soil. The observations of Broadbent et al (1971) are thus now reasonably explained and a new approach to biological control has been opened (Chapter 4).

This theory that increased plant growth with seed bacterization results from biological control of root pathogens could explain the increased plant growth obtained by early workers with seed bacterization and also why their experiments were only modestly successful and difficult to repeat (Brown, 1974). The bacteria used in their experiments were not selected for broad-spectrum inhibitory activity against other microorganisms; had they been, the results might have been more encouraging. Nevertheless, some root protection may have occurred if the bacteria were no more than aggressive root colonists; even their growth on the surface of the advancing root tips could be a deterrent to root pathogens (Chapter 6). Any improvement in root health can be expected to produce a fertilizerlike response in the tops, owing to the greater uptake of mineral nutrients by the healthier roots, and probably also to the greater synthesis of growth factors in them (Cook, *in* Bezdicek and Power, 1983). This potential effect of seed bacterization introduces another source of experimental variability, namely, the need to know the population and activity of the root pathogens to be controlled. Tests conducted in a field where root pathogens are already controlled (e.g., by crop rotation) are less likely to give positive results than tests conducted in fields where root pathogens are more numerous or more active (Chapter 4).

The success of Weller and Cook (1983) in the use of seed bacterization to control a targeted pathogen (*Gaeumannomyces graminis* var. *tritici*) can be attributed to their having combined and modified three approaches: 1) that of Broadbent et al (1971) of screening for inhibitory activity in vitro, in this case against the pathogen to be controlled; 2) that of the previously mentioned California workers, of initially isolating the candidate antagonists from roots; and 3) the suggestion of Baker and Cook (1974) that candidate antagonists be isolated exclusively from roots of the specific host infected with the pathogen in a suppressive soil (Chapter 7). The latter approach was used to help increase the chances of finding effective antagonists adapted not just to wheat roots but to wheat roots infected with *G. graminis* var. *tritici*, where their protective action is needed the most.

Bacterization got off to a poor start in the Soviet Union in the 1950s during the Lysenko era; there was a grain of truth in the theory behind bacterization but also much contradictory chaff. The reality of increased plant growth from rhizosanitizing microorganisms has now been fully confirmed and is being energetically investigated in field trials by several companies and laboratories. Even more significant, this work has widened the horizon of plant pathologists; trained to isolate parasitic pathogens from host tissue, they must also now seek nonparasitic pathogens in the rhizosphere outside the host. Already, a considerable number of such pathogens have been reported (Woltz, 1978).

Biological Control in Less Developed Countries, and Some Lessons for the Agriculture of More Developed Countries

Biological control seems to be most used in those types of agriculture where chemicals are not affordable. In the United States, biological control is used more for the relatively low-value field crops than for the high-value horticultural crops. Biological control accomplished mainly through agricultural practices is also used widely in the less developed countries, because expensive equipment and large capital investments are not required for some types of biological control practices. There is no better example than the People's Republic of China, where intercropping, mixed cropping, crop rotation, organic fertilization, intensive tillage, and flooding, among other practices that promote biological control (Chapter 10), are used on essentially all of the agricultural land. This labor-intensive culture uses large quantities of organic matter and manure and apparently has reduced root diseases of crops to unimportance (Kelman and Cook, 1977; Williams, 1979). Many of these practices were developed empirically over centuries of trial and error, and they provide an interlocking, sustainable, and stable method of crop production. A nation such as China is less able than many western nations to afford modern synthetic pesticides, except in specialized situations, and hence they continue to use agricultural practices that provide biological control. Any shift within such a country toward a western-type agriculture with more intensive use of chemical pesticides should be made only if it improves or at least does not upset the existing biological controls.

Biological control as presently practiced in the United States generally does not provide the kind of complete protection and produce the spectacular yield increases associated with chemical controls. A major objective of both researchers and growers is and should be to find ways to produce the highest economic yield. It has only been natural, therefore, to turn to greater use of chemicals. However, we can take another lesson from the accomplishments in the People's Republic of China, namely, that it is possible through a biological type of farming system to achieve yields of cereal grains as great and even greater than those in the United States. The accomplishments of the Chinese would also indicate that achieving and maintaining the highest possible yields are not in conflict with sustainability in agriculture (Chapter 1), since both appear to be happening in China.

The less developed countries characteristically depend on the large and relatively inexpensive labor force in order to carry out practices such as mixed- and intercropping and fertilization with organic manures. Obviously there are limits to the kinds of practices that can be used in the western type of agriculture, where labor is very expensive. On the other hand, western agriculture should adopt as many of these practices as possible, fitting them in wherever practical, or learning from the example itself and then finding other ways to achieve the same thing. It should not be necessary to sacrifice yield significantly in order to use biological control more widely. Furthermore, it does not necessarily follow that a technological cultural procedure used in a highly scientific agriculture such as that of the sugarcane industry of Hawaii is not economic in developing countries. The

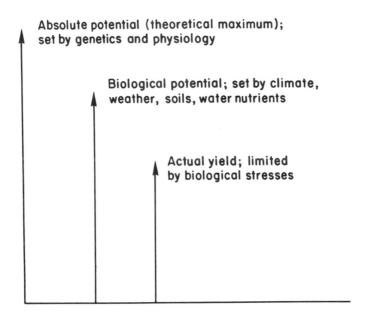

Figure 11.1. Diagrammatic representation of the absolute potential (theoretical maximum) yield, the biological potential, the actual yield, and the controlling factors of each for a crop.

crop-logging procedure developed in Hawaii has been placed in successful operation in Iran, Iraq, Ecuador, India, Venezuela, and Brazil (Clements, 1980).

The actual yield in any given field is usually well below the potential yield for that field, as determined by the cultivar, edaphic and climatic factors, and the energy-requiring inputs (Chapter 1). This indicates that significant gains are still possible with new technology, perhaps with new biotechnology. The *absolute potential* (Figure 11.1) or theoretical maximum yield would be achieved if no physical, chemical, or biological stress factors were limiting, but this situation rarely, if ever, occurs. On the other hand, we should expect to achieve yields approaching the *biological potential* of the crop, i.e., the yield determined by the cultivar, edaphic and climatic limitations, and agronomic inputs. The difference between the biological potential and actual yield can be attributed mainly or entirely to diseases, weeds, insects, and nematodes. The so-called yield plateau for any given crop is probably due as much to the presence of biological stresses that prevent the yield from increasing as to the absence of fertilizer, water, and other factors needed to help the yield increase (Figure 11.2). Unfortunately, the

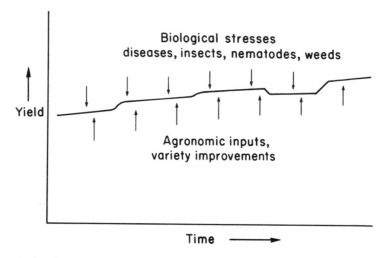

Figure 11.2. Diagrammatic representation of the factors determining the yield plateau of a crop.

tendency when attempting to increase yields has often been to increase the agronomic and energy-requiring inputs rather than to remove the biological stresses.

Soil fumigation with methyl bromide for wheat in eastern Washington has consistently increased yields by 20–25%, without increasing the amount of fertilizer or other inputs (Cook, *in* Bezdicek and Power, 1983; Cook and Haglund, 1982). This finding indicates that poor root health rather than insufficient fertilizer or water are limiting to wheat yield in the area. Soil fumigation is not economical as a commercial practice but is now used to provide a standard to measure the progress being made with experimental biological controls. The large difference that commonly exists between the absolute and the biological potentials and between the biological potential and the actual yield of important crops such as wheat makes clear that much progress in yield improvement can be made through improvements in plant health. Moreover, what may seem uneconomical or impractical today may be fully affordable tomorrow. This was the situation for control of soilborne pathogens of strawberries using soil fumigation with methyl bromide in California (Wilhelm and Paulus, 1980).

Biological control offers promising economic disease control for a wide range of crops, including: long-term, low-value forest trees; short-term, low-value cereal crops; short-term, intermediate-value vegetable crops; long-term, high-value orchard crops; and short-term high-value strawberries, floricultural crops, and horticultural nursery crops. Such versatility is not characteristic of other disease-control strategies.

A Partner at the Feast

As we stated in our first book (Baker and Cook, 1974), "Man must learn to visualize the pathogen on his crops as a partner at the feast, there before himself and . . . as much a part of the scene as he, and more likely to hold a residence permit. Each organism is as much the center of its own universe as man believes himself to be."

The late H. N. Hansen used to tell plant pathology graduate students at the University of California, Berkeley, that "to really understand fungus ecology, you have to think like a fungus." This comical advice encapsulates profound wisdom. The objective of microorganisms, as of other living things, is to survive and reproduce – that is, to enable both the individual and the population to survive (Chapter 5). The means to achieve this objective have been selected during the evolutionary process. While they may not appear superficially to be the most effective means of survival, they probably are from the standpoint of the fungus in relation to its microbial associates and microenvironments.

We may consider, for example, that it is wasteful for *Fusarum roseum* 'Culmorum' to expend energy by advancing two or three internodes up the culms of wheat to cause crown and foot rot in wheat plants; it can kill the plant by rotting only the lower few millimeters of the culms. The advantage to Culmorum is not in killing wheat plants, but rather in having a method of becoming the pioneer inhabitant of wheat stubble. This strategy of establishing in the straw while it is still part of a living plant allows Culmorum to beat the competition by being there "fustes with the mostes" and thus avoiding competition from strict saprophytes that could also live in the straw. This enables the fungus to produce more chlamydospores in plant residue and to develop a high inoculum density, thereby improving its chances of infecting plants the next year. Chances of chlamydospore survival also are improved when the fungus becomes imbedded in host tissue. A cultural practice or plant modification that prevents the fungus from progressing up the culm, if used in a field already infested with this pathogen, may not reduce crop loss during the first one or two years but eventually will reduce losses if continued for several years. As we have emphasized repeatedly, biological control often takes longer than chemical control but it is also longer lasting.

The more we learn about the mechanisms of microbial ecology, the better our chances of understanding the performance of microorganisms in the field. The recent finding that mycelia and rhizomorphs of *Armillaria mellea*, like those of other Basidiomycetes, produce antibiotics inhibitory to other microorganisms is such an example. The mycelium of *A. mellea* in colonized wood normally is not attacked by *Trichoderma* spp., but exposure of the colonized wood in soil to carbon disulfide, methyl bromide, or aerated steam, or exposure of main roots and crowns of infected trees to drying, "weakens" the mycelium and reduces its ability to produce protective antibiotics. *Trichoderma* is then able to attack and slowly eliminate the pathogen from the wood. Similarly, knowledge that *Rhizoctonia solani* is able to infect pepper, eggplant, or tomato fruit in contact with the soil and then to grow into some of the developing seeds explains how a selected virulent strain of the pathogen is seedborne to a new field or seedbed, increasing the survival chances for the fungus (Baker, *in* Parmeter, 1970).

The type of symptom produced on the host also affords valuable clues as to how a pathogen attacks its host. Thus, *Pythium spp.*, able to rapidly infect root tips before bacteria there have developed sufficiently to protect the root, characteristically cause decay of root tips. On the other hand, *Rhizoctonia solani*, developing more slowly, does not invade root tips but attacks more mature roots and stems, usually near the soil level. Its coarse, tough mycelia will hold soil particles that dangle from uprooted seedlings, but the smaller and fragile mycelia of *Pythium* does not do this.

Expectations for Biological Control

Because of the complexity of the interactions involved in biological control, it is improbable that future changes will be revolutionary in either the approaches used or the kinds of microorganisms involved. Every technique involves natural limitations imposed by long evolutionary biological balance, which, like a double-edged sword, cuts both ways. Our objective should not be to further upset the balance by making large cuts, but to modify proved existing practices by numerous small nicks so as to nudge the microbiota into configurations in line with our schemes and strategies.

Each new subject aspires to be so significant and relevant as to dominate its field, but history bears witness that each new discovery eventually falls into place, enlarging and ennobling the edifice of science but not dominating it. Two promising new areas of research will have to find their niches in the existing structure. One new area already mentioned concerns the beneficial (phytosanitizing) microorganisms that inhibit pathogens (including nonparasitic exopathogens) in the rhizosphere or the phylloplane; use of these offers a promising method of increasing crop growth and yield without increasing energy or land demands or environmental pollution. The other new area is genetic management (genetic engineering) of antagonistic microorganisms and of resistant host plants; this is a means to produce microorganisms designed to perform particular antagonistic functions and of plants designed to resist specific pathogens. This can be illustrated by a theoretical example. The prevalent xylem-inhabiting bacteria in American elms (Carter, 1945) obviously have no inhibitory effect on Dutch elm disease (caused by *Ceratocystis ulmi*), whereas some *Pseudomonas* spp. from soil that are inhibitory to *C. ulmi* in culture do not survive in elm xylem tissue. Perhaps transfer of the factor for inhibition from the pseudomonad to a normal xylem inhabitant of elm could produce an effective biological control of a disease that has so far defied control. To put this type of research into a realistic frame, however, it has been suggested that the movement of genes between microorganisms, between plants, or between microorganisms and plants could eventually select for genetic homogeneity and thus decrease the genetic variability and increase the risk of disease.

It is clear that none of the problems of biocontrol are going to be simple or easily solved. All present methods widely accepted or used have been the result of years of intensive, patient investigation. A number of workers have been actively studying biological control for more than 20 years: J. Rishbeth on *Heterbasidion annosum* on pines in England; A. S. Costa on cross protection against tristeza virus in citrus in Brazil; R. Mankau on predacious and parasitic microbiota on plant parasitic nematodes in California; J. Grente on hypovirulence of *Endothia parasitica* in France; K. F. Baker on root pathogens of nursery crops in California; C. Leben on bacteria on the phylloplane and on seeds in Ohio; G. C. Papavizas and associates on antagonists introduced into soil in Maryland. R. J. Cook studied biocontrol of *Fusarium roseum* 'Culmorum' and *Gaeumannomyces graminis* var. *tritici* in Washington for nearly 17 years. T. Kommedahl worked on application of antagonists on seeds in Minnesota, J. Ricard on introduced antagonists on various woody and herbaceous crops in Sweden, and R. R. Baker on introduced antagonists in soil in Colorado for more than 14 years. D. H. Marx has studied the effect of ectomycorrhizal fungi on root pathogens for 13 years in Georgia. A. Kerr has studied biocontrol of *Agrobacterium tumefaciens* on woody plants in Australia, and N. J. Fokkema the effect of foreign organic matter on biocontrol in the phylloplane in The Netherlands for 11 years. P. Broadbent studied biocontrol of *Phytophthora cinnamomi* and bacterization of seed for eight years in Australia. Tenacity and persistence obviously are necessary for success in this field, as is being able "to think like a fungus."

It is clear that biological control does not operate by any single mechanism; indeed, every disease situation is unique. Therefore, careful field observations often reveal the clues needed to develop effective biological control. Such control measures then must be integrated into the old standby methods and into existing cultural practices; they rarely replace the established practices. A biological balance on a new plateau eventually is established. It is a common misconception that an objective of biological control is an across-the-board replacement of previous procedures and that elimination of the pathogen is a prerequisite of effective control. Neither is true. Thus, the effective biological control of *Phytophthora cinnamomi* in avocado groves in eastern Australia does not eliminate the pathogen but curbs the production of its principal infective unit, the zoospore. This is accomplished essentially by a calculated reversion to the culture of the native rain forest there.

Concern about worldwide pollution of the environment can only intensify in the future, and chemicals inevitably will be subjected to further restrictions. If single alternative methods are substituted, the control may be too slow or the cost too high. The increased use of multiple indirect methods based on the specific ecology of the pathogen appears to be the best answer. Plant pathologists have, since the initiation of this approach by J. L. Jensen in Denmark in 1882, tended to use such integrated control and will not find such an adjustment difficult.

Control of plant disease by thermal treatment of soil has been considered economical only under greenhouse conditions. However, the development of solarization treatment in areas of intense sunlight presents a useful extension of the method. The soil is moistened and covered with clear polyethylene sheets. The temperatures attained $(37-50°C)$ are lower than those with steam $(60-100°C)$, but the times are much longer (14 days or more, compared with 30 minutes). Pathogens may be killed, or they be so stressed by the treatment that they are rendered more susceptible to antagonists (Katan, 1981). It is even possible to successfully treat soil in growing orchards by this method (Ashworth et al, 1982; 1983). Because the mechanism of action is so different from that in soil steaming, the whole subject of thermal treatment is reopened for future development. The combined use of selective chemicals and antagonists offers another promising new area for exploitation, as shown by Carter and Price (1974).

It can fairly be predicted that these and other exciting future developments will not revolutionize biological control or render obsolete those practices now used in biocontrol. However, each will modify existing methods and will improve the overall performance. With the increasing interest in and study of biocontrol, the future of the subject can only be bright and scientifically exciting.

> *Perhaps man will some day reconcile the greatness of his human creativity with the greatness of the wild that created it. If he can do this without diminishing either man's great works or nature's, then we shall indeed walk in beauty for as long as the rivers shall run and the grass shall grow.* —S. C. JETT, 1967

LITERATURE CITED

Abd-El Moity, T. H. 1979. Situation of white rot disease of onion in Egypt. Proc. First Workshop on White Rot in Onions, Inst. Hortic. Plant Breeding, Wageningen, Netherlands. pp. 14–15.

Abd-El Moity, T. H., and M. N. Shatla. 1981. Biological control of white rot disease *Sclerotium cepivorum* of onion by *Trichoderma harzianum*. Phytopathol. Zeit. 100:29–35.

Abd-El Moity, T. H., G. C. Papavizas, and M. N. Shatla. 1982. Induction of new isolates of *Trichoderma harzianum* tolerant to fungicides and their experimental use for control of white rot of onion. Phytopathology 72:396–400.

Ackerson, R. C., D. R. Krieg, T. D. Miller, and R. E. Zartman. 1977. Water relations of field grown cotton and sorghum: Temporal and diurnal changes in leaf water, osmotic, and turgor potentials. Crop Science 17:76–80.

Adams, P. B. 1975. Factors affecting survival of *Sclerotinia sclerotiorum* in soil. Plant Dis. Reptr. 59:599–603.

Adams, P. B., and W. A. Ayers. 1980. Factors affecting parasitic activity of *Sporidesmium sclerotivorum* on sclerotia of *Sclerotinia minor* in soil. Phytopathology 70:366–368.

Adams, P. B., and W. A. Ayres. 1982. Biological control of Sclerotinia lettuce drop in the field by *Sporidesmium sclerotivorum*. Phytopathology 72:485–488.

Adebayo, A. A., and R. F. Harris. 1971. Fungal growth responses to osmotic as compared to matric water potential. Soil Sci. Soc. Am. Proc. 35:465–469.

Ahmed, A. H. M., and H. T. Tribe. 1977. Biological control of white rot of onion (*Sclerotium cepivorum*) by *Coniothyrium minitans*. Plant Pathol. 26:75–78.

Akai, S., and T. Kuramoto. 1968. Microorganisms existing on leaves of rice plants and the occurance of brown leaf spot. Ann. Phytopathol. Soc. Japan 34:313–316.

Alabouvette, C., F. Rouxel, and J. Louvet. 1977. Recherches sur la résistance des sols aux maladies III. Effets du rayonnement γ sur la microflore d'un sol et sa résistance à la Fusariose vasculaires du Melon. Ann. Phytopath. 9:467–471.

Alabouvette, C., F. Rouxel, and J. Louvet. 1980a. Recherches sur la résistance des sols aux maladies VII. Etude comparative de la germination des chlamydospores de *Fusarium oxysporum* et *Fusarium solani* au contact de sols résistant et sensible aux fusarioses vasculaires. Ann. Phytopathol. 12:21–30.

Alabouvette, C., R. Tramier, and D. Grouet. 1980b. Recherches sur la résistance des sols aux maladies VIII. Perspectives d'utilisation de la résistance des sols pour lutter contre les Fusarioses vasculaires. Ann. Phytopathol. 12:83–93.

Alabouvette, C., I. Lemaitre, and M. Pussard. 1981. Densité de population de l'amibe mycophage *Thecamoeba granifera* s. sp. *minor* (Amoebida, Protozoa). Mesure et variations expérimentales dans le sol. Rev. Ecol. Biol. Sol 18:179–182.

Aldrich, J., and R. Baker. 1970. Biological control of *Fusarium roseum* f. sp. *dianthi* by *Bacillus subtilis*. Plant Dis. Reptr. 54:446–448.

Allen, D. J. 1982. *Verticillium lecanii* on the bean rust fungus, *Uromyces appendiculatus*. Trans. Brit. Mycol. Soc. 79:362–364.

Allen, M. F., M. G. Boosalis, E. D. Kerr, A. E. Muldoon, and H. J. Larsen. 1982. Sugar beets, *Rhizoctonia solani,* and *Laetisaria arvalis*. Field response to perturbation. Mycol. Soc. Amer. Newsletter 33(1):34.

Allen, R. N., and F. J. Newhook. 1973. Chemotaxis of zoospores of *Phytophthora cinnamomi* to ethanol in capillaries at soil pore dimensions. Trans. Brit. Mycol. Soc. 61:287–302.

Allen, R. N., K. G. Pegg, L. I. Forsberg, and D. J. Firth. 1980. Fungicidal control in pineapple and avocado of diseases caused by *Phytophthora cinnamomi*. Austral. Jour. Exp. Agric. Anim. Husb. 20:119–124.

Ames, L. M. 1949. New cellulose destroying fungi isolated from military material and equipment. Mycologia 41:637–648.

Anagnostakis, S. 1981. Stability of double-stranded RNA components of *Endothia parasitica* through transfer and subculture. Exper. Mycol. 5:236–242.

Anagnostakis, S. L. 1982. Biological control of chestnut blight. Science 215:466–471.

Anagnostakis, S. L., and P. R. Day. 1979. Hypovirulence conversion in *Endothia parasitica*. Phytopathology 69:1226–1229.

Anderson, E. J. 1962–64. Indirect effects of agricultural chemicals in soil. Long-term effects of soil fungicides. Proc. Annu. Conf. Control Soil Fungi, San Francisco and San Diego, Calif. 9:17; 10:13–14.

Anderson, E. J. 1966. 1–3 dichloropropene, 1–2 dichloropropane mixture found active against *Pythium arrhenomanes* in field soil. Down to Earth 22(3):23.

Anderson, J. P. E., and K. H. Domsch. 1973. Quantification of bacterial and fungal contributions to soil respiration. Arch. Mikrobiol. 93:113–127.

Anderson, J. P. E., and K. H. Domsch. 1978a. Mineralization of bacteria and fungi in chloroform-fumigated soils. Soil Biol. Biochem. 10:207–213.

Anderson, J. P. E., and K. H. Domsch. 1978b. A physiological method for the quantitative measurement of microbial biomass in soils. Soil Biol. Biochem. 10:215–221.

Anderson, T. R., and Z. A. Patrick. 1978. Mycophagous amoeboid organisms from soil that perforate spores of *Thielaviopsis basicola* and *Cochliobolus sativus*. Phytopathology 68:1618–1626.

Anderson, T. R., and Z. A. Patrick. 1980. Soil vampyrellid amoebae that cause small perforations in conidia of *Cochliobolus sativus*. Soil Biol. Biochem. 12:159–167.

Anonymous, 1982. New killer wine yeasts. American Type Culture Collection Quarterly Newsletter. 2(2):2.

Artman, J. D. 1972. Further tests in Virginia using chain saw-applied *Peniophora gigantea* in loblolly pine stump inoculations. Plant Dis. Reptr. 56:958–960.

Asher, M. J. C. 1980. Variation in pathogenicity and cultural characteristics in *Gaeumannomyces graminis* var. *tritici*. Trans. Brit. Mycol. Soc. 75:213–220.

Asher, M. J. C., and P. J. Shipton, eds. 1981. Biology and Control of Take-All. Academic Press, New York. 538 pp.

Ashworth, L. J., Jr. 1979. Polyethylene tarping of soil in a pistachio nut grove for control of *Verticillium dahliae*. Phytopathology 69:913.

Ashworth, L. J., Jr., and O. C. Huisman. 1980. New hope for *Verticillium* control in cotton. Calif. Agric. 34(10):19–20.

Ashworth, L. J., Jr., D. P. Morgan, S. A. Gaona, and A. H. McCain. 1982. Polyethylene tarping controls Verticillium wilt in pistachios. Calif. Agric. 36(5–6):17–18.

Ashworth, L. J., Jr., D. P. Morgan, S. A. Gaona, and A. H. McCain. 1983. Control of Verticillium wilt of pistachio by overall tarping, tarping of individual tree sites, and preplanting tarping of fallow soils with polyethylene mulches. Plasticulture. In press.

Austen, R. A. 1657. A Treatise of Fruit-Trees. Henry Hall, Oxford.

Austin, B., C. H. Dickinson, and M. Goodfellow. 1977. Antagonistic interactions of phylloplane bacteria with *Drechslera dictyoides* (Drechsler) Shoemaker. Can. Jour. Microbiol. 23:710–715.

Averre, C. W. III, and A. Kelman. 1964. Severity of bacterial wilt as influenced by ratio of virulent to avirulent cells of *Pseudomonas solanacearum* in inoculum. Phytopathology 54:779–783.

Ayers, W. A., and P. B. Adams. 1979. Mycoparasitism of sclerotia of *Sclerotinia* and *Sclerotium* species by *Sporidesmium sclerotivorum*. Can. Jour. Microbiol. 25:17–23.

Backman, P. A., and R. Rodriguez-Kabana. 1975. A system for the growth and delivery of biological control agents to the soil. Phytopathology 65:819–821.

Bagyaraj, D. J., A. Manjunath, and D. D. R. Reddy. 1979. Interaction of vesicular arbuscular mycorrhiza with root knot nematodes in tomato. Plant Soil 51:397–403.

Bailey, L. H., and E. Z. Bailey. 1976. Hortus Third: A Concise Dictionary of Plants Cultivated in the United States and Canada. MacMillan, New York. 1290 pp.

Baker, K. F. 1947. Seed transmission of *Rhizoctonia solani* in relation to control of seedling damping-off. Phytopathology 37:912–924.

Baker, K. F., ed. 1957. The U. C. system for producing healthy container-grown plants. Calif. Agric. Exp. Sta. Manual 23:1–332.

Baker, K. F. 1959. Factors in the standardization of container-grown nursery stock. Pac. Coast Nurseryman 18(9):31–32, 70-71, 74–75.

Baker, K. F. 1973. Biological control of soilborne plant pathogens. (First A. W. Dimock Lecture, Cornell Univ.). Cornell Univ., Ithaca. 14 pp.

Baker, K. F. 1978. Biological control of *Phytophthora cinnamomi*. Proc. Internatl. Plant Prop. Soc. 28:72–79.

Baker, K. F. 1980. Developments in plant pathology and mycology, 1930-1980. Pages 207–236 *In* Commonwealth Agricultural Bureau, Perspectives in World Agriculture. Commonwealth Agricultural Bureaux, Slough. 532 pp.

Baker, K. F., and R. J. Cook. 1974 (original ed.). Biological Control of Plant Pathogens. W. H. Freeman, San Francisco. Reprinted ed., 1982. Amer. Phytopathol. Soc., St. Paul, MN. 433 pp.

Baker, K. F., L. H. Davis, R. D. Durbin, and W. C. Snyder. 1977. Greasy blotch of carnation and flyspeck of apple diseases caused by *Zygophiala jamaicensis*. Phytopathology 67:580-588.

Baker, K. F., N. T. Flentje, C. M. Olsen, and H. M. Stretton. 1967. Effect of antagonists on growth and survival of *Rhizoctonia solani* in soil. Phytopathology 57:591–597.

Baker, K. F., and R. G. Linderman. 1979. Unique features of the pathology of ornamental plants. Annu. Rev. Phytopathol. 17:253–277.

Baker, K. F., O. A. Matkin, and L. H. Davis. 1954. Interaction of salinity injury, leaf age, fungicide application, climate, and *Botrytis cinerea* in a disease complex of column stock. Phytopathology 44:39–42.

Baker, K. F., and C. M. Olsen. 1960. Aerated steam for soil treatment. Phytopathology 50:82.

Baker, K. F., and W. C. Snyder, eds. 1965. Ecology of Soil-Borne Plant Pathogens. Prelude to Biological Control. Univ. Calif. Press, Berkeley. 571 pp.

Baker, R. 1968. Mechanisms of biological control of soil-borne pathogens. Annu. Rev. Phytopathol. 6:263–294.

Baker, R. 1980. Measures to control Fusarium and Phialophora wilt pathogens of carnations. Plant Disease 64:743–749.

Baker, R., and R. Drury. 1981. Inoculum potential and soilborne pathogens: The essence of every model is within the frame. Phytopathology 71:363–372.

Baker R., P. Hanchey, and S. D. Dottarar. 1978. Protection of carnation against fusarium stem rot by fungi. Phytopathology 68:1495–150l.

Bald, J. G. 1952. Stomatal droplets and the penetration of leaves by plant pathogens. Amer. Jour. Bot. 39:97–99.

Balis, C. 1970. A comparative study of *Phialophora radicicola*, an avirulent fungal root parasite of grasses and cereals. Ann. Appl. Biol. 66:59–73.

Baltruschat, H., and F. Schönbeck. 1975. Untersuchungen über den Einfluss der endotrophen Mycorrhiza auf den Befall von tabak mit *Thielaviopsis basicola*. Phytopathol. Zeit. 84:172–188.

Bancroft, K. 1912. A root disease of the Para rubber tree (*Fomes semitostus* Berk.) Fed. Malay States Dept. Agric. Bull. 13:1–30.

Barber, D. A., and J. K. Martin. 1976. The release of organic substances by cereal roots in soil. New Phytol. 76:69–80.

Bardzik, J. M., H. V. Marsh, Jr., and J. R. Havis. 1971. Effects of water stress on the activities of three enzymes in maize seedlings. Plant Physiol. 47:828–831.

Barea, J. M., and C. Azcón-Aguilar. 1982. Production of plant growth-regulating substances by the vesicular-arbuscular mycorrhizal fungus, *Glomus mosseae*. App. Environ. Microbiol. 43:810-813.

Bar-Joseph, M. 1978. Cross protection incompleteness: A possible cause for natural spread of citrus tristeza virus after a prolonged lag period in Israel. Phytopathology 68:1110-1111.

Barnett, H. L., and F. L. Binder. 1973. The fungal host-parasite relationship. Annu. Rev. Phytopathol. 11:273–292.

Barnett, H. L., and V. G. Lilly. 1962. A destructive mycoparasite, *Gliocladium roseum*. Mycologia 54:72–77.

Barron, G. L., and A. H. S. Onions. 1966. *Verticillium chlamydosporium* and its relationshps to *Diheterospora, Stemphyliopsis* and *Paecilomyces*. Can. Jour. Bot. 44:861–869.

Barrons, K. C. 1939. Studies on the nature of root knot nematode. Jour. Agric. Res. 58:263–271.

Bärtschi, H., V. Gianinazzi-Pearson, and I. Vegh. 1981. Vesicular-arbuscular mycorrhiza formation and root rot disease (*Phytophthora cinnamomi*) development in *Chamaecyparis lawsoniana*. Phytopathol. Zeit. 102:213–218.

Basham, J. T. 1973. Heart rot of black spruce in Ontario. II. The mycoflora in defective and normal wood of living trees. Can. Jour. Bot. 51:1379–1392.

Bateman, D. F., and S. V. Beer. 1965. Simultaneous production and synergistic action of oxalic acid and polygalacturonase during pathogenesis by *Sclerotium rolfsii*. Phytopathology 55:204–211.

Beer, S. V., J. L. Norelli, J. R. Rundle, S. S. Hodges, J. R. Palmer, J. I. Stein, and H. S. Aldwinkle. 1980. Control of fire blight by non-pathogenic bacteria. Phytopathology 70:459.

Bell, R. G. 1969. Studies on the decomposition of organic matter in flooded soil. Soil. Biol. Biochem. 1:105–116.

Benjamin, M., and F. J. Newhook. 1982. Effect of glass microbeads on *Phytophthora* zoospore motility. Trans. Brit. Mycol. Soc. 78:43–46.

Bennett, C. W., and A. S. Costa. 1949. Tristeza disease of citrus. Jour. Agric. Res. 78:207–237.

Benson, D. M., and R. Baker. 1970. Rhizosphere competition in model soil systems. Phytopathology 60:1058–1061.

Berkeley, R. C. W., and M. Goodfellow, eds. 1981. The Aerobic Endospore-Forming Bacteria. Classification and Identification. Academic Press, New York. 373 pp.

Berry, J., and O. Björkman. 1980. Photosynthetic response and adaption to temperature in higher plants. Annu. Rev. Plant Physiol. 31:491–543.

Berthelay-Sauret, S. 1973. Utilisation de mutants auxotrophes dans les recherches sur le déterminisme de "l'hypovirulence exclusive." Ann. Phytopathol. 5:318.

Beute, M. K., and R. Rodriguez-Kabana. 1979. Effect of wetting and the presence of peanut tissues on germination of sclerotia of Sclerotium rolfsii produced in soil. Phytopathology 69:869–872.

Bezdicek, D. F., and J. F. Power. 1983. Current Technology and its Role in a Sustainable Agriculture. Amer. Soc. Agron. Spec. Publ. In press.

Bhatt, D. D., and E. K. Vaughan. 1962. Preliminary investigations on biological control of gray mold (Botrytis cinerea) of strawberries. Plant Dis. Reptr. 46:342–345.

Bhatt, D. D., and E. K. Vaughan. 1963. Inter-relationships among fungi associated with strawberries in Oregon. Phytopathology 53:217–220.

Biraghi, A. 1951. Sul proposto raggruppamento di Endothia fluens (Sow.) S. et S. e di Endothia parasitica (Murr.) P. J. et H. W. And. in un uniea specie. Boll. Staz. Patol. Veg. (Florence), 9 (ser 3):133–157.

Birchfield, W. 1960. A new species of Catenaria parasitic on nematodes of sugarcane. Mycopathol. Mycol. Appl. 13:331–338.

Bird, L. S. 1982. The MAR (multi-adversity resistance) system for genetic improvement of cotton. Plant Disease 66:172–176.

Bisby, G. B. 1939. Trichoderma viride Pers. ex Fries and notes on Hypocrea. Trans. Brit. Mycol. Soc. 23:149–168.

Bitton, G., and K. C. Marshall, eds. 1980. Adsorption of Microorganisms to Surfaces. John W. Wiley & Sons, New York. 439 pp.

Blakeman, J. P. 1972. Effect of plant age on inhibition of Botrytis cinerea spores by bacteria on beet-root leaves. Physiol. Plant Pathol. 2:143–152.

Blakeman, J. P. 1978. Microbial competition for nutrients and germination of fungal spores. Ann. Appl. Biol. 89:151–155.

Blakeman, J. P., ed. 1981. Microbial Ecology of the Phylloplane. Academic Press, London. 502 pp.

Blakeman, J. P., and I. D. S. Brodie. 1977. Competition for nutrients between epiphytic microorganisms and germination of spores of plant pathogens on beetroot leaves. Physiol. Plant Pathol. 10:29–42.

Blakeman, J. P., and N. J. Fokkema. 1982. Potential for biological control of plant disease on the phylloplane. Annu. Rev. Phytopathol. 20:167–192.

Blakeman, J. P., and A. K. Fraser. 1971. Inhibition of Botrytis cinerea spores by bacteria on the surface of chrysanthemum leaves. Physiol. Plant Pathol. 1:45–54.

Blaker, N. S., and J. D. MacDonald. 1981. Predisposing effects of soil moisture extremes on the susceptibility of rhododendron to Phytophthora root and crown rot. Phytopathology 71:831–834.

Bliss, D. E. 1951. The destruction of *Armillaria mellea* in citrus soils. Phytopathology 41:665–683.

Bockus, W. W., R. K. Webster, C. M. Wick, and L. F. Jackson. 1979. Rice residue disposal influences overwintering inoculum level of *Sclerotium oryzae* and stem rot severity. Phytopathology 69:862–865.

Bollen, G. J. 1971. Resistance to benomyl and some chemically related compounds in strains of *Penicillium* species. Neth. Jour. Plant Pathol. 77:187–192.

Bolton, J. 1788–1791. An History of Fungusses Growing Around Halifax. 4 vols. Huddersfield.

Booth, C. 1971. The Genus Fusarium. Commonwealth Mycological Institute, Kew. 237 pp.

Bouhot, D. 1975. Recherches sur l'écologie des champignons parasites dans le sol. VII. Quantification de la technique d'estimation du potentiel infectieux des sols, terreaux et substrats, infestés par *Pythium* sp. Ann. Phytopathol. 7:147–154.

Bouhot, D. 1980. Le potentiel infectieux des sol (soil infectivity) un concept, un modèle pour sa mesure, quelques applications. Ph.D. Thesis, Universite de Nancy. 150 pp.

Bouhot, D., and H. Joannes. 1979. Ecologie des champignons parasites dans le sol. IX. Mesures du potentiel infectieux des sols naturellement infestés de *Pythium* sp. Soil Biol. Biochem. 11:417–429.

Bouhot, D., and R. Perrin. 1980. Mise en évidence de résistances biologiques aux *Pythium* en sol forestier. Eur. Jour. For. Pathol. 10:77–89.

Bowen, G. D., and A. D. Rovira. 1976. Microbial colonization of plant roots. Annu. Rev. Phytopathol. 14:121–144.

Boyer, J. S. 1970. Leaf enlargement and metabolic rates in corn, soybean, and sunflower at various leaf water potentials. Plant Physiol. 46:233–235.

Boyer, J. S. 1971. Nonstomatal inhibition of photosynthesis in sunflower at low leaf water potentials and high light intensities. Plant Physiol. 48:532–536.

Boyle, L. W. 1961. The ecology of *Sclerotium rolfsii* with emphasis on the role of saprophytic media. Phytopathology 51:117–119.

Brader, L. 1980. Advances in applied entomology. Ann. Appl. Biol. 94:349–365.

Brady, B. L. 1960. Occurrence of *Itersonilia* and *Tilletiopsis* on lesions caused by *Entyloma*. Trans. Brit. Mycol. Soc. 43:31-50.

Brathwaite, C. W. D. 1978. Inhibition of *Sclerotium rolfsii* by *Pseudomonas aeruginosa* and *Bacillus subtilis* and its significance in the biological control of southern blight of pigeon pea [*Cajanus cajan* (L.) Millsp.]. Abstr. 3rd Internat. Congr. Plant Pathol. München. p. 197.

Brian, P. W. 1944. Production of gliotoxin by *Trichoderma viride*. Nature 154:667–668.

Brian, P. W., and J. G. McGowan. 1945. Viridin: A highly fungistatic substance produced by *Trichoderma viride*. Nature 156:144–145.

Brian, P. W., and J. C. McGowan. 1946. Biologically active metabolic products of the mould *Metarrhizium glutinosum* S. Pope. Nature 157:334.

Broadbent, P., and K. F. Baker. 1974a. Behaviour of *Phytophthora cinnamomi* in soils suppressive and conducive to root rot. Austral. Jour. Agric. Res. 25:121–137.

Broadbent, P., and K. F. Baker. 1974b. Association of bacteria with sporangium formation and breakdown of sporangia in *Phytophthora* spp. Austral. Jour. Agric. Res. 25:139–145.

Broadbent, P., K. F. Baker, and Y. Waterworth. 1971. Bacteria and actinomycetes antagonistic to fungal root pathogens in Australian soils. Austral. Jour. Biol. Sci. 24:925–944.

Broadbent, P., K. F. Baker, N. Franks, and J. Holland. 1977. Effect of *Bacillus* spp. on increased growth of seedlings in steamed and nontreated soil. Phytopathology 67:1027–1033.

Brodie, I. D. S., and J. P. Blakeman. 1975. Competition for carbon compounds by a leaf surface bacterium and conidia of *Botrytis cinerea*. Physiol. Plant Pathol. 6:125–135.

Brooks, D. H., and M. G. Dawson. 1968. Influence of direct-drilling of winter wheat on incidence of take-all and eyespot. Ann. Appl. Biol. 61:57–64.

Brown, B. N. 1976. *Phytophthora cinnamomi* associated with patch death in tropical rain forests in Queensland. Austral. Plant Pathol. Newsletter 5:1–4.

Brown, M. E. 1974. Seed and root bacterization. Annu. Rev. Phytopathol. 12:181–197.

Brown, M. E., and D. Hornby. 1971. Behavior of *Ophiobolus graminis* on slides buried in soil in the presence of wheat seedlings. Trans. Brit. Mycol. Soc. 56:95–103.

Brown, W. 1922. Studies on the physiology of parasitism. VIII. On the exosmosis of nutrient substances from the host tissue into the infection drop. Ann. Bot. 36:101–119.

Brown, W., and N. Montgomery. 1948. Problems in the cultivation of winter lettuce. Ann. Appl. Biol. 35:161–180.

Bruehl, G. W. 1951. Root rot in grains and grasses. South Dakota Farm Home Research 2:76–79.

Bruehl, G. W. 1957. Cephalosporium stripe disease of wheat. Phytopathology 47:641–649.

Bruehl, G. W. 1968. Ecology of Cephalosporium stripe disease of winter wheat in Washington. Plant Dis. Reptr. 52:590-594.

Bruehl, G. W., ed. 1975. Biology and Control of Soil-borne Plant Pathogens. Amer. Phytopathol. Soc., St. Paul, MN. 216 pp.

Bruehl, G. W. 1976. Management of food resources by fungal colonists of cultivated soils. Annu. Rev. Phytopathol. 14:247–264.

Bruehl, G. W., and B. Cunfer. 1971. Physiologic and environmental factors that affect the severity of snow mold of wheat. Phytopathology 61:792–799.

Bruehl, G. W., and P. Lai. 1968. The probable significance of saprophytic colonization of wheat straw in the field by *Cephalosporium gramineum*. Phytopathology 58:464–466.

Bruehl, G. W., and J. B. Manandhar. 1972. Some water relations of *Cercosporella herpotrichoides*. Plant Dis. Reptr. 56:594–596.

Bruehl, G. W., R. L. Millar, and B. Cunfer. 1969. Significance of antibiotic production by *Cephalosporium gramineum* to its saprophytic survival. Can. Jour. Plant Sci. 49:235–246.

Brun, H., B. Tivoli, and F. Carpentier. 1976. Influence des caracteristiques des sols sur la manifestation de maladies d'origine tellurique et la mise en oeuvre de méthodes de lutte biologique. Sci. Agron. Rennes 1976, 37–48.

Buchanan, R. E., and N. E. Gibbons, eds. 1974. Bergey's Manual of Determinative Bacteriology. 8th ed. Williams and Wilkins, Baltimore. 1246 pp.

Bull, A. T. 1970a. Inhibition of polysaccharases by melanin: Enzyme inhibition in relation to mycolysis. Arch. Biochem. Biophys. 137:345–356.

Bull, A. T. 1970b. Chemical composition of wild-type and mutant *Aspergillus nidulans* cell walls. The nature of polysaccharide and melanin constituents. Jour. Gen. Microbiol. 63:75–94.

Buller, A. H. R. 1915. The fungus lore of the Greeks and Romans. Trans. Brit. Mycol. Soc. 5:21–66.

Bulliard, P. 1791–1812. Histoire des Champignons de la France. 4 vols. Paris.

Burdsall, H. H., Jr. 1979. *Laetisaria* (Aphyllophorales, Corticiaceae), a new genus for the teleomorph of *Isaria fuciformis*. Trans. Brit. Mycol. Soc. 72:419–422.

Burdsall, H. H., Jr., H. C. Hoch, M. G. Boosalis, and E. C. Setliff. 1980. *Laetisaria arvalis* (Aphyllophorales, Corticiaceae): A possible biological control agent for *Rhizoctonia solani* and *Pythium* species. Mycologia 72:728–736.

Burford, J. R. 1976. Effect of the application of cow slurry to grassland on the composition of the soil atmosphere. Jour. Sci. Food Agric. 27:115–126.

Burges, H. D., ed. 1981. Microbial Control of Pests and Plant Diseases, 1970-1980. Academic Press, New York. 914 pp.

Burke, D. W. 1965. Fusarium root rot of beans and behavior of the pathogen in different soils. Phytopathology 55:1122–1126.

Burke, D. W., D. E. Miller, L. D. Holmes, and A. W. Barker. 1972. Counteracting bean root rot by loosening the soil. Phytopathology 62:306–309.

Burke, M. J., L. V. Gusta, H. A. Quamme, C. J. Weiser, and P. H. Li. 1976. Freezing and injury in plants. Annu. Rev. Plant Physiol. 27:507–528.

Burr, T. J., M. N. Schroth, and T. Suslow. 1978. Increased potato yields by treatment of seedpieces with specific strains of *Pseudomonas fluorescens* and *P. putida*. Phytopathology 68:1377–1383.

Burt, E. A. 1925. The Thelephoraceae of North America. XIV. Ann. Missouri Bot. Gard. 12:213–357.

Burton, W. G., and M. J. Wigginton. 1970. The effect of a film of water upon the oxygen status of a potato tuber. Potato Res. 13:180-186.

Bushnell, W. R. 1970. Patterns in the growth, oxygen uptake, and nitrogen content of single colonies of wheat stem rust on wheat leaves. Phytopathology 60:92–99.

Butterfield, E. J., J. E. De Vay, and R. H. Garber. 1978. The influence of several crop sequences on the incidence of Verticillium wilt of cotton and on the population of *Verticillium dahliae* in field soil. Phytopathology 68:1217–1220.

Cain, R. F. 1982. Studies on Fungi Imperfecti. I. *Phialophora*. Can. Jour. Bot. 30:338–343.

California Agricultural Experiment Station. 1982. Special issue: Genetic engineering of plants. Calif. Agric. 36(8):1–36.

California Department of Agriculture. 1958. Regulations for the centrification of avocado nursery stock. Calif. Dept. Agric. Mimeo. 2 pp.

Carter, J. C. 1945. Wetwood of elms. Ill. Nat. Hist. Surv. 23:407–448.

Carter, M. V. 1971. Biological control of *Eutypa armeniacae*. Austral. Jour. Exp. Agric. Anim. Husb. 11:687–692.

Carter, M. V. 1983. Biological control of *Eutypa armeniacae*. V. Guidelines for establishing routine wound protection in commercial apricot orchards. Austral. Jour. Exp. Agric. Anim. Husb. 23:In press.

Carter, M. V., and T. V. Price. 1974. Biological control of *Eutypa armeniacae*. II. Studies of the interaction between *E. armeniacae* and *Fusarium lateritium* and their relative sensitivities to benzimidazole chemicals. Austral. Jour. Agric. Res. 25:105–109.

Caruso, F. L., and J. Kuć. 1977a. Protection of watermelon and muskmelon against *Colletotrichum lagenarium* by *Colletotrichum lagenarium*. Phytopathology 67:1285–1289.

Caruso, F. L., and J. Kuć. 1977b. Field protection of cucumber, watermelon, and muskmelon against *Colletotrichum lagenarium* by *Colletotrichum lagenarium*. Phytopathology 67:1290-1292.

Castanho, B., and E. E. Butler. 1978a. Rhizoctonia decline: A degenerative disease of *Rhizoctonia solani*. Phytopathology 68:1505–1510.

Castanho, B., and E. E. Butler. 1978b. Rhizoctonia decline: Studies on hypovirulence and potential use in biological control. Phytopathology 68:1511–1514.

Castanho, B., E. E. Butler, and R. J. Shepherd. 1978. The association of double-stranded RNA with Rhizoctonia decline. Phytopathology 68:1515–1519.

Cayrol, J.-C., and J.-P. Frankowski. 1979. Une methode de lutte biologique contre les nematodes a galles des racines appartenante au genre *Meloidogyne*. Rev. Hortic. 193:15–23.

Cayrol, J.-C., J.-P. Frankowski, A. Laniece, G. d'Hordemare, and J. P. Talon. 1978. Contre les nematodes en champignonniere. Mise au point d'une methode de lutte biologique a l'aide d'un hyphomycete predateur *Arthrobotrys robusta* souche *antipolis* (Royale 300). Rev. Hortic. 184:23–30.

Chang, I.-P., and T. Kommedahl. 1968. Biological control of seedling blight of corn by coating kernels with antagonistic microorganisms. Phytopathology 58:1395–1401.

Charudattan, R., and H. L. Walker, eds. 1982. Biological Control of Weeds with Plant Pathogens. John Wiley & Sons, New York. 293 pp.

Chatterjee, A. K., L. N. Gribbins, and J. A. Carpenter. 1969. Some observations on the physiology of *Erwinia herbicola* and its possible implication as a factor antagonistic to *Erwinia amylovora* in the "fire blight" syndrome. Can. Jour. Microbiol. 15:640-642.

Chelminski, R. 1979. A fungus beats the chestnut blight at its own game. Smithsonian 10:97–102.

Chester, K. S. 1933. The problem of acquired physiological immunity in plants. Quart. Rev. Biol. 8:129–154, 275–324.

Chet, I., and R. Baker. 1980. Induction of suppressiveness to *Rhizoctonia solani* in soil. Phytopathology 70:994–998.

Chet, I., and R. Baker. 1981. Isolation and biocontrol potential of *Trichoderma hamatum* from soil naturally suppressive to *Rhizoctonia solani*. Phytopathology 71:286–290.

Chet, I., G. E. Harman, and R. Baker. 1981. *Trichoderma hamatum*: Its hyphal interactions with *Rhizoctonia solani* and *Pythium* spp. Microb. Ecol. 7:29–38.

Chiariello, N., J. C. Hickman, and H. A. Mooney. 1982. Endomycorrhizal role for interspecific transfer of phosphorus in a community of annual plants. Science 217:941–943.

Chiu, W. F., and Y. H. Chang. 1982. Advances of science of plant protection in the People's Republic of China. Annu. Rev. Phytopathol. 20:71–92.

Chivers, A. H. 1915. A monograph of the genera *Chaetomium* and *Ascotrichia*. Mem. Torrey Bot. Club 14:155–240.

Chou, L. G., and A. F. Schmitthenner. 1974. Effect of *Rhizobium japonicum* and *Endogone mosseae* on soybean root rot caused by *Pythium ultimum* and *Phytophthora megasperma* var. *sojae*. Plant Dis. Reptr. 58:221–225.

Christensen, N. W., R. G. Taylor, T. L. Jackson, and B. L. Mitchell. 1981. Chloride effects on water potentials and yield of winter wheat infected with take-all root rot. Agron. Jour. 73:1053–1058.

Christie, J. R. 1936. The development of root-knot nematode galls. Phytopathology 26:1–22.

Clark, F. E. 1942. Experiments toward the control of the take-all disease of wheat and the Phymatotrichum root rot of cotton. U.S. Dept. Agric. Tech. Bull. 835:1–27.

Clarkson, D. T., and J. B. Hanson. 1980. The mineral nutrition of higher plants. Annu. Rev. Plant Physiol. 31:239–298.

Clarnholm, M., and T. Rosswall. 1980. Biomass and turnover of bacteria in a forest soil and a peat. Soil Biol. Biochem. 12:49–57.

Clawson, K. L., and B. L. Blad. 1982. Infrared thermometry for scheduling irrigation of corn. Agron. Jour. 74:311–316.

Clayton, E. E. 1923. The relation of soil moisture to the Fusarium wilt of tomato. Amer. Jour. Bot. 10:133–147.

Clements, H. F. 1980. Sugarcane Crop Logging and Crop Control. Principles and Practices. Univ. of Hawaii Press, Honolulu. 520 pp.

Coker, W. C., and J. N. Couch. 1928. The Gasteromycetes of the Eastern United States and Canada. Univ. North Carolina Press, Chapel Hill. 201 pp.

Colbaugh, P. F., and R. M. Endo. 1972. Drought stress as a factor stimulating the saprophytic activity of *Helminthosporium sativum* on bluegrass debris. Phytopathology 62:751.

Coley-Smith, J. R., K. Verhoeff, and W. R. Jarvis, eds. 1980. Biology of Botrytis. Academic Press, London. 318 pp.

Collins, M. A. 1982. Rust disease and development of phylloplane microflora of *Antirrhinum* leaves. Trans. Brit. Mycol. Soc. 79:117–122.

Cook, R. J. 1968. Fusarium root and foot rot of cereals in the Pacific Northwest. Phytopathology 58:127–131.

Cook, R. J. 1970. Factors affecting saprophytic colonization of wheat straw by *Fusarium roseum* f. sp. *cerealis* 'Culmorum'. Phytopathology 60:1672–1676.

Cook, R. J. 1973. Influence of low plant and soil water potentials on diseases caused by soilborne fungi. Phytopathology 63:451–458.

Cook, R. J. 1980. Fusarium foot rot of wheat and its control in the Pacific Northwest. Plant Disease 64:1061–1066.

Cook, R. J. 1981. The influence of rotation crops on take-all decline phenomenon. Phytopathology 71:189–192.

Cook, R. J. 1982. Progress toward biological control of plant pathogens, with special reference to take-all of wheat. Agric. For. Bull. 5:22–30.

Cook, R. J., and G. W. Bruehl. 1968. Ecology and possible significance of perithecia of *Calonectria nivalis* in the Pacific Northwest. Phytopathology 58:702–703.

Cook, R. J., and A. A. Christen. 1976. Growth of cereal root-rot fungi as affected by temperature-water potential interactions. Phytopathology 66:193–197.

Cook, R. J., and N. T. Flentje. 1967. Chlamydospore germination and germling survival of *Fusarium solani* f. *pisi* in soil as affected by soil water and pea seed exudation. Phytopathology 57:178–182.

Cook, R. J., and W. A. Haglund. 1982. Pythium root rot: A barrier to yield of Pacific Northwest wheat. Wash. State Univ. Agric. Res. Centr. Res. Bull. No. XB0913. 20 pp.

Cook, R. J., and Y. Homma. 1979. Influence of soil water potential on activity of amoebae responsible for perforations of fungal spores. Phytopathology 69:914.

Cook, R. J., and T. Naiki. 1982. Virulence of *Gaeumannomyces graminis* var. *tritici* from fields under short-term and long-term wheat cultivation in the Pacific Northwest, USA. Plant Pathology 31:201–207.

Cook, R. J., and R. I. Papendick. 1970. Soil water potential as a factor in the ecology of *Fusarium roseum* f. sp. *cerealis* 'Culmorum'. Plant Soil 32:131–145.

Cook, R. J., and R. I. Papendick. 1972. Influence of water potential of soils and plants on root disease. Annu. Rev. Phytopathol. 10:349–374.

Cook, R. J., and R. I. Papendick. 1978. Role of water potential in microbial growth and development of plant disease, with special reference to postharvest pathology. HortScience 13:559–564.

Cook, R. J., R. I. Papendick, and D. M. Griffin. 1972. Growth of two root-rot fungi as affected by osmotic and matric water potentials. Proc. Soil Sci. Soc. Amer. 36:78–82.

Cook, R. J., and A. D. Rovira. 1976. The role of bacteria in the biological control of *Gaeumannomyces graminis* by suppressive soils. Soil Biol. Biochem. 8:267–273.

Cook, R. J., and M. N. Schroth. 1965. Carbon and nitrogen compounds and germination of chlamydospores of *Fusarium solani* f. *phaseoli*. Phytopathology 55:254–256.

Cook, R. J., J. W. Sitton, and J. T. Waldher. 1980. Evidence for *Pythium* as a pathogen of direct-drilled wheat in the Pacific Northwest. Plant Disease 64:102–103.

Cook, R. J., and A. M. Smith. 1977. Influence of water potential on production of ethylene in soil. Can. Jour. Microbiol. 23:811–817.

Cook, R. J., and W. C. Snyder. 1965. Influence of host exudates on growth and survival of germlings of *Fusarium solani* f. *phaseoli* in soil. Phytopathology 55:1021–1025.

Cook, R. J., and J. T. Waldher. 1977. Influence of stubble-mulch residue management on *Cercosporella* foot rot and yields of winter wheat. Plant Dis. Reptr. 61:96–100.

Cook, R. J., and R. D. Watson, eds. 1969. Nature of the influence of crop residues on fungus-induced root diseases. Wash. Agric. Exp. Sta. Bull. 716:1–32.

Cooke, R. 1968. Relationships between nematode-destroying fungi and soil-borne nematodes. Phytopathology 58:909–913.

Cooksey, D. A., and L. W. Moore. 1982a. Biological control of crown gall with an agrocin mutant of *Agrobacterium radiobacter*. Phytopathology 72:919–921.

Cooksey, D. A., and L. W. Moore. 1982b. High frequency spontaneous mutations to agrocin 84 resistance in *Agrobacterium tumefaciens* and *A. rhizogenes*. Physiol. Plant Pathol. 20:129–135.

Coons, G. H. 1937. Progress in plant pathology: Control of disease by resistant varieties. Phytopathology 27:622–631.

Corke, A. T. K. 1978. Interactions between microorganisms. Ann. Appl. Biol. 89:89–93.

Corke, A. T. K., and T. Hunter. 1979. Biocontrol of *Nectria galligena* infection of pruning wounds on apple shoots. Jour. Hortic. Sci. 54:47–55.

Costa, A. S., and G. W. Müller. 1980. Tristeza control by cross protection: A U.S.-Brazil cooperative success. Plant Disease 64:538:541.

Cother, E. J., and D. M. Griffin. 1974. Chlamydospore germination in *Phytophthora drechsleri*. Trans. Brit. Mycol. Soc. 63:273–279.

Couch, J. N. 1945. Observations on the genus Catenaria. Mycologia 37:163–193.

Crowe, R. J., and D. H. Hall. 1980. Vertical distribution of sclerotia of *Sclerotium cepivorum* and host root systems relative to white rot of onion and garlic. Phytopathology 70:70-73.

Cunfer, B. M. 1975. Colonization of ergot honeydew by *Fusarium heterosporum*. Phytopathology 65:1372–1374.

Cunningham, P. C. 1967. A study of ploughing depth and foot and root rots of spring wheat. Ire. Jour. Agric. Res. 6:33–39.

Curl, E. A. 1979. Suppression of *Rhizoctonia solani* and damping-off in cotton by mycophagous insects of the order Collembola. Phytopathology 69:526.

Davet, P., M. Artigues, and C. Martin. 1981. Production en conditions non aseptiques d'inoculum de *Trichoderma harzianum* Rifai pour des essais de lutte biologique. Agronomie 1:933–936.

Davies, W. J. 1977. Stomatal responses to water stress and light in plants grown in controlled environments and in the field. Crop Sci. 17:735–740.

Davis, M. J., A. G. Gillaspie, Jr., R. W. Harris, and R. H. Lawson. 1980. Ratoon stunting disease of sugarcane: Isolation of the casual bacterium. Science 210:1365–1367.

Davis, R. M., and J. A. Menge. 1980. Influence of *Glomus fasciculatus* and soil phosphorus on Phytophthora root rot of citrus. Phytopathology 70:447–452.

Davis, R. M., J. A. Menge, and D. C. Erwin. 1979. Influence of *Glomus fasciculatus* and soil phosphorus on Verticillium wilt of cotton. Phytopathology 69:453–456.

Davis, R. M., J. A. Menge, and G. A. Zentmyer. 1978. Influence of vesicular-arbuscular mycorrhizae on Phytophthora root rot of three crop plants. Phytopathology 68:1614–1617.

Day, P. R. 1977. The Genetic Basis of Epidemics in Agriculture. New York Academy of Science, New York.

Day, P. R., J. A. Dodds, J. E. Elliston, R. A. Jaynes, and S. L. Anagnostakis. 1977. Double-stranded RNA in *Endothia parasitica*. Phytopathology 67:1393–1396.

Deacon, J. W. 1973a. Control of the take-all fungus by grass leys in intensive cereal cropping. Plant Pathol. 22:88–94.

Deacon, J. W. 1973b. Factors affecting occurrence of the Ophiobolus patch disease of turf and its control by *Phialophora radicicola*. Plant Pathol. 22:149–155.

Deacon, J. W. 1974a. Interactions between varieties of *Gaeumannomyces graminis* var. *tritici* and *Phialophora radicicola* on roots, stem bases, and rhizomes of the Gramineae. Plant Pathol. 25:85–92.

Deacon, J. W. 1974b. Further studies on *Phialophora radicicola* and *Gaeumannomyces graminis* on roots and stem bases of grasses and cereals. Trans. Brit. Mycol. Soc. 63:307–327.

Deacon, J. W. 1976a. Biological control of the take-all fungus, *Gaeumannomyces graminis*, by *Phialophora radicicola* and similar fungi. Soil Biol. Biochem. 8:275–283.

Deacon, J. W. 1976b. Studies on *Pythium oligandrum*, an aggressive parasite of other fungi. Trans. Brit. Mycol. Soc. 66:383–391.

Deacon, J. W., and C. M. Henry. 1980. Age of wheat and barley roots and infection by *Gaeumannomyces graminis* var. *tritici*. Soil Biol. Biochem. 12:113–118.

DeBach, P., ed. 1964. Biological Control of Insect Pests and Weeds. Reinhold, New York. 844 pp.

Dehne, H. W. 1982. Interaction between vesicular-arbuscular mycorrhizal fungi and plant pathogens. Phytopathology 72:1115–1119.

Dehne, H. W., and F. Schönbeck. 1975. Untersuchungen über den Einfluss der Endotrophen Mykorrhiza auf die Fusarium-Welke der Tomate. Zeit. Pflanzenkr. Pflanzensch. 82:630-632.

Dekker, J., and S. G. Georgopoulos, eds. 1982. Fungicide Resistance in Crop Protection. Pudoc, Wageningen. 265 pp.

Deleney, H. G. 1977. The Australian avocado industry. Proc. Austral. Avocado Research Workshop, Binna Burra, Queensland. pp. 7–20.

Dennis, C., and J. Webster. 1971a. Antagonistic properties of species-groups of *Trichoderma*. I. Production of non-volatile antibiotics. Trans. Brit. Mycol. Soc. 57:25–39.

Dennis, C., and J. Webster. 1971b. Antagonistic properties of species-groups of Trichoderma. II. Production of volatile antibiotics. Trans. Brit. Mycol. Soc. 57:41–48.

Dennis, R. W. G. 1956. A revision of the British Helotiaceae in the Herbarium of the Royal Botanic Gardens, Kew, with notes on related European species. Commonwealth Mycol. Inst. Mycol. Papers 62:1–216.

Dennis, R. W. G. 1968. British Ascomycetes. Cramer, Stuttgart. 453 pp.

Dhingra, O. D., and J. B. Sinclair. 1974. Effect of soil moisture and carbon:nitrogen ratio on survival of *Macrophomina phaseolina* in soybean stems in soil. Plant Dis. Reptr. 58:1034–1037.

Dickinson, C. H. 1981. Interactions of fungicides with minor pathogens on cereals. EPPO Bull. 11:311–316.

Dickinson, C. H., and T. F. Preece, eds. 1976. Microbiology of Aerial Plant Surfaces. Academic Press, New York. 669 pp.

Dickson, J. G. 1923. Influence of soil temperature and moisture on the development of the seedling-blight of wheat and corn caused by *Gibberella saubinetti*. Jour. Agric. Res. 23:837–870.

Dimock, A. W., and K. F. Baker. 1951. Effect of climate on disease development, injuriousness, and fungicidal control, as exemplified by snapdragon rust. Phytopathology 41:536–552.

Dobell, C. 1913. Observations on the life history of Cienkowski's "Arachnula." Arch. Protistkd. 31:317–353.

Dobson, R. L., R. L. Gabrielson, A. S. Baker, and L. Bennett. 1983. Effects of lime particle size and distribution and fertilizer formulation on club root disease caused by *Plasmodiophora brassicae*. Plant Disease 67:50-52.

Dodd, J. L. 1980. The role of plant stresses in development of corn stalk rots. Plant Disease 64:533–537.

Doherty, M. A., and T. F. Preece. 1978. *Bacillus cereus* prevents germination of uredospores of *Puccinia allii* and the development of rust disease of leek, *Allium porrun,* in controlled environments. Physiol. Plant Pathol. 12:123–132.

Dommergues, Y. R., and S. V. Krupa. 1978. Interactions between Non-Pathogenic Soil Microorganisms and Plants. Elsevier, Amsterdam. 475 pp.

Domsch, K. H., W. Gams, and T. H. Anderson. 1981. Compendium of Soil Fungi. 2 vols. Academic Press, New York. 1250 pp.

Doupnik, B., Jr., and M. G. Boosalis. 1980. Ecofallow—a reduced tillage system—and plant diseases. Plant Disease 64:31–35.

Drayton, F. L. 1929. Bulb growing in Holland and its relation to disease control. Sci. Agric. 9:494–509.

Drechsler, C. 1936. A *Fusarium*-like species of *Dactylella* capturing and consuming testaceous rhizopods. Jour. Wash. Acad. Sci. 26:397–404.

Drechsler, C. 1937. Some Hyphomycetes that prey on free-living terricolous nematodes. Mycologia 29:447–552.

Drechsler, C. 1940. Three new Hyphomycetes preying on free-living terricolous nematodes. Mycologia 32:448–470.

Drechsler, C. 1946. Several species of *Pythium* peculiar in their sexual development. Phytopathology 36:781–864.

Drechsler, C. 1950. Several species of *Dactylella* and *Dactylaria* that capture free-living nematodes. Mycologia 42:1–79.

Dube, A. J., R. L. Dodman, and N. T. Flentje. 1971. The influence of water activity on the growth of *Rhizoctonia solani*. Austral. Jour. Biol. Sci. 24:57–65.

Duddington, C. L. 1951. *Dactylella lobata,* predacious on nematodes. Trans. Brit. Mycol. Soc. 34:489–491.

Duniway, J. M. 1975. Formation of sporangia by *Phytophthora drechsleri* in soil at high matric potentials. Can. Jour. Bot. 53:1270-1275.

Duniway, J. M. 1976. Movement of zoospores of *Phytophthora cryptogea* in soils of various textures and matric potentials. Phytopathology 66:877–882.

Duniway, J. M. 1977. Predisposing effect of water stress on the severity of Phytophthora root rot in safflower. Phytopathology, 67:884–889.

Duniway, J. M. 1979. Water relations of water molds. Annu. Rev. Phytopathol. 17:431–460.

Dutta, B. K. 1981. Studies on some fungi isolated from the rhizosphere of tomato plants and the consequent prospect for the control of Verticillium wilt. Plant Soil 63:209–216.

Dye, D. W. 1969. A taxonomic study of the genus Erwinia. III. The "herbicola" group. N. Zeal. Jour. Sci. 12:223–236.

Ebben, M. H., and D. M. Spencer. 1978. The use of antagonistic organisms for the control of black root rot of cucumber, *Phomopsis sclerotioides*. Ann. Appl. Biol. 89:103–106.

Edmunds, L. K. 1964. Combined relation of plant maturity, temperature, and soil moisture to charcoal stalk rot development in grain sorghum. Phytopathology 54:514–517.

Elad, Y., I. Chet, and J. Katan. 1980a. *Trichoderma harzianum*: A biocontrol agent effective against *Sclerotium rolfsii* and *Rhizoctonia solani*. Phytopathology 70:119–121.

Elad, Y., J. Katan, and I. Chet. 1980b. Physical, biological, and chemical control integrated for soilborne diseases of potatoes. Phytopathology 70:418–422.

Elad, Y., I. Chet, and Y. Henis. 1981a. Biological control of *Rhizoctonia solani* in strawberry fields by *Trichoderma harzianum*. Plant Soil 60:245–254.

Elad, Y., Y. Hadar, E. Hadar, I. Chet, and Y. Henis. 1981b. Biological control of *Rhizoctonia solani* by *Trichoderma harzianum* in carnation. Plant Disease 65:675–677.

Elad, Y., Y. Hadar, I. Chet, and Y. Henis. 1982a. Prevention with *Trichoderma harzianum* Rifai aggr., of reinfestation by *Sclerotium rolfsii* Sacc. and *Rhizoctonia solani* Kühn of soil fumigated with methylbromide, and improvement of disease control in tomatoes and plants. Crop Prot. 1:199–2ll.

Elad, Y., A. Kalfon, and I. Chet. 1982b. Control of *Rhizoctonia solani* in cotton by seed-coating with *Trichoderma* spp. Plant Soil 66:279–281.

Elad, Y., I. Chet, P. Boyle, and Y. Henis. 1983. Parasitism of *Trichoderma* spp. on *Rhizoctonia solani* and *Sclerotium rolfsii* – scanning electron microscopy and fluorescence microscopy. Phytopathology 73:85–88.

Elleman, C. J., and P. F. Entwistle. 1982. A study of glands on cotton responsible for the high pH and cation concentration of the leaf surface. Ann. Appl. Biol. 100:553–558.

Ellingboe, A. H. 1981. Changing concepts in host-pathogen genetics. Annu. Rev. Phytopathol. 19:125–143.

Ellis, J. G., A. Kerr, M. Van Montagu, and J. Schell. 1979. *Agrobacterium*: Genetic studies on agrocin 84 production and the biological control of crown gall. Physiol. Plant Pathol. 15:311–319.

Ellis, M. B. 1971. Dematiaceous Hyphomycetes. Commonwealth Mycological Institute, Kew. 608 pp.

Ellis, M. B. 1976. More Dematiaceous Hyphomycetes. Commonwealth Mycological Institute, Kew. 507 pp.

Ellwood, D. C., J. N. Hedger, M. J. Latham, J. M. Lynch, and J. H. Slater, eds. 1980. Contemporary Microbial Ecology. Academic Press, New York. 438 pp.

Eplee, R. E. 1975. Ethylene: A witchwood seed germination stimulant. Weed Science 23:433–436.

Ercolani, G. L., D. J. Hagedorn, A. Kelman, and R. E. Rand. 1974. Epiphytic survival of *Pseudomonas syringae* on hairy vetch in relation to epidemiology of bacterial brown spot of bean in Wisconsin. Phytopathology 64:1330-1339.

Erwin, D. C., S. Bartnicki-Garcia, and P. H. Tsao, eds. 1983. *Phytophthora*: Its Biology, Taxonomy, Ecology, and Pathology. Amer. Phytopathol. Soc., St. Paul, MN. 400 pp.

Estores, R. A., and T. A. Chen. 1972. Interactions of *Pratylenchus penetrans* and *Meloidogyne incognita* as coinhabitants in tomato. Jour. Nematol. 4:170-174.

Etheridge, D. E. 1956. Occurrence of *Coryne sarcoides* with heart-rot fungi on spruce in Alberta, Canada. Trans. Brit. Mycol. Soc. 39:385–386.

Etheridge, D. E. 1957. Comparative studies of *Coryne sarcoides* (Jacq.) Trel. and two species of wood-destroying fungi. Can. Jour. Bot. 35:595–663.

Evans, G., W. C. Snyder, and S. Wilhelm. 1966. Inoculum increase of the Verticillium wilt fungus in cotton. Phytopathology 56:590-594.

Fan, Shêng-Chih. 1974. "Fan Shêng-Chih Shu", an Agriculturist Book of China written by Fan Shêng-Chih in the First Century B. C. (translated by S. Sheng-Han). Science Press, Peking. 67 pp.

Fawcett, H. S. 1931. The importance of investigations on the effects of known mixtures of microorganisms. Phytopathology 21:545–550.

Fellows, H., and C. H. Ficke. 1934. Wheat take-all. Kansas Agric. Exp. Sta. Ann. Rep. 1932–1934:95–96.

Férault, A. C., B. Tivoli, J. M. Lemaire, and D. Spire. 1979. Etude de l'evolution comparée du niveau d'agressivité et du contenu en particules de type viral d'une souche de *Gaeumannomyces graminis* (Sacc.) Arx et Olivier (*Ophiobolus graminis* Sacc.). Ann. Phytopathol. 11:185–191.

Ferguson, J. 1958. Reducing plant disease with fungicidal soil treatment, pathogen-free stock, and controlled microbial colonization. Ph.D. Thesis, University of California, Berkeley. 169 pp.

Ferris, R. S. 1981. Calculating rhizosphere size. Phytopathology 71:1229–1231.

Flentje, N. T., and H. K. Saksena. 1964. Pre-emergence rotting of peas in South Australia. III. Host-pathogen interaction. Austral. Jour. Biol. Sci. 17:665–675.

Fletcher, J. T., and D. Butler. 1975. Strain changes in populations of tobacco mosaic virus from tomato crops. Ann. Appl. Biol. 81:409–412.

Fletcher, J. T., and J. M. Rowe. 1975. Observations and experiments on the use of an avirulent mutant strain of tobacco mosaic virus as a means of controlling tomato mosaic. Ann. Appl. Biol. 81:171–179.

Fokkema, N. J. 1968. The influence of pollen on the development of *Cladosporium herbarum* in the phyllosphere of rye. Neth. Jour. Plant Pathol. 74:159–165.

Fokkema, N. J. 1971. The effect of pollen in the phyllosphere of rye on colonization by saprophytic fungi and on infection by *Helminthosporium sativum* and other leaf pathogens. Neth. Jour. Plant Pathol. 77(Suppl. 1):1–60.

Fokkema, N. J. 1973. The role of saprophytic fungi in antagonism against *Dreschlera sorokiniana* (*Helminthosporium sativum*) on agar plates and on rye leaves with pollen. Physiol. Plant Pathol. 3:195–205.

Fokkema, N. J. 1978. Preliminary research on biological control of *Septoria nodorum* and *Cochliobolus sativus* in wheat. Acta Bot. Neerl. 27:153.

Fokkema, N. J., and M. P. De Nooij. 1981. The effect of fungicides on the microbial balance in the phyllosphere. EPPO Bull. 11:303–310.

Fokkema, N. J., and J. W. Lorbeer. 1974. Interactions between *Alternaria porri* and the saprophytic mycoflora of onion leaves. Phytopathology 64:1128–1133.

Fokkema, N. J., and F. van der Meulen. 1976. Antagonism of yeastlike phyllosphere fungi against *Septoria nodorum* on wheat leaves. Neth. Jour. Plant Pathol. 82:13–16.

Fokkema, N. J., J. G. den Houter, Y. J. C. Kosterman, and A. L. Nelis. 1979. Manipulation of yeasts on field-grown wheat leaves and their antagonistic effect on *Cochliobolus sativus* and *Septoria nodorum*. Trans. Brit. Mycol. Soc. 72:19–29.

Fokkema, N. J., I. Riphagen, R. J. Poot, and C. de Jong. 1983. Aphid honeydew, a potential stimulant of *Cochliobolus sativus* and *Septoria nodorum* on wheat leaves, and the competitive role of the saprophytic mycoflora. Trans. Brit. Mycol. Soc. 80:In press.

Fokkema, N. J., J. A. J. van de Laar, A. L. Nelis-Blomberg, and B. Schippers. 1975. The buffering capacity of the natural mycoflora of rye leaves to infection by *Cochliobolus sativus,* and its susceptibility to benomyl. Neth. Jour. Plant Pathol. 81:176–186.

Forbes, R. J. 1965. Studies in Ancient Technology. Vol. 2, 2nd ed., Leiden: E. J. Brill. 220 pp.

Forsyth, W. 1791. Observations on the Diseases, Defects, and Injuries in all Kinds of Fruit and Forest Trees. London. 71 pp.

Foster, R. C. 1981. The ultrastructure and histochemistry of the rhizosphere. New Phytol. 89:263–273.

Foster, R. C., A. D. Rovira, and T. W. Cock, 1983. Ultrastructure of the Root-Soil Interface. Amer. Phytopathol. Soc., St. Paul, MN.

Fox, R. A. 1977. The impact of ecological, cultural and biological factors on the strategy and costs of controlling root diseases in tropical plantation crops as exemplified by *Hevea brasilensis.* Jour. Rubber Res. Inst. Sri Lanka 54:329–362.

Fraser, L. R., K. Long, and J. Cox. 1968. Stem pitting of grapefruit – field protection by the use of mild virus strains. Pages 27–31 *in* Proc. 4th Conf. Internat'l. Organ. Citrus Virologists. Univ. Florida Press, Gainesville. 404 pp.

Frick, L. J., and R. M. Lister. 1978. Serotype variability in virus-like particles from *Gaeumannomyces graminis.* Virology 85: 504–517.

Friend, J., and D. R. Threlfall, eds. 1976. Biochemical Aspects of Plant-Parasite Relationships. Academic Press, New York. 354 pp.

Furtado, I. 1969. Effect of copper fungicides on the occurrence of the pathogenic form of *Colletotrichum coffeanum.* Trans. Brit. Mycol. Soc. 53:325–328.

Furuya, H., and T. Ui. 1981. The significance of soil microorganisms on the inhibition of the macroconidial germination of *Fusarium solani* f. sp. *phaseoli* in a soil suppressive to common bean root rot. Ann. Phytopathol. Soc. Japan. 47:42–49.

Gams, W., and A. Van Zaayen. 1982. Contribution to the taxonomy and pathogenicity of fungicolous *Verticillium* species. I. Taxonomy. Neth. Jour. Plant Pathol. 88:57–78.

Garcilaso de la Vega. 1966. Royal Commentaries of the Incas (translated by H. V. Livermore). Univ. Texas Press, Austin. 1530 pp.

Garren, K. H., and G. B. Duke. 1958. The effects of deep covering of organic matter and non-dirting weed control on peanut stem rot. Plant Dis. Reptr. 42:629–636.

Garrett, S. D. 1940. Soil conditons and the take-all disease of wheat. V. Further experiments on the survival of *Ophiobolus graminis* in infected wheat stubble buried in soil. Ann. Appl. Biol. 27:199–204.

Garrett, S. D. 1948. Soil conditions and the take-all disease of wheat. III. Interaction between host plant nutrition, disease escape, and disease resistance. Ann. Appl. Biol. 35:14–17.

Garrett, S. D. 1956. Biology of Root-infecting Fungi. Cambridge Univ. Press, London. 294 pp.

Garrett, S. D. 1970. Pathogenic Root-infecting Fungi. Cambridge Univ. Press, London. 294 pp.

Garrett, S. D., and W. Buddin. 1947. Control of take-all under the Chamberlain system of intensive barley growing. Agriculture: Jour. Minist. Agric. 54:425–426.

Gates, D. A. 1980. Biophysical Ecology. Springer-Verlag, New York, Heidelberg, Berlin. 611 pp.

Georgy, N. I., and J. R. Coley-Smith. 1982. Variation in morphology of *Sclerotium cepivorum* sclerotia. Trans. Brit. Mycol. Soc. 79:534–536.

Gerdemann, J. W. 1968. Vesicular-arbuscular mycorrhiza and plant growth. Annu. Rev. Phytopathol. 6:397–418.

Gerdemann, J. W., and J. M. Trappe. 1974. The Endogonaceae in the Pacific Northwest. Amer. Mycol. Soc. Mycol. Memoir 5:1–76.

Gerlach, W., and H. Nirenberg. 1982. The genus Fusarium – a Pictorial Atlas. Mitt. Biol. Bundesanst. Land- Forstwirtsh., Berlin-Dahlem. 209:1–406.

Gerlagh, M. 1968. Introduction of *Ophiobolus graminis* into new polders and its decline. Neth. Jour. Plant Pathol. 74: (Suppl. 2):1–97.

Gessler, C., and J. Kuć. 1982. Induction of resistance to Fusarium wilt of cucumber by root and foliar pathogens. Phytopathology 73:1439–1441.

Ghaffar, A. 1969. Biological control of white rot of onion. II. Effectiveness of *Penicillium nigricass* (Bain.) Thom. Mycopathol. et Mycol. Applic. 38:113–127.

Gibbs, A., and B. Harrison. 1976. Plant Virology. John Wiley & Sons, New York. 292 pp.

Gibbs, J. N. 1980. Role of *Ceratocystis piceae* in preventing infection by *Ceratocystis fagacearum* in Minnesota. Trans. Brit. Mycol. Soc. 74:171–174.

Gibbs, J. N., and M. E. Smith. 1978. Antagonism during the saprophytic phase of the life cycle of two pathogens of woody hosts—*Heterobasidion annosum* and *Ceratocystis ulmi*. Ann. Appl. Biol. 89:125–128.

Gilbert, R. G., and G. E. Griebel. 1969. The influence of volatile substances from alfalfa on *Verticillium dahliae* in soil. Phytopathology 56:1400-1403.

Gilbert, R. G., and R. G. Linderman. 1971. Increased activity of soil microorganisms near sclerotia of *Sclerotium rolfsii* in soil. Can. Jour. Microbiol. 17:557–562.

Gilligan, C. A. 1979. Modeling rhizosphere infection. Phytopathology 69:782–784.

Gilligan, C. A. 1980. Dynamics of root colonization by the take-all fungus, *Gaeumannomyces graminis*. Soil Biol. Biochem. 12:507–512.

Gindrat, D., E. van der Hoeven, and A. R. Moody. 1977. Control of *Phomopsis sclerotioides* with *Gliocladium roseum* or *Trichoderma*. Neth. Jour. Plant Pathol. 83 (Suppl. 1):429–438.

Gintis, B. O., G. Morgan-Jones, and R. Rodriguez-Kabana. 1982. Mycoflora of young cysts of *Heterodera glycines* in North Carolina soils. Nematropica 12:295–300.

Glynne, M. D. 1935. Incidence of take-all on wheat and barley on experimental plots at Woburn. Ann. Appl. Biol. 22:225–235.

Gommers, F. J. 1973. Nematicidal principles in compositae. Meded. Landbouwhogesch. Wageningen. 73–17:1–71.

Gordon, R. E., W. C. Haynes, and C. H-N. Pang. 1913. The genus *Bacillus*. U.S.D.A. Agr. Handbk. 427:1–283.

Gosevami, B. K., and H. J. Rumpenhorst. 1978. Association of an unknown fungus with potato cyst nematodes, *Globodera rostochiensis* and *Globodera pallida*. Nematalogica 24:251-256.

Graham, J. H., and J. A. Menge. 1982. Influence of vesicular- arbuscular mycorrhizae and soil phosphorus on take-all disease of wheat. Phytopathology 72:95–98.

Graham, J. H., R. G. Linderman, and J. A. Menge. 1982. Development of external hyphae by different isolates of mycorrhizal *Glomus* spp. in relation to root colonization and growth of Troyer citrange. New Phytol. 91:183–189.

Grant, T. J., and A. S. Costa. 1951. A mild strain of the tristeza virus of citrus. Phytopathology 41:114–122.

Green, R. J., Jr. 1980. Soil factors affecting survival of microsclerotia of *Verticillium dahliae*. Phytopathology 70:353–355.

Greig, B. J. W. 1976. Biological control of *Fomes annosus* by *Peniophora gigantea*. Eur. Jour. For. Pathol. 6:65–71.

Grente, J., and S. Sauret. 1969a. L'hypovirulence exclusive phénomène original en pathologie végétale. C. R. Acad. Sci. Paris 268:2347–2350.

Grente, J., and S. Sauret. 1969b. L'hypovirulence exclusive est-elle contrôlée par des déterminants cytoplasmiques? C. R. Acad. Sci. Paris 268:3173–3176.

Grieve, P. W., and M. J. W. Povey. 1981. Evidence for the osmotic dehydration theory of freeze damage. Jour. Sci. Food Agric. 32:96–98.

Griffin, D. M. 1972. Ecology of Soil Fungi. Chapman and Hall, London. 193 pp.

Griffin, D. M. 1977. Water potential and wood-decay fungi. Annu. Rev. Phytopathol. 15:319–29.

Griffin, D. M., and G. Quail. 1968. Movement of bacteria in moist, particulate systems. Austral. Jour. Biol. Sci. 21:579–582.

Griffin, G. J. 1970a. Carbon and nitrogen requirements for macroconidial germination of *Fusarium solani*: Dependence on conidial density. Can. Jour. Microbiol. 16:733–740.

Griffin, G. J. 1970b. Exogenous carbon and nitrogen requirements for chlamydospore germination by *Fusarium solani*: Dependence on spore density. Can. Jour. Microbiol. 16:1366–1368.

Griffin, G. J., K. H. Gerren, and J. D. Taylor. 1981. Influence of crop rotation and minimum tillage on the population of *Aspergillus flavus* group in peanut field soil. Plant Disease 65:898–900.

Griffiths, E. 1981. Iatrogenic plant diseases. Annu. Rev. Phytopathol. 19:69–82.

Grigg, D. B. 1974. The Agricultural Systems of the World. An Evolutionary Approach. Cambridge Univ. Press, London. 358 pp.

Grinstein, A., J. Katan, A. A. Razik, O. Zeydan, and Y. Elad. 1979. Control of *Sclerotium rolfsii* and weeds in peanuts by solar heating of the soil. Plant Dis. Rep. 63:1056–1059.

Grogan, R. G., and G. S. Abawi. 1975. Influence of water potential on growth and survival of *Whetzelinia sclerotiorum*. Phytopathology 65:122–128.

Grogan, R. G., M. A. Sall, and Z. K. Punja. 1980. Concepts for modeling root infection by soilborne fungi. Phytopathology 70:361–363.

Grosclaude, C. 1970. Premiers essais de protection biologique des blessures de taille vis-a-vis du *Stereum purpureum* Pers. Ann. Phytopathol. 2:507–516.

Grosclaude, C., J. Ricard, and B. Dubos. 1973. Inoculation of *Trichoderma viride* spores via pruning shears for biological control of *Stereum purpureum* on plum tree wounds. Plant Dis. Reptr. 57:25–28.

Grossbard, E. 1948. Production of an antibiotic substance on wheat straw and other organic material and in the soil. Nature 161:614–615.

Grossbard, E. 1952. Antibiotic production of fungi on organic manures and in soil. Jour. Gen. Microbiol. 6:295–310.

Groves, J. W., and D. E. Wilson. 1967. The nomenclatural status of Coryne. Taxon 16:35–41.

Gudmestad, N. C., J. E. Huguelet, and R. T. Zink. 1978. The effect of cultural practices and straw incorporation into the soil on Rhizoctonia disease of potato. Plant Dis. Reptr. 62:985–989.

Habte, M., and M. Alexander. 1975. Protozoa as agents responsible for the decline of *Xanthomonas campestris* in soil. Appl. Microbiol. 29:159–164.

Hadar, Y., I. Chet, and Y. Henis. 1979. Biological control of *Rhizoctonia solani* damping-off with wheat bran culture of *Trichoderma harzianum*. Phytopathology 69:64–68.

Hadwiger, L. A., and M. E. Schwochau. 1971. Ultraviolet light-induced formation of pisatin and phenylalanine ammonia lyase. Plant Physiol. 47:588–590.

Halsey, M., and R. Powelson. 1981a. The influence of NH_4Cl with KCl on antagonism between fluorescent pseudomonads (FP) and *Gaeumannomyces graminis* var. *tritici* (GGT) on the wheat rhizoplane. Phytopathology 71:105.

Halsey, M. E., and R. L. Powelson. 1981b. "Bacterization" seed treatment and banded fertilizers for suppression of take-all root rot of wheat. Oregon Agric. Exp. Sta. Spec. Rept. 633:41–46.

Hancock, J. G. 1977. Factors affecting soil populations of *Pythium ultimum* in the San Joaquin Valley of California. Hilgardia 45:107–122.

Hancock, J. G. 1981. Longevity of *Pythium ultimum* in moist soils. Phytopathology 71:1033–1037.

Hancock, J. G., and G. S. Benham, Jr. 1980. Fungal decay of buried cotton stems. Soil Biol. Biochem. 12:35–41.

Hancock, J. G., and O. C. Huisman. 1981. Nutrient movement in host-pathogen systems. Annu. Rev. Phytopathol. 19:309–31.

Hänssler, G., and M. Hermanns. 1981. *Verticillium lecanii* as a parasite on cysts of *Heterodera schachtii*. Zeit. Pflanzenkr. Pflanzensch. 88:678–681.

Hare, R. 1970. The Birth of Penicillin and the Disarming of Microbes. Allen and Unwin., London. 236 pp.

Harlan, J. R. 1976. Diseases as a factor in plant evolution. Annu. Rev. Phytopathol. 14:31–51.

Harley, J. L. 1969. The Biology of Mycorrhiza. Leonard Hill, London. 334 pp.

Harley, J. L., and S. E. Smith. 1983. Mycorrhizal Symbiosis. Academic Press, New York. 472 pp.

Harman, G. E., I. Chet, and R. Baker. 1980. *Trichoderma hamatum* effects on seed and seedling disease induced in radish and pea by *Pythium* spp. or *Rhizoctonia solani*. Phytopathology 70:1167–1172.

Harman, G. E., I. Chet, and R. Baker. 1981. Factors affecting *Trichoderma hamatum* applied to seeds as a biocontrol agent. Phytopathology 71:569–572.

Harris, D. C. 1979. The occurrence of *Phytophthora syringae* in fallen apple leaves. Ann. Appl. Biol. 91:309–312.

Hartley, C. 1921. Damping-off in forest nurseries. U. S. Dept. Agric. Bull. 934:1–99.

Hattori, T. 1973. Microbial Life in the Soil. Marcel Dekker, New York. 427 pp.

Hayes, A. J. 1982. Phylloplane micro-organisms of Rosa cv. Picadilly following infection by *Diplocarpon rosae*. Trans. Brit. Mycol. Soc. 79:311–319.

Hayward, A. C. 1974. Latent infections by bacteria. Annu. Rev. Phytopathol. 12:87–97.

Heald, F. D. 1933. Manual of Plant Disease. McGraw-Hill, New York. 953 pp.

Heatherly, L. G., W. J. Russell, and T. M. Hinckley. 1977. Water relations and growth of soybeans in drying soil. Crop Sci. 17:381–386.

Heitefuss, R., and P. H. Williams, eds. 1976. Encyclopedia of Plant Physiology. Vol. 4. Physiological Plant Pathology. Springer-Verlag Berlin. 890 pp.

Henis, Y., and Y. Ben-Yephet. 1970. Effect of propagule size of *Rhizoctonia solani* on saprophytic growth, infectivity, and virulence on bean seedlings. Phytopathology 60:1351–1356.

Henis, Y., A. Ghaffar, and R. Baker. 1978. Integrated control of *Rhizoctonia solani* damping-off of radish: Effect of successive plantings, PCNB, and *Trichoderma harzianum* on pathogen and disease. Phytopathology 68:900-907.

Henis, Y., A. Ghaffar, and R. Baker. 1979b. Factors affecting suppressiveness to *Rhizoctonia solani* in soil. Phytopathology 69:1164–1169.

Henis, Y., Y. Elad, I. Chet, Y. Hadar, and E. Hadar. 1979a. Control of soilborne plant pathogenic fungi in carnation, strawberry and tomato by *Trichoderma harzianum*. Phytopathology 69:1031.

Henry, A. W. 1931. The natural microflora of the soil in relation to the foot-rot problem of wheat. Can. Jour. Res., C. 4:69–77.

Henry, A. W. 1932. Influence of soil temperature and soil sterilization on the reaction of wheat seedlings to *Ophiobolus graminis* Sacc. Can. Jour. Res. 7:198–203.

Heye, C. C., and J. H. Andrews. 1982. Biological control of the apple scab pathogen, *Venturia inaequalis,* under field conditions. Phytopathology 72:1136.

Hildebrand, A. A., and P. M. West. 1941. Strawberry root rot in relation to microbiological changes induced in root rot soil by the incorporation of certain cover crops. Can. Jour. Res. 19:183–198.

Hirst, J. M. 1974. What is biological control? Nature 252:147.

Hoch, H. C., and G. S. Abawi. 1979. Biological control of Pythium root rot of table beet with *Corticium* sp. Phytopathology 69:417–419.

Hoitink, H. A. J. 1980. Composted bark, a lightweight growth medium with fungicidal properties. Plant Disease 66:142–147.

Hollender, A., R. D. DeMoss, S. Kaplan, J. Kominsky, D. Savage, and R. S. Wolfe. 1982. Genetic Engineering of Microorganisms for Chemicals. Plenum Press, New York. 485 pp.

Hollings, M. 1982. Mycoviruses and plant pathology. Plant Disease 66:1106–1112.

Holton, C. S., G. W. Fischer, R. W. Fulton, H. Hart, and S. E. A. McCallan, eds. 1959. Plant Pathology Problems and Progress. 1908–1958. University Wisconsin Press, Madison. 588 pp.

Homma, Y., J. W. Sitton, R. J. Cook, and K. M. Old. 1979. Perforation and destruction of pigmented hyphae of *Gaeumannomyces graminis* by vampyrellid amoebae from Pacific Northwest wheat field soils. Phytopathology 69:1118–1122.

Hoppe, P. E. 1957. An improved technique for "cold testing" maize seed. Proc. 4th Internat'l Congr. Crop Prot., Hamburg 2:1489–1490.

Hornby, D. 1975. Inoculum of the take-all fungus: Nature, measurement, distribution and survival. EPPO Bull. 5:319–333.

Hornby, D., and C. A. I. Goring. 1972. Effects of ammonium and nitrate nutrition on take-all disease of wheat in pots. Ann. Appl. Biol. 70:225–231.

Horne, W. T. 1914. The oak fungus disease of fruit trees. Calif. Comm. Hortic. Monthly Bull. 3:275–282.

Horsfall, J. G., and E. B. Cowling, eds. 1977–1980. Plant Disease An Advanced Treatise. Vol. I-V. Academic Press, New York. 2388 pp.

Howell, C. R. 1982. Effect of *Gliocladium virens* on *Pythium ultimum, Rhizoctonia solani,* and damping-off of cotton seedlings. Phytopathology 72:496–498.

Howell, C. R., and R. D. Stipanovic. 1979. Control of *Rhizoctonia solani* on cotton seedlings with *Pseudomonas fluorescens* and with an antibiotic produced by the bacterium. Phytopathology 69:480-482.

Howell, C. R., and R. D. Stipanovic. 1980. Suppression of *Pythium ultimum* -induced damping-off of cotton seedlings by *Pseudomonas fluorescens* and its antibiotic, pyroluteorin. Phytopathology 70:712–715.

Hsiao, T. C. 1973. Plant responses to water stress. Annu. Rev. Plant Physiol. 24:519–570.

Hsu, S. C., and J. L. Lockwood. 1973. Soil fungistasis: Behavior of nutrient-independent spores and sclerotia in a model system. Phytopathology 63:334–337.

Huang, C. S., R. C. V. Tenente, F. C. C. DaSilva, and J. A. R. Lara. 1981. Effect of *Crotalaria spectabilis* and two nematicides, on numbers of *Meloidogyne incognita* and *Helicotylenchus dihystera.* Nematologica 27:1–5.

Huang, H. C. 1977. Importance of *Coniothyrium minitans* in survival of sclerotia of *Sclerotinia sclerotiorum* in wilted sunflower. Can. Jour. Bot. 55:289–295.

Huang, H. C. 1978. *Gliocladium catenulatum*: Hyperparasite of *Sclerotinia sclerotiorum* and *Fusarium* species. Can. Jour. Bot. 56:2243–2246.

Huang, H. C. 1980. Control of Sclerotinia wilt of sunflower by hyperparasites. Can. Jour. Plant Pathol. 2:26–32.

Huang, H. C., and J. A. Hoes. 1976. Penetration and infection of *Sclerotinia sclerotiorum* by *Coniothyrium minitans*. Can. Jour. Bot. 54:406–410.

Huang, H. C., and J. A. Hoes. 1980. Importance of plant spacing and sclerotial position to development of Sclerotinia wilt of sunflower. Plant Disease 64:81–84.

Hubbard, J. P., Y. Hadar, and G. E. Harman. 1982. Suppression of the biological control agent *Trichoderma hamatum* on seeds by soil-borne *Pseudomonas* spp. Phytopathology 72:1009.

Huber, D. M., C. G. Painter, H. C. McKay, and D. L. Peterson. 1968. Effect of nitrogen fertilization on take-all of winter wheat. Phytopathology 58:1470-1472.

Hubert, E. E. 1935. Observations on *Tuberculina maxima*, a parasite of *Cronartium ribicola*. Phytopathology 25:253–261.

Hughes, S. J. 1951. Studies on micro-fungi. IX Calcarisporium, Verticicladium, and Hansfordia (gen. nov.) Commonw. Mycol. Inst., Kew. Mycol. Paper No. 43:1–25.

Hungerford, R. D. 1977. Natural inactivation of blister rust cankers on western white pine. For. Sci. 23:343–350.

Hussey, R. S., and R. W. Roncadori. 1982. Vesicular-arbuscular mycorrhizae may limit nematode activity and improve plant growth. Plant Disease 66:9–14.

Hwang, S. F., R. J. Cook, and W. A. Haglund. 1982. Mechanisms of suppression of chlamydospore germination of *Fusarium oxysporum* f. sp. *pisi* in soils. Phytopathology 72:948.

Ikediugwu, F. E. O., and J. Webster. 1970a. Antagonism between *Coprinus heptemerus* and other coprophilous fungi. Trans. Brit. Mycol. Soc. 54:181–204.

Ikediugwu, F. E. O., and J. Webster. 1970b. Hyphal interference on a range of coprophilous fungi. Trans. Brit. Mycol. Soc. 54:205–210.

Ikediugwu, F. E. O., C. Dennis, and J. Webster. 1970. Hyphal interference by *Peniophora gigantea* against *Heterobasidion annosum*. Trans. Brit. Mycol. Soc. 54:307–309.

Imbriani, J. L., and R. Mankau. 1977. Ultrastructure of the nematode pathogen, *Bacillus penetrans*. Jour. Invert. Pathol. 30:337–347.

Inglis, D. A. 1982. The *Fusarium roseum* sensu Snyder and Hansen complex in wheat and wheat-field soils in eastern Washington. Ph.D. Thesis, Washington State University, Pullman. 139 pp.

Ingold, C. T. 1956. The conidial apparatus of *Trichothecium roseum*. Trans. Brit. Mycol. Soc. 39:460-464.

Ingram, D. S., and J. P. Helgeson, eds. 1980. Tissue Culture Methods for Plant Pathologists. Blackwell, Oxford. 272 pp.

Ioannou, N., R. W. Schneider, R. G. Grogan, and J. M. Duniway. 1977a. Effect of water potential and temperature on growth, sporulation, and production of microsclerotia by *Verticillium dahliae*. Phytopathology 67:637–644.

Ioannou, N., R. W. Schneider, and R. G. Grogan. 1977b. Effect of oxygen, carbon dioxide, and ethylene on growth, sporulation, and production of microsclerotia by *Verticillium dahliae*. Phytopathology 67:645–650.

Ioannou, N., R. W. Schneider, and R. G. Grogan. 1977c. Effect of flooding on the soil gas composition and the production of microsclerotia by *Verticillium dahliae* in the field. Phytopathology 67:651–656.

Itai, C., and V. Vaadia. 1971. Cytokinin activity in water-stressed shoots. Plant Physiol. 47:87–90.

Jacq, V. A., and R. Fortuner. 1979. Biological control of rice nematodes using sulfate reducing bacteria. Rev. Nematol. 2:41–50.

Jaffe, M. J. 1980. Morphogenetic responses of plants to mechanical stimuli or stress. Bioscience 30:239–243.

Jager, G., and H. Velvis. 1980. Onderzoek naar het voorkomen van *Rhizoctonia*-werende aardappel-percelen in noord-Nederland. Institut voor Bodemvruchtbaar-heid Report (1–80), 66 pp.

Jager, G., A. ten Hoopen and H. Velvis. 1979. Hyperparasites of *Rhizoctonia solani* in Dutch potato fields. Neth. Jour. Plant Pathol. 85:253–268.

Jansson, H.-B. 1982. Attraction of nematodes to endoparasitic nematophagous fungi. Trans. Brit. Mycol. Soc. 79:25–29.

Jarvis, W. R., and R. A. Shoemaker. 1978. Taxonomic status of *Fusarium oxysporum* causing foot and root rot of tomato. Phytopathology 68:1679–1680.

Jarvis, W. R., and K. Slingsby. 1977. The control of powdery mildew of greenhouse cucumber by water sprays and *Ampelomyces quisqualis*. Plant Dis. Reptr. 61:728–730.

Jaynes, R. A., and J. E. Elliston. 1980. Pathogenicity and canker control by mixtures of hypovirulent strains of *Endothia parasitica* in American chestnut. Phytopathology 70:453–456.

Jaynes, R. A., and J. E. Elliston. 1982. Hypovirulent isolates of *Endothia parasitica* associated with large American chestnut trees. Plant Disease 66:769–772.

Jenkinson, D. S. 1966. Studies on the decomposition of plant material in soil. II. Partial sterilization of soil and the soil biomass. Jour. Soil Sci. 17:280–302.

Jenkinson, D. S. 1971. Studies on the decomposition of C^{14} labeled organic matter in soil. Soil Sci. 111:64–70.

Jenkinson, D. S. 1977. Studies on the decomposition of plant material in soil. V. The effects of plant cover and soil type on the loss of carbon from ^{14}C labeled ryegrass decomposing under field conditions. Jour. Soil Sci. 28:424–434.

Jenkinson, D. S., and A. Ayanaba. 1977. Decomposition of carbon-14 labeled plant material under tropical conditions. Soil Sci. Soc. Amer., Proc. 41:912–915.

Jenkinson, D. S., and D. S. Powlson. 1976. The effects of biocidal treatments on metabolism in soil — V. A method for measuring soil biomass. Soil Biol. Biochem. 8:209–213.

Jenkyn, J. F., and R. T. Plumb, eds. 1981. Strategies for the Control of Cereal Disease. Blackwell Scientific Publications, Oxford. 219 pp.

Jenns, A. E., and J. Kuć. 1979. Graft transmission of systemic resistance of cucumber to anthracnose induced by Colletotrichum lagenarium and tobacco necrosis virus. Phytopathology 69:753–756.

Jones, D., and M. E. Solomon, eds. 1974. Biology in Pest and Disease Control. Brit. Ecol. Soc. Symp. 13. 398 pp.

Jones, D., and D. Watson. 1969. Parasitism and lysis by soil fungi of Sclerotinia sclerotiorum (Lib.) deBary a phytopathogenic fungus. Nature 224:287–288.

Jones, D., A. H. Gordon, and J. S. D. Bacon. 1974. Co-operative action by endo- and exo-β-(1–3)-glucanases from parasitic fungi in the degradation of cell-wall glucans of Sclerotinia sclerotiorum (Lib.) de Bary. Biochem. Jour. 140:47–55.

Kaiser, W. J., and G. M. Horner. 1980. Root rot of irrigated lentils in Iran. Can. Jour. Bot. 58:2549–2556.

Kallio, T., and A. M. Hallaksela. 1979. Biological control of Heterobasidion annosum (Fr.) Bref. (Fomes annosus) in Finland. Eur. Jour. For. Pathol. 9:298–308.

Kao, C. W., and W. H. Ko. 1983. Nature of suppression of Pythium splendens in a South Kohala soil in Hawaii. Phytopathology. 73:In press.

Kaper, J. M., and H. E. Waterworth. 1977. Cucumber mosaic virus associated RNA 5: Causal agent for tomato necrosis. Science 196:429–431.

Kappelman, A. J., Jr. 1980. Long-term progress made by cotton breeders in developing Fusarium wilt resistant germplasm. Crop Sci. 20:613–615.

Katan, J. 1979. Solar heating of the soil and other economical environmentally safe methods of controlling soilborne pathogens, weeds and pests. Phytopathology 69:1033–1034.

Katan, J. 1980. Solar pasteurization of soils for disease control: Status and prospects. Plant Disease 64:450-454.

Katan, J. 1981. Solar heating (solarization of soil for control of soilborne pests). Annu. Rev. Phytopathol. 19:211–236.

Katan, J., A. Greenberger, H. Alon, and A. Grinstein. 1976. Solar heating by polyethylene mulching for the control of diseases caused by soil-borne pathogens. Phytopathology 66:683–688.

Kawamoto, S. O., and J. W. Lorbeer. 1976. Protection of onion seedlings from Fusarium oxysporum f. sp. cepae by seed and soil infestation with Pseudomonas cepacia. Plant Dis. Reptr. 60:189–191.

Keener, P. D. 1934. Biological specialization in Darluca filum. Torrey Bot. Club Bull. 61:475–490.

Keim, R., and R. K. Webster. 1974. Effect of soil moisture and temperature on viability of sclerotia of Sclerotium oryzae. Phytopathology 64:1499–1502.

Kellam, M. K., and N. C. Schenck. 1980. Interactions between a vesicular-arbuscular mycorrhizal fungus and root-knot nematode on soybean. Phytopathology 70:293–296.

Kelley, W. D. 1976. Evaluation of *Trichoderma harzianum*-impregnated clay granules as a biocontrol for *Phytophthora cinnamomi* causing damping-off of pine seedlings. Phytopathology 66:1023–1027.

Kelman, A., and R. J. Cook. 1977. Plant pathology in the People's Republic of China. Annu. Rev. Phytopathol. 15:409–429.

Kerr, A. 1964. The influence of soil moisture on infection of peas by *Pythium ultimum*. Austral. Jour. Biol. Sci. 17:676–685.

Kerr, A. 1972. Biological control of crown gall: Seed inoculation. Jour. Appl. Bact. 35:493–497.

Kerr, A. 1974. Soil microbiological studies on *Agrobacterium radiobacter* and biological control of crown gall. Soil Sci. 118:168–172.

Kerr, A. 1980. Biological control of crown gall through production of agrocin 84. Plant Disease 64:25–30.

Kerr, A., and C. G. Panagopoulos. 1977. Biotypes of *Agrobacterium radiobacter* var. *tumefaciens* and their biological control. Phytopathol. Zeit. 90:172–179.

Kerry, B. R., and D. H. Crump. 1980. Two fungi parasitic on females of cyst-nematodes (*Heterodera* spp.). Trans. Brit. Mycol. Soc. 74:119–125.

Kerry, B. R., D. H. Crump, and L. A. Mullen. 1980. Parasitic fungi, soil moisture and multiplication of the cereal cyst nematode, *Heterodera avenae*. Nematologica 26:57–68.

Kerry, B. R., D. H. Crump, and L. A. Mullen. 1982a. Studies of the cereal cyst nematode, *Heterodera avenae* under continuous cereals, 1974–1978. I. Plant growth and nematode multiplication. Ann. Appl. Biol. 100:477–487.

Kerry, B. R., D. H. Crump, and L. A. Mullen. 1982b. Studies of the cereal cyst nematode, *Heterodera avenae* under continuous cereals, 1975–1978. II. Fungal parasitism of nematode females and eggs. Ann. Appl. Biol. 100:489–499.

Kimmey, J. W. 1969. Inactivation of lethal-type blister rust cankers on western white pine. Jour. For. 67:296–299.

King, C. J. 1923. Habits of the cotton root rot fungus. Jour. Agric. Res. 26:405–418.

King, C. J., C. Hope, and E. D. Eaton. 1934. Some microbiological activities affected in manurial control of cotton root rot. Jour. Agric. Res. 49:1093–1107.

King, C. J., H. F. Loomis, and C. Hope. 1931. Studies on sclerotia and mycelial strands of the cotton root rot fungus. Jour. Agric. Res. 42:827–840.

Kiyomoto, R. K., and G. W. Bruehl. 1977. Carbohydrate accumulation and depletion by winter cereals differing in resistance to *Typhula idahoensis*. Phytopathology 67:206–211.

Kiyosawa, K. 1975. Studies on the effects of alcohols on membrane water permeability of *Nitella*. Protoplasma 86:243–252.

Kloepper, J. W., and M. N. Schroth. 1979. Plant growth promoting rhizobacteria: Evidence that the mode of action involves root microflora interactions. Phytopathology 69:1034.

Kloepper, J. W., and M. N. Schroth. 1981a.Development of a powder formulation of rhizobacteria for inoculation of potato seed pieces. Phytopathology 71:590-592.

Kloepper, J. W., and M. N. Schroth. 1981b. Plant growth-promoting rhizobacteria and plant growth under gnotobotic conditions. Phytopathology 71:642-644.

Kloepper, J. W., and M. N. Schroth. 1981c. Relationship of in vitro antibiosis of plant growth-promoting rhizobacteria to plant growth and the displacement of root microflora. Phytopathology 71:1020-1024.

Kloepper, J. W., J. Leong, M. Teintze, and M. N. Schroth. 1980a. Enhanced plant growth by siderophores produced by plant growth-promoting rhizobacteria. Nature 286:885-886.

Kloepper, J. W., M. N. Schroth, and T. D. Miller. 1980b. Effects of rhizosphere colonization by plant growth-promoting rhizobacteria on potato plant development and yield. Phytopathology 70:1078-1082.

Ko, W.-h. 1971. Biological control of seedling root rot of papaya caused by *Phytophthora palmivora*. Phytopathology 61:780-782.

Ko, W. H. 1982. Biological control of Phytophthora root rot of papaya with virgin soil. Plant Disease 66:446-448.

Ko, W. H., and W. C. Ho. 1983. Screening soils for suppressiveness to *Rhizoctonia solani* and *Pythium splendens*. Ann. Phytopathol. Soc. Japan 48:In press.

Ko, W.-h., and J. L. Lockwood. 1967. Soil fungistasis: Relation to fungal spore nutrition. Phytopathology 57:894-901.

Komada, H. 1975. Behavior of pathogenic *Fusarium oxysporum* in different soil types, with special reference to the interaction of the pathogen and other microorganisms. Pages 451-452 *in* T. Hasegawa, ed. Proceedings of the First Intersectional Congress of IAMS Vol. 2, Developmental Microbial Ecology Sci. Council of Janpan. 675 pp.

Kommedahl, T., ed. 1981. Proceedings of Symposia IX International Congress of Plant Protection, Wash. D.C., 5-11 August, 1979. 2 vols., Burgess Publ. Co., Minneapolis, MN. 630 pp.

Kommedahl, T., and I. C. Mew. 1975. Biocontrol of corn root infection in the field by seed treatment with antagonists. Phytopathology 65:296-300.

Kommedahl, T., and P. H. Williams, eds. 1983. Challenging Problems in Plant Health. Amer. Phytopathol. Soc., St. Paul, MN.

Kommedahl, T., and C. E. Windels. 1978. Evaluation of biological seed treatment for controlling root diseases of pea. Phytopathology 68:1087-1095.

Kommedahl, T., C. E. Windels, G. Sarbini, and H. B. Wiley. 1981. Variability in performance of biological and fungicidal seed treatments in corn, peas, and soybeans. Prot. Ecol., 3:55-61.

Kozlowski, T. T., ed. 1968-1983. Water Deficits and Plant Growth. 6 vols. Academic Press, New York. 2061 pp.

Krasil'nikov, N. A. 1958. Soil Microorganisms and Higher Plants. Akad. Nauk. U.S.S.R. Moscow. [Engl. translation, Office Technical Services 60-21126, U.S. Dept. Commerce, Washington, D.C. 474 pp. 1961.]

Krupa, S. V., and Y. R. Dommergues, eds. 1979. Ecology of Root Pathogens. Elsevier, Amsterdam. 281 pp.

Krupa, S., and J.-E. Nylund. 1972. Studies on ectomycorrhizae of pine. III. Growth inhibitation of two root pathogenic fungi by volatile organic constituents of ectomycorrhizal root systems of *Pinus sylvestris* L. Eur. Jour. For. Pathol. 2:88–94.

Krupa, S., J. Andersson, and D. H. Marx. 1973. Studies on ectomycorrhizae of pine. IV. Volatile organic compounds in mycorrhizal and nonmycorrhizal root systems of *Pinus echinata* Mill. Eur. Jour. For. Path. 3:194–200.

Kuan, T. L., and D. C. Erwin. 1980. Predisposition effect of water saturation of soil on Phytopathora root rot of alfalfa. Phytopathology 70:981–986.

Kuć, J. 1982. Induced immunity to plant disease. Bioscience 32:854–860.

Kuć, J., and S. Richmond. 1977. Aspects of the protection of cucumber against *Colletotrichum lagenarium* by *Colletotrichum lagenarium*. Phytopathology 67:533–536.

Kuć, J., G. Shockley, and K. Kearney. 1975. Protection of cucumber against *Colletotrichum lagenarium* by *Colletotricum lagenarium*. Physiol. Plant Pathol. 7:195–199.

Kuhlman, E. G. 1981a. Mycoparasitic effects of *Scytalidium uredinicola* on aeciospore production and germination of *Cronartium quercuum* f. sp. *fusiforme*. Phytopathology 71:186–188.

Kuhlman, E. G. 1981b. Parasite interaction with sporulation by *Cronartium quercuum* f. sp. *fusiforme* on loblolly and slash pine. Phytopathology 71:348–350.

Kuhlman, E. G., and F. R. Matthews. 1976. Occurrence of *Darluca filum* on *Cronartium strobilinum* and *C. fusiforme* infecting oak. Phytopathology 66:1195–1197.

Kuhlman, E. G., J. W. Carmichael, and T. Miller. 1976. *Scytalidium uredinicola*, a new mycoparasite of *Cronartium fusiforme* on *Pinus*. Mycologia 68:1188–1194.

Kuhlman, E. G., F. R. Matthews, and H. P. Tillerson. 1978. Efficacy of *Darluca filum* for biological control of *Cronartium fusiforme* and *C. strobilinum*. Phytopathology 68:507–511.

Labruyère, R. E. 1971. Common scab and its control in seed-potato crops. Inst. Plantenziekt. Onderzoek Wageningen Med. 575:1–71.

Ladd, J. N., J. M. Oades, and M. Amato. 1981. Microbial biomass formed from ^{14}C, ^{15}N-labeled plant material decomposing in soils in the field. Soil Biol. Biochem. 13:119–126.

Lamanna, C., M. F. Malette, and L. N. Zimmerman. 1973. Basic Bacteriology; its Biological and Chemical Background. 4th ed. Williams and Wilkins, Baltimore. 1149 pp.

Langton, F. A. 1969. Interactions of the tomato with two formae speciales of *Fusarium oxysporum*. Ann. Appl. Biol. 62:413–427.

Lapierre, H. 1973. Étude de l'influence des virus sur les champignons phytopathogènes du sol. Pages 62–64 *in* Perspectives de Lutte Biologique des Champignons Parasites des Plantes Cultivées et les Pourritures des Tissus Ligneux. Station Fédérale de Recherches Agronomiques de Lausanne, Switzerland.

Lapierre, H., J. M. Lamaire, B. Jouan, and G. Molin. 1970. Mise en évidence de particules virales associées á une perte de pathogenicité chez le Piétin-échaudage des céréales, *Ophiobolus graminis* Sacc. C. R. Hebd. Séanc. Acad. Sci. Paris D, 271:1833–1836.

Lapwood, D. N., and T. F. Hering. 1970. Soil moisture and the infection of young potato tubers by *Streptomyces scabies* (common scab). Potato Res. 13:296–304.

Last, F. T., and F. C. Deighton. 1965. The non-parasitic microflora on the surfaces of living leaves. Trans. Brit. Mycol. Soc. 48:83–99.

Last, F. T., and R. C. Warren. 1972. Non-parasitic microbes colonizing green leaves: Their form and functions. Endeavour 31:143–150.

Leach, L. D. 1947. Growth rates of host and pathogen as factors determining the severity of preemergence damping-off. Jour. Agric. Res. 75:161–179.

Leach, R. 1937. Observations on the parasitism and control of *Armillaria mellea*. Proc. Roy. Soc. London, Sec. B. 121:561–573.

Leach, R. 1939. Biological control and ecology of *Armillaria mellea* (Vahl.) Fr. Trans. Brit. Mycol. Soc. 23:320-329.

Leben, C. 1963. Multiplication of *Xanthomonas vesicatoria* on tomato seedlings. Phytopathology 53:778–781.

Leben, C. 1964. Influence of bacteria isolated from healthy cucumber leaves on two leaf diseases of cucumber. Phytopathology 54:405–408.

Leben, C. 1965a. Epiphytic microorganisms in relation to plant disease. Annu. Rev. Phytopathol. 3:209–230.

Leben, C. 1965b. Influence of humidity on the migration of bacteria on cucumber seedlings. Can. Jour. Microbiol. 11:671–676.

Leben, C. 1981. How plant-pathogenic bacteria survive. Plant Disease 65:633–637.

Lemaire, J. M., B. Jouan, and B. Tivoli. 1975. Un exemple de prémunition chez le Blé contre l'agent du Piétin-échaudage: *Ophiobolus graminis*. Zeit. Pflanzenkr. Pflanzensch. 82:50-51.

Lemaire J. M., F. Carpentier, J. F. Dalle, and G. Doussinault. 1979a. Lutte biologique contre le Piétin-échaudage des céréales. Modifications physiologiques chez le Blé inoculé par une souche atténuée d'*Ophiobolus graminis*. 2. Charnement de la teneur en chlorophylle. Ann. Phytopathol. 11:193–198.

Lemaire, J. M., F. Carpentier, J. F. Dalle, G. Doussinault, B. Perraton, and C. Moule. 1979b. Lutte biologique contre le Piétin-échaudage des céréales. Modifications physiologiques chez le Blé inoculé par une souche atténuée d'*Ophiobolus graminis*. 1. Précocité accrue aux premiers stades du Blé. C. R. Acad. Agric. Fr. 65:766–772.

Lemaire, J. M., B. Jouan, B. Perraton, and M. Sailly. 1971. Perspectives de lutte biologique contre les parasites des céréales d'origine tellurique en particulier *Ophiobolus graminis* Sacc. Sci. Agron. Rennes 1971, pp 1–8.

Lemaire, J. M., H. Lapierre, B. Jouan, and G. Bertrand. 1970. Découverte de particules virales chez certaines souches d'*Ophiobolus graminis*, agent du Piétin-échaudage des céréales: Consequences agronomique prévisibles. C. R. Hebd. Séanc. Acad. Agric. Fr. 56:1134–1138.

Lemaire, J. M., B. Jouan, M. Coppenet, B. Perraton, and L. Lecorre, 1976. Lutte biologique contre le Piétin-échaudage des céréales par l'utilisation de souches hypoagressives d' *Ophiobolus graminis.* Sci. Agron. Rennes. pp. 63–65.

Lemke, P. A., ed. 1979. Viruses and Plasmids in Fungi. Marcel Dekker, Inc., New York. 653 pp.

Leonard, K. J. 1980. A reinterpretation of the mathemaical analysis of rhizoplane and rhizosphere effects. Phytopathology 70:695–696.

Levitt, J., ed. 1980. Responses of Plants to Environmental Stresses. I. Chilling, Freezing, and High Temperature Stresses. II. Water, Radiation, Salt, and Other Stresses. Academic Press, New York. 1136 pp.

Lewis, B. G. 1970. Effects of water potential on the infection of potato tubers by *Streptomyces scabies* in soil. Ann. Appl. Biol. 66:83–88.

Lewis, J. A., and G. C. Papavizas, 1980. Integrated control of Rhizoctonia fruit rot of cucumber. Phytopathology 70:85–89.

Lin, Y. S., and R. J. Cook. 1979. Suppression of *Fusarium roseum* 'Avenaceum' by soil microorganisms. Phytopathology 69:384–388.

Linderman, R. G., and R. G. Gilbert. 1969. Stimulation of *Sclerotium rolfsii* in soil by volatile components of alfalfa hay. Phytopathology 59:1366–1372.

Linderman, R. G., and R. G. Gilbert. 1973. Influence of volatile compounds from alfalfa hay on microbial activity in soil in relation to growth of *Sclerotium rolfsii.* Phytopathology 63:359–362.

Linderman, R. G., L. W. Moore, K. F. Baker, and D. A. Cooksey. 1983. Strategies for detecting and characterizing biocontrol systems. Plant Disease 67: In press.

Lindow, S. E. 1979. Frost damage to potato reduced by bacteria antagonistic to ice nucleation-active bacteria. Phytopathology 69:1036.

Lindow, S. E. 1981. Frost damage to pear reduced by antagonistic bacteria, bactericides, and ice nucleation inhibitors. Phytopathology 71:237.

Lindow, S. E. 1983. Methods of preventing frost injury caused by epiphytic ice nucleation active bacteria. Plant Disease 67:327–333.

Lindow, S. E., D. C. Arny, and C. D. Upper. 1977. Distribution of epiphytic ice nucleation-active strains of *Pseudomonas syringae.* Proc. Amer. Phytopathol. Soc. 4:107.

Lindow, S. E., D. C. Arny, and C. D. Upper. 1978a. *Erwinia herbicola:* A bacterial ice nucleus active in increasing frost injury to corn. Phytopathology 68:523–527.

Lindow, S. E., D. C. Arny, and C. D. Upper. 1978b. Distribution of ice nucleation-active bacteria on plants in nature. Appl. Environ. Microbiol. 36:831–838.

Lindow, S. E., D. C. Arny, W. R. Barchet, and C. D. Upper. 1975a. The relationship between populations of bacteria active in ice nucleation and frost sensitivity in herbaceous plants. E. O. S. Trans. Amer. Geophysical Union. 56:994.

Lindow, S. E., D. C. Arny, and C. D. Upper. 1975b. Increased frost sensitivity of maize in the presence of *Pseudomonas syringae.* Proc. Amer. Phytopathol. Soc. 2:57.

Lindow, S. F., Arny, D. C., Barchet, W. R., and Upper, C. D. 1976. *Erwinia herbicola* isolates active in ice nucleation incite frost damage to corn (*Zea mays* L.) Proc. Amer. Phytopathol. Soc. 3:224.

Linford, M. B. 1937. The feeding of root-knot nematodes in root tissue and nutrient solution. Phytopathology 27:824–835.

Linford, M. B., F. Yap, and J. M. Oliveira. 1938. Reduction of soil populations of the root-knot nematode during decomposition of organic matter. Soil Sci. 45:127–141.

Lippincott, B. B., and J. A. Lippincott. 1969. Bacterial attachment to a specific wound site as an essential stage in tumor initiation by *Agrobacterium tumefaciens*. Jour. Bact. 97:620-628.

Lippincott, B. B., M. H. Whatley, and J. A. Lippincott. 1977. Tumor induction by *Agrobacterium* involves attachment of the bacterium to a site on the host plant cell wall. Plant Physiol. 59:388–390.

Lipps, P. E., and G. W. Bruehl. 1980. Infectivity of *Pythium* spp. zoospores in snow rot of wheat. Phytopathology 70:723–726.

Littlefield, L. J. 1969. Flax rust resistance induced by prior inoculation with an avirulent race of *Melampsora lini*. Phytopathology 59:1323–1328.

Liu, S. D., and R. Baker. 1980. Mechanism of biological control in soil suppressive to *Rhizoctonia solani*. Phytopathology 70:404–412.

Liu, S.Y., and E. K. Vaughan. 1965. Control of *Pythium* infection in table beet seedlings by antagonistic microorganisms. Phytopathology 55:986–989.

Locke, J. C., J. J. Marois, and G. C. Papavizas. 1982. Biological control of Fusarium wilt of greenhouse grown chrysanthemums. Phytopathology 72:709.

Lockeretz, W., G. Shearer, and D. H. Kohl. 1981. Organic farming in the corn belt. Science 211:540-547.

Lockhart, D. A. S., V. A. F. Heppel, and J. C. Holmes. 1975. Take-all (*Gaeumannomyces graminis* (Sacc.) Arx and Olivier) incidence in continuous barley growing and effect of tillage method. EPPO Bull. 5:375–383.

Loper, J. E., M. N. Schroth, and N. J. Panopoulos. 1982. Influence of bacterial source of indole-3-acetic acid (IAA) on root elongation of sugar beet. Phytopathology 72:997.

Loutit, M. W., and J. A. R. Miles, eds. 1978. Microbiol Ecology. Springer-Verlag, Berlin.

Louvet, J., F. Rouxel, and C. Alabouvette. 1976. Recherches sur la résistance des sols aux maladies. I. Mise en évidence de la nature microbiologique de la résistance d'un sol au développement de la Fusariose vasculaire du melon. Ann. Phytopathol. 8:425–436.

Lovrekovich, L., and G. L. Farkas. 1965. Induced protection against wildfire disease in tobacco leaves treated with heat-killed bacteria. Nature 205:823–824.

Lucas, G. B. 1975. Diseases of Tobacco. Biological Consulting Associates, Raleigh, N. C. 621 pp.

Lundholm, B., and M. Stackerud, eds. 1980. Environmental Protection and Biological Forms of Control of Pest Organisms. Swedish Natural Science Research Council Ecological Bulletins 31:1–171.

Luxmoore, R. J., L. H. Stolzy, and J. Letey. 1970a. Oxygen diffusion in the soil-plant system. I. A model. Agron. Jour. 62:317–322.

Luxmoore, R. J., L. H. Stolzy, and J. Letey. 1970b. Oxygen diffusion in the soil-plant system. II. Respiration rate, permeability, and porosity of consecutive excised segments of maize and rice roots. Agron. Jour. 62:322–324.

Luxmoore, R. J., L. H. Stolzy, and J. Letey. 1970c. Oxygen diffusion in the soil-plant system. III. Oxygen concentration profiles, respiration rates, and the significance of plant aeration predicted for maize roots. Agron. Jour. 62:325–329.

Lynch, J. M. 1972. Identification of substrates and isolation of microorganisms responsible for ethylene production in the soil. Nature 240:45–46.

Lynch, J. M., and S. H. T. Harper. 1974. Formation of ethylene by a soil fungus. Jour. Gen. Microbiol. 80:187–195.

Lyons, J. M. 1973. Chilling injury in plants. Annu. Rev. Plant Physiol. 24:445–466.

MacDonald, J. D. 1982. Effect of salinity stress on the development of Phytopththora root rot of chrysanthemum. Phytopathology 72:214–219.

MacDonald, W. L., F. C. Cech, J. Luchok, and C. Smith, eds. 1979. Proceedings of the American Chestnut Symposium. West Virginia University, Morgantown. 122 pp.

Mace, M. E., A. A. Bell, and C. H. Beckman, eds. 1981. Fungal Wilt Diseases of Plants. Academic Press, New York. 640 pp.

MacNish, G. C. 1980. The use of ammonium nitrogen to reduce take-all (Gaeumannomyces graminis var. tritici) in wheat. APPS Natl. Path. Conf. 4:27.

Magie, R. O. 1980. Fusarium disease of gladioli controlled by inoculation of corms with non-pathogenic fusaria. Proc. Fla. State Hortic. Soc. 93:172–175.

Malajczuk, N., and A. J. McComb. 1979. The microflora of unsuberized roots of Eucaluptus calaphylla R. Br. and Eucalyptus marginata Donn. ex Sm. seedlings grown in soil suppressive and conducive to Phytophthora cinnamomi Rands. I. Rhizoplane bacteria, actinomycetes and fungi. II. Mycorrhizal roots and associated microflora. Austral. Jour. Bot. 27: 235–254, 255–272.

Malajczuk, N., and C. Theodorou. 1979. Influence of water potential on growth and cultural characteristics of Phytophthora cinnamomi. Trans. Brit. Mycol. Soc. 72:15–18.

Malajczuk, N., A. J. McComb, and C. A. Parker. 1977. Infection by Phytopthora cinnamomi Rands of roots of Eucalyptus calophylla R. Br. and Eucalyptus marginata Donn. ex Sm. Austral. Jour. Bot. 25:483–500.

Manandhar, J. B., and G. W. Bruehl. 1973. In vitro interactions of Fusarium and Verticillium wilt fungi with water, pH, and temperature. Phytopathology 63:413–419.

Mankau, R. 1968. Effect of nematocides on nematode-trapping fungi associated with the citrus nematode. Plant Dis. Rep. 52:851-855.

Mankau, R. 1972. Utilization of parasites and predators in nematode pest management ecology. Proc. Annu. Tall Timbers Conf. 1972:129–143.

Mankau, R. 1975. Bacillus penetrans n. comb. causing a virulent disease of plant-parasitic nematodes. Jour. Invert. Path. 26:333–339.

Mankau, R. 1975. Prokayote affinities of Duboscqia penetrans Thorne. Jour. Protozool. 22:31–34.

Mankau, R. 1980. Biological control of nematode pests by natural enemies. Annu. Rev. Phytopathol. 18:415–440.

Mankau, R., and S. Prasad. 1972. Possibilities and problems in the use of a sporozoan endoparasite for biological control of plant parasitic nematodes. Nematologica 2:7–8.

Mankau, R., and N. Prasad. 1977. Infectivity of *Bacillus penetrans* in plant parasitic-nematodes. Jour. Nematol. 9:40-45.

Mankau, R., J. L. Imbriani, and A. H. Bell. 1976. SEM observations on nematode cuticle penetration by *Bacillus penetrans*. Jour. Nematol. 8:179–181.

Marois, J. J., and D. J. Mitchell. 1981a. Effects of fumigation and fungal antagonists on the relationships of inoculum density to infection incidence and disease severity in Fusarium crown rot of tomato. Phytopathology 71:167–170.

Marois, J. J., and D. J. Mitchell. 1981b. Effects of fungal communities on the pathogenic and saprophytic activties of *Fusarium oxysporum* f. sp. *radicis-lycopersici*. Phytopathology 71:1251–1256.

Marois, J. J., D. J. Mitchell, and R. M. Sonoda. 1981. Biological control of Fusarium crown rot of tomato under field conditions. Phytopathology 71:1257–1260.

Marois, J. J., S. A. Johnson, M. T. Dunn, and G. C. Papavizas. 1982. Biological control of Verticillium wilt of eggplant in the field. Plant Disease 66:1166–1168.

Marshall, D. S. 1982. Effect of *Trichoderma harzianum* seed treatment and *Rhizoctonia solani* inoculum concentration on damping-off of snap bean in acidic soils. Plant Disease 66:788–789.

Marshall, K. C. 1976. Interfaces in Microbial Ecology. Harvard Univ. Press, Cambridge, MA. 156 pp.

Martin, F. N., and J. G. Hancock. 1981. Relationship between soil salinity and population density of *Pythium ultimum* in the San Joaquin Valley of California. Phytopathology 71:893.

Martin, F. N., and J. G. Hancock. 1982. The effects of Cl⁻ and *Pythium oligandrum* on the ecology of *Pythium ultimum*. Phytopathology 72:996.

Martin, J. K. 1977. Factors influencing the loss of organic and carbon from wheat roots. Soil Biol. Biochem. 9:1–7.

Marx, D. H. 1969. The influence of ectotrophic mycorrhizal fungi on the resistance of pine roots to pathogenic infections. I. Antagonism of mycorrhizal fungi to root pathogenic fungi and soil bacteria. II. Production, identification, and biological activity of antibiotics produced by *Leucopaxillus cerealis* var. *piceina*. Phytopathology 59:153–163, 411–417.

Marx, D. H. 1972. Ectomycorrhizae as biological deterrents to pathogenic root infections. Annu. Rev. Phytopath. 10:429–454.

Marx, D. H. 1977. Tree host range and world distribution of the ectomycorrhizal fungus *Pisolithus tinctorius*. Can. Jour. Microbiol. 23:217–223.

Marx, D. H., and W. C. Bryan. 1975. Growth and ectomycorrhizal development of loblolly pine seedlings in fumigated soil infested with the fungal symbiont *Pisolithus tinctorius*. For. Sci. 21:245–254.

Marx, D. H., and C. B. Davey. 1969. The influence of ectotrophic mycorrhizal fungi on the resistance of pine roots to pathogenic infections. III. Resistance of aseptically formed mycorrhizae to infection by *Phytophthora cinnamomi.* Phytopathology 59:549–558.

Mataré, R., and M. J. Hattingh. 1978. Effect of mycorrhizal status of avocado seedlings on root rot caused by *Phytophthora cinnamomi.* Plant Soil 49:433–435.

Matta, A. 1971. Microbial penetration and immunization of uncongenial host plants. Annu. Rev. Phytopathol. 9:387–410.

Matthysse, A. G., K. V. Holmes, and R. H. G. Gunlitz. 1982. Binding of *Agrobacterium tumefaciens* to carrot protoplasts. Physiol. Plant Pathol. 20:27–33.

Matuo, T., and W. C. Snyder. 1972. Host virulence and the *Hypomyces* stage of *Fusarium solani* f. sp. *pisi.* Phytopathology 62:731–735.

McBeth, C. W., and A. L. Taylor. 1944. Immune and resistant cover crops valuable in root-knot-infested peach orchards. Proc. Amer. Soc. Hort. Sci. 45:158–166.

McBride, R. P. 1969. A microbiological control of *Melampsora medusae.* Can. Jour. Bot. 47:711–715.

McCain, A. H., L. E. Pyeatt, T. G. Byrne, and D. B. Farnham. 1980. Suppressive soil reduces carnation disease. Calif. Agric. 34(5):9.

McCann, J., V. D. Luedders, and V. H. Dropkin. 1982. Selection and reproduction of soybean cyst nematodes on resistant soybeans. Crop Sci. 22:78–80.

McCoy, M. L., and R. L. Powelson. 1974. A model for determining spatial distribution of soil-borne propagules. Phytopathology 64:145–147.

McKeen, W. E. 1952. *Phialophora radicicola* Cain, a corn rootrot pathogen. Can. Jour. Bot. 30:344–347.

McKinney, H. H. 1929. Mosaic diseases in the Canary Islands. West Africa, and Gibrator. Jour. Agric. Res. 39:557–578.

McMichael, B. L., W. R. Jordan, and R. D. Powell. 1972. An effect of water stress on ethylene production by intact cotton petioles. Plant Physiol. 49:658–660.

Meiler, D., and A. Taylor. 1971. The effect of cochliodinol, a metabolite of *Chaetomium cochlioides,* on the respiration of microspores of *Fusarium oxysporum.* Can. Jour. Microbiol. 17:83–86.

Melouk, H. A., and C. E. Horner. 1975. Cross protection in mints by *Verticillium nigrescens* against *V. dahliae.* Phytopathology 65:767–769.

Mendgen, K. 1979. *Verticillium lecanii,* ein Hyperparasit auf dem Getreidegelbrost *(Puccinia striiformis).* Mitt. Biol. Bundesanst. Land- Forstwirtsch. Berlin Dahlem 191:301–302.

Menzies, J. D. 1959. Occurrence and transfer of a biological factor in soil that suppresses potato scab. Phytopathology 49:648–652.

Menzies, J. D., and R. G. Gilbert. 1967. Responses of the soil microflora to volatile components in plant residues. Proc. Soil Sci. Soc. Amer. 31:495–496.

Menzies, J. D., and G. E. Griebel. 1967. Survival and saprophytic growth of *Verticillium dahliae* in uncropped soil. Phytopathology 57:703–709.

Merriman, P. R. 1976. Survival of sclerotia of *Sclerotinia sclerotiorum* in soil. Soil Biol. Biochem. 8:385–389.

Merriman, P. R., and S. Isaacs. 1978. Evaluation of onions as a trap crop for *Sclerotium cepivorum*. Soil Biol. Biochem. 10:339–340.

Merriman, P. R., I. M. Samson, and B. Schippers. 1981. Stimulation of germination of sclerotia of *Sclerotium cepivorum* at different depths in soil by artificial onion oil. Neth. Jour. Plant Pathol. 87:45–53.

Merriman, P. R., M. Pywell, G. Harrison, and J. Nancarrow. 1979. Survival of sclerotia of *Sclerotinia sclerotiorum* and effects of cultivation practices on disease. Soil Biol. Biochem. 11:567–570.

Merriman, P. R., R. D. Price, F. Kollmorgen, T. Piggott, and E. H. Ridge. 1974. Effect of seed inoculation with *Bacillus subtilis* and *Streptomyces griseus* on the growth of cereals and carrots. Austral. Jour. Agric. Res. 25:219–226.

Mew, I. C., and T. Kommedahl. 1972. Interaction among microorganisms occurring naturally and applied to pericarps of corn kernels. Plant Dis. Reptr. 56:861–863.

Meyer, J. A., and H. Maraite. 1971. Multiple infection and symptom mitigation in vascular wilt diseases. Trans. Brit. Mycol. Soc. 57:371–377.

Meyers, J. A., and R. J. Cook. 1972. Induction of chlamydospore formation in *Fusarium solani* by abrupt removal of the organic carbon substrate. Phytopathology 62:1148–1153.

Michael, A. H., and P. E. Nelson. 1972. Antagonistic effect of soil bacteria on *Fusarium roseum* 'Culmorum' from carnation. Phytopathology 62:1052–1056.

Micheli, P. A. 1729. Nova Plantarum Genera Juxta Tournefortii Methodum Disposita. Florence.

Middleton, J. T. 1943. The Taxonomy, Host Range and Geographic Distribution of the Genus Pythium. Mem. Torrey Bot. Club 20:1–171.

Miki, N. K., K. J. Clarke, and M. E. McCully. 1980. A histological and histochemical comparison of the mucilages on the root tips of several grasses. Can. Jour. Bot. 58:2581–2593.

Millar, R. L., and R. Hemphill. 1978. β-Glucosidase associated with cyanogenesis in *Stemphylium* leafspot of birdsfoot trefoil. Physiol. Plant Pathol. 13:259–270.

Millar, R. L., and V. J. Higgins. 1970. Association of cyanide with infection of birdsfoot trefoil by *Stemphylium loti*. Phytopathology 60:104–110.

Millard, W. A., and C. B. Taylor. 1927. Antagonism of micro-organisms as the controlling factor in the inhibition of scab by green-manuring. Ann. Appl. Biol. 14:202–216.

Miller, D. E., and D. W. Burke. 1974. Influence of soil bulk density and water potential on Fusarium root rot of beans. Phytopathology 64:526–529.

Miller, D. E., and D. W. Burke. 1975. Effect of soil aeration on Fusarium root rot of beans. Phytopathology 65:519–523.

Miller, J. H., J. E. Giddens, and A. A. Foster. 1957. A survey of the fungi of forest and cultivated soils of Georgia. Mycologia 49:779–808.

Minter, D. W. 1980. Possible biological control of *Lophodermium seditiosum*. Pages 75–80 *in* C. S. Millar, ed. Current Research on Conifer Needle Diseases. Proc. IUFRO W. P. on Needle Diseases, Sarajevo, Bosnia, Yugoslavia, 15–19 Sept., 1980.

Minter, D. W., and C. S. Millar. 1980. Ecology and biology of three *Lophodermium* species on secondary needles of *Pinus sylvestris*. Eur. Jour. For. Pathol. 10:169–181.

Misaghi, I. J., L. J. Stowell, R. G. Grogan, and L. C. Spearman. 1982. Fungistatic activity of water-soluble fluorescent pigments of fluorescent pseudomonads. Phytopathology 72:33–36.

Mitchell., D. T., and R. C. Cooke. 1968. Some effects of temperature on germination and longevity of sclerotia in *Claviceps purpurea*. Trans. Brit. Mycol. Soc. 51:721–729.

Molitoris, H. P., M. Hollings, and H. A. Woodi, eds. 1979. Fungal Viruses. Springer-Verlag, Berlin. 194 pp.

Moody, A. R., and D. Gindrat. 1977. Biological control of cucumber black root rot by *Gliocladium roseum*. Phytopathology 67:1159–1162.

Moore, K. J. 1978. The influence of no-tillage on take-all of wheat. Ph.D. Thesis. Washington State University, Pullman, 62 pp.

Moore, L. W. 1977. Prevention of crown gall on Prunus roots by bacterial antagonists. Phytopathology 67:139–144.

Moore, L. W., and D. A. Cooksey. 1981. Biology of *Agrobacterium tumefaciens*: Plant interactions. Internatl. Review Cytology, Suppl. 13:15–46.

Moore, W. D. 1949. Flooding as a means of destroying the sclerotia of *Sclerotinia sclerotiorum*. Phytopathology 39:920-927.

Morgan-Jones, G., and R. Rodriguez-Kabana. 1981. Fungi associated with cysts of *Heterodera glycines* in an Alabama soil. Nematropica 11:69–74.

Morgan-Jones, G., G. Godoy, and R. Rodriguez-Kabana. 1981. *Verticillium chlamydosporium,* fungal parasite of *Meloidogyne arenaria* females. Nematropica 11:115–120.

Morquer, R., G. Viala, J. Rouch, J. Fayret, and G. Bergé. 1963. Contribution à l'étude morphogénique du genre *Gliocladium*. Bull. Soc. Mycol. Fr. 79:137–241.

Mortvedt, J. J., P. M. Giordano, and W. L. Lindsey, eds. Micronutrients in Agriculture. Soil Sci. Soc. Amer., Madison, WI. 666 pp.

Mosse, B. 1962. The establishment of vesicular-arbuscular mycorrhiza under aseptic conditions. Jour. Gen. Microbiol. 27:509–520.

Mosse, B. 1973. Advances in the study of vesicular-arbuscular mycorrhiza. Annu. Rev. Phytopathology 11:171–196.

Mossop, D. W., and C. H. Procter. 1975. Cross protection of glasshouse tomatoes against tobacco mosaic virus. N. Zeal. Jour. Exp. Agric. 3:343–348.

Mount, M. S., and G. H. Lacy, eds. 1982–83. Phytopathogenic Prokaryotes. 2 vols. Academic Press, New York. 1056 pp.

Mower, R. L., W. C. Snyder, and J. G. Hancock. 1975. Biological control of ergot by *Fusarium*. Phytopathology 65:5–10.

Muirhead, I. F., and B. J. Deverall. 1981. Role of appressoria in latent infection of banana fruits by *Colletotrichum musae*. Physiol. Plant Pathol. 19:77–84.

Muller, H. G. 1958. The constricting ring mechanism of two predacious hyphomycetes. Trans. Brit. Mycol. Soc. 41:341–364.

Müller, K. O., and H. Börger. 1940. Experimentelle untersuchungen über die Phytophthora-resistenz der kartoffel. Arb. Biol. BundAnst. Land- Fortswirts. 23:189–231.

Munnecke, D. E., and P. A. Chandler. 1957. A leaf spot of *Philodendron* related to stomatal exudation and to temperature. Phytopathology 47:299– 303.

Munnecke, D. E., W. Wilbur, and E. F. Darley. 1976. Effect of heating or drying on *Armillaria mellea* or *Trichoderma viride* and the relation to survival of *A. mellea* in soil. Phytopathology 66:1363–1368.

Myers, D. F., and G. A. Strobel. 1981. *Pseudomonas syringae* as an antagonist: Laboratory and greenhouse effectiveness against Dutch elm disease. Phytopathology 71:1006.

Naiki, T., and R. J. Cook. 1983a. Heterokaryosis as a factor in loss of pathogenicity in *Gaeumannomyces graminis* var. *tritici*. Phytopathology 73:In Press.

Naiki, T., and R. J. Cook. 1983b. Relationship between self-inhibition and inability of isolates of *Gaeumannomyces graminis* var. *tritici* to cause disease. Phytopathology 73:In Press.

Nair, N. G., and P. C. Fahy. 1972. Bacteria antagonistic to *Pseudomonas tolaasii* and their control of brown blotch of the cultivated mushroom *Agaricus bisporus*. Jour. Appl. Bact. 35:439–442.

Nair, N. G., and P. C. Fahy. 1976. Commercial application of biological control of mushroom bacterial blotch. Austral. Jour. Agric. Res. 27:415– 422.

Nash, S. M., and W. C. Snyder. 1962. Quantitative estimations by plate counts of propagules of the bean root rot *Fusarium* in field soils. Phytopathology 52:567– 572.

Neal, J. L., Jr., R. I. Larson, and T. G. Atkinson. 1973. Changes in rhizosphere populations of selected physiological groups of bacteria related to substitution of specific pairs of chromosomes in spring wheat. Plant Soil 39:209– 212.

Nelson, E. E. 1975. Survival of *Poria weirii* on paired plots in alder and conifer stands. Microbios 12:155–158.

Nelson, P. E., T. A. Toussoun, and R. J. Cook, eds. 1981. Fusarium: Diseases, Biology, and Taxonomy. Penn. State Univ., University Park, PA. 457 pp.

Nelson, R. R., ed. 1973. Breeding Plants for Disease Resistance. Penn. State Univ. Press, Univ. Park, PA. 40l pp.

New, P. B., and A. Kerr. 1972. Biological control of crown gall: Field measurements and glasshouse experiments. Jour. Appl. Bact. 35:279–287.

Newhook, F. J. 1951. Microbiological control of *Botrytis cinerea* Pers. I. The role of pH changes and bacterial antagonism. II. Antagonism by fungi and actinomycetes. Ann. Appl. Biol. 38:169–184, 185–202.

Newhook, F. J. 1957. The relationship of saprophytic antagonism to control of *Botrytis cinerea* Pers. on tomatoes. N. Zeal. Jour. Sci. Technol., Sect. A. 38:473–481.

Newhook, F. J., and F. D. Podger. 1972. The role of *Phytophthora cinnamomi* in Australian and New Zealand forests. Annu. Rev. Phytopathol. 10:299–326.

Nilsson, H. C. 1969. Studies on root and foot rot diseases of cereals and grasses. I. On resistance to *Ophiobolus graminis* Sacc. Lanbruks. Annaler 35:275–807.

Noble, P. S. 1974. Introduction to Biophysical Plant Physiology. W. H. Freeman, San Francisco. 488 pp.

Novacky, A., G. Acedo, and R. N. Goodman. 1973. Prevention of bacterially induced hypersensitive reaction by living bacteria. Physiol. Plant Pathol. 3:133–136.

Odvody, G. N., M. G. Boosalis, and E. D. Kerr. 1980. Biological control of *Rhizoctonia solani* with a soil-inhabiting Basidiomycete. Phytopathology 70:655–658.

Ohr, H. D., and D. E. Munnecke. 1974. Effects of methyl bromide on antibiotic protection by *Armillaria mellea*. Trans. Brit. Mycol. Soc. 62:65–72.

Old, K.M. 1977. Giant soil amoebae cause perforation of conidia of *Cochliobolus sativus*. Trans. Brit. Mycol. Soc. 68:277–320.

Old, K. M., and J. F. Darbyshire. 1978. Soil fungi as food for giant amoebae. Soil Biol. Biochem. 10:93–100.

Old, K. M., and Z. A. Patrick. 1976. Perforation and lysis of spores of *Cochliobolus sativus* and *Thielaviopsis basicola* in natural soils. Can. Jour. Bot. 54:2798–2809.

Old, R. W., and S. B. Primrose. 1981. Principles of Gene Manipulation. An Introduction to Genetic Engineering. 2nd ed. Univ. California Press, Berkeley. 214 pp.

Olivier, J. M., and J. Guillaumes. 1981. Essais de lutte biologique contre la tache bacterienne (agent: *Pseudomonas fluorescens* = *tolaasii*). Proc. 11th Internatl. Congr. Sci. Cultiv. Edible Fungi, Sydney. 16 pp.

Olivier, J. M., J. Guillaumes, and D. Martin. 1978. Study of a bacterial disease of mushroom caps. Proc. 11 Internatl. Conf. Plant Path. Bacteria, Angers, 1978:903–916.

Olsen, C. M. 1964. Antagonistic effects of microorganisms on *Rhizoctonia solani* in soil. Ph.D. Thesis, University of California, Berkeley. 152 pp.

Olsen, C. M., and K. F. Baker. 1968. Selective heat treatment of soil, and its effect on the inhibition of *Rhizoctonia solani* by *Bacillus subtilis*. Phytopathology 58:79–87.

Olsen, C. M., N. T. Flentje, and K. F. Baker. 1967. Comparative survival of monobasidial cultures of *Thanatephorus cucumeris* in soil. Phytopathology 57:598–601.

Olson, S. 1982. Why is the sea constant? Science82 3(9):112.

Omar, M., and W. A. Heather. 1979. Effect of saprophytic phylloplane fungi on germination and development of *Melampsora larici-populina*. Trans. Brit. Mycol. Soc. 72:225–231.

Oschwald, W. R., ed. 1978. Crop Residue Management Systems. Amer. Soc. Agron., Madison. Spec. Publ. 31:1–248.

Owens, L. D., R. G. Gilbert, G. E. Griebel, and J. D. Menzies. 1969. Identification of plant volatiles that stimulate microbial respiration and growth in soil. Phytopathology 59:1468–1472.

Paleg, L. G., and D. Aspinall, eds. 1982. Physiology and Biochemistry of Drought Resistance in Plants. Academic Press, New York. 492 pp.

Palleroni, N. J., and M. Doudoroff. 1972. Some properties and taxonomic subdivisions of the genus *Pseudomonas*. Annu. Rev. Phytopathol. 10:73–100.

Panopoulos, N. J., ed. 1981. Genetic Engineering in the Plant Sciences. New York: Praeger 271 pp.

Papacostas, G., and J. Gaté. 1928. Les associations microbiennes, leurs applications thérapeutique. Paris:G. Doin.

Papavizas, G. C., ed. 1981. Biological Control in Crop Production. Beltsville Symposium in Agricultural Research 5. Allanheld, Osmun Co., London. 461 pp.

Papavizas, G. C. 1982. Survival of *Trichoderma harzianum* in soil and in pea and bean rhizospheres. Phytopathology 72:121–125.

Papavizas, G. C., and J. A. Lewis. 1981. Induction of new biotypes of *Trichoderma harzianum* resistant to benomyl and other fungicides. Phytopathology 71:247–248.

Papavizas, G. C., and R. D. Lumsden. 1980. Biological control of soilborne fungal propagules. Annu. Rev. Phytopathol. 18:389–413.

Papavizas, G. C., J. A. Lewis, and T. H. Abd-El Moity. 1982. Evaluation of new biotypes of *Trichoderma harzianum* for tolerance of benomyl and enhanced biocontrol capabilities. Phytopathology 72:126–132.

Papavizas, G. C., S. B. Morris, and J. J. Marois. 1982. Isolation and enumeration of propagules of *Laetisaria arvalis* from soil. Phytopathology 72:709.

Papendick, R. I., and R. J. Cook. 1974. Plant water stress and development of Fusarium foot rot in wheat subjected to different cultural practices. Phytopathology 64:358–363.

Papendick, R. I., L. L. Boersma, D. Colacicco, J. M. Kla, C. A. Kraenzle, P. B. Marsh, A. S. Newman, J. F. Parr, J. B. Swan, and I. G. Youngberg. 1980. U.S. Department of Agriculture Report and Recommendations on Organic Farming. A Special Report Prepared for the Secretary of Agriculture. U. S. Government Printing Office, Washington, D. C. 164 pp.

Parke, J. L., and R. G. Linderman. 1980. Association of vesicular-arbuscular mycorrhizal fungi with the moss *Funaria hygrometrica*. Can. Jour. Bot. 58:1898–1904.

Parkinson, D., and J. S. Waid, eds. 1960. The Ecology of Soil Fungi. Liverpool Univ. Press, Liverpool. 324 pp.

Parmeter, J. R. Jr., ed. 1970. *Rhizoctonia solani*: Biology and Pathology. Univ. California Press, Berkeley. 255 pp.

Parr, J. F., W. R. Gardner, and L. F. Elliott, eds. 1981. Water Potential Relations in Soil Microbiology. Soil Sci. Soc. Amer., Madison, WI. 151 pp.

Paul, E. A., and J. N. Ladd, eds. 1981. Soil Biochemistry, Vol. 5. Marcel Dekker, New York. 504 pp.

Pavlica, D. A., T. S. Hora, J. J. Bradshaw, R. K. Skogerboe, and R. Baker. 1978. Volatiles from soil influencing activities of soil fungi. Phytopathology 68:758–765.

Pegg, K. G. 1977a. Biological control of *Phytophthora cinnamomi* root rot of avocado and pineapple in Queensland. Austral. Nurs Assoc. Ann. Conf. Seminar Papers 1977:7–12.

Pegg, K. G. 1977b. Soil application of elemental sulphur as a control of *Phytophthora cinnamomi* root and heart rot of pineapple. Austral. Jour. Exp. Agric. Anim. Husb. 17:859–865.

Pegg, K. G. 1978. Disease-free avocado nursery trees. Queensl. Agric. Jour. 104:134–136.

Pérombelon, M. C. M., and A. Kelman. 1980. Ecology of the soft rot erwinias. Annu. Rev. Phytopathol. 18:361–387.

Philipp, W. D., and G. Crüger. 1979. Parasitismus von *Ampelomyces quisqualis* auf Echten Mehltaupilzen an Gurken und anderen Gemüsearten. Zeit. Pflanzenkr. Pflanzensch. 86:129–142.

Phillips, D. J. 1965. Ecology of plant pathogens in soil. IV. Pathogenicity of macroconidia of *Fusarium roseum* f. sp. *cerealis* produced on media of high or low nutrient content. Phytopathology 55:328–329.

Phillips, D. J., and S. Wilhelm. 1971. Root distribution as a factor influencing symptom expression of Verticillium wilt of cotton. Phytopathology 61:1312–1313.

Phillips, D. V., C. Leben, and C. C. Allison. 1967. A mechanism for the reduction of Fusarium wilt by a *Cephelosporium* species. Phytopathology 57:916–919.

Phipps, P. M., and M. K. Beute. 1979. Population dynamics of *Cylindrocladium crotalariae* microsclerotia in naturally-infested soil. Phytopathology 69:240-243.

Pimentel, D., ed. 1981. Handbook of Pest Management in Agriculture, Vol. II. CRC Press, Boca Raton, FL. 528 pp.

Plenchette, C., V. Furlan, and J. A. Fortin. 1981. Growth stimulation of apple trees in unsterilized soil under field conditions with VA mycorrhiza inoculation. Can. Jour. Bot. 59:2003–2008.

Potter, M. C. 1908. On a method of checking parasitic diseases in plants. Jour. Agric. Sci. 3:102–107.

Pottle, H. W., A. L. Shigo, and R. O. Blanchard. 1977. Biological control of wound hymenomycetes by *Trichoderma harzianum*. Plant Dis. Reptr. 61:687–690.

Powell, C. C. 1982. Some new technologies in greenhouse crop production. Plant Disease 66:525.

Pratt, R. G. 1978. Germination of oospores of *Sclerospora sorghi* in the presence of growing roots of host and nonhost plants. Phytopathology 68:1606–1613.

Preece, T. F., and C. H. Dickinson, eds. 1971. Ecology of Leaf Surface Microorganisms. Academic Press, New York. 640 pp.

Prescott, J. A. 1920. A note on the sharqi soils of Egypt. A study in partial sterilisation. Jour. Agric. Sci. 10:177–181.

Preston, N. C. 1943–1961. Observations on the genus *Mycothecium Tode*. I. The three classic species. II. *Myrothecium gramineum* Lib. and two new species. III.

The cylindrical-spored species of *Myrothecium* known in Britain. Trans. Brit. Mycol. Soc. 26:158–168; 31:271–276; 44:31–41.

Primrose, S. B. 1976. Ethylene-forming bacteria from soil and water. Jour. Gen. Microbiol. 97:343–346.

Pullman, G. S., and J. E. DeVay. 1981. Effect of soil flooding and paddy rice culture on the survival of *Verticillium dahliae* and incidence of Verticillium wilt of cotton. Phytopathology 72:1285–1289.

Pullman, G. S., J. E. DeVay, R. H. Garber, and A. R. Weinhold. 1981a. Soil solarization: Effects on Verticillium wilt of cotton and soilborne populations of *Verticillium dahliae, Pythium* spp., *Rhizoctonia solani,* and *Thielaviopsis basicola.* Phytopathology 71:954–959.

Pullman, G. S., J. E. DeVay, and R. H. Garber. 1981b. Soil solarization and thermal death: A logarithmic relationship between time and temperature for four soilborne plant pathogens. Phytopathology 71:959–964.

Punja, Z. K., R. G. Grogan, and T. Unruh. 1982. Comparative control of *Sclerotium rolfsii* on golf greens in northern California with fungicides, inorganic salts, and *Trichoderma* spp. Plant Disease 66:1125–1128.

Purkayastha, B. P., and B. Bhattacharyya. 1982. Antagonism of micro-organisms from jute phyllosphere towards *Colletotrichum corchori.* Trans. Brit. Mycol. Soc. 78:509–513.

Pusey, P. L., and C. L. Wilson. 1982. Detection of double-stranded RNA in *Ceratocystis ulmi.* Phytopathology 72:423–428.

Raabe, R. D. 1962. Host list of the root rot fungus, *Armillaria mellea.* Hilgardia 33:25–88.

Rabb, R. L., and F. E. Guthrie, ed. 1970. Concepts of Pest Management. N. C. State Univ., Raleigh. 242 pp.

Rahe, J. E., and R. M. Arnold. 1975. Injury-related phaseollin accumulation in *Phaseolus vulgaris* and its implications with regard to specificity of host-parasite interaction. Can. Jour. Bot. 53:921–928.

Rana, G. L., G. H. Kaloostian, G. N. Oldfield, A. L. Granett, E. C. Calavan, H. D. Pierce, I. M. Lee, and D. J. Gumpf. 1975. Acquisition of *Spiroplasma citri* through membranes by homopterous insects. Phytopathology 65:1143–1145.

Rao, N. N. R., and Pavgi, M. S. 1976. A mycoparasite on *Sclerospora graminicola* Can. Jour. Bot. 54:220-223.

Raper, K. B., and C. Thom. 1949. A Manual of the Penicillia. Williams and Wilkins, Baltimore. 875 pp.

Rast, A. T. B. 1972. MII-16, an artificial symptomless mutant of tobacco mosaic virus for seedling inoculation of tomato crops. Neth. Jour. Plant Pathol. 78:110-112.

Rast, A. T. B. 1979. Infection of tomato seed by different strains of tobacco mosaic virus with particular reference to the symptomless mutant, MII-16. Neth. Jour. Plant Pathol. 85:223–233.

Rathaiah, Y., and M. S. Pavgi. 1973. *Fusarium semitectum* mycoparasitic on *Cercosporae.* Phytopathol. Zeit. 77:278–281.

Rawlinson, C. J. 1975. Role of fungal viruses in pathogenicity of fungi. Abstracts of Third International Congress for Virology, Madrid, p. 147.

Rawlinson, C. J., D. Hornby, V. Pearson, and J. M. Carpenter. 1973. Virus-like particles in the take-all fungus, *Gaeumannomyces graminis*. Ann. Appl. Biol. 74:197–209.

Reinking, O. A., and M. M. Manns. 1933. Parasitic and other fusaria counted in tropical soils. Zeit. Parasitenkr. 6:23–75.

Reis, E. M., R. J. Cook, and B. L. McNeal. 1982. Effect of mineral nutrition on take-all of wheat. Phytopathology 72:224–229.

Reis, E. M., R. J. Cook, and B. L. McNeal. 1983. Elevated pH and associated reduced trace-nutrient availability as factors contributing to increased take-all of wheat upon soil liming. Phytopathology 73:411–413.

Rennie, R. J. 1981. Potential use of induced mutations to improve symbioses of crop plants with N_2-fixing bacteria. IAEH-SM-251/45:293–321.

Ricard, J. L. 1970. Biological control of *Fomes annosus* in Norway spruce (*Picea abies*) with immunizing commensals. Studia Forest. Suecica 84:1–50.

Ricard, J. 1977. Experience with immunizing commensals. Neth. Jour. Plant Pathol. 83 (Suppl. 1):443–448.

Ricard, J. L. 1981. Commercialization of a *Trichoderma* based mycofungicide: Some problems and solutions. Biocontrol News Inform. 2(2):95–98.

Ricard, J. L., and W. B. Bollen. 1968. Inhibition of *Poria carbonica* by *Scytalidium* sp., an imperfect fungus isolated from Douglas-fir poles. Can. Jour. Bot. 46:643–647.

Ricard, J. L., and P. Laird. 1968. Current research in the control of *Fomes annosus* with *Scytalidium* sp., an immunizing commensal. Pages 104–109 *in* IUFRO, Proc. 3rd Interational Conference on *Fomes annosus,* Aarbus, Denmark.

Ricard, J. L., M. M. Wilson, and W. B. Bollen. 1969. Biological control of decay in Douglas-fir poles. Forest Prod. Jour. 19:41–45.

Ridge, E. H. 1976. Studies on soil fumigation. II. Effects on bacteria. Soil Biol. Biochem. 8:249–253.

Rifai, M. A. 1969. A revision of the genus *Trichoderma*. Commonw. Mycol. Inst., Mycol. Papers 116:1–56.

Rifai, M. A., and R. C. Cooke. 1966. Studies on some didymosporous genera of nematode-trapping Hyphomycetes. Trans. Brit. Mycol. Soc. 49:147–168.

Riggle, J. H., and E. J. Klos. 1972. Relationship of *Erwinia herbicola* to *Erwinia amylovora*. Can. Jour. Bot. 50:1077–1083.

Riggs, R. D., D. A. Slack, M. L. Hamblen, and L. Rakes. 1980. Nematode control studies in soybeans. Ark. Agric. Exp. Sta. Report series 252. 32 pp.

Rishbeth, J. 1963. Stump protection against *Fomes annosus*. III. Inoculation with *Peniophora gigantea*. Ann. Appl. Biol. 52:63–77.

Rishbeth, J. 1979. Modern aspects of biological control of *Fomes* and *Armillaria*. Eur. Jour. For. Pathol. 9:331–340.

Roberts, W. 1874. Studies on biogenesis. Phil. Trans. Roy. Soc. London 164:457–477.

Roberts, W. P., M. E. Tate, and A. Kerr. 1977. Agrocin 84 is a 6-N-phosphoramidate of an adenine nucleotide analogue. Nature 265:379–381.

Robertson, N. F. 1958. Observations of the effect of water on the hyphal apices of *Fusarium oxysporum*. Ann. Bot. N. S. 22:159–173.

Robertson, N. F. 1959. Experimental control of hyphal branching and branch form in hyphomycetous fungi. Linn. Soc. London Jour. (Bot.) 56:207–211.

Romanos, M. A., C. J. Rawlinson, M. R. Almond, and K. W. Buck. 1980. Production of fungal growth inhibitors by isolates of *Gaeumannomyces graminis* var. *tritici*. Trans. Brit. Mycol. Soc. 74:79–88.

Ross, D. J., K. R. Tate, A. Cairns, and E. A. Pansier. 1980. Microbial biomass estimations in soils from tussock grasslands by three biochemical procedures. Soil Biol. Biochem. 12:375–383.

Ross, E. W., and D. H. Marx. 1972. Susceptibility of sand pine to *Phytophthora cinnamomi*. Phytopathology 62:1197–1200.

Ross, J. P. 1972. Influence of Endogone mycorrhizae on Phytophthora rot of soybean. Phytopathology 62:896–897.

Rouxel, F., C. Alabouvette, and J. Louvet. 1977. Recherches sur la résistance des sols aux maladies. II. Incidence de traitements thermiques sur la résistance microbiologique d'un sol à la Fusariose vasculaire du Melon. Ann. Phytopathol. 9:183–192.

Rovira, A. D. 1959. Root excretions in relation to the rhizosphere effect. IV. Influence of plant species, age of plant, light, temperature and calcium nutrition on exudation. Plant Soil 11:53–64.

Rovira, A. D., and A. Simon. 1982. Integrated control of *Heterodera avenae*. EPPO Bull. 12:517–523.

Rovira, A. D., E. I. Newman, H. J. Bowen, and R. Campbell. 1974. Quantitative assessment of the rhizoplane microflora by direct microscopy. Soil Biol. Biochem. 6:211–216.

Rowe, R. C., and J. D. Farley. 1978. Control of Fusarium crown and root rot of greenhouse tomatoes by inhibiting recolonization of steam-disinfested soil with a captafol drench. Phytopathology 68:1221–1224.

Rowe, R. C., J. D. Farley, and D. L. Coplin. 1977. Airborne spore dispersal and recolonization of steamed soil by *Fusarium oxysporum* in tomato greenhouses. Phytopathology 67:1513–1517.

Ruinen, J. 1961. The phyllosphere. I. An ecologically neglected milieu. Plant Soil 15:81–109.

Ruinen, J. 1966. The phyllosphere. IV. Cuticle decomposition by microorganisms in the phyllosphere. Ann. Inst. Pasteur (Paris) 111 (3 Suppl.):342–346.

Runia, W. Th., and D. Peters. 1980. The response of plant species used in agriculture and horticulture to viroid infections. Neth. Jour. Plant Pathol. 86:135–146.

Russell, R. S. 1977. Plant Root Systems: Their Function and Interaction with the Soil. McGraw-Hill, London. 298 pp.

Saksena, S. B. 1960. Effect of carbon disulfide fumigation on *Trichoderma viride* and other soil fungi. Trans. Brit. Mycol. Soc. 43:111–116.

Sanders, F. E., B. Mosse, and P. B. Tinker. 1975. Endomycorrhizas. Academic Press, New York. 626 pp.

Sanford, G. B. 1926. Some factors affecting the pathogenicity of *Actinomyces scabies*. Phytopathology 16:525–547.

Sanford, G. B., and W. C. Broadfoot. 1931. A note on the biological control of root rots of cereals. Studies of the effects of other soil-inhabiting microorganisms on the virulence of *Ophiobolus graminis* Sacc. Sci. Agric. 11:460,512–528.

Sargent, J. A., I. C. Tommerup, and D. S. Ingram. 1973. The penetration of a susceptible lettuce variety by the downy mildew fungus *Bremia lactucae* Regel. Physiol. Plant Pathol. 3:231–239.

Sarkar, A. N., and R. G. Wyn Jones. 1982. Effect of rhizosphere pH on the availability and uptake of Fe, Mn, and Zn. Plant Soil 66:361–372.

Sayre, R. M. 1980. Promising organisms for biocontrol of nematodes. Plant Disease 64:526–532.

Schenck, N. C., ed. 1982. Methods and Principles of Mycorrhizal Research. Amer. Phytopathol. Soc., St. Paul. 256 pp.

Schenck, N. C., and M. K. Kellam. 1978. The influence of vesicular arbuscular mycorrhizae on disease development. Florida Agric. Exp. Sta. Bull. 798 (Technical):1–16.

Scher, F. M., and R. Baker. 1980. Mechanism of biological control in a *Fusarium*-suppressive soil. Phytopathology 70:412–417.

Scher, F. M., and R. Baker. 1982. Effect of *Pseudomonas putida* and a synthetic iron chelator on induction of soil supressive to Fusarium wilt pathogens. Phytopathology 72:1567–1573.

Scherff, R. H. 1973. Control of bacterial blight of soybean by *Bdellovibrio bacteriovorus*. Phytopathology 63:400-402.

Scherff, R. H., J. E. DeVay, and T. W. Carroll. 1966. Ultrastructure of host-parasite relationships involving reproduction of *Bdellovibrio bacteriovorus* in host bacteria. Phytopathology 56:627–632.

Schippers, B., and W. Gams, eds. 1979. Soil-borne Plant Pathogens. Academic Press, New York. 686 pp.

Schippers, B., and A. K. F. Schermer. 1966. Effect of antifungal properties of soil on dissemination of the pathogen and seedling infection originating from Verticillium-infected achenes of *Senecio*. Phytopathology 56:549–552.

Schippers, B., D. J. Boerwinkel, and H. Konings. 1978. Ethylene not responsible for inhibition of conidium germination by soil volatiles. Neth. Jour. Plant Pathol. 84:101–107.

Schippers, B., J. W. Meijer, and J. I. Liem. 1982. Effect of ammonia and other soil volatiles on germination and growth of soil fungi. Trans. Brit. Mycol. Soc. 79:253–259.

Schmidt, E. L. 1979. Initiation of plant root-microbe interactions. Annu. Rev. Microbiol. 33:355–376.

Schnathorst, W. E., and D. E. Mathre. 1966. Cross-protection in cotton with strains of *Verticillium albo-atrum*. Phytopathology 56:1204–1209.

Schneider, R. W., ed. 1982. Suppressive Soils and Plant Disease. Amer. Phytopathol. Soc., St. Paul, MN. 96 pp.

Schoeneweiss, D. F. 1975. Predisposition, stress, and plant disease. Annu. Rev. Phytopathol. 13:193–211.

Schönbeck, F., and H. W. Dehne. 1977. Damage to mycorrhizal and nonmycorrhizal cotton seedlings by *Thielaviopsis basicola*. Plant Dis. Reptr. 61:266–267.

Schönbeck, F., and H. W. Dehne. 1979. Untersuchuingen zum Einfluss der endotrophen Mykorrhiza auf Pflanzenkrankheiten. Zeit. Pflanzenkr. Pflanzensch. 86:103–112.

Schoulties, C. L., K. F. Baker, and C. Sabersky-Lehmann. 1980. Factors influencing zoospore production by *Phytophthora cinnamomi* in axenic culture. Can. Jour. Bot. 58:2117–2122.

Schroth, M. N., and J. G. Hancock. 1981. Selected topics in biological control. Annu. Rev. Microbiol. 35:453–476.

Schroth, M. N., and J. G. Hancock. 1982. Disease-suppressive soil and root-colonizing bacteria. Science 216:1376–1381.

Schroth, M. N., and F. F. Hendrix, Jr. 1962. Influence of nonsusceptible plants on the survival of *Fusarium solani* f. *phaseoli* in soil. Phytopathology 52:906–909.

Schroth, M. N., and W. J. Moller. 1976. Crown gall controlled in the field with a nonpathogenic bacterium. Plant Dis. Reptr. 60:275–278.

Schroth, M. N., Thomson, S. V., and Moller, W. J. 1979. Streptomycin resistance in *Erwinia amylovora*. Phytopathology 69:565–568.

Schroth, M. N., A. R. Weinhold, A. H. McCain, D. C. Hildebrand, and N. Ross. 1971. Biology and control of *Agrobacterium tumefaciens*. Hilgardia 40:537–552.

Scott, P. R. 1970. *Phialophora radicicola,* an avirulent parasite of wheat and grass roots. Trans. Brit. Mycol. Soc. 55:163–167.

Sequeira, L. 1958. Bacterial wilt of bananas: Dissemination of the pathogen and control of the disease. Phytopathology 48:64–69.

Sequeira, L., and T. L. Graham. 1977. Agglutination of avirulent strains of *Pseudomonas solanacearum* by potato lectin. Physiol. Plant Pathol. 11:43–54.

Sequeira, L., and L. M. Hill. 1974. Induced resistance in tobacco leaves: The growth of *Pseudomonas solanacearum* in protected tissues. Physiol. Plant Pathol. 4:447–455.

Setlow, J. K., and A. Hollaender, eds. 1979–1982. Genetic Engineering: Principles and Methods. 4 vols. Plenum Press, New York.

Shawish, O., and R. Baker. 1982. Thigmomorphogenesis and predisposition of hosts to Fusarium wilts. Phytopathology 72:63–68.

Shea, S. R. 1979. Forest management and *Phytophythora cinnamomi* in Australia. Pages 73–100 *In* K. M. Old, ed. Phytophthora and Forest Management in Australia. CSIRO, Melbourne. 114 pp.

Shea, S. R., and N. Malajczuk. 1977. Potential for control of eucalypt dieback in Western Australia. Austral. Nurs. Assoc. Ltd. Annu. Conf. Seminar Papers 1977:13–19.

Shepherd, C. J. 1975. *Phytophthora cinnamomi*—An ancient immigrant to Australia. Search 6:484–490.

Shigo, A. L. 1979. Tree decay. An expanded concept. U.S. Dept. Agric., Agric. Inform. Bull. 419:1–73.

Shigo, A. L. 1982. Tree decay in our urban forests: What can be done about it? Plant Disease 66:763–768.

Shigo, A. L., and H. G. Marx. 1977. Compartmentalization of decay in trees. U. S. Dept. Agric., Agric. Inform. Bull. 405:1–73.

Shipton, P. J., R. J. Cook, and J. W. Sitton. 1973. Occurrence and transfer of a biological factor in soil that suppresses take-all of wheat in eastern Washington. Phytopathology 63:511–517.

Shishiyama, J., F. Araki, and S. Akai. 1970. Studies on cutin esterase. II. Characteristics of cutin esterase from *Botrytis cinerea* and its activity on tomato-cutin. Plant Cell Physiol. 11:937–945.

Shokes, F. M., S. D. Lyda, and W. R. Jordan. 1977. Effect of water potential on the growth and survival of *Macrophomina phaseolina*. Phytopathology 67:239–241.

Shortle, W. C. 1979. Mechanisms of compartmentalization of decay in living trees. Phytopathology 69:1147–1151.

Sifton, H. B. 1957. Air-space tissues in plants. II. Bot. Rev. 23:303–312.

Sinclair, W. A., D. P. Cowles, and S. M. Hee. 1975. Fusarium root rot of Douglas-fir seedlings: Suppression by soil fumigation, fertility management, and inoculation with spores of the fungal symbiont *Laccaria laccata*. For. Sci. 21:390–399.

Sinden, S. L., J. E. DeVay, and P. A. Backman. 1971. Properties of syringomycin, a wide spectrum antibiotic and phytotoxin produced by *Pseudomonas syringae*, and its role in the bacterial canker disease of peach trees. Physiol. Plant Pathol. 1:199–213.

Singer, R. 1975. The Agaricales in Modern Taxonomy. 2nd ed. Cramer, Weinheim. 912 pp.

Sitton, J. W., and R. J. Cook. 1981. Comparative morphology and survival of chlamydospores of *Fusarium roseum* 'Culmorum' and 'Graminearum'. Phytopathology 71:85–90.

Sivasithamparam, K., C. A. Parker, and C. S. Edwards. 1979. Bacterial antagonists to the take-all fungus and fluorescent pseudomonads in the rhizosphere of wheat. Soil Biol. Biochem. 11:161–165.

Skotland, C. B. 1971. Pathogenic and nonpathogenic *Verticillium* species from south central Washington. Phytopathology 61:435–436.

Smiley, R. W. 1978a. Antagonists of *Gaeumannomyces graminis* from the rhizoplane of wheat in soils fertilized with ammonium- or nitrate-nitrogen. Soil Biol. Biochem. 10:169–174.

Smiley, R. W. 1978b. Colonization of wheat roots by *Gaeumannomyces graminis* inhibited by specific soils, microorganisms and ammonium-nitrogen. Soil Biol. Biochem. 10:175–179.

Smiley, R. W., and R. J. Cook. 1973. Relationship between take-all of wheat and rhizosphere pH in soils fertilized with ammonium vs. nitrate-nitrogen. Phytopathology 63:882–890.

Smith, A. M. 1972. Drying and wetting sclerotia promotes biological control of *Sclerotium rolfsii* Sacc. Nutrient leakage promotes biological control of dried sclerotia of *Sclerotium rolfsii* Sacc. Biological control of fungal sclerotia in soil. Soil Biol. Biochem. 4:119–123, 125–129, 131–134.

Smith, A. M. 1973. Ethylene as a cause of soil fungistasis. Nature 246:311–313.

Smith, A. M. 1976a. Ethylene production by bacteria in reduced microsites in soil and some implications to agriculture. Soil Biol. Biochem. 8:293–298.

Smith, A. M. 1976b. Ethylene in soil biology. Annu. Rev. Phytopathol. 14:53–73.

Smith, A. M., and R. J. Cook. 1974. Implications of ethylene production by bacteria for biological balance of soil. Nature 252:703–705.

Smith, J. D. 1980. Is biologic control of *Marasmius oreades* fairy rings possible? Plant Disease 64:348–354.

Smith, K. A. 1978. Ineffectiveness of ethylene as a regulator of soil microbial activity. Soil Biol. Biochem. 10:269–272.

Smith, K. A., and S. W. F. Restall. 1971. The occurrence of ethylene in anaerobic soil. Jour. Soil Sci. 22:430-443.

Smith, K. A., and R. S. Russell. 1969. Occurrence of ethylene, and its significance in anaerobic soil. Nature 222:769–771.

Smith, K. T., R. O. Blanchard, and W. C. Shortle. 1979. Effects of spore load of *Trichoderma harzianum* on wood-invading fungi and volume of discolored wood associated with wounds in *Acer rubrum*. Plant Dis. Reptr. 63:1070-1071.

Smith, N. R., R. E. Gordon, and F. E. Clark. 1946. Aerobic mesophilic sporeforming bacteria. U.S. Dept. Agric. Misc. Publ. 559:1–112.

Smith, R. A. 1982. Nutritional study of *Pisolithus tinctorius*. Mycologia 74:54–58.

Smith, S. N. 1977. Comparison of germination of pathogenic *Fusarium oxysporum* chlamydospores in host rhizosphere soils conducive and suppressive to wilts. Phytopathology 67:502–510.

Smith, S. N., and W. C. Snyder. 1971. Relationship of inoculum density and soil types to severity of Fusarium wilt of sweet potato. Phytopathology 61:1049–1051.

Smith, S. N., and W. C. Snyder. 1972. Germination of *Fusarium oxysporum* chlamydospores in soils favorable and unfavorable to wilt establishment. Phytopathology 62:273–277.

Sneh, B., S. J. Humble, and L. J. Lockwood. 1977. Parasitism of oospores of *Phytophthora megasperma* var. *sojae*, *P. cactorum, Pythium* sp., and *Aphanomyces euteiches* in soil by oomycetes, chytridiomycetes, hyphomycetes, actinomycetes, and bacteria. Phytopathology 67:622–628.

Snyder, W. C., H. N. Hansen, and J. W. Oswald. 1957. Cultivars of the fungus, *Fusarium*. Jour. Madras Univ., B. 27:185–195.

Snyder, W. C., M. N. Schroth, and T. Christou. 1959. Effect of plant residues on root rot of bean. Phytopathology 49:755–756.

Sparling, G. P. 1981. Microcalorimetry and other methods to assess biomass and activity in soil. Soil Biol. Biochem. 13:93–98.

Speakman, J. B., and B. G. Lewis. 1978. Limitation of *Gaeumannomyces graminis* by wheat root responses to *Phialophora radicicola*. New Phytol. 80:373–380.

Spencer, D. M. 1980. Parasitism of carnation rust (*Uromyces dianthi*) by *Verticillium lecanii*. Trans. Brit. Mycol. Soc. 74:191–194.

Spencer, D. M., and P. T. Atkey. 1981. Parasitism of carnation rust and brown rust of wheat by *Verticillium lecanii*. Glasshouse Crops Res. Inst. Ann. Rept. 1980:134

Sprague, R. 1950. Diseases of Cereals and Grasses in North America. Ronald Press, New York. 538 pp.

Spurr, H. W., Jr. 1979. Ethanol treatment—A valuable technique for foliar biocontrol studies of plant disease. Phytopathology 69:773–776.

Spurr, H. W., Jr., and M. Sasser. 1982. Distribution of *Pseudomonas cepacia*, a broad-spectrum antagonist to plant pathogens in North Carolina. Phytopathology 72:710.

Stall, R. E., and A. A. Cook. 1979. Evidence that bacterial contact with the plant cell is necessary for the hypersensitive reaction but not the susceptible reaction. Physiol. Plant Pathol. 14:77–84.

Stanghellini, M. E., and T. J. Burr. 1973. Effect of soil water potential on disease incidence and oospore germination of *Pythium aphanidermatum*. Phytopathology 63:1496–1498.

Stanghellini, M. E., L. J. Stowell, W. C. Kromland, and P. von Bretzel. 1983. Distribution of *Pythium aphanidermatum* in rhizosphere soil and factors affecting expression of the absolute inoculum potential. Phytopathology 73:In press.

Starr, M. P., and N. L. Baigent. 1966. Parasitic interaction of *Bdellovibrio bacteriovorus* with other bacteria. Jour. Bact. 91:2006–2017.

Steadman, J. R. 1979. Control of plant diseases caused by *Sclerotinia* species. Phytopathology 69:904–907.

Sterne, R. E., G. A. Zentmyer, and M. R. Kaufmann. 1977a. The effect of matric and osmotic potential of soil on Phytophthora root disease of *Persea indica*. Phytopathology 67:1491–1494.

Sterne, R. E., G. A. Zentmyer, and M. R. Kaufmann. 1977b. The influence of matric potential, soil texture, and soil amendment on root disease caused by *Phytophthora cinnamomi*. Phytopathology 67:1495–1500.

Stevens, R. J., and I. S. Cornforth. 1974. The effect of pig slurry applied to a soil surface on the composition of the soil atmosphere. Jour. Sci. Food Agric. 25:1263–1272.

Stillwell, M. A., R. E. Wall, and G. M. Strunz. 1973. Production, isolation, and antifungal activity of scytalidin, a metabolite of *Scytalidium* species. Can. Jour. Microbiol. 19:597–602.

Stirling, G. R. 1981. Effect of temperature on infection of *Meloidogyne javanica* by *Bacillus penetrans*. Nematologica 27:458–462.

Stirling, G. R., and R. Mankau. 1978. *Dactylella oviparasitica*, a new fungal parasite of *Meloidogyne* eggs. Mycologia 70:774–783.

Stirling, G. R., and R. Mankau. 1979. Mode of parasitism of *Meloidogyne* and other nematode eggs by *Dactylella oviparasitica*. Jour. Nematol. 11:282–288.

Stirling, G. R., and M. F. Wachtel. 1980. Mass production of *Bacillus penetrans* for the biological control of root-knot nematodes. Nematologica 26:308–312.

Stirling, G. R., M. V. McKenry, and R. Mankau. 1979. Biological control of root-knot nematodes (*Meloidogyne* spp.) on peach. Phytopathology 69:806–809.

Stolp, H., and M. P. Starr. 1963. *Bdellovibrio bacteriovorus* gen. et sp. n., a predaory ectoparasitic, and bacteriolytic microorganism. Antonie van Leeuwenhoek Jour. Microbiol. Serol. 29:217–248.

Stolzy, L. H., J. Letey, L. J. Klotz, and C. K. Labanauskas. 1965. Water and aeration as factors in root decay of *Citrus sinensis*. Phytopathology 55:270–275.

Stone, I. 1980. The Origin: A Biographical Novel of Charles Darwin. Doubleday, Garden City, New York. 743 pp.

Stoner, W. N., and W. D. Moore. 1953. Lowland rice farming, a possible cultural control for *Sclerotinia sclerotiorum* in the Everglades. Plant Dis. Reptr. 37:181–186.

Stotzky, G., and R. T. Martin. 1963. Soil mineralogy in relation to the spread of Fusarium wilt of banana in Central America. Plant Soil 18:317–337.

Stotzky, G., and L. T. Rem. 1966. Influence of clay minerals on microorganisms. I. Montmorillonite and kaolinite on bacteria. Can. Jour. Microbiol. 12:547–563.

Strange, R. N., and H. Smith. 1978. Specificity of choline and betaine as stimulants of *Fusarium graminearum*. Trans. Brit. Mycol. Soc. 70:187–192.

Streets, R. B. 1969. Diseases of the Cultivated Plants of the Southwest. Univ. Ariz. Press, Tucson. 390 pp.

Strobel, G. A., and D. F. Myers. 1981. *Pseudomonas syringae* as an antagonist: Field tests of its effectiveness against Dutch elm disease. Phytopathology 71:1007.

Strobel, N. E., R. S. Hussey, and R. W. Roncadori. 1982. Interactions of vesicular-arbuscular mycorrhizal fungi, *Meloidogyne incognita,* and soil fertility on peach. Phytopathology 72:690–694.

Strunz, G. M., M. Kakushima, and M. A. Stillwell. 1972. Scytalidin: A new fungitoxic metabolite produced by a *Scytalidium* species. Jour. Chem. Soc. Perkin Trans. 1:2280–2283.

Stumbo, C. R., P. L. Gainey, and F. E. Clark. 1942. Microbiological and nutritional factors in the take-all disease of wheat. Jour. Agric. Res. 64:653–665.

Stumpf, P. K., ed. 1980. The Biochemistry of Plants, A Comprehensive Treatise. Vol. 4. Lipids: Structure and Function. Academic Press, New York. 650 pp.

Stumpf, P. K., and E. E. Conn, eds. 1981. The Biochemistry of Plants, A Comprehensive Treatise. Vol. 2. Metobolism and Respiration. Academic Press, New York. 687 pp.

Süle, S., and C. I. Kado. 1980. Agrocin resistance in virulent derivatives of *Agrobacterium tumefaciens* harboring the pTi plasmid. Physiol. Plant Pathol. 17:347–356.

Sundheim, L. 1977. Attempts at biological control of *Phomopsis sclerotioides* in cucumber. Neth. Jour. Plant Path. 83 (Suppl. 1):439–442.

Sundheim, L. 1982a. Control of cucumber powdery mildew by the hyperparasite *Ampelomyces quisqualis* and fungicides. Plant Pathol. 31:209–214.

Sundheim, L. 1982b. Effects of four fungi on conidial germination of the hyperparasite *Ampelomyces quisqualis*. Acta Agric. Scand. 32:341–347.

Sundheim, L., and T. Amundsen. 1982. Fungicide tolerance in the hyperparasite *Ampelomyces quisqualis* and integrated control of cucumber powdery mildew. Acta Agric. Scand. 32:349–355.

Sung, J. M., and R. J. Cook. 1981. Effect of water potential on reproduction and spore germination by *Fusarium roseum* 'Graminearum', 'Culmorum', and 'Avenaceum'. Phytopathology 71:499–504.

Suslow, T. V. 1982. Soil-borne bacteriophage as a limiting factor in root colonization by beneficial rhizobacteria. Phytopathology 72:997.

Suslow, T. V., and M. N. Schroth. 1982a. Role of deleterious rhizobacteria as minor pathogens in reducing crop growth. Phytopathology 72:111–115.

Suslow, T. V., and M. N. Schroth. 1982b. Rhizobacteria of sugar beets: Effects of seed application and root colonization on yield. Phytopathology 72:199–206.

Sutherland, J. B., and R. J. Cook. 1980. Effects of chemical and heat treatments on ethylene production in soil. Soil Biol. Biochem. 12:357–362.

Sutton, B. C. 1980. The Coelomycetes: Fungi imperfecti with pycnidia, acervuli and stromata. Common. Mycol. Inst., Kew, England. 696 pp.

Swendsrud, D. P., and L. Calpouzos. 1970. Rust uredospores increase the germination of pycnidiospores of *Darluca filum*. Phytopathology 60:1445–1447.

Swendsrud, D. P., and L. Calpouzos. 1972. Effect of inoculation sequence and humidity on infection of *Puccinia recondita* by the mycoparasite *Darluca filum*. Phytopathology 62:931–932.

Swinburne, T. R. 1973. Microflora of apple leaf scars in relation to infection by *Nectria galligena*. Trans. Brit. Mycol. Soc. 60:389–403.

Swinburne, T. R. 1978. Post-infection antifungal compounds in quiescent or latent infections. Ann. Appl. Biol. 89:322–325.

Swinburne, T. R., J. G. Barr, and A. E. Brown. 1975. Production of antibiotics by *Bacillus subtilis* and their effect on fungal colonists of apple leaf scars. Trans. Brit. Mycol. Soc. 65:211–217.

Szaniszlo, P. J., P. E. Powell, C. P. P. Reid, and G. R. Cline. 1981. Production of hydroxamate siderophore iron chelators by ectomycorrhizal fungi. Mycologia 73:1158–1174.

Taber, R. A. 1982a. *Gigaspora* spores and associated hyperparasites in weed seeds in soil. Mycologia 74:1026–1031.

Taber, R. A. 1982b. Occurrence of *Glomus* spores in weed seeds in soil. Mycologia 74:515–520.

Taber, R. A., and R. E. Pettit. 1981. Potential for biological control of Cercosporidium leafspot of peanuts by *Hansfordia*. Phytopathology 71:260.

Tainter, F. H., and W. D. Gubler. 1973. Natural biological control of oak wilt in Arkansas. Phytopathology 63:1027–1034.

Taylor, J. B., and E. M. Guy. 1981. Biological control of root-infecting basidiomycetes by species of *Bacillus* and *Clostridium*. New Phytol. 87:729–732.

Taylor, P. A., and P. J. LeB. Williams. 1975. Theoretical studies on the coexistence of competing species under continuous flow conditions. Can. Jour. Microbiol. 21:90-98.

Teintze, M., and J. Leong. 1981. Structure of pseudobactin A, a second siderphore from plant growth promoting *Pseudomonas* B10. Biochemistry 20:6457–6462.

Teintze, M., M. B. Hossain, C. L. Barnes, J. Leong, and D. van der Helm. 1981. Structure of ferric pseudobactin, a siderophore from a plant growth promoting *Pseudomonas*. Biochemistry 20:6446–6457.

Templeton, G. F., R. J. Smith, Jr., and W. Klomparens. 1980. Commercialization of fungi and bacteria for biological control. Biocontrol News Inform. 1(4):291–294.

Thatcher, F. S. 1939. Osmotic and permeability relations in the nutrition of fungus parasites. Amer. Jour. Bot. 26:449–458.

Thomson, S. V., M. N. Schroth, W. J. Moller, and W. O. Reil. 1976. Efficacy of bactericides and saprophytic bacteria in reducing colonization and infection of pear flowers by *Erwinia amylovora*. Phytopathology 66:1457–1459.

Thorne, G. 1961. Principles of Nematology. McGraw-Hill, New York. 553 pp.

Thorpe, T. A., ed. 1981. Plant Tissue Culture. Methods and Application to Agriculture. Academic Press, New York. 379 pp.

Thresh, J. M., ed. 1981. Pests, Pathogens, and Vegetation. Pitman, Boston. 517 pp.

Thresh, J. M. 1982. Cropping practices and virus spread. Annu. Rev. Phytopathol. 20:193–218.

Tivoli, B., J. M. Lemaire, and B. Jouan. 1974. Prémunition du Blé contre *Ophiobolus graminis* Sacc. Par des souches peu agressives du même parasite. Ann. Phytopathol. 6:395–406.

Toussoun, T. A., and P. E. Nelson. 1976. A Pictorial Guide to the Identification of *Fusarium* Species According to the Taxonomic System of Snyder and Hansen. 2nd ed. Penn. State Univ. Press, Univ. Park, Penn. 43 pp.

Toussoun, T. A., R. V. Bega, and P. E. Nelson, eds. 1970. Root Diseases and Soil-borne Pathogens. Univ. Calif. Press, Berkeley. 252 pp.

Toussoun, T. A., W. Menzinger, and R. S. Smith, Jr. 1969. Role of conifer litter in ecology of *Fusarium*: Stimulation of germination in soil. Phytopathology 59:1396–1399.

Toussoun, T. A., S. M. Nash, and W. C. Snyder. 1960. Effect of nitrogen sources and glucose on the pathogenesis of *Fusarium solani* f. *phaseoli*. Phytopathology 50:137–140.

Toyama, N., N. Fuji, and K. Ogawa. 1970. Cellulose, cell separating enzyme and mycolytic enzyme. Report, Applied Microbial Laboratory, Faculty of Agriculture, Miyazaki Univ., Miyayaki, Japan.

Trappe, J. M. 1977. Selection of fungi for ectomycorrhizal inoculation in nurseries. Annu. Rev. Phytopathol. 15:203–222.

Trappe, J. M. 1982. Synoptic keys to the genera and species of zygomycetous mycorrhizal fungi. Phytopathology 72:1102–1108.

Triantaphyllou, A. C. 1973. Environmental sex differentiation of nematodes in relation to pest management. Annu. Rev. Phytopathol. 11:441–462.

Tribe, H. T. 1957. On the parasitism of *Sclerotinia trifoliorum* by *Coniothyrium minitans*. Trans. Brit. Mycol. Soc. 40:489–499.

Tribe, H. T. 1977. A parasite of white cysts of *Heterodera: Catenaria auxiliaris*. Trans. Brit. Mycol. Soc. 69:367–376.

Trolldenier, G. 1971. Bodenbiologie. Franckh'sche Verlagehandlung, Stuttgart.

Trolldenier, G. 1981. Influence of soil moisture, soil acidity and nitrogen source on take-all of wheat. Phytopathol. Zeit. 102:163–177.

Trolldenier, G. 1982. Influence of potassium nitrition and take-all on wheat yield in dependence of inoculum density. Phytopathol. Zeit. 103:340-382.

Tronsmo, A., and C. Dennis. 1977. The use of *Trichoderma* species to control strawberry fruit rots. Neth. Jour. Plant Pathol. 83 (Suppl. l):449–455.

Tronsmo, A., and J. Raa. 1977. Antagonistic action of *Trichoderma pseudokonigii* against the apple pathogen *Botrytis cinerea*. Phytopathol. Zeit. 89:216–220.

Trujillo, E. E., and R. B. Hine. 1965. The role of papaya residues in papaya root rot caused by *Pythium aphanidermatum* and *Phytophthora parasitica*. Phytopathology 55:1293–1298.

Trujillo, E. E., and W. C. Snyder. 1963. Uneven distribution of *Fusarium oxysporum* f. *cubense* in Honduras soils. Phytopathology 53:167–170.

Trutmann, P., P. J. Keane, and P. R. Merriman. 1980. Reduction of sclerotial inoculum of *Sclerotinia sclerotiorum* with *Coniothyrium minitans*. Soil Biol. Biochem. 12:461–465.

Trutmann, P., P. J. Keane, and P. R. Merriman. 1982. Biological control of *Sclerotinia sclerotiorum* on aerial parts of plants by the hyperparasite *Coniothyrium minitans*. Trans. Brit. Mycol. Soc. 78:521–529.

Tsuneda, A., Y. Hiratsuka, and P. J. Maruyama. 1980. Hyperparasitism of *Scytalidium uredinicola* on western gall rust, *Endocronartium harknessii*. Can. Jour. Bot. 58:1154–1159.

Tu, J. C. 1980. *Gliocladium virens,* a destructive mycoparasite of *Sclerotinia sclerotiorum*. Phytopathology 70:670-674.

Tu, J. C., and O. Vaartaja. 1981. The effect of the hyperparasite (*Gliocladium virens*) on *Rhizoctonia solani* and on *Rhizoctonia* root rot of white beans. Can. Jour. Bot. 59:22–27.

Tulloch, M. 1972. The genus *Myrothecium* Tode ex Fr. Commonw. Mycol. Inst. Mycol. Papers 130:1–44.

Turner, G. J., and H. T. Tribe. 1975. Preliminary field plot trials on biological control of *Sclerotinia trifoliorum* by *Coniothyrium minitans*. Plant Pathol. 24:109–113.

Turner, G. J., and H. T. Tribe. 1976. On *Coniothyrium minitans* and its parasitism of *Sclerotinia* species. Trans. Brit. Mycol. Soc. 66:97–105.

Turner, N. C. 1974. Stomatal behavior and water status of maize, sorghum, and tobacco under field conditions. Plant Physiol. 53:360-365.

Tveit, M., and M. B. Moore. 1954. Isolates of *Chaetomium* that protect oats from *Helminthosporium victoriae*. Phytopathology 44:686–689.

Tyler, J. 1933. Reproduction without males in aseptic root culture of the root-knot nematode. Hilgardia 7:373–388.

Uecker, F. A., W. A. Ayers, and P. B. Adams. 1978. A new Hyphomycete on sclerotia of *Sclerotinia sclerotiorum*. Mycotaxon 7:275–282.

Underköfler, L. A. 1976. Microbial Enzymes in Industrial Microbiology. McGraw-Hill, New York. 465 pp.

Utkhede, R. S., and J. E. Rahe. 1980. Biological control of onion white rot. Soil Biol. Biochem. 12:101–104.

Utkhede, R. S., and J. E. Rahe. 1983. Chemical and biological control of onion white rot in muck and mineral soils. Plant Disease 67:153–155.

Utkhede, R. S., J. E. Rahe, and D. J. Ormrod. 1978. Occurrence of *Sclerotium cepivorum* sclerotia in commercial onion farm soils in relation to disease development. Plant Dis. Reptr. 62:1030-1034.

Van Alfen, N. K. 1982. Biology and potential for disease control of hypovirulence of *Endothia parasitica*. Annu. Rev. Phytopathol. 20:349–362.

Van Alfen, N. K., and V. Allard-Turner. 1979. Susceptibility of plants to vascular disruption of large molecules. Plant Physiol. 63:1072–1075.

Van Alfen, N. K., and B. D. McMillan. 1982. Macromolecular plant-wilting toxins: Artifacts of the bioassay method? Phytopathology 72:132–135.

Van Alfen, N. K., R. A. Jaynes, S. L. Anagnostakis, and P. R. Day. 1975. Chestnut blight: Biological control by transmissible hypovirulence in *Endothia parasitica*. Science 189:890-891.

Van der Laan, P. A. 1954. Nader onderzoek over het aaltjesvangende amoeboide organisme *Theratromyxa weberi* Zwillenberg. Tijdschr. Plantenziekt. 60:139–145.

Van der Plaats-Niterink, A. J. 1981. Monograph of the genus *Pythium*. Studies in Mycology No. 21. Centraalbur. Schimmelcult. Baarn, The Netherlands. 244 pp.

Van der Plank, J. E. 1963. Plant Diseases: Epidemics and Control. Academic Press, New York. 349 pp.

Van der Plank, J. E. 1968. Disease Resistance in Plants. Academic Press, New York and London. 206 pp.

Van der Plank, J. E. 1975. Principles of Plant Infection. Academic Press, New York. 216 pp.

Vanderplank, J. E. 1978. Genetic and Molecular Basis of Plant Pathogenesis. Springer-Verlag, Berlin. 167 pp.

Vanderplank, J. E. 1982. Host-Pathogen Interactions in Plant Pathogenesis. Academic Press, New York. 202 pp.

Van Rheenan, H. A., O. E. Hasslebach, and S. G. S. Muigai. 1981. The effect of growing beans together with maize on the incidence of bean diseases and pests. Neth. Jour. Plant Pathol. 87:193–199.

Veselý, D. 1978. Parasitic relationships between *Pythium oligandrum* Drechsler and some other species of the Oomycetes class. Zentralbl. Bakt. Par. Infect. Hyg. Abt. II, 133:341–349.

Vidaver, A. K. 1976. Prospects for control of phytopathogenic bacteria by bacteriophages and bacteriocins. Annu. Rev. Phytopathol. 14:451–465.

Vojinović, Ž. D. 1972. Antagonists from soil and rhizosphere to phytopathogens. Final Tech. Rep. Inst. Soil Sci., Beograd, Yugoslavia 130 pp.

Vojinović, Ž. D. 1973. The influence of microorganisms following *Ophiobolus graminis* Sacc. on its further pathogenicity. Org. Eur. Med. Prot. Plantes Bull. 9:91–101.

Wadi, J. A. 1982. Biological control of *Verticillium dahliae* on potato. Ph.D. Thesis. Washington State University, Pullman. 66 pp.

Waksman, S. A. 1947. What is an antibiotic or an antibiotic substance? Mycologia 39:565–569.

Waksman, S. A. 1967. The Actinomycetes. A Summary of Current Knowledge. Ronald Press, New York. 280 pp.

Waksman, S. A., and E. Bugie. 1944. Chaetomin, a new antibiotic substance produced by *Chaetomium cochlioides*. I. Formation and properties. Jour. Bacteriol. 48:527–530.

Walker, J. A., and R. B. Maude. 1975. Natural occurrence and growth of *Gliocladium roseum* on the mycelium and sclerotia of *Botrytis allii*. Trans. Brit. Mycol. Soc. 65:335–338.

Walker, J. C. 1969. Plant Pathology. McGraw-Hill, New York. 819 pp.

Walker, J. C., and R. Smith. 1930. Effect of environmental factors upon resistance of cabbage to yellows. Jour. Agric. Res. 41:1–15.

Walker, J. C., and W. C. Snyder. 1933. Pea wilt and root rots. Wisconsin Agric. Exp. Sta. Bull. 424:1–16.

Warcup, J. H. 1976. Studies on soil fumigation. IV. Effects on fungi. Soil Biol. Biochem. 8:261–266.

Warner, R. 1982. Metamorphosis. Science82 3(10):43–46.

Warren, R. C. 1972. The effect of pollen on the fungal leaf microflora of *Beta vulgaris* L. and on infection of leaves by *Phoma betae*. Neth. Jour. Plant Pathol. 78:89–98.

Waterhouse, G. M. 1968. The Genus *Pythium* Pringsheim. Commonw. Mycol. Inst. Mycol. Papers 110:1–71.

Waterworth, H. E., J. M. Kaper, and M. E. Tousignant. 1979. CARNA 5, the small cucumber mosaic-dependent replicating RNA regulates disease expression. Science 204:845–847.

Wearing, A. H., and L. W. Burgess. 1979. Water potential and the saprophytic growth of *Fusarium roseum* 'Graminearum.' Soil Biol. Biochem. 11:661–667.

Weber, J. 1981. A natural biological control of Dutch elm disease. Nature 292:449–451.

Webster, J., and N. Lomas. 1964. Does *Trichoderma viride* produce gliotoxin and viridin? Trans. Brit. Mycol. Soc. 47:535–540.

Weindling, R. 1932. *Trichoderma lignorum* as a parasite of other soil fungi. Phytopathology 22:837–845.

Weindling, R., and O. H. Emerson. 1936. The isolation of a toxic substance from the culture filtrate of *Trichoderma*. Phytopathology 26:1068–1070.

Weindling, R., and H. S. Fawcett. 1936. Experiments in the control of Rhizoctonia damping-off of citrus seedlings. Hilgardia 10:1–16.

Weinhold, A. R., and T. Bowman. 1968. Selective inhibition of the potato scab pathogen by antagonistic bacteria and substrate influence on antibiotic production. Plant Soil 28:12–24.

Weinhold, A. R., J. W. Oswald, T. Bowman, J. Bishop, and D. Wright. 1964. Influence of green manures and crop rotation on common scab of potato. Am. Potato Jour. 41:265–273.

Weller, D. M. 1983. Colonization of wheat roots by a fluorescent pseudomonad suppressive to take-all. Phytopathology 73:In press.

Weller, D. M., and R. J. Cook. 1982. Pseudomonads from take-all conducive and suppressive soils. Phytopathology 72:264.

Weller, D. M., and R. J. Cook. 1983. Suppression of take-all of wheat by seed-treatment with fluorescent pseudomonads. Phytopathology 73:463–469.

Wells, H. D., and D. K. Bell. 1979. Variable antagonistic reaction in vitro of *Trichoderma harzianum* against several pathogens. Phytopathology 69:1048–1049.

Wells, H. D., D. K. Bell, and C. A. Jaworski. 1972. Efficacy of *Trichoderma harzianum* as a biocontrol for *Sclerotium rolfsii*. Phytopathology 62:442–447.

Whatley, M. H., J. B. Margot, J. Schell, B. B. Lippincott, and J. A. Lippincott. 1978. Plasmid and chromosomal determination of *Agrobacterium* adherence specificity. Jour. Gen. Microbiol. 107:395–398.

Wheeler, H. 1975. Plant Pathogenesis. Springer-Verlag, Berlin. 106 pp.

White, K. D. 1970. Roman Farming. Cornell Univ. Press, Ithaca, NY. 536 pp.

White, N. H. 1954. Decoy crops. The use of decoy crops in the eradication of certain soil-borne plant diseases. Austral. Jour. Sci. 17:18–19.

White, W. L., R. T. Darby, G. M. Stechart, and K. Sanderson. 1948. Assay of cellulolytic activity of molds isolated from fabrics and related items exposed in the tropics. Mycologia 40:34–84.

Whittington, W. J., ed. 1969. Root Growth. Butterworth, London. 450 pp.

Whittington, J., and D. J. Read. 1982. Vascular-arbuscular mycorrhizae in natural vegetative systems. III. Nutrient transfer between plants with mycorrhizal interconnections. New Phytol. 90:277–284.

Wicker, E. F. 1968. Toxic effects of cycloheximide and phytoactin on *Tuberculina maxima*. Phytoprotection 41:91–98.

Wicker, E. F. 1981. Natural control of white pine blister rust by *Tuberculina maxima*. Phytopathology 71:997–1000.

Wicker, E. F., and J. Y. Woo. 1973. Histology of blister rust cankers parasitized by *Tuberculina maxima*. Phytopathol. Zeit. 76:356–366.

Wicklow, D. T., and G. C. Carroll, eds. 1981. The Fungal Community. Marcel Dekker, New York. 855 pp.

Wiebe, H. H., G. S. Campbell, W. H. Gardner, S. L. Rawlings, J. W. Cary, and R. W. Brown. 1971. Measurement of plant and soil water status. Utah Agric. Exp. Sta. Bull. 484:1–71.

Wildermuth, G. B., and A. D. Rovira. 1977. Hyphal density as a measure of suppression of *Gaeumannomyces graminis* var. *tritici* on wheat roots. Soil Biol. Biochem. 9:203–205.

Wilhelm, S. 1965. *Pythium ultimum* and the soil fumigation growth response. Phytopathology 55:1016–1020.

Wilhelm, S., and A. O. Paulus. 1980. How soil fumigation benefits the California strawberry industry. Plant Disease 64:264–270.

Wilkinson, H. T., J. R. Alldredge, and R. J. Cook. 1982a. Estimated distances for infection of wheat roots by *G. graminis* var. *tritici* in take-all suppressive and conducive soils. Phytopathology 72:949.

Wilkinson, H. T., D. M. Weller, and J. R. Alldredge. 1982b. Enhanced biological control of wheat take-all when inhibitory pseudomonas strains are introduced on inoculum or seed as opposed to directly into soil. Phytopathology 72:948–949.

Willcox, J., and H. T. Tribe. 1974. Fungal parasitism in cysts of *Heterodera*. I. Preliminary investigations. Trans. Brit. Mycol. Soc. 62:585–594.

Williams, P. H. 1979. Vegetable crop protection in the People's Republic of China. Annu. Rev. Phytopathol. 17:311–24.

Williams, T. D. 1969. The effects of formalin, nabam, irrigation and nitrogen on *Heterodera avenae* Woll., *Ophiobolus graminis* Sacc. and the growth of spring wheat. Ann. Appl. Biol. 64:325–334.

Williamson, R., ed. 1981. Genetic Engineering. Academic Press, New York.

Wilson, J. M., and D. M. Griffin. 1975. Respiration and radial growth of soil fungi at two osmotic potentials. Soil Biol. Biochem. 7:269–274.

Windels, C. E. 1981. Growth of *Penicillium oxalicum* as a biological seed treatment on pea seed in soil. Phytopathology 71:929–933.

Windels, C. E., and T. Kommedahl. 1978. Factors affecting *Penicillium oxalicum* as a seed protectant against seedling blight of pea. Phytopathology 68: 1656–1661.

Windels, C. E., and T. Kommedahl. 1982a. Rhizosphere effects of pea seed treatment with *Penicillium oxalicum*. Phytopathology 72: 198–194.

Windels, C. E., and T. Kommedahl. 1982b. Pea cultivar effect on seed treatment with *Penicillium oxalicum* in the field. Phytopathology 72:541–543.

Wolfe, M. S., and J. A. Barrett. 1980. Can we lead the pathogen astray? Plant Disease 64:148–155.

Woltz, S. S. 1978. Nonparasitic plant pathogens. Annu. Rev. Phytopathol. 16:403–430.

Wong, P. T. W. 1975. Cross-protection against the wheat and oat take-all fungi by *Gaeumannomyces graminis* var. *graminis*. Soil Biol. Biochem. 7:189–194.

Wong, P. T. 1980. Effect of temperature on growth of avirulent fungi and cross-protection against wheat take-all. APPS Natl. Plant Pathol. Conf. 4:37.

Wong, P. T. W., and D. M. Griffin. 1976. Bacterial movement at high matric potentials. I. In artificial and natural soils. Soil Biol. Biochem. 8:215–218.

Wong, P. T. W., and T. R. Siviour. 1979. Control of *Ophiobolus* patch in *Agrostis* turf using avirulent fungi and take-all suppressive soils in pot experiments. Ann. Appl. Biol. 92:191–197.

Wong, P. T. W., and R. J. Southwell. 1980. Field control of take-all by avirulent fungi. Ann. Appl. Biol. 94:41–49.

Wood, R. K. S., and M. Tveit. 1955. Control of plant diseases by use of antagonistic organisms. Bot. Rev. 21:441–492.

Wright, J. M. 1956. The production of antibiotics in soil. III. Production of gliotoxin in wheatstraw buried in soil. Ann. Appl. Biol. 44:461–466.

Wu, R., ed. 1979. Methods in Enzymology. Vol. 68. Recombinant DNA. Academic Press, New York. 555 pp.

Wymore, L. A., and R. Baker. 1982. Factors affecting cross-protection in control of Fusarium wilt of tomato. Plant Disease 66:908–910.

Yarwood, C. E. 1956. Cross protection with two rust fungi. Phytopathology 46:540-544.

Yin, S. Y., J. K. Chang, and P. C. Xun. 1965. Studies in the mechanisms of antagonistic fertilizer "5406" IV. The distribution of the antagonist in soil and its influence on the rhizosphere. Acta Microbiol. Sin. 11:259–288.

Yin, S. Y., D. C. Keng, K. Y. Yang, and D. Cheu. 1957. A further study on the biological control of Verticillium wilt of cotton. Acta Phytopathol. Sin. 3:55–61.

Yost, D., and S. McGill. 1982. Solar energy cleans up the soil. The Furrow. January, pp. 6–7.

Yu, S. Q., and E. J. Trione. Biochemical detrmination of fungal spore viability. Phytopathology 72:957.

Zentmyer, G. A. 1980. *Phytophthora cinnamomi* and the diseases it causes. Monogr. 10. Amer. Phytopathol. Soc. St. Paul, MN. 96 pp.

Zentmyer, G. A., K. F. Baker, and W. A. Thorn. 1952. The role of nursery stock in the dissemination of soil pathogens. Phytopathology 42:478–479.

Zopf, W. 1980. Die Pilze in Morphologischer, Physiologisher und Systematicher Beziehung. Breslau.

Zuck, M. G., F. Hyland, and F. L. Caruso. 1982. Possible hyperparasitism of *Venturia inaequalis* perithecia by *Cladosporium* spp. and *Hyalodendron* spp. Phytopathology 72:268.

Zuckerman, B. M., and R. A. Rohde. 1981. Plant Parasitic Nematodes. Academic Press, New York. 712 pp.

INDEX

Principal references are given in **boldface**.